Geographical Information
SCIENCE

Geographical Information
SCIENCE
Case Studies in Earth and Environmental Monitoring

Edited by

GEORGE P. PETROPOULOS
Geoinformatics, Department of Geography,
Harokopio University of Athens, Athens, Greece

CHRISTOS CHALKIAS
Applied Geography and GIS, Harokopio
University of Athens, Athens, Greece

ELSEVIER

Elsevier
Radarweg 29, PO Box 211, 1000 AE Amsterdam, Netherlands
125 London Wall, London EC2Y 5AS, United Kingdom
50 Hampshire Street, 5th Floor, Cambridge, MA 02139, United States

Notices
Knowledge and best practice in this field are constantly changing. As new research and experience broaden our understanding, changes in research methods, professional practices, or medical treatment may become necessary.

Practitioners and researchers must always rely on their own experience and knowledge in evaluating and using any information, methods, compounds, or experiments described herein. In using such information or methods they should be mindful of their own safety and the safety of others, including parties for whom they have a professional responsibility.

To the fullest extent of the law, neither the Publisher nor the authors, contributors, or editors, assume any liability for any injury and/or damage to persons or property as a matter of products liability, negligence or otherwise, or from any use or operation of any methods, products, instructions, or ideas contained in the material herein.

ISBN: 978-0-443-13605-4

For information on all Elsevier publications visit our website at
https://www.elsevier.com/books-and-journals

Publisher: Candice Janco
Acquisitions Editor: Peter Llewellyn
Editorial Project Manager: Aleksandra Packowska
Production Project Manager: Paul Prasad Chandramohan
Cover Designer: Christian Bilbow

Typeset by TNQ Technologies

Working together
to grow libraries in
developing countries

www.elsevier.com • www.bookaid.org

Contents

4. Spatiotemporal patterns of vegetation regeneration dynamics in a natural Mediterranean ecosystem using EO imagery and Google Earth Engine cloud platform   69

Ioannis Lemesios, Spyridon E. Detsikas and George P. Petropoulos

5. Understanding and monitoring the dynamics of Arctic permafrost regions under climate change using Earth Observation and cloud computing: The contribution of EO-PERSIST project   91

George P. Petropoulos, Vassilia Karathanassi, Kleanthis Karamvasis, Aikaterini Dermosinoglou and Spyridon E. Detsikas

SECTION 2   Agricultural environment

6. Development of algorithms based on the integration of vegetation indices and meteorological data for the identification of low productivity agricultural areas   111

M. Lanfredi, R. Coluzzi, M. D'Emilio, V. Imbrenda, L. Pace, C. Samela, T. Simoniello, L. Salvati and J. Mughini Gras

SECTION 3   Coastal/aquatic environment

SECTION 5   Atmosphere/air environment

List of contributors

Hamaad Raza Ahmad
Institute of Soil and Environmental Sciences, University of Agriculture, Faisalabad, Pakistan

Cinzia Albertini
Dipartimento di Scienze del Suolo, della Pianta e degli Alimenti, Università degli Studi di Bari Aldo Moro, Bari, Italy; Dipartimento di Ingegneria Civile, Ambientale, del Territorio, Edile e di Chimica, Politecnico di Bari, Bari, Italy

Gourgiotis Anestis
Department of Planning and Regional Development, University of Thessaly, Volos, Greece

Gleyce Kelly Dantas Araújo Figueiredo
Faculty of Agricultural Engineering, UNICAMP - Campinas, Brazil

Muhammad Hassan Bashir
Institute of Soil and Environmental Sciences, University of Agriculture, Faisalabad, Pakistan

Dimitrios-Vasileios Batzakis
Department of Geography, School of Environment, Geography and Applied Economics, Harokopio University of Athens, Athens, Greece

Sadia Bibi
Institute of Soil and Environmental Sciences, University of Agriculture, Faisalabad, Pakistan

Christos Chalkias
Department of Geography, School of Environment, Geography and Applied Economics, Harokopio University, Athens, Greece

Zenaida Chitu
National Meteorological Administration, Bucharest, Romania

R. Coluzzi
Institute of Methodologies for Environmental Analysis, National Research Council of Italy (IMAA-CNR), Tito Scalo, Italy; NBFC, National Biodiversity Future Center, Palermo, Italy

M. D'Emilio
Institute of Methodologies for Environmental Analysis, National Research Council of Italy (IMAA-CNR), Tito Scalo, Italy

Daniela Fernanda da Silva Fuzzo
Department of Agricultural and Biological Sciences, University of the State of Minas Gerais, Frutal Unit, Brazil

Avgoustina I. Davri
Department of Geography, Harokopio University of Athens, Athens, Greece

Aikaterini Dermosinoglou
Department of Geography, Harokopio University of Athens, Athens, Greece

Spyridon E. Detsikas
Department of Geography, Harokopio University of Athens, Athens, Greece

Vassilis Detsis
Department of Economics and Sustainable Development, School of Environment, Geography and Applied Economics, Harokopio University, Athens, Greece

Fran Domazetović
University of Zadar, Department of Geography, Center for Geospatial Technologies, Zadar, Croatia

Bruno Enrique Fuzzo
Master's Program in Environmental Sciences, University of the State of Minas Gerais, Fruit Unit, Brazil

Antigoni Faka
Department of Geography, School of Environment, Geography and Applied Economics, Harokopio University, Athens, Greece

Triantafyllos Falaras
Department of Geography, School of Environment, Geography and Applied Economics, Harokopio University, Athens, Greece

João Alberto Fischer Filho
Department of Agricultural and Biological Sciences, University of the State of Minas Gerais, Frutal Unit, Brazil

Carmen Garcia
Department of Geography and Land Planning, Universidad de Castilla-La Mancha, Albacete, Spain

Andrea Gioia
Dipartimento di Ingegneria Civile, Ambientale, del Territorio, Edile e di Chimica, Politecnico di Bari, Bari, Italy

Ioannis Z. Gitas
Laboratory of Forest Management and Remote Sensing, School of Forestry and Natural Environment, Aristotle University of Thessaloniki, Thessaloniki, Greece

Georgios Gkatzios
Department of Geography, Harokopio University of Athens, Athens, Greece

Abdulghani Hasan
Department of Landscape Architecture, Planning and Management, Swedish University of Agricultural Sciences, Alnarp, Sweden

Vito Iacobellis
Dipartimento di Ingegneria Civile, Ambientale, del Territorio, Edile e di Chimica, Politecnico di Bari, Bari, Italy

V. Imbrenda
Institute of Methodologies for Environmental Analysis, National Research Council of Italy (IMAA-CNR), Tito Scalo, Italy; NBFC, National Biodiversity Future Center, Palermo, Italy

Radu Irimia
Faculty of Geography, University of Bucharest, Bucharest, Romania

Kleomenis Kalogeropoulos
Department of Surveying and Geoinformatics Engineering, University of West Attica, Athens, Greece

Kleanthis Karamvasis
School of Rural, Surveying and Geoinformatics Engineering, National Technical University of Athens, Athens, Greece

Vassilia Karathanassi
School of Rural, Surveying and Geoinformatics Engineering, National Technical University of Athens, Athens, Greece

Efthimios Karymbalis
Department of Geography, School of Environment, Geography and Applied Economics, Harokopio University of Athens, Athens, Greece

Ourania Kounadi
University of Vienna, Department of Geography and Regional Research, Vienna, Austria

Nektarios N. Kourgialas
Water Recourses-Irrigation & Environmental Geoinformatics Laboratory, Institute of Olive Tree, Subtropical Crops and Viticulture, Hellenic Agricultural Organization (ELGO Dimitra), Chania, Greece

M. Lanfredi
Institute of Methodologies for Environmental Analysis, National Research Council of Italy (IMAA-CNR), Tito Scalo, Italy; NBFC, National Biodiversity Future Center, Palermo, Italy

Ioannis Lemesios
Department of Geography, Harokopio University of Athens, Athens, Greece

Panagiota Louka
Department of Natural Resources Development and Agricultural Engineering, Agricultural University of Athens, Athens, Greece

Salvatore Manfreda
Dipartimento di Ingegneria Civile, Edile e Ambientale, Università degli Studi di Napoli Federico II, Naples, Italy

Ivan Marić
University of Zadar, Department of Geography, Center for Geospatial Technologies, Zadar, Croatia

Swati Maurya
Department of Environmental Science, Integral University, Lucknow, Uttar Pradesh, India

Khalid Mehmood
Institute of Environmental Health and Ecological Security, School of the Environment and Safety Engineering, Jiangsu University, Zhenjiang, Jiangsu, China

Ioanna Merkouriadi
Finnish Meteorological Institute, Helsinki, Finland

Smrutisikha Mohanty
Department of Life Sciences, School of Natural Sciences, Shiv Nadar Institution of Eminence (Deemed to be University), Greater Noida, Uttar Pradesh, India

J. Mughini Gras
Octopus S.R.L., Rome, Italy

L. Pace
Department of Earth and Geoenvironmental Sciences, University of Bari "Aldo Moro", Bari, Italy

Varsha Pandey
Department of Civil Engineering, Indian Institute of Technology Bombay, Mumbai, Maharashtra, India

Prem C. Pandey
Department of Life Sciences, School of Natural Sciences, Shiv Nadar Institution of Eminence (Deemed to be University), Greater Noida, Uttar Pradesh, India

Apostolos G. Papadopoulos
Department of Geography, School of Environment, Geography and Applied Economics, Harokopio University, Athens, Greece

Andreas Persson
Lund University GIS Centre, Lund, Sweden; Department of Physical Geography and Ecosystem Science, Lund University, Lund, Sweden

George P. Petropoulos
Department of Geography, Harokopio University of Athens, Athens, Greece

Triantafyllia Petsini
Department of Geography, Harokopio University of Athens, Athens, Greece

Maria Pigaki
Department of Geography and Regional Planning, National Technical University of Athens, Athens, Greece

Petter Pilesjö
Lund University GIS Centre, Lund, Sweden; Department of Physical Geography and Ecosystem Science, Lund University, Lund, Sweden

Gustavo Rodrigues Pereira
Faculty of Agricultural Engineering, UNICAMP – Campinas, Brazil

Jorge Jorge Ruiz
Finnish Meteorological Institute, Helsinki, Finland

Saifullah
Institute of Soil and Environmental Sciences, University of Agriculture, Faisalabad, Pakistan

L. Salvati
Department of Methods and Models for Economics, Territory and Finance (MEMOTEF), University of Rome "La Sapienza", Rome, Italy

C. Samela
Institute of Methodologies for Environmental Analysis, National Research Council of Italy (IMAA-CNR), Tito Scalo, Italy

Ionut Sandric
Faculty of Geography, University of Bucharest, Bucharest, Romania

Ayesha Siddique
Institute of Soil and Environmental Sciences, University of Agriculture, Faisalabad, Pakistan

Ante Šiljeg
University of Zadar, Department of Geography, Center for Geospatial Technologies, Zadar, Croatia

T. Simoniello
Institute of Methodologies for Environmental Analysis, National Research Council of Italy (IMAA-CNR), Tito Scalo, Italy; NBFC, National Biodiversity Future Center, Palermo, Italy

Anjali Kumari Singh
Department of Energy and Environment, Faculty of Science and Environment, Mahatma Gandhi Chitrakoot Gramodaya Vishwavidayalya, Satna, Madhya Pradesh, India

Prashant K. Srivastava
Remote Sensing Laboratory, Institute of Environment and Sustainable Development, Banaras Hindu University, Varanasi, Uttar Pradesh, India

Demetris Stathakis
Department of Planning and Regional Development, University of Thessaly, Volos, Greece

Nikolaos Stathopoulos
Operational Unit "BEYOND Centre for Earth Observation Research and Satellite Remote Sensing", Institute for Astronomy, Astrophysics, Space Applications and Remote Sensing, National Observatory of Athens, Athens, Greece

Dimitris Stavrakoudis
Laboratory of Forest Management and Remote Sensing, School of Forestry and Natural Environment, Aristotle University of Thessaloniki, Thessaloniki, Greece

Alexandra Stefanidou
Laboratory of Forest Management and Remote Sensing, School of Forestry and Natural Environment, Aristotle University of Thessaloniki, Thessaloniki, Greece

Swati Suman
Department for Innovations in Biological, Agri-food and Forest System, University of Tuscia, Viterbo, Italy

Konstantinos Tsanakas
Department of Geography, School of Environment, Geography and Applied Economics, Harokopio University of Athens, Athens, Greece

Andreas Tsatsaris
Department of Surveying and Geoinformatics Engineering, University of West Attica, Athens, Greece

Demetrios E. Tsesmelis
Laboratory of Technology and Policy of Energy and Environment, School of Science and Technology, Hellenic Open University, Patra, Greece

Chrysovalantis Tsiakos
Department of Geography, School of Environment, Geography and Applied Economics, Harokopio University, Athens, Greece

Kanella Valkanou
Department of Geography, Harokopio University, Athens, Greece

Eleni Vasileiou
Department of Geological Sciences, School of Mining Engineering and Metallurgy, National Technical University of Athens, Athens, Greece

Mariana Vallejo Velázquez
University of Vienna, Department of Geography and Regional Research, Vienna, Austria

Md. Wasim
Department of Life Sciences, School of Natural Sciences, Shiv Nadar Institution of Eminence (Deemed to be University), Greater Noida, Uttar Pradesh, India

Preface

This book, *Geographical Information Science: Case Studies in Earth and Environmental Monitoring*, explores the diverse applications of Geographical Information Systems (GIS) and Earth Observation (EO) to understanding our planet's environment. Each chapter takes a comprehensive journey through different environmental fields and presents in a systematic way the practical use of those technologies for earth and environmental monitoring in various disciplines.

The anthology begins with an insight into the natural environment and covers techniques for identifying geomorphic features, assessing vegetation dynamics after forest fires, and quantitatively analyzing the Earth's landscapes. In agriculture, case studies estimate productivity in Mediterranean regions, identify low-productivity agricultural areas using GIS-based approaches, and integrate vegetation indices with meteorological data.

The chapters on coastal and marine environments describe methods for vulnerability assessment, delta monitoring, and the study of soil erosion and coastal change, emphasizing the role of GIS/EO in understanding dynamic environmental systems.

The focus then shifts to the urban milieu, with an emphasis on mapping noise levels in historic centers, studying urbanization through advanced EO operational products, and assessing air quality with GIS tools—a meticulous exploration of evolving urban landscapes.

In the atmospheric section, chapters focus on mapping nighttime light pollution, assessing air quality using remote sensing and GIS, and scientific understanding of Earth's atmospheric conditions.

The final section of this book includes chapters on the use of GIS and remote sensing in natural disasters, covering topics such as operational mapping of burned areas, vulnerability assessment of coastal areas, and agricultural drought monitoring—a collective effort to understand and mitigate the effects of Earth's unpredictable events.

This collaboration of researchers, scientists, and experts, centered on scientific exploration, contributes to our knowledge of the Earth's complex systems. The collection serves as a learning tool, providing detailed insights into the use of synergistic GIS and EO methods and their role in understanding Earth's intricate environment.

Users are encouraged to adapt this book to their own needs to gain a deep understanding of the possibilities and potential of both GIS and EO technologies in environmental monitoring. Any errors, suggestions, or comments can be communicated to the editor at gpetropoulos@hua.gr and chalkias@hua.gr.

Special thanks go to the reviewers of the chapters and to the members of the Elsevier Editorial Team for their support throughout the preparation of the present volume. Closing, we are grateful to the different contributors for sharing with us extracts of their research allowing to produce this book.

Editors

George P. Petropoulos and Christos Chalkias

SECTION 1

Natural environment

CHAPTER 1

Identification of geomorphic and morphotectonic features using geographic information system techniques

Kanella Valkanou[1], Efthimios Karymbalis[2] and Konstantinos Tsanakas[2]

[1]Department of Geography, Harokopio University, Athens, Greece; [2]Department of Geography, School of Environment, Geography and Applied Economics, Harokopio University of Athens, Athens, Greece

1. Introduction

Geomorphology is the study of the Earth's surface development and evolution (Keller & Pinter, 2002) aiming to establish the relationship between landforms and the processes (geological, physical, chemical, and biological) that create and modify them (Morisawa, 1958; Evans, 1998). The evolution of relief is based on process-response models, which provide both quantitative and qualitative representations of how processes impact the landscape. It is often necessary to consider past events and processes that have influenced the landscape since many landforms cannot be fully explained solely by current natural processes (Summerfield, 1991). The "geomorphological record," encompassing land-forms and Quaternary deposits, often dating from several thousand to 2.6 million years old, serves as a valuable tool in neotectonic studies (Keller & Pinter, 2002) as it indirectly assists in drawing conclusions about the recent tectonic activity in a tectonically active area (Pavlidis, 2003; Scheidegger, 2004).

Earth's surface-forming processes have diverse effects on human activities; for example, earthquakes, large landslides, or floods can pose natural hazards to communities. Therefore, the study of landforms can assist in the identification of high-risk areas. Furthermore, applied geomorphology plays a role in effective environmental management and supports technical construction issues (Summerfield, 1991). In tectonically active areas, present topography results from the interplay between tectonics and erosional processes (Bishop, 2007). Tectonic activity gives rise to distinct landforms, such as mountain ranges, horsts and grabens, fluvial terraces, uplifted marine terraces, incised V-shaped valleys, gorges, etc. It's worth noting that the type of fault responsible for the landscape formation determines the specific landforms created (e.g., horizontal displacement faults create different features than normal faults) (Slemmons & Depolo, 1986; Karymbalis et al., 2016; Leeder & Jackson, 1993). Additionally, neotectonic activity directly impacts fluvial systems (drainage networks and watersheds), as well as orographic fronts. The quantitative analysis

Geographical Information Science
ISBN 978-0-443-13605-4
https://doi.org/10.1016/B978-0-443-13605-4.00009-6

of these landforms utilizing geographic information system (GIS) techniques provides valuable insights and the results are often considered in research aimed at assessing the tectonic activity of an area (López-Ramos et al., 2022).

The aim of this review chapter is to emphasize the utility of geoinformation technology in research related to quantitative geomorphological analysis of landscapes in tectonically active regions. Methods of automatic classification of the relief, as well as methodologies of identification and mapping of specific landforms, are presented. In addition, the most commonly used morphotectonic indices both individual and composite are discussed. Examples of the application of such approaches in the tectonically active Evia Island in Greece, as well as visualization of the results are also presented.

2. Geomorphometry

The study of the processes that affect the Earth's surface and landforms apart from qualitative observations requires the quantitative characterization of the topography (Wilson & Bishop, 2013). This is essential because the role played by topography in shaping the Earth's dynamic systems, including atmospheric, geomorphological, hydrological, geological, and ecological processes, is fundamental. Usually, there is either a strong or weak relationship between landforms and processes. Therefore, the study of land surface features provides crucial evidence for investigating the nature and intensity of their forming processes (Wilson & Bishop, 2013). Consequently, it is necessary to quantify the morphological characteristics of the Earth's surface and categorize the topography as it reflects the interaction of various systems. Quantitative analysis of the land surface, or the quantitative representation of topography, can be achieved through the calculation of morphometric parameters and indices using software and models that manage elevation information (Pike et al., 2009). This quantitative description of the landforms is a subject of geomorphometry (Tagil & Jenness, 2008), which is a multidisciplinary field encompassing geoscience (mainly geomorphology), mathematics, and geoinformatics. The established terms associated with geomorphometry, also known as morphometry, include relief analysis, quantitative geomorphology, digital relief modeling, digital relief analysis, landform modeling, land surface modeling, surface topography, surface metrology, interpretation of surface relief, and topographical analysis (Pike, 2002).

Landform analysis uses elevational data, particularly digital terrain models that represent the "bare" surface of the Earth, and it is closely related to spatial analysis and constitutes a fundamental component of geographic information science (Wilson & Deng, 2008). The quantitative analysis of the Earth's surface and landforms, in combination with other information such as geology and vegetation, supports the understanding of natural processes and relief features, making it particularly useful for spatial planning and land use zoning. Furthermore, the utility of morphometry in geomorphological and morphotectonic studies is evident, as recently active neotectonic structures can influence the Earth's surface morphology and interact with geomorphological processes during relief evolution (Pedrera et al., 2009).

Even the simplest morphological analysis of an area includes the study of altimetric data and its derivatives (Jordan et al., 2005). According to Pike et al. (2009), quantitative geomorphological analyses encompass the following five stages: (1) elevation measurements, (2) creation of a surface model using sampled data, (3) correction of errors in the surface model, (4) extraction of surface parameters, and (5) application of the resulting parameters and "objects." The foundation of geomorphometry and quantitative geomorphological analysis is the digital elevation model (DEM). Apart from obtaining elevation information and presenting geographic data in three dimensions, DEMs are also used to calculate many other morphological parameters and indicators through their derivatives such as morphological slope, slope orientation, relief shading, and curvatures. The potential for GIS use was recognized, and it has since been widely employed in relief modeling and mapping, both on a local and regional scale, especially in studies concerning processes affecting the Earth's surface and in automated methods for extracting topographic and morphometric parameters.

2.1 Relief classification

As mentioned above, geomorphometry primarily focuses on extracting surface parameters, which are topographic features of the relief, and subsequently classifying the surface into spatial entities through digital topography (Wilson & Bishop, 2013).

Topographical parameters can be categorized into two groups: primary and secondary. Primary parameters are computed directly from altimetric data and encompass variables such as elevation, slope, and curvatures. Secondary parameters, on the other hand, such as the topographic wetness index or stream power index, consist of combinations of the primary parameters and serve as indicators that describe or characterize the spatial variability of specific processes that are responsible for shaping the relief or act on its surface (Moore et al., 1991; Wilson & Deng, 2008).

The automatic classification of landforms holds particular significance for geomorphology research as there is a readily discernible relationship between topography, geological structures, tectonic activity, and surface processes such as erosion and fluvial incision (Florinsky, 2017).

2.1.1 Topographic position index

Classifying relief features is a fundamental scientific approach for extracting information and knowledge on the landscape of an area from vast datasets. In contrast to perceiving continuous representations, the classification and categorization of things simplifies one's ability to comprehend and distinguish patterns (Strobl, 2008). Hence, the classification of landforms or the morphological characteristics of the Earth's surface follows a similar principle, albeit with recent advancements in semiautomated techniques. This process involves reducing complexity into a smaller number of easily identifiable functional units, from which valuable information about tectonic activity and active geomorphic processes can be derived (Burrough et al., 2000).

User manuals for topographic position index (TPI) calculation tools make it evident that using TPI at various scales in combination with slope allows for the classification of the relief based on both topographic positions (e.g., peak, valley) and landform types (e.g., narrow canyon with steep slopes, valley with gently sloping sides). This scale-dependent classification using different neighborhood radii and DEM cell sizes has been applied in various approaches worldwide (e.g., in Turkey by Tagil and Jenness (2008) or in Greece by Ilia et al. (2013)).

The initial phase of classification involves determining the position on the slope. The TPI index values are utilized for this purpose, combining two properties: the extremeness of the index values and the slope at each point. High index values are expected on crests, while low values correspond to midvalleys (i.e., valleys), and values near zero characterize flat or moderately sloping surfaces. The slope value (Tagil & Jenness, 2008) helps differentiate flat surfaces from those of moderate slopes. While the slope location classification grid may vary, most applications use either four (e.g., valley, flat surface, midslope, peak) or six (e.g., valley, low slope, flat area, medium slope, high slope, ridge) categories. Specific parameters are defined, such as the TPI index threshold (for identifying peaks and valleys) and slope threshold (for distinguishing flat areas from midslope areas), along with neighborhood type, radius, and shape. The TPI index type is frequently used due to its normalized elevation, which is the ratio of the index value to the neighborhood standard deviation. In this case, units are in "standard deviation," where a normalized elevation value of one (1) signifies that the cell is one standard deviation higher than the average elevation of its neighborhood (Jenness et al., 2013). Based on the described criteria, the topographic location index thresholds are set to minus one (-1) standard deviation and plus one ($+1$) standard deviation (i.e., $\pm SD$), while the slope threshold is typically set at ten (10) degrees for classification into four (4) classes. For classification into six (6) classes, additional thresholds of -0.5 and 0.5 can be employed.

The final phase of classification involves categorizing landforms and relies on using different cell sizes of the TPI at varying scales to identify diverse and more complex landform types. To achieve up to ten (10) landform categories, criteria proposed by Weiss (2001) can be employed (Fig. 1.1).

2.1.2 Classification of relief according to Hammond and Dikau

Another methodology for the automatic classification of large-scale landforms using DEMs is a process initially developed by Dikau (1989) and subsequently refined by Dikau et al. (1991). This methodology streamlines Hammond's manual procedures (Hammond, 1954, 1964a, 1964b). According to Dikau (1989), large-scale landforms are those whose

Figure 1.1 Classification of landforms in northern Evia Island, Greece, based on the categories proposed by Weiss (2001), combining TPI at a small and large scale (50 and 150 m, respectively).

dimensions range from 1000 to 10 km^2. Dikau et al. (1991) conducted tests of this model in New Mexico and demonstrated that their approach can effectively classify extensive areas in terms of their geomorphic characteristics.

Hammond's method, which has gained wide acceptance (Drăgut & Blaschke, 2006), incorporates quantitative factors such as slope, relative relief, and profile type to identify distinct landforms. Using the automated method, it is possible to recognize as many as 96 different landforms, although these are typically further grouped. Such categorization is proposed by Brabyn (1998), where he begins with level one (1) encompassing twenty two (22) classes and ultimately reaches level six (6) with only two (2) classes (plains and nonflat areas). In the five-level grouping system, there are four (4) main classes: plains, hills, mountains, and plateaus. Even in this case, there is a possibility that not all of them are identified (Fig. 1.2).

2.2 Lineament analysis

Another crucial methodological component of geomorphometry involves the analysis of lineaments within an area. The term "lineament" was initially introduced by O'Leary et al. (1976) to describe a mappable, linear surface feature, which can be simple or complex. Lineaments are considered segments that align in a rectilinear or slightly curvilinear way and have a pattern distinct from adjacent surface features. Masoud and Koike (2006) describe lineaments as physical structures representing zones of structural weakness. Gupta (2003) provides an overview of lineation interpretation in various geological

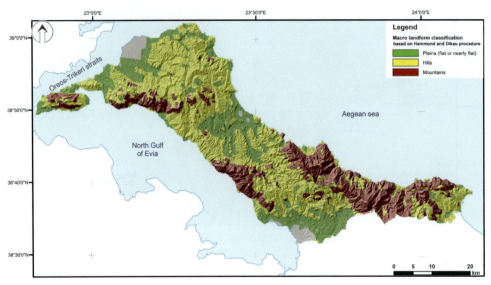

Figure 1.2 Macro landform classification of the north part of Evia Island (Greece) based on Hammond and Dikau procedure.

features, including shear zones, faults, trenches, geological formation contacts, fold axes, discontinuities, fracture traces, and changes in topography, vegetation, soil type, and more. Faults and fracture zones often appear as lines on the Earth's surface. Hence, lineaments are typically associated with faults or linear zones of fracture and deformation (Florinsky, 2012) and their identification is very important in neotectonic studies. Morphological analysis of topographical features is widely used for identifying such tectonic lines (mainly faults). Aerial photographs and satellite images have traditionally been the primary tools for this purpose (Jordan et al., 2005). It's worth noting that automated photogram extraction methods, while providing information on both negative and positive lineaments, may not distinguish between linear forms resulting from anthropogenic activities (such as roads or irrigation canals) and those formed by geological or geomorphological processes. This can result in a high number of lineaments (in terms of frequency and length) compared to other methods for identifying similar linear features (Abdullah et al., 2010). Positive lineaments encompass features interpreted as "positive topography," including ridges, fronts, dips, or craters, while negative lineaments correspond to low topography and represent discontinuities, faults, and shear zones (Abdullah et al., 2010; Raj, 1983).

However, in recent years, DEMs, shaded relief maps, slope maps, as well as orientation maps or curvatures have also been widely used (Onorati et al., 1992; Ganas et al., 2005). Visual interpretation, often performed through the study of DEMs or shaded relief maps, remains a commonly used methodology for mapping these features. It's important to note that this approach has faced criticism due to the subjectivity of researchers and their skills in performing the mapping.

Nonetheless, with or without automated methods, shaded relief analysis is the most widely adopted technique for detecting and mapping fault zones and has been proven to be more effective than satellite images' interpretation (Florinsky, 2012). Shaded relief analysis can identify up to 90% of faults that have been recognized and mapped using geological and geophysical methods (Onorati et al., 1992). Shaded relief maps obtained from DEMs with a resolution of 20 m and a light source at a height of 45 degrees, or variations of the light angle from 0 to 360 degrees per 45-degree increments, have been extensively employed to yield satisfactory results (e.g., Abdullah et al., 2010; Alhirmizy, 2015; Muhammad & Awdal, 2012). This method involves creating two shaded relief images, each resulting from a combination of shaded relief maps with light sources at specific angles. These two images can be used to extract lineaments utilizing automated techniques through software like PCI Geomatica. The algorithm typically includes edge detection, thresholding, and curve extraction steps.

Statistical analysis of linear tectonic structures begins with assessing their spatial distribution, particularly their frequency (calculated using the Kernel function), which is crucial as it can indicate zones where tectonic control is more pronounced (Fig. 1.3). Additionally, the orientation of tectonic lines is a vital aspect of statistical analysis, often depicted through corresponding rose diagrams (Fig. 1.4).

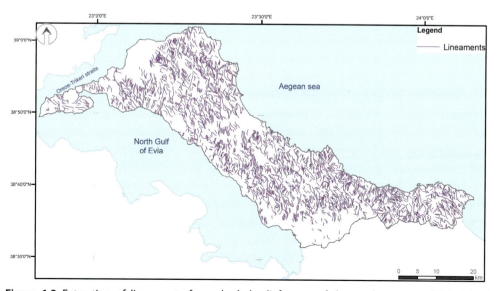

Figure 1.3 Extraction of lineaments from shaded relief maps of the northern part of Evia Island (Greece) based on different angles of light source (0, 45, 90, 135 degrees).

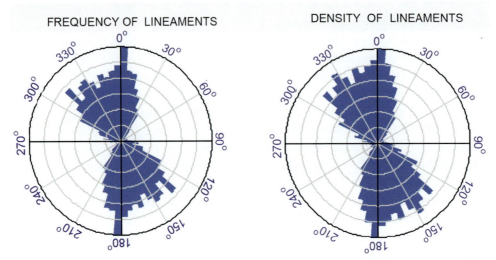

Figure 1.4 Rose diagrams of lineaments as derived from automatic extraction. The example is from the north part of Evia Island (Greece).

3. Identification and mapping of landforms—specific geomorphometry

In addition to general geomorphometry, there is specific geomorphometry which entails the measurement and analysis of specific landforms with distinct boundaries, such as planation surfaces, terraces (both fluvial and marine), alluvial fans, etc (Evans, 1972). To perform measurements on a landform accurately, it must be mapped as precisely as possible (Goudie, 2004) and this can be achieved utilizing GIS techniques.

3.1 Planation surfaces-inclined surfaces—morphological discontinuities

Planation surfaces are a subject of ongoing debate within the realm of relief forms (Veselský et al., 2015). Despite their seemingly straightforward definition, mapping and analyzing them can be quite complex (Goudie, 2004). The formation, characteristics, and preservation of planation surfaces are also influenced by the lithology of the geological formations they are developed on, the climatic conditions, and the tectonic activity. These landforms hold significance because they provide insights into the evolution of terrain over time (Posea, 1997; Karymbalis et al., 2013) and often serve as records of mantle dynamics (Guillocheau et al., 2018). They signify periods of tectonic calm and a humid, warm climate during the development of mountain masses. They are indications of the frequency of tectonic movements, with surfaces at higher elevations generally considered older (Pavlopoulos et al., 2009). To study them effectively, accurate and detailed mapping is required.

A comprehensive analysis of planation surfaces involves the examination of their morphological characteristics, the application of appropriate visualization techniques using DEMs and derived maps, on-site fieldwork where feasible, and the study of Google Earth Pro images when fieldwork is not possible. Ultimately, a geographical distribution map of the planation surfaces, grouped based on their altitudes of development, should be created. The initial step for creating such maps is to define typical horizontal surfaces, which are large, easily identifiable surfaces characterized by an average slope lower than that of neighboring areas.

The maps of morphological gradients and slope orientation can be valuable tools for studying the planation surfaces of an area and serve as fundamental analysis maps, either individually or in combination with each other or with geological maps. They aid in locating and initially identifying horizontal or near-horizontal surfaces, surfaces with low slopes, but also morphological discontinuities often associated with tectonic structures. A useful methodology for mapping these surfaces has been developed by Vassilakis (2006) and was tested in Evia Island Greece (Fig. 1.5).

This includes the identification of surfaces with slope of 0—1 degree, of surfaces with lower degree of flattening 2—5 degrees, as well as the identification of morphological discontinuities with slope >45 degrees, which constitute areas of intense erosion or topographic prominence. The methodology uses the map of morphological slopes, the conversion of raster data into vectors, and the calculation of selected surface parameters like area, perimeter, maximum, minimum, mean elevation, and range. The last step of

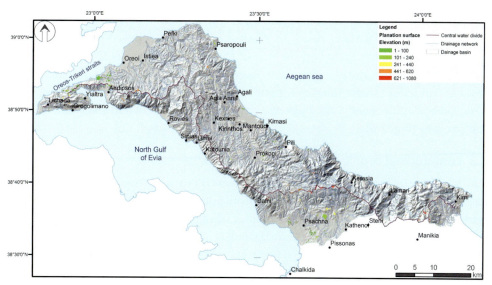

Figure 1.5 Distribution and classification of the identified planation surfaces (0—1 degree) in the northern part of Evia Island, Greece.

the process includes the selection of the most reliable and representative surfaces based on their area in combination with the cell size of the DEM and finally, the classification of these planation surfaces.

3.2 Fluvial terraces

Fluvial terraces, whether depositional or rocky, can be found either individually or in pairs on either side of a riverbed. They represent a geomorphic response of rivers to climatic changes, tectonic activity, and alterations in base level, which can be influenced by both climate and tectonics. These changes in base level, along with variations in river supply and sediment load, result in the formation of terraces of varying ages, with the higher-altitude terraces typically being the oldest. Their characteristic morphology includes a flat surface bounded by the valley walls and a steeply sloping front that descends toward the riverbed. The original abandoned plain remains as a flat bench or terrace, separated from newer terraces by steep slopes (Summerfield, 1991).

The advancement of high-resolution DEMs has facilitated the development of several methods for identifying and mapping terraces. These methods are cost-effective and efficient when compared to detailed field surveys and can lead to a preliminary map of fluvial terraces which is very useful for the fieldwork. In the initial stages of research, geometric approaches can be employed to locate fluvial terraces. For example, Demoulin et al. (2007) utilized local slope and elevation data from each cell compared to the bed elevation, focusing on topographic features along longitudinal profiles of river channels. However, the necessity of creating terrace maps on a broader scale prompted further advancements in these techniques. Stout and Belmon (2014) developed a semiautomated toolbox known as TerEx. This toolbox is designed to identify potential terrace surfaces based on detailed DEMs, the digitized drainage network, and various parameters including the size of the focal window, local relief, minimum area, and maximum distance from the riverbed. Following the initial identification, manual correction of the surfaces is performed to enhance the accuracy of the result. The toolbox subsequently calculates the extent of the terraces and their height above the riverbed (Fig. 1.6).

3.3 Longitudinal profile analysis—knickpoints

The study of the longitudinal profile of the rivers is crucial in studies related to the tectonic activity of an area. The longitudinal profile of a river is a diagram derived from the relationship between elevation (y) and distance (along the bed) downstream of the mouth (x) and represents the gradient of the river from its headwaters to its outlet. There are two types of profiling, those in which the horizontal axis is on a numerical scale and those in which it is on a logarithmic scale, while the vertical axis is numerically scaled in both cases. Balanced rivers—those whose bed slope, width, and depth are in dynamic equilibrium with the discharge and sediment load transported along the bed—typically develop

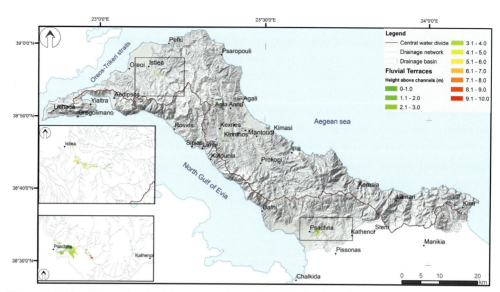

Figure 1.6 Map illustrating fluvial terraces along channels of selected drainage networks, obtained after implementing the described methodology in the northern part of Evia Island, Greece.

a smooth concave profile. So, the semilog plot of these, as originally proposed by Hack (1957, pp. 45–95) for areas with common lithology, will be a straight line of the equation:

$$H = C - K \ln(L).$$

where H is the elevation above the datum, L is the length from the highest point of the riverbed, C is the elevation of the highest point of the riverbed, and K is the slope of the line (the slopes are such that erosion and deposition processes to be in equilibrium).

The concave profile results in part from the well-known inverse deposition—slope relationship whereby as discharge increases and the grain size of the sedimentary load decreases downstream, smaller gradients are required to transport material and simultaneously to erode bedrock (Pazzaglia et al., 1998).

The concave equilibrium profile of a river as well as the steepness of the profile depends on a number of factors related to geology, geomorphology, sediment transport (e.g., Hack, 1957, pp. 45–95), as well as the tectonic history of the area (e.g., Snyder et al., 2000), and provides important information on neotectonic activity (e.g., Molin et al., 2004; Ramírez-Herrera, 1998). A variation from equilibrium is the curvature of the profile as a whole, which is expressed by a low gradient in the upstream part of the river (Gilbert, 1877) and a relatively steep gradient in the downstream part (Hanks & Webb, 2006). This form may result from the presence of more erodible formations along the river bed upstream (smooth slopes) and more erosion resistant formations (resulting in steeper slopes) downstream. Locally abrupt changes (discontinuities) in the profile of a river can be due to the presence of different lithology formations along the bed. Often such changes are not associated with equilibrium conditions but are due to local faults that vertically cross the bed (Goldrick & Bishop, 1995; Kirby & Whipple, 2012). The curved section is called knickpoint (it is a dip in base level or a discontinuity in the longitudinal profile of the river), and can take the form of a vertical knickpoint step or a knickpoint slope change. Depending on their length compared to the total length of the river, it is possible to draw important conclusions about their relationship with tectonic movements at various scales (Molin et al., 2004). For example, studying the sharpness of knickpoints is useful because it gives information about how recent active tectonics or capture-piracy-river events took place. Generally, the steeper an inflection point, the more recently it was created (Wobus et al., 2006). Also, the deviation from the state of equilibrium is usually determined by the way the river system reacts to a temporal and spatial disturbance.

River profiles are usually studied individually by conventional methods such as analysis of normalized diagrams of river length (Li/L where $L =$ the total length of the riverbed) in relation to altitude ($\Delta Hi/\Delta H$ where $\Delta H =$ the total difference in altitude from the highest point of the bed to the estuary). This analysis allows profile comparisons between rivers of different lengths and gradients and highlights variations in river gradient (Demoulin, 1998). When analyzing a river profile, its curvature is the change in its slope, so calculating this curvature is useful in identifying possible anomalies, the stage of bed erosion, or any elevation of the area through which the river passes. It should be noted that according to Zaprowski et al. (2005), the steam concavity index Sci covaries with the concavity "θ" derived from the slope of the regression line in the logarithmic slope—area plots (SA plots). Both of these indices effectively measure the curvature of the profile and can be easily calculated utilizing GIS functionalities. Often, the profiles are studied alongside the stream length-gradient index (SL). It is intended that the "new knickpoints", that are not associated with any already recorded fault (compared to the fault fabric of a study area), in sections with a high SL index, shall be coevaluated with the lineaments and in this way, the new lineaments (possible faults) that affect the profiles will emerge (Fig. 1.7).

Knickpoint mapping is essential in assessing the neotectonic activity in an area. The "Knickpoint finder" software (Queiroz et al., 2015) is a valuable tool for identifying knickpoints. It was developed as an extension to ESRI's ArcGIS platform to expedite analyses in neotectonics studies. This tool is based on the methods of Hack (1973) and Etchebehere et al. (2004) and utilizes the DEM along with settings like elevation equidistance to locate knickpoints along the channels of the drainage network of an area. For each detected point, it calculates the relative drainage energy (RDE), which is a derivative of the Hack index (SL) indicating the current energy within a specific part of the

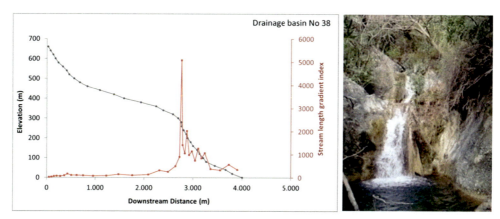

Figure 1.7 (a) Example of analysis using longitudinal profile and SL values of a drainage basin in Evia Island, Greece, and (b) photo of field observation at SL perturbations.

drainage network's bed. Additionally, it calculates RDEt, which considers the total length of a river and the total slope between its source and estuary. The ratio RDEs/RDEt reveals the degree of anomaly in the analyzed drainage network. When this ratio is ≥2, it suggests an anomaly in the network's development. Specifically, anomalies between 2 and 10 are considered second order, while those exceeding 10 are considered first order (Fig. 1.8).

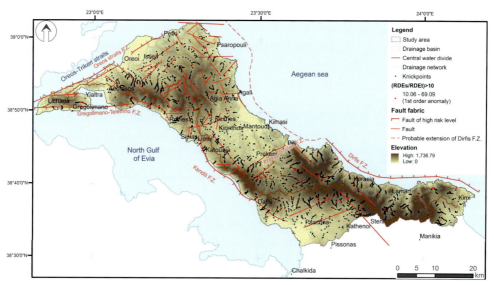

Figure 1.8 Map illustrating the results obtained after implementing the methodology for mapping knickpoints in the northern part of Evia Island, Greece.

4. Morphometric indices of active tectonics

Tectonic geomorphology studies utilize an integrated approach, employing specific geomorphological indicators that are closely tied to the tectonics of a region. The goal is to decipher the tectonic history responsible for the surface features we observe today. In some cases, the reverse approach is also used, where information about the tectonic history helps predict the landforms that will emerge. These studies can unveil the nature, timing, and distribution of faulting and seismic activity that occurred tens of thousands of years ago (Maroukian et al., 1995; Mayer, 1986; Papanastassiou & Gaki-Papanastassiou, 1993). One of the primary challenges in tectonic geomorphology lies in developing methodologies to extract tectonic information (in the form of morphometric indices) from existing topography and geomorphology in an area (Wang et al., 2009). Morphometric indices of active tectonics are widely used tools for establishing links between tectonics and surface morphology on regional or local scales, often used to determine recent geological deformations (Burbank & Anderson, 2008; Keller & Pinter, 2002). These indicators reveal how landforms respond to recent tectonic processes. Among the most significant landforms sensitive to tectonic movements, such as uplift and tilting, are fluvial systems (Leeder et al., 1991; Gaki-Papanastassiou et al., 2007). The characteristics of these systems are directly tied to tectonic movements, making them valuable records of tectonic deformation (Schumm, 1986). Therefore, examining drainage networks and associated catchments using morphometric indices can yield valuable insights into the tectonic history of a region (Toudeshki & Arian, 2011a, 2011b). The practical applications of geomorphological analysis in assessing active tectonics have been demonstrated in numerous studies across various tectonically active regions worldwide (e.g., Pérez-Peña et al., 2010; Valkanou et al., 2020; Valkanou et al., 2021). These studies have contributed to quantifying the impact of these processes on surface morphology in tectonically active areas.

Regardless of the study area's size, there are geomorphological indicators related to fluvial systems that can be linked to uplift rates (Kirby & Whipple, 2001). These geomorphological indicators are often closely associated with a region's tectonic regime and can help define it. The initial step involves constructing and analyzing longitudinal profiles of the main rivers and their tributaries. Subsequently, various simple geomorphological indices can be computed, including the asymmetry index (AF), the transverse topographic symmetry coefficient (T), the stream length–gradient index (SL), the valley height to width ratio (Vf), the hypsometric curve and integral (Hi), and the mountain front sinuosity index (Smf) (Table 1.1). Researchers often compare the values of these indices and display their spatial variations. The calculation of all these morphometric variables and the visualization of the spatial distribution of their values can be easily achieved with the use of GIS techniques.

Table 1.1 Morphometric indices that are commonly used for the estimation of relative tectonic activity.

Asymmetry factor (AF)	$AF = (Ar/A)*100$, Ar = the area of the basin in right-hand of the main channel (facing downstream) and A = the total area of the drainage basin	Hare and Gardner (1985)
Transverse topographic symmetry factor (T)	$T = Da/Dd$, Da = the distance from the midline of the drainage basin to the midline of the active meander belt and Dd = the distance from the basin midline to the basin divide	Cox (1994)
Stream length–gradient index (SL)	$SL = (\Delta h/\Delta I)*L$, $\Delta h/\Delta I$ = the channel slope or gradient of the reach and L = the total channel length from the midpoint of the reach of interest upstream to the highest point on the channel	Hack, 1973; Keller & Pinter, 2002
Valley height-width ratio (Vf)	$Vf = 2Vfw/(Eld + Erd - 2Esc)$, Vfw = the width of valley floor, Eld and Erd = the elevations of the left and right valley divides, respectively, and Esc = the elevation of the valley floor	Bull and McFadden (1977)
Hypsometric integral (Hi)	$Hi = (H_{mean} - H_{min})/(H_{max} - H_{min})*100$, H_{mean} = mean elevation, H_{min} = minimum elevation, and H_{max} = maximum elevation of the drainage basin	Pike and Wilson (1971), Mayer (1990)
Mountain front sinuosity (Smf)	$Smf = Lmf/Ls$, Lmf = the length of the mountain front along the foot of the mountain, at the break in slope and Ls = the straight-line length of the mountain front	Bull and McFadden (1977)

4.1 Composite morphometric indices

While individual morphotectonic indices are valuable in identifying specific characteristics of an area and determining the complex tectonic regime, combining multiple indicators yields more meaningful results (Keller & Pinter, 2002). This is particularly important when individual indices produce conflicting results. An early example of this approach is the classification of the Garlock fault area in southern California by Bull and McFadden (1977), who used indices like Vf, Smf, Bs, slope inclination, and alluvial fan quantitative characteristics to categorize tectonic activity. Similarly, McKeown et al. (1988, pp. 1—39) aimed to identify geomorphological evidence of recent local or regional tectonic deformations associated with seismic activity in the Ozark Mountains of Arkansas. They compared fluvial profiles, SL index values, and lithology, which led them to a similar classification.

In subsequent studies (conducted after 2003, as shown in Table 1.2), to mitigate potential uncertainties arising from applying individual morphometric indices and parameters, several researchers combined two or more indicators to extract semiquantitative information about the relative tectonic activity in specific areas (e.g., Greece, Fig. 1.9).

In most cases, their findings were corroborated by field observations. Others went further by creating new composite indices from numerical means (El Hamdouni et al., 2008; Mahmood & Gloaguen, 2012). Alipoor et al. (2011) adopted a different approach, using simple analytical ranking instead of numerical means. They argued that their method could be applied universally, even in regions beyond the Zagros Mountain range where they tested it. This contrasts with the methods of other researchers, which rely on criteria for classification that may not be applicable in areas outside the study area of the respective composite index.

After computing these indices, with the use of GIS techniques, the results are often compared with field geomorphological observations (e.g., mapping of deep valleys, landslides, changes in river flow, evidence of tectonic uplift, etc.). Additionally, they cross-reference their findings with existing literature reports on recent tectonic activity in the area to validate the results of quantitative geomorphological analysis. This approach also involves comparing results with the SL index, which has been used individually in many cases, such as in the Apennine mountains in the Marche province of central Italy by Troiani and Della Seta (2008) and in the state of Pernambuco in NE Brazil by Monteiro et al. (2010).

Table 1.2 Morphometric parameters that are often combined to create composite indices for the estimation of relative tectonic activity.

	References	Year	Study area	Vf	Smf	Af	Hi	SL	T	Bs	Rc	Rb	D	Sb	Rh	M	Slk	Hc	FD	F%	Long. R. Prof.	Mean slope of Tr. Facet	Param. Of Long. R. Prof.	Top. Prof. Of al. Fans
1	Silva et al.	2003	Southeast Spain	�	▩																			
2	Zovoili et al.	2004	Kompotades and Nea Anchalos faults (Greece)	▩	▩			▩																
3	Verrios et al.	2004	Eliki fault zone, Gulf of Corinth (Greece)	▩	▩				▩															
4	Tsimi et al.	2007	Western Corinth Rift (central Greece)	▩	▩	▩	▩																	
5	Tsimi et al.	2007	Sfakia normal fault, SouthWestern Crete (Greece)		▩	▩	▩															▩		
6	Vassilakis et al.	2007	SE coastal area of the Gulf of Corinth (Greece)			▩	▩															▩	▩	
7	Fountoulis et al.	2007	Central – Western Peloponnese (Greece)	▩	▩																			
8	El-Hamdouni et al.	2008	Sierra Nevada (southern Spain)		▩	▩	▩	▩		▩														

Continued

Table 1.2 Morphometric parameters that are often combined to create composite indices for the estimation of relative tectonic activity.—cont'd

References	Year	Study area	Vf	Smf	Af	Hi	SL	T	Bs	Rc	Rb	D	Sb	Rh	M	Slk	Hc	FD	F%	Long. R. Prof.	Mean slope of Tr. Facet	Param. Of Long. R. Prof.	Top. Prof. Of al. Fans
Tsodoulos et al.	2008	Beotia (Central Greece)	▨	▨	▨		▨		▨							▨	▨						
Pedrera et al.	2009	Eastern Betic Cordillera (Spain)	▨																				▨
Sboras et al.	2010	Beotia (Central Greece)	▨	▨	▨	▨		▨															
Pérez-Peña et al.	2010	Sierra Nevada, Betic Cordillera (SE Spain)	▨	▨	▨												▨			▨			
Dehbozorgi et al.	2010	Sarvestan area, central Zagros (Iran)	▨	▨	▨	▨	▨		▨														
Alipoor et al.	2011	High Zagros Belt (SW of Iran)	▨	▨	▨	▨	▨	▨	▨														
Toudeshki and Arian	2011a, 2011b	Ghezel Ozan River Basin (NW Iran)	▨	▨					▨														
Sarp et al.	2011	Yenicaga basin area (Turkey)	▨	▨																			
Jayappa et al.	2012	Western Ghats (India)	▨	▨	▨		▨		▨														
Mahmood and Gloaguen	2012	Hindu Kush (NW Pakistan and NE Afghanistan)	▨	▨														▨					
Michail and Chatzipetros	2013	Sperchios basin Frhiotis (Central Greece)	▨	▨	▨	▨	▨	▨	▨														

#	Author	Year	Location
20	Kokinou et al.	2013	Heraklion Basin (Crete Greece)
21	Arian and Aram	2014	Kermanshah area (Western Iran)
22	Matos et al.	2014	Mt. Medvednica (NW Croatia)
23	Dar et al.	2014	Karewa Basin of Kashmir Valley (Northwest India)
24	Fountoulis et al.	2015	NW Peloponnese (Greece)
25	Matos et al.	2016	Bilogora Mt. Area (NE Croatia)
26	Eynoddin et al.	2017	Bozgoush Basin, SW of Caspian Sea (NW of Iran)
27	Koukouvelas et al.	2017	Kenchreai fault, Gulf of Corinth (Greece)
28	Zygouri et al.	2018	Corinth graben (Greece)
29	Valkanou et al.	2020	Evia Island (Greece)

Vf, valley height-width ratio index; Smf, mountain front sinuosity index; Af, asymmetry factor; Hi, hypsometric integral; SL, stream length–gradient index; T, transverse topographic symmetry factor; Bs, basin shape; Rc, circulation ratio; Rb, bifurcation ratio; D, drainage density; Sb, drainage basin slope; Rh, relief ratio; M, Melton ruggedness number; Slk, SL index normalized with graded river gradient (k); Hc, hypsometric curve; FD, fractal dimension; $F\%$, percentage of faceting along mountain front; $Lon. R. Prof.$, longitudinal river profile; $Tr. Facet$, triangular facet; $Param. of Long. R. Prof.$, parameters of longitudinal river profiles -SI, Cf, Cmax, $\Delta I/L$, Ksn, θ, and ks-; $Top. Prof. of Al. Fans$, topographic profiles of alluvial fans.

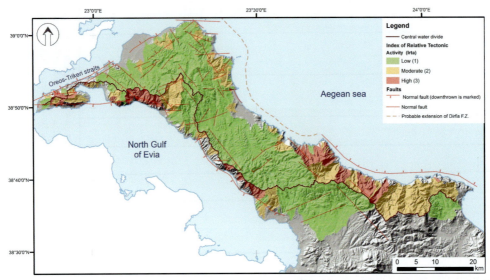

Figure 1.9 Spatial distribution map of index of relative tectonic activity (IRTA), applied in Evia Island, Greece.

5. Conclusions

Geomorphology is crucial for understanding the evolution of Earth's surface over geological time scales. It is a multidisciplinary field that explores the complex relationship between landforms and the processes that shape them.

This chapter illustrates the significance of geomorphological studies and the role of geomorphometry, quantitative morphometric analysis, and GIS technology in comprehending, characterizing, and classifying landscapes in tectonically active regions. To evaluate the prevailing tectonic regime and neotectonic deformation, a comprehensive analysis of the study area's topography is essential, encompassing quantitative assessments at various scales and employing multiple methodologies. The initial step in recognizing the primary landforms of a tectonically active region involves the automated categorization of the relief. This is achieved through techniques such as employing the TPI and the Hammond and Dikau's landform classification. It is also crucial to examine the lineaments that correspond to faults or linear zones of fracture using the DEM of the area and its derived data. Morphological and tectonic studies greatly benefit from this analysis. GIS software and tools are valuable for the identification, and mapping, of specific geographical features, such as planation surfaces, fluvial terraces, and knickpoints. Evaluating the spatial distribution of these geographical features, in combination with the values of individual or composite morphotectonic indices (such as asymmetry index (AF), transverse topographic symmetry coefficient (T), stream length-gradient index

(SL), valley height to width ratio (Vf), hypsometric curve and integral (Hi), and mountain front sinuosity index (Smf)) that can be readily computed using GIS techniques, can facilitate the assessment of relative neotectonic activity.

6. Key takeaway and future research

A thorough comprehension of landform dynamics is attained through a detailed analysis of landscapes utilizing advanced GIS techniques, which involves automated relief categorization and the calculation of morphometric indices. The application of landform classification methodological approaches, along with DEM analysis, enables the accurate delineation of landforms and the assessment of neotectonic activity. This interdisciplinary approach enhances the understanding of Earth's processes and underlines the significant role of geoinformation technologies in geomorphological research in tectonically active regions. However, it's essential to conduct field observations to validate the findings of such approaches. Potential areas for future research include the implementation of advanced remote sensing techniques, machine learning applications, predictive modeling, and the integration of geospatial data, such as satellite imagery. These methods can provide highly accurate monitoring and measurements of active tectonic processes and deformation along with automated detection of landform changes. Furthermore, geochronological studies, involving sampling and dating, as well as investigations into paleoseismology and paleogeography, can prove helpful in depicting the recent geomorphological evolution, and in determining the timing and duration of past geological events.

References

Abdullah, A., Akhir, J. M., & Abdullah, I. (2010). *Automatic mapping of lineaments using shaded relief images derived from digital elevation model (DEMs) in the Maran—Sungi Lembing area* (Vol 15, pp. 949—957). Malaysia: EJGE.

Alhirmizy, S. (2015). Automatic mapping of lineaments using shaded relief images derived from digital elevation model (DEM) in Kirkuk northeast Iraq. *International Journal of Science and Research, 4*(5), 2228—2233.

Alipoor, R., Poorkermani, M., Zare, M., & El Hamdouni, R. (2011). Active tectonic assessment around Rudbar Lorestan dam site, high Zagros Belt (SW of Iran). *Geomorphology, 128*, 1—14.

Arian, M., & Aram, Z. (2014). Relative tectonic activity classification in the Kermanshah area, western Iran. *Solid Earth, 5*, 1277—1291.

Bishop, P. (2007). Long-term landscape evolution: Linking tectonics and surface processes. *Earth Surface Processes and Landforms, 32*, 329—365.

Brabyn, L. (1998). GIS analysis of Macro landform. New Zealand, 1998. In *10th Colloquium of the spatial information research Centre, University of Otago*.

Bull, W. B., & McFadden, L. (1977). Tectonic geomorphology north and South of the Garlock fault, California. In D. O. Dohering (Ed.), *Geomorphology in arid regions. Binghamton: Publ. In geomorphology* (pp. 115—138). State University of New York.

Burbank, D. W., & Anderson, R. S. (2008). *Tectonic geomorphology*. UK: Blackwell Science.

Burrough, P. A., Van Gaans, P. F. M., & MacMillan, R. A. (2000). High resolution landform classification using fuzzy-k means. *Fuzzy Sets and Systems, 113*, 37—52.

Cox, R. T. (1994). Analysis of drainage-basin symmetry as a rapid technique to identify areas of possible Quaternary tilt-block tectonics: An example from the Mississippi Embayment. *Geological Society of America Bulletin, 106*, 571–581.

Dar, R. A., Romshoo, S. A., Chandra, R., & Ahmad, I. (2014). Tectono-geomorphic study of the Karewa basin of Kashmir valley. *Journal of Asian Earth Sciences, 92*, 143–156.

Dehbozorgi, M., Pourkermani, M., Arian, M., Matkan, A. A., Motamedi, H., & Hosseiniasl, A. (2010). Quantitative analysis of relative tectonic activity in the Sarvestan area, central Zagros, Iran. *Geomorphology, 121*, 329–341.

Demoulin, A. (1998). Testing the tectonic significance of some parameters of Longitudinal River Profiles: The case of the Ardenne (Belgium, NW Europe). *Geomorphology, 24*, 189–208.

Demoulin, A., Bovy, B., Rixhon, G., & Cornet, Y. (2007). An automated method to extract fluvial terraces from digital elevation models: The Vesdre valley, a case study in eastern Belgium. *Geomorphology, 91*(1–2), 51–64.

Dikau, R., Brabb, E. E., & Mark, R. M. (1991). Landform classification of New Mexico by computer. *U.S. Geological Survey Open File Report, 91–634*, 1–15.

Dikau, R. (1989). The application of a digital relief model to landform analysis. In J. F. Raper (Ed.), *Three dimensional applications in geographical information systems* (pp. 51–77). London: Taylor and Francis.

Drăgut, L., & Blaschke, T. (2006). Automated classification of landform elements using object-based image analysis. *Geomorphology, 81*, 330–344.

El Hamdouni, R., Irigaray, C., Fernández, T., Chacón, J., & Keller, E. A. (2008). Assessment of relative active tectonics, southwest border of the Sierra Nevada (southern Spain). *Geomorphology, 96*, 150–173.

Etchebehere, M. L. C., Saad, A. R., Fulfaro, V. J., & Perinotto, J. A. J. (2004). Aplicação do Índice "Relação Declividade-Extensão—RDE" na Bacia do Rio do Peixe (SP) para Detecção de Deformações Neotectônicas. *Revista do Instituto de Geociências—USP Série Científica, São Paulo, 4*(2), 43–56.

Evans, I. S. (1972). General geomorphometry, derivatives of altitude, and descriptive statistics. In R. J. Chorley (Ed.), *Spatial analysis in geomorphology* (pp. 17–90). New York: Harper and Row.

Evans, I. S. (1998). What do terrain statistics really mean? In S. Lane, K. Richards, & J. Chandler (Eds.), *Landform monitoring, modelling and analysis* (pp. 119–138). Chichester: Wiley.

Eynoddin, E., Solgi, A., Pourkermani, M., Matkan, A., & Arian, M. (2017). Assessment of relative active tectonics in the Bozgoush basin (SW of Caspian Sea). *Open Journal of Marine Science, 7*, 211–237.

Florinsky, I. V. (2017). An illustrated introduction to general geomorphometry. *Progress in Physical Geography: Earth and Environment, 41*(6), 723–752. https://doi.org/10.1177/0309133317733667

Florinsky, I. V. (2012). *Digital terrain analysis in soil science and geology* (1st ed.). Oxford: Academic Press is an imprint of Elsevier.

Fountoulis, I., Vassilakis, E., Mavroulis, S., Alexopoulos, J., Dilalos, S., & Erkeki, A. (2015). Synergy of tectonic geomorphology, applied geophysics and remote sensing techniques reveals new data for active extensional tectonism in NW Peloponnese (Greece). *Geomorphology, 237*, 52–64.

Gaki-Papanastassiou, K., Papanastassiou, D., & Maroukian, H. (2007). Recent uplift rates at Perachora Peninsula, east Gulf of Corinth, Greece, based on geomorphological—archaeological evidence and radiocarbon dates. *Hellenic Journal of Geosciences, 42*, 45–56.

Ganas, A., Pavlides, S., & Karastathis, V. (2005). DEM-based morphometry of range-front escarpments in Attica, central Greece, and its relation to fault slip rates. *Geomorphology, 65*, 301–319.

Gilbert, G. K. (1877). *Report on the geology of the Henry mountains*. Washington, D.C.: United States Government Printing Office.

Goldrick, G., & Bishop, P. (1995). Differentiating the roles of lithology and uplift in the steepening of bedrock river Long profiles: An example from Southeastern Australia. *The Journal of Geology, 103*(2), 227–231.

Goudie, A. S. (2004). *Encyclopedia of geomorphology* (Vol. 1). London: Routledge Ltd. A–I.

Guillocheau, F., Simon, B., Baby, G., Bessin, P., Robin, C., & Dauteuil, O. (2018). Planation surfaces as a record of mantle dynamics: The case example of Africa. *Gondwana Research, 53*(8298). https://doi.org/10.1016/j.gr.2017.05.015

Gupta, R. P. (2003). *Remote sensing geology*. Berlin: Springer Berlin Heidelberg.

Hack, J. T. (1973). Stream-profile analysis and stream-gradient index. *Journal of Research of the U. S. Geological Survey, 1*(4), 421–429.

Hack, J. T. (1957). *Studies of longitudinal stream profiles in Virginia Maryland* (pp. 45–95). U.S. Geological Survey Professional Paper.

Hammond, E. H. (1964a). Analysis of properties in landform geography: An application to broadscale landform mapping. *Annals of the Association of American Geographers, 54*(1), 11–19.

Hammond, E. H. (1964b). Classes of land surface form in the forty-eight states, U.S.A. *Annals of the Association of American Geographers, 54*(1) (p.map supplement).

Hammond, E. H. (1954). Small scale continental landform maps. *Annals of the Association of American Geographers, 44*, 32–42.

Hanks, T. C., & Webb, R. H. (2006). Effects of tributary debris on the longitudinal profile of the Colorado River in Grand Canyon. *Journal of Geophysical Research, 111*(F02020).

Hare, P. W., & Gardner, T. W. (1985). Geomorphic indicators of vertical neotectonism along converging plate margins, Nicoya Peninsula, Costa Rica. In M. Morisawa, & J. T. Hack (Eds.), *Tectonic geomorphology: Proceedings of the 15th geomorphology symposia series, Binghamton* (pp. 75–104). Boston: Allen and Unwin.

Ilia, I., Rozos, D., & Koumantakis, I. (2013). Landform classification using GIS techniques. The case of Kimi municipality area, Euboea Island, Greece. Chania, 2013. *Bulletin of the Geological Society of Greece, XLVII.*

Jayappa, K. S., Markose, V. J., & Nagaraju, M. (2012). Identification of geomorphic signatures of neotectonic activity using DEM in the Precambrian terrain of Western Ghats, India. In *International archives of the photogrammetry, remote sensing and spatial information sciences, XXII ISPRS congress, Melbourne, Australia, XXXIX(B8)* (pp. 215–220).

Jenness, J., Brost, B., & Beler, P. (2013). *Land facet corridor designer, revision 1.2.884 (including tools for topographic position index): Extension for ArcGIS. [Online] Arizona: USDA forest service rocky mountain research station.* Available at: http://www.jennessent.com/arcgis/land_facets.htm. (Accessed 2 February 2017).

Jordan, G., Meijninger, B. M. L., van Hinsbergen, D. J. J., Meulenkamp, J. E., & van Dijk, P. M. (2005). Extraction of morphotectonic features from DEMs: Development and applications for study areas in Hungary and NW Greece. *International Journal of Applied Earth Observation and Geoinformation, 7*, 163–182.

Karymbalis, E., Papanastassiou, D., Gaki-Papanastassiou, K., Ferentinou, M., & Chalkias, C. (2016). Late Quaternary rates of stream incision in Northeast Peloponnese, Greece. *Frontiers in Earth Science, 10*, 455–478.

Karymbalis, E., Papanastassiou, D., Gaki-Papanastassiou, K., Tsanakas, K., & Maroukian, H. (2013). Geomorphological study of Cephalonia Island, Ionian Sea, Western Greece. *Journal of Maps, 9*, 121–134.

Keller, E. A., & Pinter, N. (2002). *Active tectonics, earthquake, uplift and landscape* (2nd ed.). New Jersey: Prentice Hall.

Kirby, E., & Whipple, K. (2001). Quantifying differential rock-uplift rates via stream profile analysis. *Geology, 29*(5), 415–418.

Kirby, E., & Whipple, K. X. (2012). Expression of active tectonics in erosional landscapes. *Journal of Structural Geology, 44*, 54–75.

Kokinou, E., Skilodimou, H., & Bathrellos, G. (2013). Morphotectonic analysis of Heraklion Basin (Crete, Greece). *Bulletin of the Geological Society of Greece, 47*, 285–294.

Koukouvelas, I. K., Zygouri, V., Papadopoulos, G. A., & Verroios, S. (2017). Holocene record of slip-predictable earthquakes on the Kenchreai Fault, Gulf of Corinth, Greece. *Journal of Structural Geology, 94*, 258–274.

López-Ramos, A., Medrano-Barboza, J. P., Martínez-Acosta, L., Acuña, G. J., Remolina López, J. F., & López-Lambraño, A. A. (2022). Assessment of Morphometric Parameters as the Basis for Hydrological Inferences in Water Resource Management: A Case Study from the Sinú River Basin in Colombia. *ISPRS International Journal of Geo-Information, 11*, 459.

Leeder, M. R., & Jackson, J. A. (1993). The interaction between normal faulting and drainage in active extensional basins, with examples from the western United States and central Greece. *Basin Research, 5*, 79–102.

Leeder, M. R., Seger, M. J., & Stark, C. P. (1991). Sedimentation and tectonic geomorphology adjacent to major active and inactive normal faults, southern Greece. *Journal of the Geological Society of London, 148*, 331–343.

Mahmood, S. A., & Gloaguen, R. (2012). Appraisal of active tectonics in Hindu Kush: Insights from DEM derived geomorphic indices and drainage analysis. *Geoscience Frontiers, 3*(4), 407–428.

Maroukian, H., Papanastassiou, D., & Gaki-Papanastassiou, K. (1995). Geomorphological and Seismotectonic study of the broader area of the lake Vouliagmeni, peninsula of Perachora. In *Proceedings of the 4th panhellenic geographical congress of the Greek geographical society* (pp. 162–175).

Masoud, A., & Koike, K. (2006). Tectonic architecture through Landsat-7 ETM+/SRTM DEM- derived lineaments and relationship to the hydrogeologic setting in Siwa region, NW Egypt. *Journal of African Earth Sciences, 45*, 467–477.

Matoš, B., Pérez-Peña, J. V., & Tomljenović, B. (2016). Landscape response to recent tectonic deformation in the SW Pannonian Basin: Evidence from DEM-based morphometric analysis of the Bilogora Mt. area, NE Croatia. *Geomorphology, 263*, 132–155.

Matoš, B., Tomljenović, B., & Trenc, N. (2014). Identification of tectonically active areas using DEM: a quantitative morphometric anal y sis of Mt. Medvednica, NW Croatia. *Geological Quaterly, 58*(1), 51–70.

Mayer, L. (1990). *Introduction to quantitative geomorphology: An exercise manual.* Englewood Cliffs, NJ: Prentice Hall.

Mayer, L. (1986). Tectonic Geomorphology of Escarpments and Mountain Fronts. In Geophysics Study Committee (Ed.), *Active tectonics: Impact on society* (pp. 125–135). Washington, D.C.: National Academy Press.

McKeown, F. A., Jones-Cecil, M., Askew, B. L., & McGrath, M. B. (1988). *Analysis of stream-profile data and inferred tectonic activity, Eastern Ozark mountains region* (pp. 1–39). U.S. Geological Survey Bulletin, 1807.

Michail, M., & Chatzipetros, A. (2013). Morphotectonic analysis of faults in Sperchios basin (Fthiotis, Central Greece). *Bulletin of the Geological Society of Greece, XLVII*(1), 295–304.

Molin, P., Pazzaglia, F. J., & Dramis, F. (2004). Geomorphic expression of active tectonics in a rapidly-deforming forearc Sila Massif, Calabria, Southern Italy. *American Journal of Science, 304*, 539–589.

Monteiro, K.d. A., Missura, R., & Correa, A. C.d. B. (2010). Application of the Hack index—Or stream length-gradient index (Sl index)—To the Tracunhaém river watershed, Pernambuco, Brazil. *Geociencias, 29*(4), 533–539.

Moore, I. D., Grayson, R. B., & Ladson, A. R. (1991). Digital terrain modelling: a review of hydrological, geomorphological and biological applications. *Hydrological Processes, 5*, 3–30.

Morisawa, M. (1958). Measurement of Drainage-Basin Outline Form. *The Journal of Geology, 66*(5), 587–591.

Muhammad, M. M., & Awdal, A. H. (2012). Automatic Mapping of Lineaments Using Shaded Relief Images Derived from Digital Elevation Model (DEM) in Erbil-Kurdistan, northeast Iraq. *Advances in Natural and Applied Sciences, 6*(2), 138–146.

O' Leary, D. W., Friedman, J. D., & Pohn, H. A. (1976). Lineament, linear, lineation: Some proposed new standards for old terms. *Geological Society of America Bulletin, 87*, 1463–1469.

Onorati, G., Ventura, R., Poscolieri, M., Chiarini, V., & Crucilla, U. (1992). The Digital Elevation Model of Italy for geomorphology and structural geology. *Catena, 19*(2), 147–178.

Pérez-Peña, J. V., Azor, A., Azañón, J. M., & Keller, E. A. (2010). Active tectonics in the Sierra Nevada (Betic Cordillera, SE Spain): Insights from geomorphic indexes and drainage pattern analysis. *Geomorphology, 119*, 74–87.

Papanastassiou, D., & Gaki-Papanastassiou, K. (1993). Geomorphological Observations at the area of Kehries-Ancient Corinth and their correlation with seismological data. In *Proceedings of the 3rd panhellenic geographical congress of the Greek geographical society, B* (pp. 210–223).

Pavlides, S.B. (2003). *Geology of earthquakes.* Thessaloniki: University Studio Press.

Pavlopoulos, K., Evelpidou, N., & Vassilopoulos, A. (2009). *Mapping geomorphological environments.* Berlin: Springer.

Pazzaglia, F. J., Gardner, T. W., & Merritts, D. (1998). Bedrock fluvial incision and longitudinal profile development over geologic time scales determined by fluvial terraces. *Washington DC American Geophysical Union Geophysical Monograph Series, 107*, 207–235.

Pedrera, A., Pérez-Peña, J. V., & Galindo-Zaldívar, J. (2009). Testing the sensitivity of geomorphic indices in areas of low-rate active folding (eastern Betic Cordillera, Spain). *Geomorphology, 105*, 218−231.

Pike, R. J. (2002). *A bibliography of terrain modeling (geomorphometry), the quantitative representation of topography; supplement 4.0: Open-file report 02-465, 2002*. U.S. Department of the Interior, U.S. Geological Survey.

Pike, R. J., Evans, I., & Hengl, T. (2009). Geomorphometry: A brief guide. In T. Hengl, & H. I. Reuter (Eds.), *Geomorphometry—Concepts, software, applications, series developments in soil science* (Vol 33, pp. 3−30). Amsterdam: Elsevier.

Pike, R. J., & Wilson, S. E. (1971). Elevation-relief ratio, hypsometric integral, and geomorphic area-altitude analysis. *Geological Society of America Bulletin, 82*(4), 1079−1084.

Posea, G. (1997). Suprafetele și nivelele de eroziune. *Revista de Geomorfologie, I*, 11−30.

Queiroz, G. L., Salamuni, E., & Nascimento, E. R. (2015). Knickpoint finder: A software tool that improves neotectonic analysis. *Computers & Geosciences, 76*, 80−87.

Raj, J. K. (1983). Negative lineaments in the granitic bedrock areas of NW Peninsular Malaysia. *Geological Society of Malaysia Bulletin, 16*, 61−70.

Ramírez-Herrera, M. T. (1998). Geomorphic assessment of active tectonics in the Acambay Graben, Mexican Volcanic Belt. *Earth Surface Processes and Landforms, 23*, 317−332.

Sarp, G., Geçen, R., Toprak, V., & Duzgun, S. (2011). *Morphotectonic properties of Yeniçağa basin area in Turkey* (p. 5). Ankara: METU, Geodetic and Geographic Information Technologies.

Sboras, S., Ganas, A., & Pavlides, S. (2010). Morphotectonic analysis of the Neotectonic and Active Faults of Beotia (Central Greece), using G.I.S. techniques. *Bulletin of the Geological Society of Greece, Proceedings of the 12th International Congress, Patras, 43*(3), 1607−1618.

Scheidegger, A. E. (2004). *Morphotectonics*. Berlin: Springer.

Schumm, S. A. (1986). Alluvial river response to active tectonics. In N. R. Council (Ed.), *Studies in geophysics, active tectonics* (pp. 80−94). Washington: National Academy Press.

Silva, P. G., Goy, J. L., Zazo, C., & Bardají, T. (2003). Fault-generated Mountain fronts in southeast Spain: geomorphologic assessment of tectonic and seismic activity. *Geomorphology, 50*, 203−225.

Slemmons, D. B., & Depolo, C. M. (1986). Evaluation of Active Faulting and Associated Hazards. In Geophysics Study Committee (Ed.), *Active tectonics: Impact on society* (pp. 45−62). Washington, D.C.: National Academy Press.

Snyder, N. P., Whipple, K. X., Tucker, G. E., & Merritts, D. J. (2000). Landscape response to tectonic forcing: Digital elevation model analysis of stream profiles in the Mendocino triple junction region, northern California. *GSA Bulletin, 112*(8), 1250−1263.

Stout, J. C., & Belmon, P. (2014). TerEx Toolbox for semi-automated selection of fluvial terrace and floodplain features from lidar. *Earth Surface Processes and Landforms, 39*(5), 569−580.

Strobl, J. (2008). Segmentation-based Terrain Classification. In Q. Zhou, B. Lees, & G. Tang (Eds.), *Advances in digital terrain analysis* (pp. 125−141). Berlin Heidelberg: Springer-Verlag.

Summerfield, M. A. (1991). *Global geomorphology*. Harlow: Longman.

Tagil, S., & Jenness, J. (2008). GIS-Based Automated Landform Classification and Topographic Landcover and Geologic Attributes of Landforms Around the Yazoren Polje, Turkey. *Journal of Applied Sciences, 8*(6), 910−921.

Toudeshki, V. H., & Arian, M. (2011a). Morphotectonic Analysis in the Ghezel Ozan River Basin, NW Iran. *Journal of Geography and Geology, 3*(1), 258−265.

Toudeshki, V. H., & Arian, M. (2011b). Morphotectonic Analysis in the Ghezel Ozan River Basin, NW Iran. *Journal of Geography and Geology, 3*(1), 1−21.

Troiani, F., & Della Seta, M. (2008). The use of the Stream Length−Gradient index in morphotectonic analysis of small catchments: A case study from Central Italy. *Geomorphology, 102*, 159−168.

Tsimi, C., Ganas, A., Ferrier, G., Drakatos, G., Pope, R. J., & Fassoulas, C. (2007a). Morphotectonics of the Sfakia normal fault, southwestern Crete. In , *1. Proceedings of 8th pan-hellenic geographical conference of the Greek geographical society, Athens, Greece* (pp. 180−188).

Tsimi, C., Ganas, A., Soulakellis, N., Kairis, O., & Valmis, S. (2007b). Morphotectonics of the Psathopyrgos active fault, western Corinth rift, central Greece. *Bulletin of the Geological Society of Greece, 40*, 500−511.

Tsodoulos, I. M., Koukouvelas, I. K., & Pavlides, S. (2008). Tectonic geomorphology of the easternmost extension of theGulf of Corinth (Beotia, Central Greece). *Tectonophysics, 453*, 211−232.

Valkanou, K., Karymbalis, E., Papanastassiou, D., Soldati, M., Chalkias, C., & Gaki-Papanastassiou, K. (2021). Assessment of neotectonic landscape deformation in Evia Island, Greece, using GIS-based multi-criteria analysis. *ISPRS International Journal of Geo-Information, 10*, 118.

Valkanou, K., Karymbalis, E., Papanastassiou, D., Soldati, M., Chalkias, C., & Gaki-Papanastassiou, K. (2020). Morphometric analysis for the assessment of relative tectonic activity in Evia Island, Greece. *Geosciences, 10*(7), 264.

Vassilakis, E. M. (2006). *Study of the tectonic structure of Messara basin, Central Crete, with the aid of remote sensing techniques and geographic information systems*. PhD Thesis. Athens: National and Kapodistrian University of Athens, Department of Geology and Geoenvironment, Department of Dynamics, Tectonics and Applied Geology.

Vassilakis, E., Skourtsos, E., & Kranis, H. (2007). Tectonic uplift rate estimation with the quantification of morphometric indices. In *Proceedings of the 8th panhellenic geographical congress, Athens, Greece, 4—7 October 2007* (pp. 17—26).

Verrios, S., Zygouri, V., & Kokkalas, S. (2004). Morphotectonic analysis in the Eliki fault zone (Gulf of Corinth, Greece). *Bulletin of the Geological Society of Greece, XXXVI*(4), 1706—1715.

Veselský, M., Bandura, P., Burian, L., Harciníková, T., & Bella, P. (2015). Semi-automated recognition of planation surfaces and other flat landforms: a case study from the Aggtelek Karst, Hungary. *Open Geosciences, 1*, 799—811.

Wang, A., Wang, G., Zhang, K., Xiang, S., Li, D., & Liu, D. (2009). Late Neogene mountain building of eastern Kunlun orogen: Constrained by DEM analysis. *Journal of Earth Sciences, 20*(2), 391—400.

Weiss, A. D. (2001). *Topographic position and landforms analysis*. San Diego, CA: Poster Presentation, ESRI User Conference.

Wilson, J. P., & Bishop, M. P. (2013). Geomorphometry (Volume-Editor). In J. Shroder, (Editor-in-chief), & M. P. Bishop (Eds.), *Treatise on geomorphology, Vol 3, remote sensing and GIScience in geomorphology* (pp. 162—186). San Diego: Academic Press.

Wilson, J. P., & Deng, Y. (2008). Terrain Analysis. In K. K. Kemp (Ed.), *Encyclopedia of geographic information science* (pp. 465—467). Thousand Oaks: SAGE Publications.

Wobus, C., Whipple, K. X., Kirby, E., Snyder, N., Johnson, J., Spyropolou, K., Crosby, B., & Sheehan, D. (2006). Tectonics from topography: Procedures, promise, and pitfalls. In S. D. Willett, N. Hovius, M. T. Brandon, & D. M. Fisher (Eds.), *Tectonics, climate, and landscape evolution* (pp. 55—74). Penrose Conference Series: Geological Society of America Special Paper 398.

Zaprowski, B. J., Pazzaglia, F. J., & Evenson, E. B. (2005). Climatic influences on profile concavity and river incision. *Journal of Geophysical Research, 110*(F03004).

Zovoili, E., Konstantinidi, E., & Koukouvelas, I. K. (2004). Tectonic geomorphology of escarpments: the cases of Kompotades and Nea Anchialos faults. *Bulletin of the Geological Society of Greece, 36*, 1716—1725.

Zygouri, V., Verroios, S., Anninou, A., & Koukouvelas, I. (2018). Fuzzy cognitive mapping vs. statistic approach of geomorphological analysis of active faults: Preliminary results from a case study within the Corinth Graben, Greece. *IFAC-Papers On Line, 51*(30), 372—377.

CHAPTER 2

Monitoring snow by combining physical snow models and microwave remote sensing techniques: two case studies from northern Finland

Ioanna Merkouriadi and Jorge Jorge Ruiz
Finnish Meteorological Institute, Helsinki, Finland

1. Introduction

Knowing how much snow is accumulated every winter over the land, the glaciers, and the sea ice is crucial for numerous reasons. Snow is a medium of unique physical properties with a strong influence on our climate. From the moment snow starts to form inside the clouds until it reaches the Earth's surface, it experiences atmospheric forces that shape its properties. Processes such as wind force and air temperature and moisture keep influencing the evolution of snow after it has been deposited and continue shaping it further. For example, strong winds tend to break the snow crystals into smaller fragments and result in densely packed snow. Even though the characteristics of snow can vary drastically under the forces of the atmosphere, broadly speaking, snow has strong insulating and reflective properties. The high air volume within its crystal structure enacts the insulation properties of snow, keeping the ground and its vegetation relatively warm under cold atmospheric conditions. Snow is also characterized by high albedo values, reflecting most of the solar radiation back to space, and keeping the surface of the Earth cool. In addition to its importance in the surface energy balance, and therefore the Earth's climate, snow is an important freshwater reservoir. One-sixth of the world's population relies on snow as a primary source of freshwater (Barnett et al., 2005). Mountain snowpacks provide a slow release of fresh water into rivers and waterways. Knowing how much snow is out there and when does it melt is therefore crucial for numerous reasons spanning from climate projections to industrial applications. Freshwater management, hydropower production estimates, weather forecasts, and climate monitoring are a few examples among many.

SWE describes the amount of water that is contained within the snowpack. It is related to snow depth and density, and it is commonly used to characterize snow mass. Although there are great scientific and technological advances in the field, global monitoring of SWE remains a challenge. With the current technology, monitoring of SWE over large areas can be either achieved by earth observations, that is, satellites or

Geographical Information Science
ISBN 978-0-443-13605-4
https://doi.org/10.1016/B978-0-443-13605-4.00001-1

airborne campaigns, or by physical snow models. Both methods come with caveats, and understanding their limitations is important in order to use them properly. Active and passive microwave systems, two distinct techniques in microwave remote sensing, have been proposed for SWE monitoring from space. Active microwaves involve the transmission of microwaves by the sensor itself, providing higher resolution. Passive microwaves rely on detecting microwave radiation that is emitted naturally from any object due to their thermal energy, and they do not require active transmission.

Currently, passive microwave sensors provide the only operational satellite products on SWE, and they offer global daily coverage for almost half a century. The main downside of the existent microwave sensors is their low spatial resolution (in the order of tens of kilometers) that limits their ability to capture the high spatial variability of snow over heterogeneous landscapes. Active microwave sensors, such as synthetic aperture radars (SARs), can mitigate the coarse resolution challenges; however, currently no SAR system operating at high frequencies required for snow mass observations exists. Recently, radar instruments for suitably high frequencies have been proposed for development (Derksen et al., 2021). A limitation that concerns equally passive and active microwave systems is related to the retrieval algorithms that are required to interpret microwave signatures into SWE relative information.

When microwave signatures interact with the snow, they are affected by scattering and absorption in the snow medium. These mechanisms are controlled by several properties and structural characteristics of the snowpack, including snow depth and density, layering of the snow, snow microstructure (the size and type of the snow grains), temperature and moisture, hereafter referred to as *snowpack states*. Several microwave models have been developed to describe theoretically and empirically the relationship between microwave signatures and the snowpack states. Thus, to retrieve SWE from microwave signatures, a priori information of the snowpack states is crucial. The challenge simply lies in the fact that microwave models are faced with too many unknowns and insufficient observations. As an alternative, the InSAR (Interferometric Synthetic Aperture Radar) method, only requires snow density as prior information, and the retrieval algorithm is not too sensitive to it (Leinss et al., 2015). InSAR technology is able to exploit the interferometric phase to retrieve ΔSWE, as the phase is affected by the snow accumulation between passes (Guneriussen et al., 2001). Moreover, low-frequency SAR has recently drawn the attention of space agencies, with the on-going development of NISAR (NASA-IRSO), ROSE-L (ESA), or ALOS-4 (JAXA). However, this method presents several drawbacks and challenges of different kinds. Firstly, repeat-pass interferometry requires acquisitions from the same viewing geometry with typical temporal baselines from satellites ranging between 6 and 14 days. In retrievals based on the interpretation of phase, coherence is a critical parameter, as it is a measure of the amount of noise present in the interferogram. The snowpack is prone to decorrelation due to meltdown events, wind and precipitation, with higher frequencies being more susceptible than low frequencies

(Jorge-Ruiz et al., 2022). Relatively long temporal baselines suffer from greater decorrelation. Secondly, satellite-borne SAR observations suffer from atmospheric phase screening, which is a component added to the phase due to the interaction of the wave with the atmosphere. Additionally, the InSAR phase is ambiguous by nature and requires calibration. Typically, this is addressed by using reference points such as corner reflectors (CRs).

Physical snow models are essential tools in snow science, providing simulations of the complex behavior of snow physics within the snowpack and its interactions with the surrounding environment. Snow models are based on fundamental principles of physics and thermodynamics, allowing us to simulate and predict various aspects of snow accumulation, melt, and transformation. Typically, snow models consider various factors and processes related to snow accumulation and melt, such as:

- Meteorological data: Snow models use meteorological data (air temperature, precipitation, wind speed and direction, etc.) to simulate snowfall, snowpack accumulation, and snowmelt.
- Snowpack physics: Snow models consider the physical properties of snow, such as density, temperature, and water content. They simulate how these properties change over time due to processes like compaction and melting.
- Topography: The terrain and slope of the land can affect snow distribution and accumulation. Snow models may account for these factors to simulate the spatial variability of snowpacks.
- Vegetation: Forest cover and vegetation can influence snow accumulation and melt rates. Some models include vegetation parameters to account for these effects.
- Assimilation: Some snow models can use snow observations from the field to ensure their accuracy.

The strength of snow models lies in their adaptability to different spatial and temporal scales, making them versatile tools for a wide range of applications. Recent advances in computational power, together with the improved quality and accessibility of topographical, land cover, and meteorological data, position physical snow models as powerful tools for monitoring snow on a global scale. However, even though they can mitigate passive microwave challenges related to spatial resolution, snow models come with their own weaknesses and limitations. The main challenge snow models confront is uncertainties in the input data. Snow models rely on input data such as temperature, precipitation, wind, and humidity. Uncertainties in these inputs can lead to severe biases in the model predictions. In remote regions, where automatic weather stations (AWSs) are sparse, or entirely absent, snow models rely on input data informed by atmospheric reanalysis. Atmospheric reanalysis data are generated using advanced numerical weather prediction models that assimilate various observations from satellites, weather stations, and other sources to create a detailed and consistent representation of the Earth's atmosphere. Despite advancements in atmospheric reanalysis products, accurately determining winter precipitation amounts remains problematic. This challenge has the potential to introduce significant uncertainties in global-scale SWE simulations by snow models.

There has been a growing interest in developing synergies between microwave remote sensing and physical snow modeling for global monitoring of SWE, in an effort to mitigate challenges in both. Inspired by the potential of synergies, two case studies were developed in Northern Finland. The first one aimed to identify challenges associated with utilizing physical snow models to inform microwave models, and to offer potential solutions to these challenges (Merkouriadi et al., 2021). A local experiment was designed, utilizing tower-based remote sensing instruments, field observations of the snowpack properties, a state-of-the-art snow evolution model, and a semiempirical radiative transfer model. The second one aimed to evaluate the retrieval of satellite InSAR ΔSWE and examine the potential for assimilation into physical snow models.

2. Materials and methods

2.1 The study area

Both experiments took place in the municipality of Sodankylä, Northern Finland. Sodankylä lies in the region of Finnish Lapland (Fig. 2.1). Each year, this area experiences a seasonal snow cover that typically lasts for about 200 days, spanning from October through May (Merkouriadi et al., 2017). The highest snow depth of the year is typically reached in March, and it is on average 80 cm. During the winter months, the soil freezes down to 2 m (Rautiainen et al., 2014), and air temperatures can plummet to below −30°C. The snow conditions in Sodankylä are characteristic of snowpacks found in boreal forests (Sturm & Liston, 2021). A campaign called NoSREx (Lemmetyinen et

Figure 2.1 Location of the study area in Northern Finland.

al., 2016) took place in Sodankylä from 2009 until 2013. It was organized by the Arctic Space Center of the Finnish Meteorological Institute that is located in the area (67.368°N, 26.633°E). NoSREx focused on investigating the interaction of boreal snowpack states with active and passive microwave instruments.

2.1.1 Remote sensing observations

Two tower-based systems, an active microwave (SnowScat) and a passive microwave (SodRad), were deployed in the study area and they were measuring continuously during the NoSREx campaign. SnowScat is a fully polarimetric radar that operates on frequencies from X to Ku bands (Werner et al., 2010). It was installed approximately 10 m above the ground, and it was scanning at four elevation angles (60 degrees, 50 degrees, 40 degrees, and 30 degrees from Nadir) every 3 hours. SodRad is a dual-polarization radiometer system with four modular, dual-polarization receivers operating from X to W bands. It was installed at a different location, approximately 4 m above the ground. SodRad used receivers at 10.65, 18.7, 37, and 90 GHz from 2009 to 2012. The 90 GHz receiver was replaced by a 21 GHz receiver in 2012. It was scanning at elevation angles between 30 degrees and 70 degrees from Nadir at a 5 degrees interval.

The ALOS-2/PALSAR-2 is a SAR satellite sensor that has been operating since 2014 by JAXA. It operates at L-band (1257.5 MHz) and has a repeat pass of 14 days. In the Fine Stripmap imaging mode, the sensor offers a range and azimuth resolution of 10 m over a swath of 70 km. Although the ALOS2 acquisitions included VV and VH polarizations, the latter was used as is proven that co-pol perform better over snow. The SAR images presented here were acquired in an agreement between ESA and JAXA during 2019 and 2021.

2.1.2 In situ observations

Snowpits, also known as snow profiles, were performed by creating an elongated vertical excavation wall in the snowpack that allows for detailed examination and measurement of the snow's physical and structural properties. Snowpit measurements are conducted frequently at the Arctic Research Center as part of their monitoring program. During the NoSREx campaign, snowpit measurements took place near the microwave instrument platforms once or twice a week. Several snow properties and characteristics were documented weekly, including total snow depth and SWE, alongside detailed profiles of temperature, density, moisture, stratigraphy (layering), and snow grain size and type. Temperature and density values were recorded every 10 cm from the snow surface to the ground. Distinct snow layers within the snowpack, including old snow layers, new snow, wind-deposited layers, and crusts were identified visually. The size and type of snow grains present in each layer such as facets, rounds, or depth hoar were measured and photographed. Postprocessing involved analyzing grain size and type from macro photographs of the snow samples taken against a reference grid. Grain size and type are the snowpack's structural properties with the largest effect in the microwave signatures.

2.2 The models

2.2.1 The physical snow model

SnowModel (Liston & Elder, 2006a) was used to simulate the snowpack states that are required for the microwave models. SnowModel is a state-of-the-art collection of snow distribution and snow evolution modeling tools, suitable for various environments that experience snow. These tools are highly adaptable, working effectively across diverse spatial domains (from equatorial mountain tops to polar regions), temporal domains (from past to future climates, spanning from hours to centuries), and a wide range of spatial and temporal resolutions (from 1 m to 100 km and from 1 hour to daily intervals). They are versatile in addressing various aspects of snowpack analysis, from identifying snowpack layers to assessing rain-on-snow-related icing events. SnowModel simulates key processes, including snow precipitation, the redistribution and sublimation of blowing snow, changes in snow density, snowpack development, and melt. Moreover, SnowModel is coupled to a high-resolution meteorological model named MicroMet (Liston & Elder, 2006b). MicroMet spatially disseminates meteorological data across the simulation area and provides the essential inputs to drive SnowModel. Both Micro-Met and SnowModel rely on meteorological data, which can be derived from meteorological stations or gridded atmospheric datasets, including information such as air temperature, relative humidity, precipitation, wind speed, and wind direction. Additionally, SnowModel is integrated with SnowAssim (Liston & Hiemstra, 2008), a numerical tool designed to incorporate available field or remote sensing observations into the modeling process. These observations encompass critical snow properties and structural characteristics, including SWE, snow depth, snow density and temperature profiles, snow grain size, and various other snow-related features.

At each time step, SnowModel performs an energy balance calculation given by,

$$(1 - \alpha_s)Q_{si} + Q_{li} + Q_{le} + Q_h + Q_e + Q_c = Q_m, \tag{2.1}$$

where α_s is the albedo at the surface, Q_{si} is the incoming solar (shortwave) radiation reaching the surface, Q_{li} is the incoming longwave radiation reaching the surface, Q_{le} is the emitted longwave radiation from the surface to the atmosphere, Q_h is the turbulent sensible heat flux, Q_e is the turbulent latent heat flux, Q_c is the conductive heat flux, and Q_m is the energy available for melt. The only unknown is the surface temperature T_0 that is solved iteratively after setting Q_m to zero. If snow is present and $T_0 > 0$, T_0 is set at zero and the equation is solved for Q_m, to estimate the amount of energy available for melting.

A mass budget calculation is also performed at each time step of the model. SWE changes in response to the precipitation and melt rates defined by the energy balance calculations, as well as water-equivalent sublimation and blowing-snow processes. The snow density increases due to compaction following Anderson (1976),

$$\frac{\partial \rho_s}{\partial t} = A_1 h_w^* \rho_s \exp\left[-B\left(T_f - T_s\right)\right] \exp\left(-A_2 \rho_s\right), \tag{2.2}$$

where ρ_s (kg m^{-3}) is the snow density, $h_w^* = \frac{1}{2} h_w$ (m) is the weight of snow defined as half of the SWE, T_f (K) is the freezing temperature of water, and T_s (K) is the temperature of the snow taken as a linear average of the surface and bottom (snow/ground interface) snow temperatures. A_1, A_2, and B are constants based on Kojima (1967).

SWE is defined as,

$$h_w = \frac{\rho_s}{\rho_w} \zeta_s, \tag{2.3}$$

where ρ_s and ρ_w are the densities of snow and water, respectively, and ζ_s is the snow depth.

SnowModel simulates the snow grain size by applying a grain growth parameterization based on the SNTHERM model (Jordan, 1991). Snow grains grow in response to vapor pressure gradients that are triggered by temperature gradients, that is, the temperature difference between the surface and the bottom of the snowpack. The grain diameter (D) evolution is given by:

$$\frac{\partial D}{\partial t} = \frac{g_1 |U_v|}{D}, \tag{2.4}$$

SnowModel has a SWE assimilation component, for adapting and correcting model simulations. If SWE observations are available at a specific time step, SnowModel first calculates the difference between the observed and the modeled SWE at that time step. Then it calculates and applies adjustment factors to the precipitation and melting over the whole snow accumulation period, in order to recreate the observed SWE at that specific time step. This process slows down the computation time of the simulations as they have to be repeated over the whole season.

In the first case study, SnowModel was ran over a single point. In the second case study, SnowModel ran over a larger spatial domain. Therefore, topography and land cover (vegetation) data were also used to force the model, in order to simulate the spatial evolution of the snow properties.

2.2.2 The microwave emission model

MEMLS (microwave emission model of layered snowpacks; MEMLS3&a, Proksch et al., 2015) was the radiative transfer model used in this study. It is a computational model used in remote sensing and snow science to simulate the microwave emission and scattering properties of snowpacks. Specifically, MEMLS is designed to model how microwave radiation interacts with snowpacks composed of multiple layers with varying properties, such as snow density, grain size, temperature, and liquid water content. It is therefore especially useful for analyzing microwave remote sensing data collected by satellites or ground-based sensors, to examine snow properties and structural characteristics. MEMLS takes into account a wide range of physical parameters that influence microwave emission and scattering in snow, including the dielectric properties of snow components, such as ice and liquid water. Input data required to run MEMLS are typically layered snow thickness, temperature, density, moisture, and a parameter called snow exponential correlation length (l_{exp}). l_{exp} is a parameter used in microwave remote sensing to describe the spatial variability of snow microstructure. It provides information about the size and distribution of snow grains within a snowpack and plays a crucial role in determining how microwave radiation interacts with the snow. It is a crucial snow parameter that is particularly useful in remote sensing, because it strongly affects the interactions of snow with electromagnetic radiation, such as microwaves. MEMLS produces microwave emissions and scattering properties that can be later compared to the signatures received by remote sensing sensors.

In practical implementations of remote sensing monitoring of SWE, the following is happening. Remote sensing sensors detect the microwave signatures. Microwave models, such as MEMLS, use a cost function to recreate the observed signatures. A cost function is a mathematical function that quantifies the discrepancy between the predicted (model output) and actual (ground truth or observed) values in optimization problems. The aim is to minimize the cost function, in order to make the model fit an observation, by giving controlled freedom to the parameters involved in the cost function (and in the model). The reason cost functions are used in microwave applications, as already mentioned in the introduction, simply results from the fact that they involve too few equations and too many unknowns. Can we mitigate this challenge by having a physical snow model to provide the snow information (or snowpack states) to MEMLS? This is what we aim to examine in our first study case.

2.2.3 The InSAR SWE retrieval model

The interferometric phase is computed by calculating the complex correlation between two coregistered SAR images. In practice, this is calculated over an assembly of pixels, called estimation window. In repeat-pass interferometry, the SAR images are acquired at different times, thus the phase can be related to changes in the scatterers over the estimation window. The absolute value of this correlation is referred to as coherence, and it is a measure of the amount of noise present in the interferogram. Several sources of temporal decorrelation are present in interferograms, with snow meltdown being the most significant over snow-covered surfaces (Jorge-Ruiz et al., 2022). The retrieval of SWE using repeat-pass InSAR is based on the increase in the travel time a wave experiences in the presence of snow, in comparison to the snow-free scenario, since snow has higher permittivity than air. Similarly, when both acquisitions have snow, changes in the snowpack will change the interferometric phase. Dry snow is a mixture of air and ice, having a permittivity of little dependence to frequency between 10 MHz and 100 GHz. The real part of the complex permittivity, which determines the propagation speed through the medium, can be approximated from the snow density (Wiesmann et al., 1999). Absorption, which is related to the complex part of the permittivity, is small for dry snow in these frequencies. However, for wet snow this is no longer true due to the presence of liquid water. Liquid water in the snowpack modifies the main backscatter interface from the ground-snow to the snow-air or the snow volume, depending on the frequency and the amount. For this reason, the model cannot be confidently applied over wet snow. Note that this effect becomes more dominant in higher frequencies and high percentages of liquid water in the snowpack. Fig. 2.1 illustrates a schematic of the retrieval method of ΔSWE. It can be shown that the relation between the interferometric phase component induced by snow and the change in SWE or snow depth, follows (Fig. 2.2):

$$\Delta R = \left(R_s + \Delta R_s \cdot \sqrt{\epsilon}\right) - (R_a + \Delta R_a), \tag{2.5}$$

$$\Delta \Phi_s = 2\kappa_i \cdot \Delta R = -2 \cdot \kappa_i \cdot \Delta Z_s \cdot \left(\cos \theta - \sqrt{\epsilon_s - \sin^2 \theta}\right)$$
$$= -2 \cdot \kappa_i \cdot \frac{\Delta SWE}{\rho_s} \cdot \left(\cos \theta - \sqrt{\epsilon_s - \sin^2 \theta}\right), \tag{2.6}$$

with

$$\epsilon_s = 1 + 1.5995e - 3 \cdot \rho_s + 1.861e - 3 \cdot \rho_s^3, \tag{2.7}$$

where Δ Rs is wave path in presence of snow, Δ Ra is wave path in absence of snow, κ_i is the incoming wavenumber, ΔSWE and ΔZ_s are the changes in SWE and the changes in snow depth between acquisitions, respectively, ϵ_s is the real permittivity of the snow, and

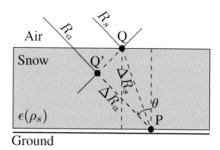

Figure 2.2 Schematic representation of the ΔSWE InSAR retrieval technique (Leinss et al., 2015).

θ is the local incidence angle. The interferometric phase has modulo 2 π. Since κ_i decreases as frequency increases, the changes in the ΔSWE required for the interferometric phase to exceed 2π are smaller. For this reason, low frequency bands, such as L- or even C-band are preferred, as they allow retrievals of larger increments of SWE. Furthermore, lower frequency bands exhibit better coherence conservation properties.

In forested areas, the forest will introduce another component to the phase due to its interaction with the signal. However, this component is expected to be small for the low frequency on which ALOS2 operates, and even smaller when the air temperature is below 0°C, meaning that the snowpack is dry, which is a precondition for the retrieval (Lemmetyinen et al., 2022).

2.3 Case study A

2.3.1 Setting up the synergy between SnowModel and MEMLS

SnowModel was driven with meteorological data from an AWS that is located in the study area (67.366618°N, 26.628976°E). The AWS was deployed by the Arctic Research Center in 2006, and it has provided continuous weather measurements ever since. All instruments and sensors at the AWS are calibrated annually. SWE observations collected in the field were used to assimilate SnowModel once a year, around mid-March. SnowModel was ran at a 3-hour time step from July 1, 2009 through June 30, 2013. The maximum number of snow layers was defined to be 12. At each time-step SnowModel produced snow depth, density, SWE, temperature, and grain diameter for each snow layer.

As explained in section 2.2, SnowModel simulates snow grain diameter (D) based on SNTHERM. MEMLS, however, as explained in section 2.3, requires another parameter to describe the microstructure, the snow exponential correlation length (l_{exp}). An equation proposed by Wiesmann et al. (2000) was used to convert D to l_{exp},

$$l_{exp} = 0.16*D, \tag{2.8}$$

The impact of the ground surface on microwave signatures was estimated by comparing early season MEMLS simulations with observations from the tower-based instruments. For brightness temperature simulations, the ground reflectivity was adjusted to match the observed values, assuming a ground temperature of 272 K. For backscatter simulations, we used realistic soil reflectivity values with a scatter component to achieve a match. We calculated downwelling sky brightness temperatures using a statistical model based on typical winter atmospheric conditions in Finland.

2.3.2 Identifying the challenges—setting up a sensitivity experiment

As already mentioned earlier, one of the main challenges snow models encounter is the uncertainties in the forcing data, and most importantly the precipitation. No matter how sophisticated models are, they will fail to produce accurate results when driven with inaccurate input data. To examine how uncertainties in the forcing data would affect a snow model, and how this will in turn affect the synergy with microwave models, a sensitivity experiment was designed.

In addition to the SWE-assimilated SnowModel run, hereafter referred to as *base run*, 10 more SnowModel simulations were performed, hereafter referred to as *SWE-perturbed runs*. In each of the SWE-perturbed runs, the SWE observation from the field used for assimilating the model was perturbed by ±10 to ±50%, in 10% increments. Thus, as commonly happens during the assimilation process, adjustment factors were calculated and applied to the precipitation inputs, resulting in either increasing or decreasing the original precipitation to achieve the perturbed SWE observation. In this manner, one could examine how the SnowModel results would be affected if precipitation inputs were biased.

All SnowModel runs, the base run and the SWE-perturbed runs, produced 12-layer information of snow depth, snow density, snow temperature, and grain size at a 3-hour time step. Information from the base run was ingested in MEMLS to produce a synthetic set of microwave observations, and assumed them to be the "ground-truth." The reason a synthetic set was used instead of the real microwave observations collected at the tower-based systems was to gain control over the experiment and reduce uncertainties resulting from the real observations or from the model parameterizations. Here, we assume that models and instruments work perfectly.

At a following step, a cost function was minimized to achieve a match between the "ground-truth," that is, the synthetic set of microwave observations guided by the base run, to the snowpack states introduced to MEMLS from the SWE-perturbed runs. Snow depth was treated as a free parameter, while all other parameters, including the microstructure (l_{exp}), were kept fixed to the information from the SWE-perturbed SnowModel runs. This method resembles an approach commonly used in microwave-based SWE retrievals, where the microstructure is being fixed and the cost function is minimized by adjusting only the snow depth (e.g., Takala et al., 2011). This sensitivity experiment enabled in-depth examination of the way common challenges in snow modeling would be demonstrated in microwave-based SWE retrievals, when snow model information gets integrated in microwave models.

2.4 Case study B

For this case study, a digital elevation model (DEM) was used as a topography input to SnowModel, as well as for the InSAR processing and postprocessing. The DEM was provided by the National Land Survey of Finland (NLS) and had a 10 m pixel resolution. A Canopy Cover map with a pixel spacing of 16 m from the Finnish Environment Institute (SYKE) was used in the analysis. A Corine Land Class map from 2018 with a pixel resolution of 20 m was used.

SnowModel ran over an area of 100×140 km, with a spatial resolution of 25 m and a temporal resolution of 3 hours. The starting time and the output maps from SnowModel were set at 12:00 UTC, matching approximately the time ALOS2 over-passes the study area. The forcing data for SnowModel was the NORA10 atmospheric reanalysis, provided by the Norwegian Meteorological Institute (Reistad et al., 2011). Snow depth was set to zero on October 1, 2020, guided by in situ observations from the area. Assimilation was performed using in situ observations of snow during late March 2020.

The interferogram presented here expands 14 days between October 11, 2020 and October 25, 2020. All the processing was performed using the open-source software ISCE2. The output resolution was set to 25 m to match SnowModel. The interfero-metric phase is affected by the ionosphere because its composition changes between the acquisitions. Although the ISCE2 processing graph for ALOS2 includes correc-tion of this component, this process is prone to fail over images with large water bodies and/or low coherence. Another important step is the calculation of the local incidence angle θ. This is computed for each pixel by calculating the angle between the gradient vector from the DEM and the incidence vector between that pixel and the radar.

The maps of ΔSWE from SnowModel were generated by differentiating the output from the two SnowModel SWE maps closest to the ALOS2 acquisition times. The aim of this case study was to compare ALOS2 retrievals with SnowModel simulations of ΔSWE. Note that SnowModel simulations were performed using in situ SWE observations for assimilation, which are typically not available in remote regions. Snow models are suscep-tible to suffering from errors due to uncertainties in the precipitation data, and therefore, they could benefit from assimilation with remote sensing data. For this reason, the com-parison between the ALOS2 retrieved and the SnowModel simulated ΔSWE is evalu-ated after calibrating the interferometric phase with known points from SnowModel. For this we follow a multipoint calibration strategy where pixels from nonforested areas are set to match the ΔSWE from SnowModel. The snow density for the InSAR calcu-lation of ΔSWE was assumed to be 250 kg/m^{-3}. This snow density is typically observed in fresh snow over the area.

3. Results and discussion

3.1 Case study A

3.1.1 Sensitivity experiment

Results from the sensitivity experiment indicated that when the snowpack states from the SWE-perturbed SnowModel runs were introduced into MEMLS, and if all parameters were fixed except the snow depth in the cost function, microwave retrievaled SWE was more biased than the simulated perturbed SWE. But why is this happening? Why were the microwave SWE retrievals consistently pulled away from valid solutions?

The answer lies in the relationship between SWE and grain size in microwave models, versus their relationship in physical snow models. In microwave models, both SWE and grain size are directly related to the microwave signatures, that is, the brightness temperature and the backscatter. Specifically, brightness temperature decreases with both increasing SWE and grain size, and backscatter increases with both increasing SWE and grain size (Fig. 2.3). It is the grain size though that has the strongest effect in both the brightness temperature and the backscatter.

In physical snow models, on the other hand, snow grains are growing in response to vapor fluxes within the snowpack that are triggered by temperature gradients, that is, the temperature difference between the top and the bottom of the snowpack. Under the presence of a thicker snowpack (larger SWE), temperature gradients are dampened, and vapor fluxes inside the snowpack decrease, resulting in slower grain growth. In simpler words, in physical snow models that resemble natural processes, SWE and grain size are not directly but inversely related (Fig. 2.4).

When a snow model underestimates SWE, it will also overestimate grain size and vice versa. Since microwave signatures are more sensitive to the grain size than to SWE, grain size biases govern the microwave simulations of SWE. When the cost function is constrained to a biased grain size and fails to consider the inherent relationship between SWE and grain size, it modifies the free parameter (snow depth) to the wrong direction in order to match the observation, leading to further deterioration of the SWE retrieval. This same pattern was observed in all experiments, and it was consistent across all years, even under small changes in the modeled SWE (within $\pm10\%$) (Fig. 2.5). Specifically, the mean absolute error (MAE) of SWE from both passive and active microwave retrievals was consistently greater than the MAE of SWE from the SWE-perturbed SnowModel runs.

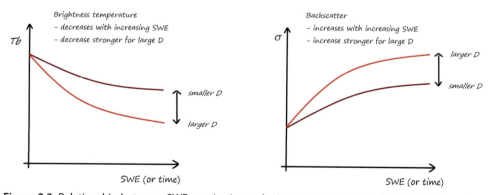

Figure 2.3 Relationship between SWEs, grain size and microwave signatures, that is, brightness temperature (left) and backscatter (right).

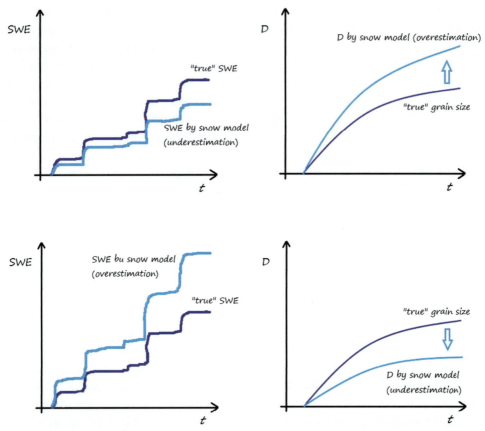

Figure 2.4 Grain size response when SWE is underestimated (upper plots) or underestimated (lower plots) in a physical snow model.

3.1.2 Proposing solutions to the challenges
3.1.2.1 The look-up table

Look-up tables are data structures that facilitate quick retrievals of data associated with specific input parameters. In our case, these are sets of snowpack states derived from physical model simulations. They were created by perturbing the precipitation inputs in a controlled manner. The microwave models were guided to select sets of the snowpack states suggested by the look-up tables. It is worth mentioning that in previous works utilizing look-up tables for microwave SWE retrievals, these tables were typically generated by applying random combinations of snowpack states (e.g., Zhu et al., 2018), an approach that could be described as unsupervised look-up tables (Fig. 2.6). Essentially, any solution matching the observation can be selected in the retrieval, leading potentially to ambiguities if multiple SWE/microstructure combinations provide a model result close to the observation. Our approach guarantees that the inherent relationships among

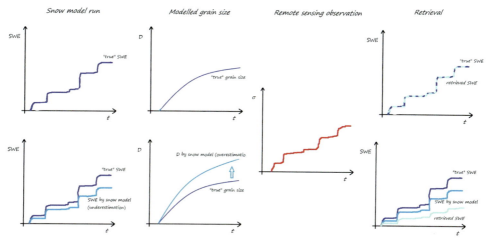

Figure 2.5 Modeled SWE and grain size, and the subsequent microwave-based SWE retrieval when physical models perform ideally (upper plots) and the same when physical models underestimate SWE (lower plots).

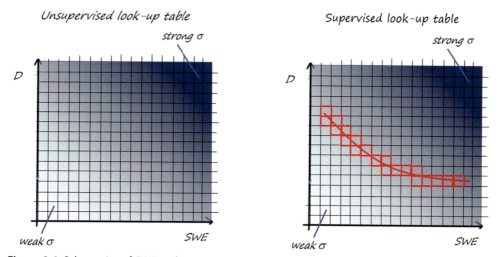

Figure 2.6 Schematics of SWE and grain size combinations provided by unsupervised (left) and supervised (right) look-up tables. The red boxes correspond to solutions that are physically possible.

the snowpack states are maintained, in a "supervised" look-up table configuration, which encourages the retrieval algorithm to select only physically realistic model solutions (Fig. 2.6).

To test this method, 10 SWE-perturbed SnowModel runs were conducted, each varying within a range of ±50% of the base run SWE, with ±5% intervals. Look-up tables were generated containing sets of snowpack states that resulted from these 20 SnowModel runs. Information from the look-up tables was ingested into MEMLS, and a cost function was applied to identify the set that yielded the smallest difference to the "ground-truth." Similarly to the sensitivity experiment, the synthetic microwave observations that derived from the SnowModel base run were considered to be the ground-truth.

The downside of the look-up table solution is that it requires several snow model runs. Therefore, it becomes computationally expensive when considering large spatial scales. In addition to the computational cost, our results indicated that when the look-up tables were applied strictly without room for adjustments in the snowpack states, the range of solutions they provided was narrow. In real-life applications, this could be mitigated by introducing a degree of freedom to the sets of snowpack states, by adjusting the cost function variances of the reference values.

3.1.2.2 The nudging method

An alternative solution was examined to reduce the computational costs of the look-up table approach. Driven by the results of the sensitivity experiment, a bias correction nudging algorithm was introduced to the SnowModel. Knowing that a consistent deterioration of the microwave SWE retrieval occurs under minor biases in the modeled SWE, the following condition was introduced: if at $T = t$ retrieved SWE(t) is greater than the modeled SWE(t) by 5%, then SnowModel SWE(t) is assimilated with SWE(t) nudged by 10% to the opposite direction (Fig. 2.7):

$$SWE'(t) = 1.1 \times SWE(t).$$

Implementing the bias correction nudging algorithm once every month led to significant improvements in the SWE-perturbed SnowModel runs. This approach proves to be cost-effective for large-scale applications and calls for further investigation.

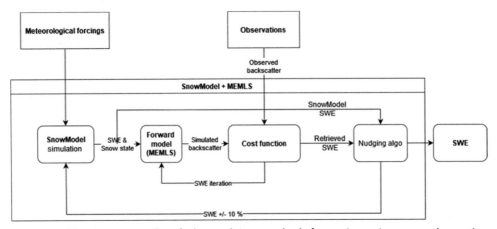

Figure 2.7 Flowchart example of the nudging method for active microwave observations (backscatter).

One reason that the nudging method was successful in this case study was the large impact of changes in SWE on the grain growth rates of the snowpack. This is common in boreal snow climates where the air temperature is well below zero for most of the winter season. However, it might not be the case in regions with milder snow climates. Would the nudging algorithm be invalid, if changes in SWE had a milder effect on the grain growth rates of the snowpack? To investigate this, an artificial meteorological data-set was generated in which air temperature was constrained at $-1°C$ throughout the snow season to ensure mild temperature gradients within the snowpack. The results of this experiment indicated that even in these milder air temperatures, small variations in the modeled SWE consistently led the retrieval away from valid solutions. Thus, it was concluded that the bias correction nudging algorithm could remain valid even in milder snow climate conditions.

3.2 Case study B

This case study was focused in the early winter of 2020–21, capturing the first snow of the season. On the first SAR image, corresponding to October 11, 2020, no snow was present over the imaged area and air temperature exceeded $0°C$. For the second SAR image forming the interferogram, dry snow was present over the image, with air temperature slightly below $0°C$. Fig. 2.8 presents the coherence for the interferogram. Some areas, such as water bodies, open bogs (also dominated by water), or cities and roads, have been masked and left out of the analysis.

Note that coherence was generally conserved considering the relatively long temporal baseline of 14 days. Some areas without significant vegetation exhibit a coherence of up to 0.75, whereas vegetated areas suffer from decreased coherence. This is particularly noticeable for the lower part of the image, where denser forest is present. In these areas, coherence is around and below 0.3. The interferometric phase is considered unreliable

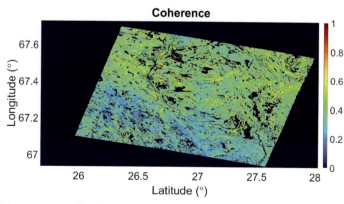

Figure 2.8 Coherence map for the InSAR pair between October 11, 2020 and October 25, 2020.

for coherence values in that range. Fig. 2.9 presents the retrieved ALOS2 ΔSWE and the simulated ΔSWE from SnowModel.

When comparing the maps in Fig. 2.9, we can see that there are significant differences in ΔSWE, particularly for the forested areas with low coherence. However, both ALOS2 and SnowModel captured similar features over certain land types. For example, the area around latitude 67.1 and longitude 27 corresponds to a barren with low vegetation. Although the coherence was relatively low, the high increase in SWE captured by ALOS2 was in line with the SnowModel simulation. Additionally, both methods captured the higher snow accumulation in the south part of the area. Fig. 2.10 presents a histogram comparing ΔSWE from ALOS2 and SnowModel, with color representing the number of points. We can see that most of the points lie close to 1:1 line. However, there is significant dispersion in both. The root mean square error (RMSE) including all pixels was 7.53 mm, which, considering the magnitude of the increment of SWE, is a moderate error. Note that this was a particularly challenging scenario, where air

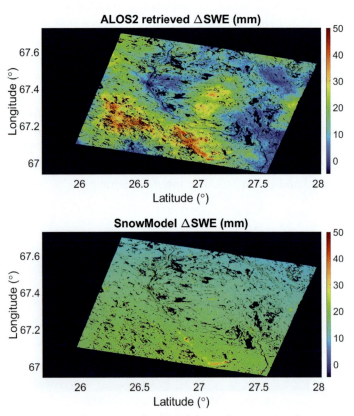

Figure 2.9 (Up) ALOS2 retrieval ΔSWE map for the InSAR pair between October 11, 2020 and October 25, 2020. (Down) SnowModel ΔSWE map between October 11, 2020 and October 25, 2020 UTC 12.

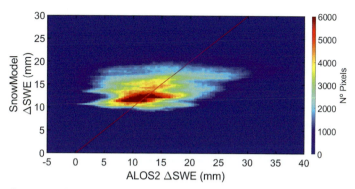

Figure 2.10 2D histogram between ALOS Δ*SWE*, on x-axis, and SnowModel Δ*SWE*, y-axis. Color represents the number of pixels.

temperature was high during the first SAR acquisition, and moderate during the second SAR acquisition. Furthermore, soil moisture changes introduced another component to the phase that was not compensated for. Later in the season, soil moisture changes between acquisitions are negligible as snow acts as a thermal and moisture insulator.

4. Conclusions

Microwave signatures are greatly affected by various snow properties and structural characteristics (snowpack states), especially by the snow microstructure, on which the lack of information poses challenges in monitoring of SWE by microwave remote sensing. Currently, information on the snowpack states is either derived from limited on-site measurements (Takala et al., 2011) or from simplified models (Kelly et al., 2003). Deriving snowpack states from physical snow models is, in theory, an appealing alternative for improving the a priori snow information required in microwave modeling, and subsequently enhancing global retrievals of SWE.

On the other hand, physical snow models come with challenges of their own, commonly related to uncertainties in the meteorological data that are required as inputs to these models, and they are naturally more sensitive to uncertainties in the precipitation forcing. Biases in the precipitation forcing do not only result in errors related to SWE but they also affect all snowpack states, including the snow microstructure that is a crucial parameter in microwave modeling. In the first case study, we explored the challenges that arise when assimilating microwave remote sensing signatures with physical snow models, and when biased precipitation inputs are introduced.

A controlled experiment was designed and conducted over a small area in Northern Finland, to address the objectives of the first case study. In situ observations of the evolution of snow properties, together with passive and active microwave observations of the snow cover were available from 2009 through 2013. First, a SnowModel run was conducted for each year, driven by meteorological data from the site and assimilated with in situ SWE observations. Then, SnowModel results of layered snow temperature, snow depth, snow density, and grain size from the base run where integrated into MEMLS to create a set of synthetic microwave observations (brightness temperature and backscatter). The quality of the synthetic observations was evaluated by comparing them to in situ observations from the ground-truth systems. Additional SnowModel runs were performed, this time by deliberately introducing errors in the simulated SWE, to resemble real-world challenges of uncertainties in the precipitation inputs. The corrupted SnowModel outputs were also ingested into MEMLS to examine how biases in snow depth, and consequently in the rest of snowpack states that are interrelated, affect the microwave simulations.

The experiment showed that even small biases in the SWE simulations led to significant errors in the microwave-based SWE retrievals. These errors were more pronounced when additional biases in snow depth predictions were introduced. The errors were not only resulting from the snow depth biases but even more so from the consequent biases in the snow microstructure. This poses a significant challenge when trying to assimilate microwave models with physical snow models. To mitigate this challenge, the natural relationship between snow depth and the snow microstructure needs to be considered.

Look-up tables derived from several SWE-perturbed SnowModel runs were examined as a possible solution to the challenges emerged at the sensitivity experiment. These look-up tables included combinations of snowpack states, ensuring that their natural relationship would be preserved. The microwave models were then guided to look for solutions among these sets. This method showed potential, but it would be computationally expensive if applied over larger spatial scales.

A bias correction nudging algorithm was explored as an alternative, cost-effective method for practical applications. By applying a condition that was derived from the sensitivity experiment, this method utilized microwave SWE retrievals guided initially by SnowModel simulations, to improve SnowModel simulations during the snow accumulation season. The nudging algorithm approach showed significant potential for correcting biases in SWE originating from the precipitation inputs, and it owes to be explored further in larger scale applications.

The second case study involved radar interferometry. InSAR is a promising tool for the retrieval of ΔSWE that has drawn the attention of the scientific community recently. Although this method has been successfully applied, it presents some limitations regarding the calibration, loss of coherence between acquisitions, or even ambiguous retrievals due to ΔSWE inducing phase changes greater than 2π. The example presented in this chapter was affected by many sources of errors and uncertainties, such as the effect of the vegetation, soil moisture, or possible uncompensated atmospheric disturbances. However, the overall trend of the ΔSWE captured by ALOS2 was generally matching the SnowModel simulation. In scenarios with less uncertainty, the InSAR retrieval could be used for assimilation into SnowModel. For example, information from the residuals between the InSAR and the SnowModel ΔSWE over a certain area could be used to correct the precipitation inputs in snow model simulations. With the new generation of low-frequency SAR satellites, the accuracy of the retrieval is expected to improve significantly, due to the reduced temporal baseline (6 days for ROSE-L and 12 days for NISAR), and for the combination of multiple frequency bands (NISAR will account with L- and S-bands, ROSE-L may orbit in a tandem formation with C-band Sentinel-1).

Further research is required to develop a cost-effective system that combines physical snow models and microwave models for large-scale applications, as well as to explore other microwave and physical snow models for this purpose. In summary, combining physical snow models, such as SnowModel, with microwave models for SWE retrievals, such as MEMLS, involves challenges, particularly when there are uncertainties in the modeled SWE. However, by considering the natural relationships among the snowpack states, there is strong potential to improve SWE retrievals through synergies between microwave remote sensing and physical snow models. Regarding InSAR, in boreal applications where the landscape is dominated by forest, further development of retrieval algorithms is required to account for the effect of the canopy. However, the InSAR retrieval algorithm is generally accurate over certain conditions, potentially enabling its employment for monitoring seasonal SWE accumulation, and hence, its ingestion in snow model assimilation protocols.

References

Anderson, E. A. (1976). *A point energy and mass balance model of a snow cover*. NOAA Technical Report.

Barnett, T., Adam, J., & Lettenmaier, D. (2005). Potential impacts of a warming climate on water availability in snow-dominated regions. *Nature, 438*, 303–309. https://doi.org/10.1038/nature04141

Derksen, C., King, J., Belair, S., Garnaud, C., Vionnet, V., Fortin, V., Lemmetyinen, J., Crevier, Y., Plourde, P., Lawrence, B., van Mierlo, H., Burbidge, G., & Siqueira, P. G. (2021). Development of the terrestrial snow mass mission. In *IEEE conference proceedings 2021 IEEE international geoscience and remote sensing symposium IGARSS, Brussels, Belgium, 2021* (pp. 614–617). https://doi.org/10.1109/IGARSS47720.2021.9553496.

Guneriussen, T., Hogda, K. A., Johnsen, H., & Lauknes, I. (2001). InSAR for estimation of changes in snow water equivalent of dry snow. *IEEE Transactions on Geoscience and Remote Sensing, 39*(10), 2101–2108. https://doi.org/10.1109/36.957273

Jordan, R. (1991). A one-dimensional temperature model for snow cover. Technical documentation for SNTHERM. *US Army Corps of Engineers, 89*.

Jorge-Ruiz, J., Lemmetyinen, J., Kontu, A., Tarvainen, R., Vehmas, R., Pulliainen, J., & Praks, J. (2022). Investigation of environmental effects on coherence loss in SAR interferometry for snow water equivalent retrieval. *IEEE Transactions on Geoscience and Remote Sensing, 60*, 1–15. https://doi.org/10.1109/TGRS.2022.3223760. Article 4306715.

Kelly, R., Chang, A., Tsang, L., & Foster, J. (2003). A prototype AMSR-E global snow area and snow depth algorithm. *IEEE Transactions on Geoscience and Remote Sensing, 41*(2), 230–242. https://doi.org/10.1109/tgrs.2003.809118

Kojima, K. (1967). Densification of seasonal snow cover. In H. Ôura (Ed.), *Physics of snow and ice* (Vol. 1, pp. 929–952). Institute of Low Temperature Science, Hokkaido University.

Leinss, S., Wiesmann, A., Lemmetyinen, J., & Hajnsek, I. (2015). Snow water equivalent of dry snow measured by differential interferometry. *IEEE Journal of Selected Topics in Applied Earth Observations and Remote Sensing, 8*(8), 3773–3790. https://doi.org/10.1109/JSTARS.2015.2432031

Lemmetyinen, J., Jorge-Ruiz, J., Cohen, J., Haapamaa, J., Kontu, A., Pulliainen, J., & Praks, J. (2022). Attenuation of radar signal by a boreal forest canopy in winter. *IEEE Geoscience and Remote Sensing Letters, 19*(1–5). https://doi.org/10.1109/LGRS.2022.3187295. Art no. 2505905.

Lemmetyinen, J., Kontu, A., Pulliainen, J., Vehviläinen, J., Rautiainen, K., Wiesmann, A., Mätzler, C., Werner, C., Rott, H., Nagler, T., Schneebeli, M., Proksch, M., Schüttemeyer, D., Kern, M., & Davidson, M. W. J. (2016). Nordic snow radar experiment. *Geoscientific Instrumentation, Methods, and Data Systems, 5*, 403–415. https://doi.org/10.5194/gi-5-403-2016

Liston, G. E., & Elder, K. (2006a). A distributed snow-evolution modeling system (SnowModel). *Journal of Hydrometeorology, 7*, 1259–1276. https://doi.org/10.1175/jhm548.1

Liston, G. E., & Elder, K. (2006b). A meteorological distribution system for high-resolution terrestrial modeling (MicroMet). *Journal of Hydrometeorology, 7*, 217–234. https://doi.org/10.1175/jhm486.1

Liston, G. E., & Hiemstra, C. A. (2008). A simple data assimilation scheme for complex snow distributions (SnowAssim). *Journal of Hydrometeorology, 9*, 989–1004. https://doi.org/10.1175/2008jhm871.1

Merkouriadi, I., Lemmetyinen, J., Liston, G. E., & Pulliainen, J. (2021). Solving challenges of assimilating microwave remote sensing signatures with a physical model to estimate snow water equivalent. *Water Resources Research, 57*. https://doi.org/10.1029/2021WR030119. e2021WR030119.

Merkouriadi, I., Leppäranta, M., & Järvinen, O. (2017). Interannual variability and trends in winter weather and snow conditions in Finnish Lapland. *Estonian Journal of Earth Sciences, 66*(1), 4757. https://doi.org/10.3176/earth.2017.03

Proksch, M., Mätzler, C., Wiesmann, A., Lemmetyinen, J., Schwank, M., Löwe, H., & Schneebeli, M. (2015). MEMLS3&a: Microwave emission model of layered snowpacks adapted to include backscattering. *Geoscientific Model Development, 8*, 2611–2626. https://doi.org/10.5194/gmd-8-2611-2015

Rautiainen, K., Lemmetyinen, J., Schwank, M., Kontu, A., Menard, C. B., Mätzler, C., Drusch, M., Wiesmann, A., Ikonen, J., & Pulliainen, J. (2014). Detection of soil freezing from L-band passive microwave observations. *Remote Sensing of Environment, 147*, 206–218. https://doi.org/10.1016/j.rse.2014.03.007

Reistad, M., Breivik, Ø., Haakenstad, H., Aarnes, O. J., Furevik, B. R., & Bidlot, J.-R. (2011). A high-resolution hindcast of wind and waves for the North sea, the Norwegian sea, and the Barents sea. *Journal of Geophysical Research: Oceans, 116*(C5). https://doi.org/10.1029/2010JC006402

Sturm, M., & Liston, G. E. (2021). Revisiting the global seasonal snow classification: An updated dataset for earth system applications. *Journal of Hydrometeorology, 22*, 2917–2938. https://doi.org/10.1175/JHM-D-21-0070.1

Takala, M., Luojus, K., Pulliainen, J., Derksen, C., Lemmetyinen, J., Kärnä, J.-P., Koskinen, J., & Bojkov, B. (2011). Estimating Northern hemisphere snow water equivalent for climate research through assimilation of space-borne radiometer data and ground-based measurements. *Remote Sensing of Environment, 115*(12), 3517–3529. https://doi.org/10.1016/j.rse.2011.08.014

Werner, C., Wiesmann, A., Strozzi, T., Schneebeli, M., & Mätzler, C. (2010). The SnowScat ground-based polarimetric scatterometer: Calibration and initial measurements from Davos Switzerland. In *IEEE international geoscience and remote sensing symposium (IGARSS), July 25–30, 2010* (pp. 2363–2366). https://doi.org/10.1109/igarss.2010.5649015

Wiesmann, A., Fierz, C., & Mätzler, C. (2000). Simulation of microwave emission from physically modeled snowpacks. *Annals of Glaciology, 31*, 397–405. https://doi.org/10.3189/172756400781820453

Wiesmann, A., & Mätzler, C. (1999). Microwave emission model of layered snowpacks. *Remote Sensing of Environment, 70*(3), 307–316.

Zhu, J., Tan, S., King, J., Derksen, C., Lemmetyinen, J., & Tsang, L. (2018). Forward and inverse radar modeling of terrestrial snow using SnowSAR data. *IEEE Transactions on Geoscience and Remote Sensing, 56*(12), 7122–7132. https://doi.org/10.1109/TGRS.2018.2848642

CHAPTER 3

Is there any pure nature left? A geospatial analysis approach of perceived landscape naturalness

Antigoni Faka[1], Triantafyllos Falaras[1], Christos Chalkias[1], Apostolos G. Papadopoulos[1] and Vassilis Detsis[2]

[1]Department of Geography, School of Environment, Geography and Applied Economics, Harokopio University, Athens, Greece; [2]Department of Economics and Sustainable Development, School of Environment, Geography and Applied Economics, Harokopio University, Athens, Greece

1. Introduction

The concept of naturalness is variously used in the literature and there is a variety of different views and perceptions of what constitutes natural and what criteria define naturalness. Baiamonte et al. (2009) report that the more intense human activities are, the lower the degree of naturalness. Angermeier (2000) considers that something is natural when it is not made or influenced by humans and mainly by technology, and defines naturalness in the following way: "*the degree to which a thing is natural, is represented by a continuous gradient between extremes of entirely natural and entirely artificial*" (p. 375). However, we recognize there is a diversity of knowledges about nature and various authoritative interpretations of what nature is (Castree, 2005). Landscape in the context of the present study is understood as defined by the European Landscape Convention: "Landscape" means an area, as perceived by people, whose character is the result of the action and interaction of natural and/or human factors." Regarding the landscapes, they are characterized by high naturalness when natural processes prevail (Baiamonte et al., 2009). On the other hand, human activities take over the natural processes in landscapes with lower naturalness.

Landscape patterns have been used to assess naturalness and ecological sustainability, especially at the national level, without limitation to administrative units, where ecological quality and landscape pattern are closely linked (Renetzeder et al., 2010). The constant landscape changes based on anthropogenic influence led to considerable environmental impacts such as biophysical transformations, alterations of habitats, disturbance of ecosystem functioning, loss of biotic diversity, even degradation of natural and cultural landscape beauty (Fischer & Lindenmayer, 2007; Frank et al., 2013; Frondoni et al., 2011; Lindenmayer et al., 2008).

Natural landscapes are also linked to health and social benefits. Human interaction with nature triggers positively reactions in the human body physiology and psychology

Geographical Information Science
ISBN 978-0-443-13605-4
https://doi.org/10.1016/B978-0-443-13605-4.00014-X

(Hamilton & Daily, 2012; Millenium Ecosystem Assessment, 2005; ten Brink et al., 2016). Many studies have reported the beneficial effects of viewing natural landscapes, including stress levels reduction, mood improvement, sense of pleasure, etc. (Velarde et al., 2007). On the other hand, loss of natural landscapes and culturally valued ecosystems is associated with negative effects on societal quality of life, including social disruptions and societal marginalization (Millenium Ecosystem Assessment, 2005; ten Brink et al., 2016).

The assessment of landscape naturalness is a useful tool in landscape management and nature conservation planning (Cooper & Murray, 1992; Ridder, 2007; Verhoog et al., 2007). A scaled index of naturalness provides information about conservation state, human intervention, and the impact of fragmentation, while it can be valuable in land use and natural resource planning (Ode et al., 2009; Pinto-Correia, 2000). According to these, subsystems conservation, areas suitability, and leisure options can be assessed leading to the establishment of related priorities (Adem Esmail & Geneletti, 2018; Raunikar & Buongiorno, 2001). Such an index can also contribute to the zonation of protected areas and the environment restoration, necessary for defining initiatives, baseline states, and targets (Adem Esmail & Geneletti, 2018).

Many studies have been carried out to evaluate naturalness, but only few of them have clearly defined criteria for its assessment (Adem Esmail & Geneletti, 2018; Kerebel et al., 2019; Machado, 2004; Ode et al., 2009; Palmer, 2019; Purcell & Lamb, 1998). In some cases, a well-defined set of criteria have been proposed, but without referring applications in which they could be implemented (Anderson, 1991; Angermeier, 2000). Naturalness has been also examined by a different point of view, by investigating the human impact (Grant, 1995; Hill et al., 2002), while some studies focus on land use and vegetation changes (Ferrari et al., 2008; Gimmi & Radeloff, 2013).

A quantitative index was presented by Machado (2004) to describe the state of naturalness in spatial units. The index was composed of diagnostic criteria based on aspects of the ecosystem that can be measured. Specifically, Machado's criteria constituted of a number of parameters related to the elements of the ecosystem, the addition of energy and matter by humans to the system, the physical alteration, the extraction of elements from the system, the fragmentation by infrastructures, and the dynamics (Machado, 2004). As a result, a scale of naturalness was constructed ranging from low to high values, each of which was defined by a set of descriptive conditions. Recognizing that man's relationship with nature is complex, containing connections and disconnections, this study seeks to distance itself from dualisms between positive and negative human impacts on the environment that are isolated from cultural and/or political contexts (Beery et al., 2023).

A similar methodology has been proposed to map landscape tranquility (Pafi et al., 2018). In this research, three composite criteria are used to assess tranquility in northern Greece: the noise regarding artificial noise, the visual-esthetic about light pollution, and the perceptual one which was related to ruralness and naturalness, utilizing open geospatial data and geographic information systems (GIS) technology.

This study aims to develop a GIS-based methodology for the assessment of landscape naturalness as it is perceived by humans, at national scale. Focusing on the perception of naturalness disturbance, human interventions in land cover, visual and noise disturbances, as well as the aesthetic value of natural landscapes were investigated. Perceived naturalness was evaluated by an integrated geographical index (Landscape Naturalness index, LNi), aiming at classifying landscapes from totally natural, without any human intervention, to transformed landscapes with clear dominance of artificial elements. The construction of LNi was based on multicriteria decision analysis (MCDA) and GIS technology which contribution in naturalness assessment has been noticed in many studies (among others Appleton & Lovett, 2003; Brabyn, 2005; Ode et al., 2009). MCDA is an effective method for dealing with problems that involve a set of discrete alternatives which are evaluated by conflicting and incommensurate criteria (Malczewski, 1999, p. 81). MCDA is also popular in spatial dimension problems, and its combined use with GIS makes it a valuable method to support spatial-related decision-making (Malczewski, 1996), such as land use management and nature conservation (Adem Esmail & Geneletti, 2018; Beinat & Nijkamp, 1998).

The methodology was designed for use at national scale and implemented in Greece. In Greece, naturalness mapping is quite limited (Boteva et al., 2004; Pafi et al., 2018) and focuses either on the ecological dimension of naturalness or on mapping specific areas. GIS technology was used to construct secondary variables and the final index, as well as to map naturalness. Landscape naturalness mapping provides valuable spatial information to the stakeholders in environmental and planning management for setting policies' objectives and priorities, and LNi constitutes a useful management tool for landscape planning, rural planning, and nature conservation.

2. Materials and methods

In the proposed methodology, four composite criteria were defined and evaluated through a set of geographical variables in order to assess landscape naturalness. Using free open spatial datasets and specialized GIS functions, landscape naturalness is depicted by evaluating LNi (Fig. 3.1).

2.1 Naturalness categories, composite criteria, and data

At the first stage of the methodology, landscape naturalness categories were defined. Five distinct categories of ordinal scale, ranked from very low (category 1) to very high (category 5) naturalness, were created, enabling the assessment of landscape naturalness in extensive geographical areas. In order to set the stage for elaborating on the criteria to assign spatial units to each category, an exercise was undertaken at low spatial level to

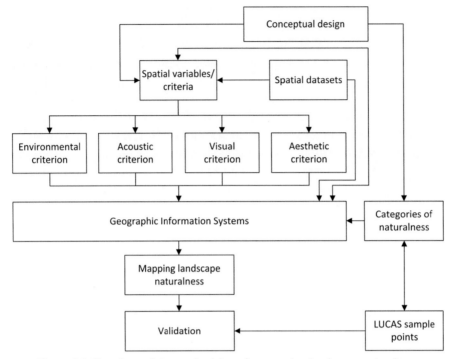

Figure 3.1 Flowchart of the methodology for assessing landscape naturalness.

develop a conceptual understanding of the continuum from very low to very high naturalness. Small islands are well suited therefore due to the undisputed definition of borders and spatial extent to be taken into account.

Afterward, the four composite criteria that assess landscape naturalness were defined (Table 3.1). The first criterion is related to the natural environment and was evaluated by the land cover factor. Human interventions in land cover were assessed in terms of degrading naturalness and not of restoration or enhancement of natural landscapes. Land cover data are widely used to assess naturalness and examine landscape patterns (Baiamonte et al., 2009; Geri et al., 2010; Fry et al., 2009; Serra et al., 2008).

The acoustic criterion is associated with noise disturbance, a very important variable which directly affects naturalness. Noise from human activity reduces naturalness depending on its level (Jackson et al., 2008). To estimate this criterion, the distance from major noise sources, such as roads, railways, and airports (Pafi et al., 2018), was analyzed.

Visually pleasing or disturbing landscapes generate emotions to the observers. Natural landscapes are linked with attractiveness and preference (Purcell & Lamb, 1998), while constant visual landscape experience implicating nature contributes to health benefits and well-being (Bratman et al., 2012; Velarde et al., 2007). The assessment

Table 3.1 Criteria that constitute LNi and the datasets that were used to evaluate each criterion.

Criteria	Variables	Data	Type	Spatial resolution	Primary source
Environmental	Land cover	Corine Land Cover 2018/CLC 2018	Raster	100 m	Copernicus Land Monitoring Service
Noise	Distance from roads	Road Network	Vector	—	Open Street Map
	Distance from railways	Railway Network	Vector	—	Open Street Map
	Distance from airports	Airports	Vector	—	Hellenic Civil Aviation Authority
Visual	Visibility on freeways	Digital Surface Model	Raster	90 m	SRTM 90 GDEM
	Visibility on railways	Road Network	Vector	—	Open Street Map
	Visibility on areas of very low naturalness regarding land cover	Railway Network	Vector	—	Open Street Map
		Corine Land Cover 2018/CLC 2018	Raster	100 m	Copernicus Land Monitoring Service
Aesthetic	Areas of special environmental interest	Digital Surface Model	Raster	90 m	SRTM 90 GDEM
		Natura 2000	Vector	—	National Geospatial Repository
		National Parks	Vector	—	National Geospatial Repository and GBWC
		National Woodland Parks	Vector	—	National Geospatial Repository
		Aesthetic Forests	Vector	—	National Geospatial Repository

of this aspect of naturalness requires identifying landscape elements perceived as disturbing. The visual contact with human constructions, activities, and infrastructure contributes to naturalness disturbance. In this way, the visual criterion was evaluated by the perceived visual naturalness of an observer depending on visibility to freeways, railways, and artificial surfaces, which correspond to areas of minimum naturalness regarding land cover (according to the first criterion) (Kerebel et al., 2019; Ode et al., 2008).

Areas of special environmental interest and aesthetic value have also been proved that contribute positively to landscape naturalness (Cole et al., 2008; Fagerholm et al., 2012). The aesthetic criterion included such areas (National Parks, National Woodland Parks, Aesthetic Forests, areas of Natura 2000 network). The development of this criterion was based on the assumption that aesthetics is related to natural environment of high quality.

All datasets (Table 3.1) were obtained from open databases, in order to be widely available and cost effective. Corine Land Cover 2018 (CLC 2018) provides geographical information on the land cover of the Member States of the European Community. The CLC 2018 was obtained from the Copernicus Land Monitoring Service database (https://land.copernicus.eu/pan-european/corine-land-cover/clc2018?tab=download) in raster format of 100 m resolution. The SRTM 90 global digital elevation models (GDEMs), produced by NASA, have spatial resolution of 90m and are provided in mosaiced 5 × 5 degree tiles. The SRTM 90 (version 4) tiles covering the Greek territory were downloaded from the CGIAR-CSI GeoPortal (http://srtm.csi.cgiar.org/srtmdata/). The road network (freeways and primary road network) and the active national railway network were obtained from Open Street Map (using Geofabrik—https://download.geofabrik.de/). Airports were digitized as a point layer, based on location information of the Hellenic Civil Aviation Authority (http://www.ypa.gr/en/our-airports). Natura 2000 dataset was obtained from EEA (https://www.eea.europa.eu/data-and-maps/data/natura-10). Natura 2000 is a network of nature protection areas all over European Union, constituted of Special Protection Areas (SPAs) and Sites of Community Importance (SCI) defined from European legislation. In Greece, there are 202 SPA and 241 SCI, overlapping each other in many cases. The delimitation of Natura 2000 sites was obtained from geodata.gov.gr. which provides open geospatial data and services for Greece, serving as a national geospatial repository and an INSPIRE-conformant Spatial Data Infrastructure. National Parks, National Woodland Parks, and Aesthetic Forests, as polygons layers, were obtained from geodata.gov.gr, which provides open geospatial data and services for Greece, serving as a national open data catalog and an INSPIRE-conformant Spatial Data Infrastructure, as well as form the Catalog of Geospatial Data of the Greek Biotope/Wetland Center (GBWC) (http://ekbygis.biodiversity-info.gr/geonetwork/srv/eng/main.home). Greece has 17 National Parks, 10 National Woodland Parks, and 19 Aesthetic Forests.

2.2 Estimating secondary variables and composite criteria

All datasets described previously were stored in a spatial database in GIS environment enabling the calculation of secondary variables, the variables' values reclassification into the five distinct categories (1-very low to 5-very high naturalness) according to the degree of landscape naturalness they are attributed to, and the final composite criteria estimation. The raster layer's spatial resolution was set at 100 m.

The values of the environmental criterion were assigned by reclassifying the CLC 2018 categories. The reclassification was based on the contribution of the categories' characteristics (human activity, vegetation) to different levels of natural systems (Baiamonte et al., 2009) and the CLC classes' degree of naturalness indicated by the hemeroby scale (European Environment Agency, 2014) (Table 3.2).

The acoustic criterion was evaluated by distance-based indicators. Although using a model to assess noise disturbance is less accurate than in situ measurements, it is a quick, low cost, and simple method suitable for national scale analysis (Votsi et al., 2012). The noise criterion variables were calculated by creating buffer zones of specific distance from each major source of noise. Specifically, the buffer zones were defined at 800 m from the freeways, 600 m from the primary road network, 650 m from the railway network, 1500 m from the highest daily traffic load airports, and 900 m from the other major airports (Votsi et al., 2012). Each buffer zone was divided into four subzones of equal distance. The subzones were reclassified into the four naturalness categories (1: very low to 4: high), with the minimum naturalness to be assigned to the closest subzone to the noise source, while the maximum naturalness (category 5) was assigned to areas in distance greater than the buffer zone limit. In cases of road network intersection (freeways and primary road network), lowest level of

Table 3.2 Categorization of the CLC 2018 classes into the five categories of landscape naturalness.

Naturalness	CLC
1 Very low	**1.1** Urban fabric
	1.2 Industrial, commercial, and transport units
	1.3 Mine, dump and construction sites
2 Low	**1.4** Artificial, nonagricultural vegetated areas
	2.1 Arable land
	2.2 Permanent crops
	2.4 Heterogeneous agricultural areas
3 Moderate	**2.3** Pastures
	3.3 Open spaces with little or no vegetation
4 High	**3.2** Scrub and/or herbaceous vegetation associations
	4.1 Inland wetlands
	4.2 Maritime wetlands
5 Very high	**3.1** Forests
	5.1 Inland waters
	5.2 Marine waters

naturalness was taken into account. To evaluate the final acoustic criterion, the vector layers of the three variables were merged and converted to a 100 m raster layer.

Regarding the visual criterion, variables related to the visibility on freeways and railways were calculated using viewshed analysis which identifies the areas that can be seen from observation locations. In this case, freeways and railways were defined as two separate sets of observation locations and a viewshed layer was created for each one set. It has to be reported that visibility distance is affected by earth curvature, observation altitude, and atmospheric conditions. Earth curvature and altitude parameters were calculated in viewshed algorithm (http://desktop.arcgis.com/en/arcmap/10.3/tools/spatial-analyst-toolbox/viewshed.htm), while the absence of atmospheric effects was assumed. The viewshed layers were limited in a zone of 40 km, considering it the maximum horizontal visibility distance in clean atmosphere for an observer. The two final layers were reclassified into the five categories of naturalness by attributing maximum naturalness to areas with no visibility to freeways and railways, respectively, while the rest four categories were based on natural breaks method (Jenks, 1977).

The visibility on areas of minimum naturalness regarding land cover was based on the minimum naturalness category of CLC layer (environmental criterion—Table 3.2), by excluding road network and railway, as the visibility on them was assessed previously. A viewshed layer was also created by defining CLC minimum naturalness category as observation location. The five categories of naturalness were defined in the same way as the visibility on freeways and railways, by attributing maximum naturalness to areas with no visibility to the specific CLC category, and classifying the rest areas using the natural breaks method. The raster layers of the three variables were combined, attributing to the final criterion layer the lowest level of naturalness of each cell combination.

In the aesthetic criterion, two landscape naturalness categories were defined; maximum naturalness which was assigned to areas of special environmental interest, and minimum to the rest of Greece.

ArcGIS version 10.2 (ESRI Inc., Redlands, California, USA) software was used for the creation of the spatial database and mapping, as well as for local statistical modeling.

2.3 Landscape naturalness index construction

The LNi is calculated as the square root of the product of the ranked criteria divided by the total number of criteria as:

$$LNi = \sqrt{\frac{C_i * \ldots * C_n}{n}} \tag{3.1}$$

where:
  LNi = Landscape Naturalness index
  C = criterion
  n = number of criteria

Table 3.3 Categorization of LNi values into five naturalness classes.

	1	2	3	4	5
Categories of naturalness	**Very low**	**Low**	**Moderate**	**High**	**Very high**
LNi values	0.5—1.0	1.0—3.0	3.0—6.0	6.0—10.0	10.0—12.5

This method allows variables/criteria to be related in a quantifiable manner and it has also been employed to indexes using classified variables to assess environmental risk (Gornitz et al., 1994; Thieler & Hammar-Klose, 1999). The product is highly sensitive to small changes in variables ranking and the square root compresses the extreme range of scores highlighting the regions where the disturbance of landscape naturalness may be either great or inconsiderable.

The environmental, acoustic, visual, and aesthetic criteria were used in Eq. (3.1) to calculate an overall value to asses landscape naturalness. Since the values of the composite criteria range from 1 to 5, LNi values range from 0.5 to 12.5. The categorization of LNi values into five classes (1: very low naturalness to 5: very high naturalness) was based on the combination of the values of the four criteria and the corresponding resulting thresholds (Table 3.3).

3. Results

The distribution of the values assigned to each composite criterion in Greece is shown in Fig. 3.2. With respect to the environmental criterion, 21.4% of the Greek territory was characterized by landscapes with very high naturalness, while areas with very low naturalness (urban and industrial areas) corresponded only to 3.0% (Fig. 3.2a). Landscapes of high and very high naturalness were mostly located in mountainous areas, whereas low naturalness was noticed in agricultural areas. Noise disturbance is limited because very high naturalness prevails across the country (89.8%), as the major noise sources affected only neighboring areas (Fig. 3.2b). According to the visual criterion, almost 76% of Greece was characterized by high and very high naturalness (60.0% and 15.7% accordingly). Landscapes of very low and low naturalness were concentrated in north-central Greece, due to the high altitude of Mount Olympus that maximizes visibility to areas that affect landscape naturalness (freeways, railways, land cover of minimum naturalness) along with the mountains surrounding the plain of Thessaloniki. Urban areas, such as Athens and Thessaloniki, were also characterized by low naturalness based on this criterion (Fig. 3.2c). For the stricter criterion, the aesthetic one, almost 70% of Greece consisted of landscapes with a very low degree of naturalness (Fig. 3.2d), while only the areas of special environmental interest were characterized by a very high degree of naturalness (30.7%).

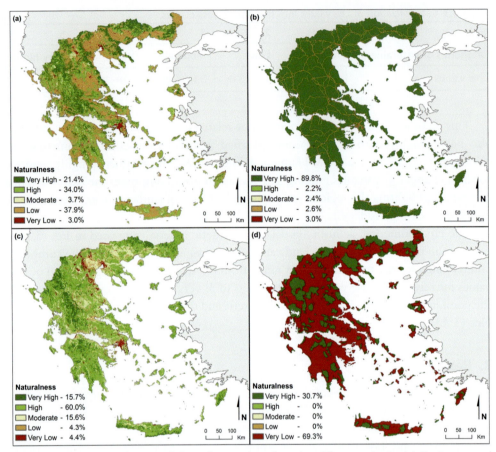

Figure 3.2 Spatial distribution of the values assigned to the different criteria. (a) Environmental, (b) noise, (c) visual, and (d) aesthetic.

Fig. 3.3 illustrates LNi results across Greece. In Fig. 3.3a, overall landscape naturalness is represented based on the LNi's continuous values (0.5–12.5). Regarding the final outcome, reclassified into five categories (Fig. 3.3b), landscapes of Greece were characterized mostly by moderate naturalness (56.7%). Scattered areas of high and very high naturalness were found throughout the country, together accounting for more than a quarter (18.9% and 9.7%, respectively) of the Greek territory. To a lesser extent (13.1%), landscapes with low naturalness were also observed throughout the country. On the other hand, landscapes with very low naturalness, located mainly in urban areas and along the road network, occupy only 1.6% of the total area of the country.

Both maps illustrate a situation where the mountainous, arid and border areas are considered of very high naturalness. This illustration implies that naturalness in the case of Greece is connected to elevation, aridity, and frontier zones (Fig. 3.3).

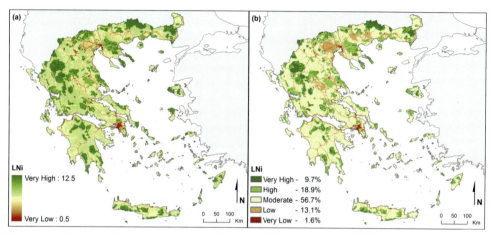

Figure 3.3 Overall landscape naturalness of Greece, (a) based on the LNi's continuous values, (b) reclassified into five categories.

4. Validation

A supervisory validation was applied to examine if the results of the proposed methodology were consistent with the perceived landscape state. The proposed validation method was based on sample points of the Land Use/Cover Area frame Survey (LUCAS) 2015.

LUCAS is a field survey of Eurostat with in situ sample collection. Plenty of information is collected regarding land use and land cover, environmental information, photographs, samples from soil surface, while it also studies grasslands (Eurostat, 2013). LUCAS points are surpassing one million in European Union constituting a 2×2 km grid of points. In 2015, field samples included 273,400 points, 7852 of which located in Greece (Gallego et al., 2015). LUCAS data are free open and provide information for several research sectors (https://ec.europa.eu/eurostat/web/lucas).

Data of LUCAS 2015 were compared to the final outcome. LUCAS sample points were spatially matched with the final landscape naturalness layer. 10 points of each naturalness category all over Greece (50 points in total) were spatially randomly selected, 48 of which were accompanied with photos. LUCAS points were characterized by the five categories of naturalness according to their land cover and photos information. Afterward, a buffer zone of 100 m radius, based on the raster layer's spatial resolution, was created for each point. Average naturalness value was estimated based on the results and compared with the naturalness category of each LUCAS point. No deviation was defined as no difference or difference in only one naturalness category, slight deviation when difference was detected in two naturalness categories and considerable deviation when a difference was found in three or more naturalness categories.

Figure 3.4 Randomly selected LUCAS 2015 sample points for supervisory validation.

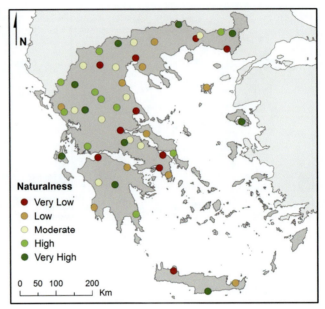

The supervisory validation, based on 50 LUCAS points (Fig. 3.4), confirmed the accuracy of the proposed methodology. Slight deviation was noticed in five out of fifty points, while only two revealed considerable deviation, resulting in 86% accuracy.

5. Discussion

In this study, we aimed to develop a methodology for the assessment of perceived landscape naturalness, by constructing a composite index. Naturalness evaluation has been already examined by many studies (Adem et al., 2018; Ferrari et al., 2008; Gimmi & Radeloff, 2013; Machado, 2004). Relative indices have been developed for assessing naturalness, but to our knowledge, this is the first study to propose an integrated index and definitive scale for landscape naturalness that is applicable on a national scale. In this study, an integrated geographic approach was used to assess the perceived naturalness of landscapes at a national scale, which is critical for formulating large-scale, evidence-based planning related to conservation and landscape management. The proposed index was designed to cover multiple aspects of naturalness, assign a name and/or definition to each level of naturalness, and adjust the levels of naturalness according to geographic scale.

The spatial patterns demonstrated by overall LNi values are in line with the Greek landscape typology suggested by Tsilimigkas & Kizos (2014). The final output revealed that almost half the area of Greece was characterized by moderately natural landscapes

(56.7%), mainly rural areas with human interventions but limited noise and visual disturbances. More than a quarter of Greece was characterized by landscapes of high and very high naturalness (28.6%), mainly in mountainous areas where there is no disturbance of the landscape. Most areas of low naturalness in Greece were located in suburban and agricultural areas, along the road network and in parts of the coastal area, because they were directly related to areas of intense human activity. For the same reason, mountain areas bordering the coastal zone and in direct visual contact with areas where human activity is prevalent received lower naturalness scores than inland mountain areas. Intense human influence was reflected in the category of very low naturalness found in landscapes with man-made facilities, infrastructure, urban and industrial areas.

The cost-effectiveness of the proposed methodology is another strength of this study, as we obtained accurate results by using open and free data. The validation procedure is also a new cost-effective method, very close to the field research. Furthermore, the implementation of GIS technology provided specialized functions and methods of analysis to create the composite criteria and the final LNi, and promoted the results mapping (Appleton & Lovett, 2003; Brabyn, 2005). The methodology that was developed, in conjunction with the use of GIS, produced naturalness models in intelligible and comprehensive representations.

The design of the LNi, focusing on the implementation at national level, leads to the construction of simple criteria, and limited the use of variables and data to avoid overlaps. For example, only highways and the main road network were considered instead of the entire road network in Greece. However, the proposed index is not limited in terms of the number of criteria, variables and data, and although it was applied at the national level, it can serve as a basis for assessing the naturalness of the landscape at different geographical scales.

Another limitation is related to the assessment of human activity by using only land cover data. For instance, solar panels and wind turbines installations were not included in the analysis due to lack of available data. Such human interventions affect naturalness, not only by means of aesthetic value but also of noise, visual, and environmental disturbances as well. The installation of wind turbines and/or solar panels transform the naturalness of landscapes in the name of environmental sustainability and substituting carbon fuels. With respect to this example, significant discussions need to take place in terms of landscape planning, rural planning, and nature conservation with regard to the prioritization of the type(s) of interventions that need to be pursued. Qualitative information gathered from individuals who have their own visions of naturalness and their own interpretation of human presence could be integrated into this index to come in terms with similar situations.

In addition, the supervisory validation enabled the comparison between the LNi results and the available photographs of LUCAS survey, but did not allow the noise validation, as the LUCAS points do not provide noise data. Finally, the proposed methodology and the use of LNi, like any national mapping attempt, should only be auxiliary when used in local context that require more detailed data and analysis.

6. Conclusions

In the present study, a methodology for the evaluation of naturalness was proposed, which is easy to apply and to interpret the results. Naturalness is assessed using measurable variables that are commonly available and can be integrated into a database GIS. Promising results from the validation process indicate that the method can be safely applied in large geographic regions with limited availability of detailed data. Naturalness mapping is a functional tool for decision makers and planners in terms of landscape change monitoring and conservation planning. LNi can be applied to different regions and can form the basis for assessing the naturalness of landscapes at different geographic scales. In future work, criteria could be evaluated by more variables, especially related to human activity. In addition, empirical information from case studies can enrich the proposed methodology by using more nuanced variables based on observations, qualitative constructs, and collective representations of landscapes. Emotions, aesthetic values, and imaginations of people may well be included to enhance the subjective aspect of the methodology. Furthermore, LNi can be applied over time periods to identify landscape naturalness changes. All in all, our modeling exercise can assist the articulation of a policy framework for natural landscapes and cooptation in landscape and rural planning.

References

Adem Esmail, B., & Geneletti, D. (2018). Multi-criteria decision analysis for nature conservation: A review of 20 years of applications. *Methods in Ecology and Evolution, 9*(1), 42—53. https://doi.org/10.1111/2041-210X.12899

Anderson, J. E. (1991). A conceptual framework for evaluating and quantifying naturalness. *Conservation Biology, 5*, 347—352. https://doi.org/10.1111/j.1523-1739.1991.tb00148.x

Angermeier, P. L. (2000). The natural imperative for biological conservation. *Conservation Biology, 14*(2), 373—381. https://doi.org/10.1046/j.1523-1739.2000.98362.x

Appleton, K., & Lovett, A. (2003). GIS-based visualisation of rural landscapes: Defining 'sufficient' realism for environmental decision-making. *Landscape and Urban Planning, 65*(3), 117—131. https://doi.org/10.1016/S0169-2046(02)00245-1

Baiamonte, G., Bazan, G., & Raimondo, M. F. (2009). *Land mosaic naturalness evaluation: A proposal for European landscapes*. Salzburg, Austria: European IALE Conference 2009. https://doi.org/10.13140/2.1.1236.4489

Beery, T., Stahl Olafsson, A., Gentin, S., Maurer, M., Stålhammar, S., Albert, C., Bieling, C., Buijs, A., Fagerholm, N., Garcia-Martin, M., Plieninger, T., & Raymond, C. (2023). Disconnection from nature: Expanding our understanding of human—nature relations. *People and Nature, 5*, 470—488. https://doi.org/10.1002/pan3.10451

Beinat, E., & Nijkamp, P. (1998). *Multicriteria analysis for land-use management*. Dordrecht: Springer Netherlands.

Boteva, D., Griffiths, G., & Dimopoulos, P. (2004). Evaluation and mapping of the conservation significance of habitats using GIS: An example from Crete, Greece. *Journal for Nature Conservation, 12*(4), 237–250. https://doi.org/10.1016/j.jnc.2004.09.002

Brabyn, L. (2005). Solutions for characterising natural landscapes in New Zealand using geographical information systems. *Journal of Environmental Management, 76*(1), 23–34. https://doi.org/10.1016/j.jenvman.2005.01.005

Bratman, G. N., Hamilton, J. P., & Daily, G. C. (2012). The impacts of nature experience on human cognitive function and mental health. *Annals of the New York Academy of Sciences, 1249*(1), 118–136. https://doi.org/10.1111/j.1749-6632.2011.06400.x

Castree, N. (2005). *Nature*. London: Routledge.

Cole, D. N., Yung, L., Zavaleta, E. S., Aplet, G. H., Chaplin, F. S., III, Graber, D. M., Higgs, E. S., Hobbs, R. J., Landres, P. B., Millar, C. I., Parsons, D. J., Randall, J. M., Stephenson, N. L., Tonnessen, K. A., White, P. S., & Woodley, S. (2008). Naturalness and beyond: Protected area stewardship in an era of global environmental change. *George Wright Forum, 25*(1), 36–56.

Cooper, A., & Murray, R. (1992). A structured method of landscape assessment and countryside management. *Applied Geography, 12*(4), 319–338. https://doi.org/10.1016/0143-6228(92)90012-C

European Environment Agency. (2014). *Good practice guide on quiet areas. Technical report No 4/2014.* Luxembourg: Publications Office of the European Union. https://www.eea.europa.eu/publications/good-practice-guide-on-quiet-areas. (Accessed 12 August 2023).

Eurostat. (2013). *LUCAS, EU's land use and land cover survey*. Luxembourg: Publications Office of the European Union. https://ec.europa.eu/eurostat/documents/4031688/5931504/KS-03-13-587-EN.PDF/4ee08a33-36ee-40c3-bf59-3b2f5baa28e1?version=1.0. (Accessed 18 September 2023).

Fagerholm, N., Käyhkö, N., Ndumbaro, F., & Khamis, M. (2012). Community stakeholders' knowledge in landscape assessments - mapping indicators for landscape services. *Ecological Indicators, 18*, 421–433. https://doi.org/10.1016/j.ecolind.2011.12.004

Ferrari, C., Pezzi, G., Diani, L., & Corazza, M. (2008). Evaluating landscape quality with vegetation naturalness maps: An index and some inferences. *Applied Vegetation Science, 11*(2), 243–250. https://doi.org/10.3170/2008-7-18400

Fischer, J., & Lindenmayer, D. B. (2007). Landscape modification and habitat fragmentation: A synthesis. *Global Ecology and Biogeography, 16*, 265–280. https://doi.org/10.1111/j.1466-8238.2007.00287.x

Frank, S., Fürst, C., Koschke, L., Witt, A., & Makeschin, F. (2013). Assessment of landscape aesthetics - validation of a landscape metrics-based assessment by visual estimation of the scenic beauty. *Ecological Indicators, 32*, 222–231. https://doi.org/10.1016/j.ecolind.2013.03.026

Frondoni, R., Mollo, B., & Capotorti, G. (2011). A landscape analysis of land cover change in the Municipality of Rome (Italy): Spatio-temporal characteristics and ecological implications of land cover transitions from 1954 to 2001. *Landscape and Urban Planning, 100*(1–2), 117–128. https://doi.org/10.1016/j.landurbplan.2010.12.002

Fry, G., Tveit, M. S., Ode, A., & Velarde, M. D. (2009). The ecology of visual landscapes: Exploring the conceptual common ground of visual and ecological landscape indicators. *Ecological Indicators, 9*(5), 933–947. https://doi.org/10.1016/j.ecolind.2008.11.008

Gallego, F. J., Palmieri, A., & Ramos, H. (2015). *Sampling system for LUCAS 2015*. Luxembourg: Publications Office of the European Union. https://ec.europa.eu/eurostat/documents/205002/6786255/LUCAS+2015+sampling_20160922.pdf. (Accessed 18 September 2023).

Geri, F., Amici, V., & Rocchini, D. (2010). Human activity impact on the heterogeneity of a Mediterranean landscape. *Applied Geography, 30*(3), 370–379. https://doi.org/10.1016/j.apgeog.2009.10.006

Gimmi, U., & Radeloff, V. C. (2013). Assessing naturalness in northern great lakes forests based on historical land-cover and vegetation changes. *Environmental Management, 52*(2), 481–492. https://doi.org/10.1007/s00267-013-0102-0

Gornitz, V., Daniels, R. C., White, T. W., & Birdwell, K. R. (1994). The development of a coastal risk assessment database: Vulnerability to sea-level rise in the U.S. Southeast. *Journal of Coastal Research, 12*, 327–338.

Grant, A. (1995). Human impacts on terrestrial ecosystems. In T. O'Riordan (Ed.), *Environmental science for environmental management* (pp. 66–79). Singapore: Longman.

Hill, M. O., Roy, D. B., & Thompson, K. (2002). Hemeroby, urbanity and ruderality: Bioindicators of disturbance and human impact. *Journal of Applied Ecology, 39*, 708–720. https://doi.org/10.1046/j.1365-2664.2002.00746.x

Jackson, S., Fuller, D., Dunsford, H., Mowbray, R., Hext, S., MacFarlane, R., & Haggett, C. (2008). *Tranquillity mapping: Developing a robust methodology for planning support. Report to the campaign to protect rural england.* Centre for Environmental & Spatial Analysis, Northumbria University, Bluespace environments and the University of Newcastle upon on Tyne.

Jenks, G. F. (1977). *Optimal data classification for choropleth maps. Occasional Paper no. 2.* Department of Geography, University of Kansas.

Kerebel, A., Gélinas, N., Déry, S., Voigt, B., & Munson, A. (2019). Landscape aesthetic modelling using Bayesian networks: Conceptual framework and participatory indicator weighting. *Landscape and Urban Planning, 185*, 258–271. https://doi.org/10.1016/j.landurbplan.2019.02.001

Lindenmayer, D., Hobbs, R. J., Montague-Drake, R., Alexandra, J., Bennett, A., Burgman, M., Cale, P., Calhoun, A., Cramer, V., Cullen, P., Driscoll, D., Fahrig, L., Fischer, J., Franklin, J., Haila, Y., Hunter, M., Gibbons, P., Lake, S., Luck, G., … Zavaleta, E. (2008). A checklist for ecological management of landscapes for conservation. *Ecology Letters, 11*(1), 78–91. https://doi.org/10.1111/j.1461-0248.2007.01114.x

Machado, A. (2004). An index of naturalness. *Journal for Nature Conservation, 12*(2), 95–110. https://doi.org/10.1016/j.jnc.2003.12.002

Malczewski, J. (1996). A GIS-based approach to multiple criteria group decision- making. *International Journal of Geographical Information Science, 10*(8), 955–971. https://doi.org/10.1080/02693799608902119

Malczewski, J. (1999). *GIS and multicriteria decision analysis.* New York: John Wiley and Sons.

Millennium Ecosystem Assessment. (2005). *Ecosystems and human well-being: Synthesis.* Washington: Island Press.

Ode, A., Fry, G., Tveit, M. S., Messager, P., & Miller, D. (2009). Indicators of perceived naturalness as drivers of landscape preference. *Journal of Environmental Management, 90*(1), 375–383. https://doi.org/10.1016/j.jenvman.2007.10.013

Ode, Å., Tveit, M., & Fry, G. (2008). Capturing landscape visual character using indicators: Touching base with landscape aesthetic theory. *Landscape Research, 33*(1), 89–117. https://doi.org/10.1080/01426390701773854

Pafi, M., Chalkias, C., & Stathakis, D. (2018). A cost-effective method for tranquility mapping using open environmental data. *Environment and Planning B: Urban Analytics and City Science, 0*(0), 1–20. https://doi.org/10.1177/2399808318779084

Palmer, J. F. (2019). The contribution of a GIS-based landscape assessment model to a scientifically rigorous approach to visual impact assessment. *Landscape and Urban Planning, 189*, 80–90. https://doi.org/10.1016/j.landurbplan.2019.03.005

Pinto-Correia, T. (2000). Future development in Portuguese rural areas: How to manage agricultural support for landscape conservation? *Landscape and Urban Planning, 50*(1–3), 95–106. https://doi.org/10.1016/S0169-2046(00)00082-7

Purcell, A. T., & Lamb, R. J. (1998). Preference and naturalness: An ecological approach. *Landscape and Urban Planning, 42*(1), 57–66. https://doi.org/10.1016/S0169-2046(98)00073-5

Raunikar, R., & Buongiorno, J. (2001). Valuation of forest amenities: A macro approach. In M. H. Pelkki (Ed.), *Proceedings of the 2000 southern forest economics workshop, March 23-25, 2000, Lexington, Kentucky* (pp. 118–123). University of Kentucky.

Renetzeder, C., Schindler, S., Peterseil, J., Prinz, M. A., Mücher, S., & Wrbka, T. (2010). Can we measure ecological sustainability? Landscape pattern as an indicator for naturalness and land use intensity at regional, national and European level. *Ecological Indicators, 10*(1), 39–48. https://doi.org/10.1016/j.ecolind.2009.03.017

Ridder, B. (2007). An exploration of the value of naturalness and wild nature. *Journal of Agricultural and Environmental Ethics, 20*(2), 195–213. https://doi.org/10.1007/s10806-006-9025-6

Serra, P., Pons, X., & Saurí, D. (2008). Land-cover and land-use change in a Mediterranean landscape: A spatial analysis of driving forces integrating biophysical and human factor. *Applied Geography, 28*(3), 189–209. https://doi.org/10.1016/j.apgeog.2008.02.001

ten Brink, P., Mutafoglu, K., Schweitzer, J. P., Kettunen, M., Twigger-Ross, C., Baker, J., Kuipers, Y., Emonts, M., Tyrväinen, L., Hujala, T., & Ojala, A. (2016). *The health and social benefits of nature and biodiversity protection. A report for the European commission (ENV.B.3/ETU/2014/0039).* London/Brussels: Institute for European Environmental Policy. https://op.europa.eu/en/publication-detail/-/publication/89612982-1e4d-11e6-ba9a-01aa75ed71a1/language-en/format-PDF. (Accessed 18 September 2023).

Thieler, E. R., & Hammar-Klose, E. S. (1999). National assessment of coastal vulnerability to sea-level rise, U.S. Atlantic Coast. *U.S. Geological Survey. Open-File Report*, 99−593.

Tsilimigkas, G., & Kizos, T. (2014). Space, pressures and the management of the Greek landscape. *Geografiska Annaler Series B Human Geography, 96*(2), 159−175. https://doi.org/10.1111/geob.12043

Velarde, M. D., Fry, G., & Tveit, M. (2007). Health effects of viewing landscapes—Landscape types in environmental psychology. *Urban Forestry and Urban Greening, 6*(4), 199−212. https://doi.org/10.1016/j.ufug.2007.07.001

Verhoog, H., Van Bueren, E. T. L., Matze, M., & Baars, T. (2007). The value of 'naturalness' in organic agriculture. *NJAS - Wageningen Journal of Life Sciences, 54*(4), 333−345. https://doi.org/10.1016/S1573-5214(07)80007-8

Votsi, N. E., Drakou, E., Mazaris, A., Kallimanis, A., & Pantis, J. (2012). Distance-based assessment of open country Quiet Areas in Greece. *Landscape and Urban Planning, 104*(2), 279−288. https://doi.org/10.1016/j.landurbplan.2011.11.004

CHAPTER 4

Spatiotemporal patterns of vegetation regeneration dynamics in a natural Mediterranean ecosystem using EO imagery and Google Earth Engine cloud platform

Ioannis Lemesios, Spyridon E. Detsikas and George P. Petropoulos
Department of Geography, Harokopio University of Athens, Athens, Greece

1. Introduction

In recent decades, wildfires have emerged as one of the most perilous natural disasters affecting global forest ecosystems, resulting in adverse environmental and economic consequences, as well as substantial human casualties (Brown et al., 2018; IPCC, 2022). Among European nations, Mediterranean countries bear the brunt of forest fires, with the majority of the total burned area concentrated during the summer season (JRC, 2020). The distinct climatic conditions, diverse vegetation, and human activities collectively render the European Mediterranean region exceptionally susceptible to forest fires (Moreira et al., 2020). The issue is further exacerbated by the pressures of climate change, which heightens the problem by escalating the occurrence of extreme weather events (Bridges et al., 2019; IPCC, 2022). Projections indicate that the Mediterranean region will witness a rise in both the frequency and severity of droughts due to climate change, subsequently leading to a heightened risk of wildfires (Karali et al., 2023; Rovithakis et al., 2022).

In this context, understanding the spatial and temporal patterns of postfire vegetation recovery dynamics holds paramount significance. This knowledge serves essential roles across various aspects of policy making and decision making, as well as in shaping the composition and structure of plant and animal communities within impacted ecosystems (Gouveia et al., 2010; Said et al., 2015). Gaining insights into the dynamics of vegetation recovery subsequent to a wildfire outbreak is vital for assessing the fire's impact and comprehending the underlying forces steering changes in postfire environments (Grissino Mayer & Swetnam, 2000; Ireland & Petropoulos, 2015; Petropoulos et al., 2014). In Mediterranean ecosystems, the process of vegetation regrowth following a fire is intricate and challenging to generalize due to an array of influencing factors. These factors encompass heightened spatial variability, ecotype variations, fire duration and

Geographical Information Science
ISBN 978-0-443-13605-4
https://doi.org/10.1016/B978-0-443-13605-4.00002-3

intensity, preexisting vegetation conditions, as well as localized topographical, climatic, and soil characteristics (Amos et al., 2018; Moreira et al., 2009; Pausas & Vallejo, 1999; Veraverbeke et al., 2010).

Concerning topography, it assumes a substantial role in the process of vegetation regeneration. Existing research has illuminated the pivotal role this factor plays in shaping an area's microclimate and its susceptibility to fire (Ireland & Petropoulos, 2015). South facing slopes, for instance, receive more intense solar radiation and subsequently undergo heightened rates of evapotranspiration compared to their north facing counterparts. This differential solar exposure leads to notably reduced soil moisture content on south facing slopes in the northern hemisphere, thereby impeding the growth of vegetation. Conversely, north facing slopes offer more favorable moisture conditions, resulting in increased vegetation growth and regrowth dynamics (Ireland & Petropoulos, 2015; Lou-haichi et al., 2021; Petropoulos et al., 2014). Consequently, south facing slopes generally exhibit elevated burn severity due to their heightened solar exposure, which diminishes soil moisture levels and creates a drier environment, consequently elevating the risk of ignition (Ozelkan et al., 2011).

When employed alongside Geographic Information Systems (GIS) techniques, Earth Observation (EO) data have demonstrated a significant potential as a valuable solution toward evaluating the fire impacts and continually in monitoring spatiotemporally affected ecosystems (Chen et al., 2021; Fragou et al., 2020; Llorens et al., 2021; Petropoulos et al., 2014). Numerous studies concerning the monitoring of fire affected regions have been conducted using satellite data in Mediterranean ecosystems (Kalivas et al., 2013; Nioti et al., 2015; Petropoulos et al., 2014; Viana-Soto et al., 2017). Furthermore, vegetation indices like the Normalized Difference Vegetation Index (NDVI) have been extensively employed to track changes and trends in vegetation following fires (Katagis et al., 2014; Smith-Ramírez et al., 2022; Wang et al., 2022). In addition, many studies have effectively utilized high spatial resolution (30 m) NDVI time series data from Landsat satellites to estimate postfire vegetation dynamics (Chen et al., 2016; García-Llamas et al., 2019; Ireland & Petropoulos, 2015; López García, 2020; Pérez et al., 2022; Petropoulos et al., 2014; Viana-Soto et al., 2017).

Nowadays, the abundance of accessible EO datasets has introduced a set of inevitable challenges in EO data processing (Amos et al., 2018). Consequently, cloud based platforms like Google Earth Engine (GEE) have arisen as highly valuable tools for managing the wealth of extensive EO datasets. These cloud based platforms offer a cost effective and computationally efficient means to process large scale EO data within a cloud environment. They concurrently facilitate the application of various levels of complexity in image processing techniques, which prove instrumental in the examination of vegetation regrowth dynamics (Yang et al., 2022).

Within this framework, the principal objective of this study is the evaluation of the dynamics of vegetation recovery after a wildfire event over a 15 year period. This assessment is conducted through a multitemporal analysis of Landsat imagery (TM and OLI), leveraging the capabilities of GEE. The study focuses on a representative Mediterranean location in Greece that had experienced a wildfire in 2007. The specific study objectives were twofold. Firstly, it aimed to delineate the spatiotemporal patterns governing vegetation regrowth dynamics within the fire affected area, a process monitored by the NDVI response. Secondly, it sought to examine the impact of topographical variables on these regrowth dynamics. Furthermore, the study made efforts to establish regression models to characterize the regenerative process quantitatively. Additionally, it aimed to investigate, through EO data and GEE analysis, the correlation between postfire recovery and topographic factors such as aspect.

2. Study area

The selected study site, Mount Parnitha, is situated approximately 30 km north of the Greek capital, Athens (Fig. 4.1). Covering an area of about 200 km^2, it spans an elevation range from 200 to 1400 m above sea level (m.a.s.l). The landscape is characterized by a predominant presence of Aleppo Pine (*Pinus halepensis*) forests, occupying the elevations between 300 and 900 m, while Greek Fir (*Abies cephalonica*) forests thrive on slopes above 900 m up to 1400 m altitude (Ganatsas et al., 2012). Below 300 m, the terrain transitions to a mix of shrubland as well as farmlands in the northern section, and suburban housing in the eastern sector. The climate is continental, marked by cold winters and relatively warmer summers. Summer temperatures typically stay under 18°C, while winter temperatures often hover around 0°C, with an annual average

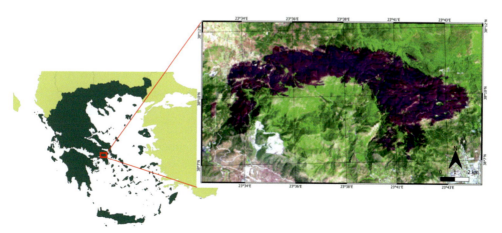

Figure 4.1 Study area location. The satellite image displays a false color composite of the first postfire LANDSAT TM image (using bands 5, 4, and 3), which was acquired on July 3, 2007, few days after fire suppression. This image clearly highlights the burn scar (visible as a dark area).

temperature of 11°C (Arianoutsou et al., 2010). The average annual rainfall in this region reaches 822 mm (at 1000 m elevation), accompanied by approximately 70 rainy days each year. Snowfall is also frequent, with an average of 33 snowy days annually and an average snow depth of 120 cm (Ganatsas et al., 2012).

The broader study area holds significant ecological value not only for Attica region but also for Greece as a whole. It is recognized as one of the 10 National Woodland Parks in Greece, established in 1961 with the primary aim of conserving the local flora and fauna. Furthermore, the National Park has been designated as part of the Natura 2000 Network (GR 3000009) in accordance with the Habitats Directive (92/43/EEC). Additionally, it is designated as a crucial area for the protection of wild birds (SPA) under the Birds Directive (2009/147/EC) (Ganatsas et al., 2012).

On the afternoon of June 27, 2007, in the Dervenohoria area, near a village named Stefani, a fire ignited due to sparks from an overloaded power line. The fire rapidly spread during the night, driven by strong winds and exacerbated by an exceptionally intense and prolonged heatwave in Athens. The following day, propelled by a moderate westerly wind, the fire advanced into the forested western slopes and canyons of the mountain, eventually reaching the summit and leaving behind a landscape of charred trees. Its primary advance ceased upon encountering sparse vegetation on the eastern slope of the mountain on the morning of June 29th. Thanks to aerial firefighting support, the fire was brought under control two days later, on July 1st, 2007. According to the official estimates, a total of 48 km^2 of land were scorched in the region, with approximately 36 km^2 falling within the Parnitha National Park. In addition, roughly 18% of the Natura 2000 area in Parnitha (about 27 km^2) was affected by the fire, while 62% of the renowned Parnitha fir forest was burnt. The impact on the local fauna was also severe, resulting in the loss of numerous animals, including turtles, various mammal species, birds, and reptiles, during the multiday blaze.

From a scientific point of view, Mount Parnitha proves to be an ideal study site for studying vegetation regeneration. First of all, is the main green lung of Attica, but also the main mechanism for reducing the temperature of the air flowing into Athens from the north. It is also a natural monument and a reserve of preserved Mediterranean mountain environment adjacent to the capital, as some 1116 plant species have been recorded in the area. In other words, Parnitha is an area of great environmental importance, ensuring better living conditions and environmental balance in the wider urban fabric. In addition, the area is characterized by a very varied topography, which includes lowland and mountainous areas with slopes ranging from 3% to 90%, with considerable differences in altitude and aspect. In particular, the altitude ranges from 200 m.a.s.l. to over 1400 m.a.s.l. (the highest altitude being 1413 m.a.s.l.). As a result, the study of the influence of the aspect on the regeneration of vegetation is therefore of great scientific interest.

Table 4.1 Acquisition dates and type of Landsat images used in the present study.

	Dates	Satellite
Prefire image	June 17, 2007	Landsat 5 TM
Postfire images	July 3, 2007	
	August 12, 2010	
	July 19, 2013	Landsat 8 OLI
	July 11, 2016	
	July 4, 2019	
	June 27, 2022	Landsat 9 OLI 2

3. Datasets

To explore the dynamics of vegetation regeneration within the burn scar of the chosen study region over a 15 year period, an analysis was conducted on seven Landsat images. Specifically, from 2007 to 2022, at three year intervals, three Landsat 5 Thematic Mapper (TM), two Landsat 8 Operational Land Imager (OLI) images, and one Landsat 9 OLI 2 image were selected. Table 4.1 provides details about the acquisition dates and type of Landsat for each image. These images acquired around the same dates ("anniversary dates") to avoid the influence of seasonal variations in both spectral radiation (e.g., due to meteorological conditions, the distance between the Sun and Earth, and Sun elevation angle) and surface reflection (Lillesand et al., 2015). Furthermore, for topographical information within the area of interest, the Shuttle Radar Topography Mission (SRTM) 1 arc second (30 m) version 3 dataset (Farr et al., 2007) was obtained. All satellite image and elevation data were acquired using the GEE platform, which provides convenient access to the complete Landsat and SRTM datasets through the Earth Engine Data Catalog.

4. Methods

The analysis of vegetation regrowth in the burned area was conducted using the GEE platform, while the statistical analysis was carried out using the R programming language. An overview of the methodology employed to satisfy the study objectives is depicted in Fig. 4.2.

4.1 Data preprocessing

All preprocessing of data used was carried out in GEE using the appropriate commands. Firstly, a polygon of the broader study area was constructed and used for masking the area of interest from the entire Landsat image. After that, the selection of preferable bands was carried out. The blue, green, red, and near infrared (NIR) bands were selected in all Landsat 5, 8, and 9 images. In addition, the scaling factor was applied to all the

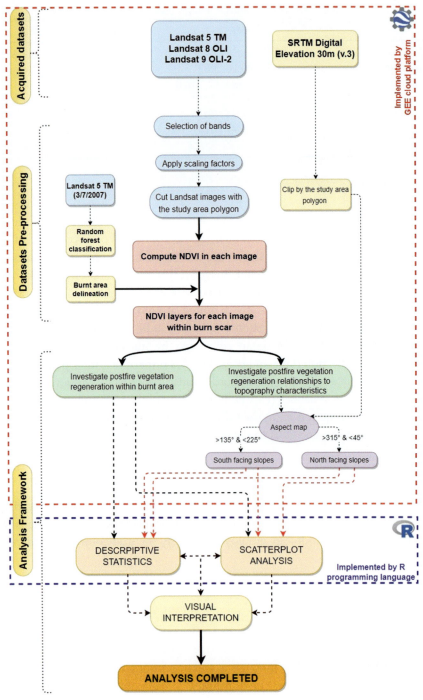

Figure 4.2 The methodological framework applied to assess vegetation regeneration dynamics within the study area.

selected bands. Landsat Collection 2 surface reflectance has a scale factor of 0.0000275 and an additional offset of −0.2 per pixel. To apply that, all bands were multiplied with the 0.0000275 and the −0.2 was added to them to take the real value of surface reflectance. No atmospheric or topographic correction was carried out since all the LANDSAT images which were acquired were already atmospherically and terrain corrected (at Level 1T).

4.2 Burnt area detection

To create the burnt area layer for the study region, the random forest (RF) supervised classifier (Breiman, 2001; Pandey et al., 2021) was utilized on the Landsat TM image captured immediately after fire suppression on July 3, 2007. The RF classification process was conducted within the GEE cloud platform. This classification method stands out as one of the most commonly employed machine learning classifiers in remote sensing, chosen for its distinct assumptions and well established track record. Specifically, RF is a sophisticated variation of pixel based classification techniques, known for consistently delivering promising results across various remote sensing datasets (Liu et al., 2023; Tselka et al., 2023). The RF method achieves this by forming an ensemble of multiple decision trees, introducing randomness in the selection of training samples and variables, which contributes to its ability to make accurate predictions and produce high quality classifications (Sheykhmousa et al., 2020).

Concerning the classification scheme, it comprised two distinct classes: "Burnt area" and "Nonburnt area". To create the training dataset representing these two classes, a total of 560 pixels were randomly selected. The parameterization for each classifier was determined through a series of trial and error iterations, following a standard approach commonly employed in similar studies (Brown et al., 2018; Petropoulos et al., 2014). For the RF classifier, the optimal number of random trees, which yielded the highest overall accuracy, was ultimately employed.

4.3 Vegetation regeneration mapping

The assessment of postfire vegetation regrowth dynamics involved a multitemporal analysis of the NDVI. NDVI was computed from the red (R) and near infrared (NIR) bands of each preprocessed image, applying the formula initially introduced by Rouse et al. (1973):

$$NDVI = \frac{\rho_{NIR} - \rho_R}{\rho_{NIR} + \rho_R} \tag{4.1}$$

where ρ_{NIR} and ρ_R represent the near infrared and the red surface spectral reflectance, respectively.

The NDVI formulation is based in the observation that thriving vegetation tends to absorb a significant portion of the red and blue bands while strongly reflecting the

NIR and green bands within the electromagnetic spectrum. The outcome of this calculation yields values theoretically ranging from -1 to $+1$. Values approaching $+1$, indicating significantly higher NIR reflectance than red, are indicative of healthy vegetation with robust photosynthetic activity. Therefore, NDVI serves as a metric closely tied to the quantity of photosynthetically active vegetation detected by the sensor in each pixel. Typically, NDVI values for vegetated areas are notably higher than 0.1 (Jensen, 2000; Petropoulos & Kalaitzidis, 2011). To provide more detail, NDVI values falling between 0.2 and 0.6 are associated with semiarid vegetation and lush green grass, whereas values ranging from 0.6 to 0.9 are characteristic of forested areas (Jensen, 2015).

The dynamics of the regrowth process were subsequently examined by comparing postfire NDVI spatial patterns to the prefire pattern within the burnt area. This procedure aimed to determine the extent to which the prefire pattern was reestablished and the rate of this recovery. Furthermore, descriptive statistics of NDVI within the studied region were computed from each image. These statistics, in combination with scatter plots and nonparametric correlation analysis, were employed to assess the NDVI variability within the burn scar (Ireland & Petropoulos, 2015; Petropoulos et al., 2014). Descriptive statistics and scatterplot analysis were conducted using R programming language version 4.2.2. To visualize the density of overlapping values within the scatterplots, the "ggpointdensity" package Version 0.1.0 was utilized (https://github.com/LKremer/ggpointdensity).

Next, the connection between vegetation regrowth dynamics and aspect was examined. More precisely, the disparities in vegetation recovery dynamics between north facing and south facing slopes are investigated. North facing slopes encompassed pixels with orientations ranging from NW (315 degrees) to NE (45 degrees), while south facing slopes included those with orientations from SE (135 degrees) to SW (225 degrees) (Petropoulos et al., 2014). Pixels falling outside of these orientations were excluded from this analysis.

5. Results

5.1 Burn scar delineation

Fig. 4.3 illustrates the burnt area layer obtained using the random forest classification. The assessment of the classification map accuracy was conducted based on the error matrix, and the results indicate an overall accuracy (OA) of approximately 98% and a Kappa coefficient (Kc) approaching 0.97. The total area burned was estimated at 46.8 km^2.

5.2 Vegetation regeneration evaluation

The outcomes related to the evaluation of spatiotemporal trends in vegetation regrowth within the burned area using NDVI can be found in Fig. 4.4. Additionally,

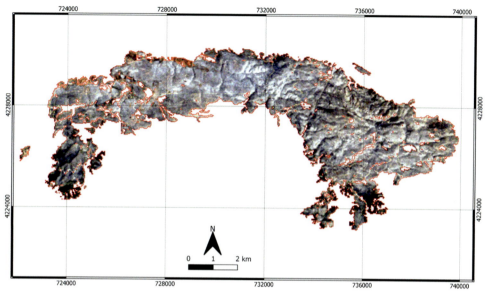

Figure 4.3 The burnt area layer resulting from the application of the Radom Forest classification method.

Table 4.2 offers the associated descriptive statistics. To begin with, upon visually comparing the NDVI map before the fire with the initial postfire NDVI map (Fig. 4.4), the extensive damage inflicted on the vegetation by the fire becomes evident. Furthermore, these adverse effects of the fire on the study area's vegetation are corroborated by the alterations in NDVI descriptive statistics between the prefire and initial postfire images (Table 4.2). To elaborate, the average NDVI within the study area prior to the fire event stands at approximately 0.58, while following the fire suppression, the corresponding NDVI plummets to 0.12. In terms of the maximum NDVI, it decreases from 0.87 before the fire to 0.25 afterward. This decline underscores the destructive nature of the fire, which incinerated the majority of the predominantly forested vegetation.

In terms of vegetation regeneration dynamics within the burned area, visual inspection of the postfire NDVI maps in conjunction with the corresponding NDVI descriptive statistics (Table 4.2) and the box plot (Fig. 4.5) reveal a gradual recovery process within the fire affected region. More specifically, the mean and maximum NDVI values in postfire images exhibit a progressive increase, indicative of vegetation regeneration in the area. Additionally, the results show that the highest rate of regeneration occurs during the initial three years following the fire (2007–10), followed by the subsequent three years (2010–13) characterized by a slower but still substantial regeneration. In the years from 2013 onwards until 2022, vegetation regeneration is evident based on NDVI results

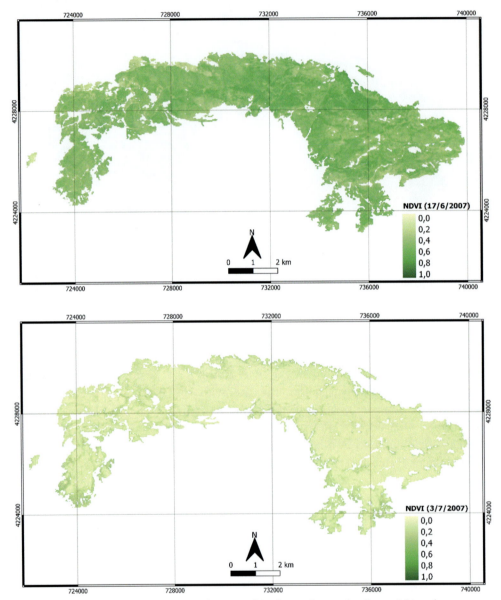

Figure 4.4 Spatial distribution of NDVI within the study area in all acquisition dates.

but with a slower rate of increase. According to the average and maximum NDVI values, along with the information from the box plot, it is evident that by 2022, NDVI levels have closely approached those from before the fire. This suggests that the vegetation in the burnt area has almost fully recovered to its prefire state.

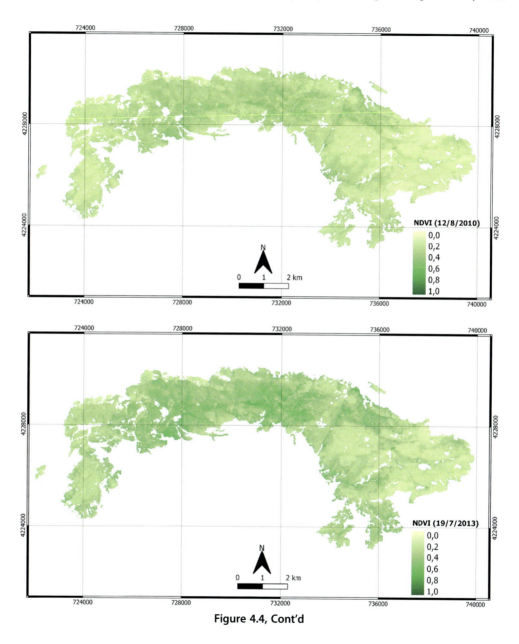

Figure 4.4, Cont'd

Furthermore, when examining the spatial distribution of postfire NDVI from 2007 onwards, depicted in Fig. 4.4 along with the difference NDVI maps shown in Fig. 4.6, it becomes evident that there are varying regeneration dynamics within the study area. Notably, more pronounced regeneration dynamics are observed in the central and western portions of the fire affected area compared to the remaining areas.

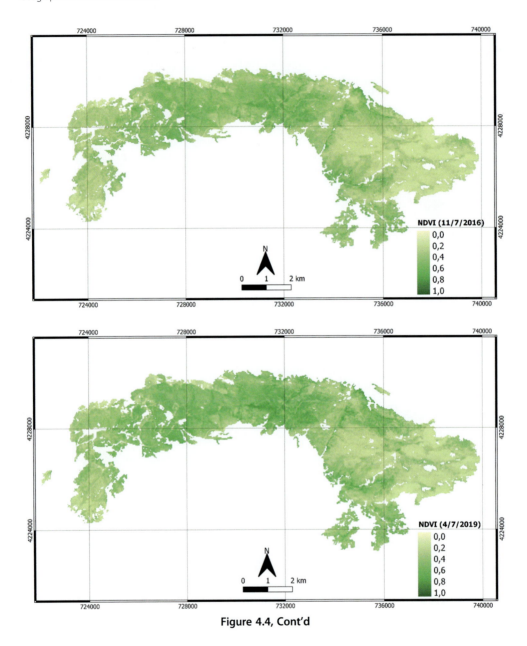

Figure 4.4, Cont'd

Furthermore, upon examining the difference map between the prefire image and the corresponding one from 2022, it's apparent that in the eastern mountainous regions of the study area, vegetation regeneration, while making progress, has not yet returned to prefire levels, even as of 2022.

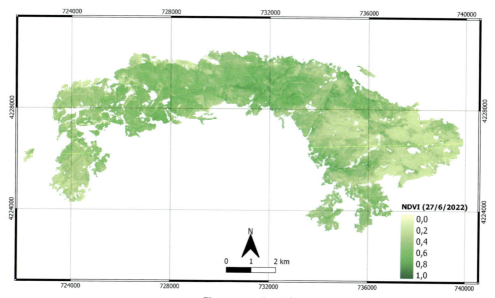

Figure 4.4, Cont'd

Table 4.2 Descriptive statistics of NDVI in different dates withing the burn scar.

Landsat image date	NDVI			
	Min	Max	Mean	Std dev
June 17, 2007	−0.32	0.87	0.58	0.09
July 3, 2007	−0.07	0.54	0.14	0.07
August 12, 2010	0.10	0.68	0.31	0.10
July 19, 2013	0.13	0.76	0.41	0.13
July 11, 2016	0.07	0.81	0.45	0.14
July 4, 2019	0.10	0.85	0.50	0.14
June 27, 2022	0.06	0.85	0.52	0.14

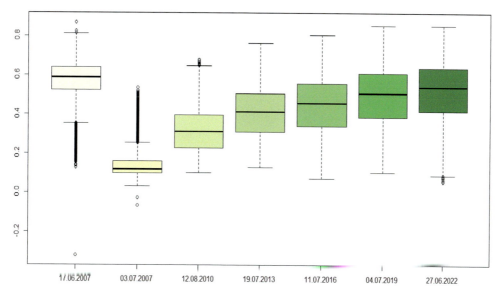

Figure 4.5 Box plot displaying the variation in NDVI within the burnt area for all acquisition images.

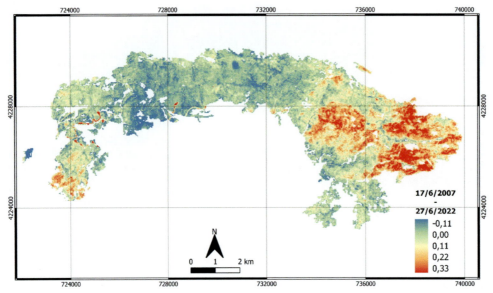

Figure 4.6 Difference map between the prefire image in 2007 and the postfire image in 2022.

5.3 Regression analysis of regeneration dynamics

To delve deeper into the dynamics of vegetation regeneration within the burned area under study, an attempt was made to develop regression models for this process. Building on prior research (Hope et al., 2007; Petropoulos et al., 2014), scatterplots were created to depict the NDVI values between prefire conditions and all subsequent postfire dates. These scatterplots offer insights into the regeneration dynamics by examining how data points are distributed in relation to the 1:1 line, which signifies the return of the burned area to its prefire state. Furthermore, calculations were performed for the slope, intercept, and R^2 of the regression line. Fig. 4.7 displays the generated scatterplots, while Table 4.3 summarizes the statistical findings related to these scatterplots.

The results obtained from analyzing the scatterplots concerning the regeneration dynamics within the study area closely mirror the earlier NDVI analysis. Notably, there's a discernible trend of data points moving toward the 1:1 line after three years from fire suppression (August 2010), signifying the initial stages of regeneration. Between 2010 and 2016, there is a gradual increase in the regeneration process, with a substantial recovery in NDVI levels observed. Starting from 2016, the regression analysis demonstrates that the majority of data points align closely with the 1:1 line, suggesting that NDVI levels are progressively approaching those observed before the fire, especially from 2019 onwards. Furthermore, the increase in the slope and R^2 value for the entire burn scar, which reflect the advancement of the regeneration process, is also evident (Table 4.3).

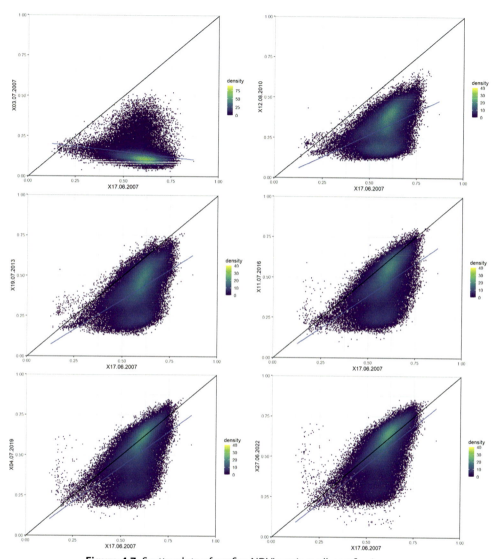

Figure 4.7 Scatterplots of prefire NDVI against all postfire images.

Table 4.3 Regression analysis between the prefire NDVI and the corresponding NDVI values for all subsequent postfire dates within the burned area.

Period	Slope	Intercept	R^2
June 2007—July 2007	−0.13	0.22	0.03
June 2007—August 2010	0.56	−0.01	0.25
June 2007—July 2013	0.74	−0.02	0.29
June 2007—July 2016	0.80	−0.01	0.29
June 2007—July 2019	0.87	−0.01	0.32
June 2007—June 2022	0.95	−0.03	0.38

5.4 Vegetation regeneration and topographic characteristics

The correlation between topographic attributes, specifically aspect, and the dynamics of vegetation regeneration (assessed through NDVI), was examined within the burnt area. Table 4.4 displays the descriptive statistics for NDVI on both north and south facing slopes. The findings indicate that the occurrence of the fire had adverse effects on both aspects, resulting in a significant reduction in both mean and maximum NDVI immediately after the fire suppression. Concerning the process of vegetation regeneration, the results demonstrated a steady rise in mean NDVI over the years in both aspects. However, it's noteworthy that the mean NDVI values in north facing slopes consistently remained approximately 0.1 higher throughout the study period. The highest regeneration rate occurred three to six years after the fire event (August 2010 to July 2013) with an increase of approximately 0.5 and 0.3 in mean NDVI in north and south facing slopes, respectively. From 2013 onwards, up to 2019, the vegetation regeneration continues to increase gradually, at a slower rate, with an increase in the NDVI about 0.1 NDVI every three years. From 2019 until the end of the study period in 2022, the increase in NDVI is even smaller, approximately 0.03 in the north facing areas and 0.01 in the corresponding south facing areas.

Table 4.4 Descriptive statistics of NDVI withing the burned area separately for north facing and south facing slopes across all dates.

Landsat image date	NDVI			
	Min	Max	Mean	Std dev
	North facing slopes			
June 17, 2007	−0.32	0.88	0.59	0.10
July 3, 2007	−0.07	0.49	0.13	0.06
August 12, 2010	0.10	0.68	0.34	0.10
July 19, 2013	0.13	0.76	0.44	0.12
July 11, 2016	0.07	0.80	0.48	0.13
July 4, 2019	0.11	0.85	0.53	0.14
June 27, 2022	0.08	0.85	0.56	0.14
	South facing slopes			
June 17, 2007	0.21	0.79	0.57	0.08
July 3, 2007	0.03	0.46	0.13	0.06
August 12, 2010	0.11	0.59	0.29	0.10
July 19, 2013	0.14	0.71	0.38	0.12
July 11, 2016	0.13	0.78	0.42	0.14
July 4, 2019	0.13	0.82	0.47	0.14
June 27, 2022	0.07	0.84	0.56	0.14

6. Discussion

The findings of this study highlight significant spatial variability and underscore the destructive impact of wildfires on the environment. They emphasize the role of wildfires, such as the one on Mount Parnitha, in altering the landscape and causing vegetation damage. Moreover, the results clearly demonstrate that the process of vegetation recovery in the affected area is time consuming. The analysis revealed substantial growth in vegetation regeneration within the burned study area, even in the early years of the study (2010–13). However, NDVI values had approached prefire levels by the study's conclusion in 2022. This indicates a significant recovery in the burned region, a process spanning 15 years. Furthermore, in certain areas, NDVI values even exceeded prefire levels, indicating an enhancement in vegetation compared to the period before the fire outbreak. These results align with earlier research that documented vegetation regeneration occurring within a timeframe of five years in Mount Carmen, Israeil (Wittenberg et al., 2007) to ten years in Chilean Mediterranean forested areas (Smith-Ramírez et al., 2022) after wildfires.

However, vegetation regeneration within the study area exhibits varying dynamics across different regions. As previously mentioned, there is substantial regeneration in the central and western sections of the burned area, where postfire NDVI values exceed prefire values in some areas. Conversely, in the eastern regions, vegetation dynamics are less pronounced, even after a 15 year study period. Indeed, as illustrated in Fig. 4.6, NDVI values in the red areas are 0.3 points lower than prefire levels. This suggests that the response of vegetation varies spatially, likely influenced by localized factors, one of which could be the composition of the vegetation itself. Studies conducted by Arianoutsou et al. (2010) and Christopoulou et al. (2018) have indicated that the slow recovery of vegetation, particularly the endemic Greek fire (*A. cephalonica*), in Mount Parnitha Nation Park after a fire is partially due to the fact that these trees are obligate seeders, and their seeds can be destroyed during summer fires, hindering their regeneration (Christopoulou et al., 2018). Consequently, in forests predominantly populated by these species, the ecosystem's recovery is likely to be led by shrub species which exhibit lower NDVI values, rather than native forest species which present higher NDVI values (Jensen, 2015). On the other hand, other species such as black pine (*Pinus nigra)* are well adapted to low intensity surface fires (Fulé et al., 2008) and their regeneration dynamics are faster (Christopoulou et al., 2018).

Furthermore, Christopoulou et al. (2018) highlighted shifts in the floristic composition and plant species richness within the regenerating fir forest on Mount Parnitha after a wildfire. In the first 2 to 8 years following the fire, the burned area exhibited high species richness, primarily due to a substantial presence of herbaceous species. However, after a decade, species richness declined and became comparable to unburned fir communities. It's during this phase that the initial Greek fir seedlings managed to establish supporting the recovery of the forest. These findings align closely with the results of the current study, which reveal the pronounced dynamics of vegetation regeneration shortly after the fire, particularly during the period from 2010 to 2013, and the subsequent reduction in dynamics in the period from 2016 to 2022.

The analysis delving into the intricate relationship between vegetation regeneration and aspect reveals that north facing slopes within our study area exhibit a slightly more robust regeneration rate when compared to their south facing counterparts. This observation is consistent with the findings of several other studies (Christopoulou et al., 2018; Fox et al., 2008; Ireland & Petropoulos, 2015; Petropoulos et al., 2014). However, this variance in regeneration is not a random occurrence but rather a reflection of the profound influence of aspect and other key topographical variables. These variables include elevation, slope, and geographical positioning, all of which collectively mold the microenvironments at play in these landscapes. The intricate interplay of these factors results in the development of diverse microclimates, distinct soil properties, and unique hydrological processes. For instance, north facing slopes typically receive differential solar radiation compared to their south facing counterparts, which significantly impacts the local environmental conditions. This discrepancy in solar exposure can lead to variations in temperature, precipitation patterns, and moisture retention in the soil. Such differences, in turn, influence the ability of vegetation to recover and flourish in the wake of a fire. Researchers have long recognized these intricate relationships. Studies by Daws et al. (2002), Moeslund et al. (2013), and Jucker et al. (2018) have all shed light on how these topographical features, especially aspect, can play a pivotal role in shaping the trajectory of postfire vegetation regeneration.

7. Conclusions

This study conducted an extensive analysis of vegetation regeneration dynamics within a Mediterranean ecosystem located in Attica, Greece, which had undergone a substantial wildfire. The research harnessed EO datasets and geospatial data analysis methods to investigate the spatiotemporal variability of vegetation recovery over a 15 year period postfire. Landsat TM, OLI, and OLI 2 satellite images, in conjunction with GEE and the R programming language, were instrumental in assessing the extent of vegetation's return to its prefire state. The widely accepted NDVI served as the primary indicator in this endeavor. Furthermore, the research delved into the relationship between vegetation regeneration dynamics and topographical elements, with a particular focus on aspect.

The analysis uncovers a notable pattern in vegetation recovery dynamics. There is a distinct phase characterized by robust regrowth immediately after the fire, notably from 2010 to 2013, followed by a gradual reduction in dynamics between 2016 and 2022. However, it's important to note that vegetation recovery is not uniform across the study area. While some regions experience substantial regrowth, surpassing their prefire NDVI values, others demonstrate less pronounced vegetation dynamics. In certain areas, even after 15 years of observation, full recovery to prefire conditions has not been achieved. In examining the connection between vegetation regrowth and aspect, the analysis reveals slightly higher regeneration rates on north facing aspects. This discrepancy can be attributed to the presence of microclimatic and hydrological conditions that are more favorable for vegetation growth in these north facing regions.

In conclusion, it is imperative to recognize that the present study, despite its contributions, carries inherent limitations. An important constraint is the absence of differentiation between various vegetation types within the regeneration dynamics. Such differentiation is vital for a comprehensive grasp of the scope and characteristics of wildland fire damage (Milne, 1986). By distinguishing between different vegetation types, it becomes feasible to discern the distinct responses of various plant species to fire, thereby enhancing our understanding of the ecological implications. Nonetheless, the selection of the NDVI as the primary analytical tool remains justified. NDVI delivers a quantifiable measure of green biomass in vegetation, a critical parameter for forest management and planning (Abdel Malak & Pausas, 2006). Analyzing changes in NDVI offers essential insights into vegetation recovery and biomass restoration, both of which are pivotal for decision making in land management and ecological restoration efforts. Moreover, this investigation underscores the capacity of EO technology, in conjunction with geospatial data analysis methodologies, and the utilization of cloud based platforms like GEE, to facilitate the evaluation of vegetation recovery. Equally significant, the methodological blueprint outlined here exhibits substantial promise for complete automation within the GEE platform, offering the prospect of operational utilization to generate robust outcomes offering valuable support for policy making and decision making processes. It is essential to note that this constitutes an area of future exploration.

References

Abdel Malak, D., & Pausas, J. (2006). Fire regime and post-fire normalized difference vegetation index changes in the eastern Iberian peninsula (Mediterranean basin). *International Journal of Wildland Fire, 15*. https://doi.org/10.1071/WF05052

Amos, C., Petropoulos, G. P., & Ferentinos, K. P. (2018). Determining the use of Sentinel-2A MSI for wildfire burning & severity detection. *International Journal of Remote Sensing, 40*(3), 905—930. https://doi.org/10.1080/01431161.2018.1519284

Arianoutsou, M., Christopoulou, A., Kazanis, D., Tountas, T., Ganou, E., Bazos, I., & Kokkoris, Y. (2010). Effects of fire on high altitude coniferous forests of Greece. In *Sixth international conference of wildland fire, Coimbra, 12.*

Breiman, L. (2001). Random forests. *Machine Learning, 45*(1), 5—32. https://doi.org/10.1023/A:1010933404324

Bridges, J. M., Petropoulos, G. P., & Clerici, N. (2019). Immediate changes in organic matter and plant available nutrients of haplic luvisol soils following different experimental burning intensities in Damak forest, Hungary. *Forests, 10*(5). https://doi.org/10.3390/f10050453

Brown, A. R., Petropoulos, G. P., & Ferentinos, K. P. (2018). Appraisal of the Sentinel-1 & 2 use in a large-scale wildfire assessment: A case study from Portugal's fires of 2017. *Applied Geography, 100*, 78–89. https://doi.org/10.1016/j.apgeog.2018.10.004

Chen, D., Fu, C., Hall, J. V., Hoy, E. E., & Loboda, T. V. (2021). Spatio-temporal patterns of optimal Landsat data for burn severity index calculations: Implications for high northern latitudes wildfire research. *Remote Sensing of Environment, 258*, 112393. https://doi.org/10.1016/j.rse.2021.112393

Chen, D., Loboda, T. V., Krylov, A., & Potapov, P. V. (2016). Mapping stand age dynamics of the Siberian larch forests from recent Landsat observations. *Remote Sensing of Environment, 187*, 320–331. https://doi.org/10.1016/j.rse.2016.10.033

Christopoulou, A., Kazanis, D., Fyllas, N., & Arianoutsou, M. (2018). Post-fire recovery of *Abies cephalonica* forest communities: The case of Mt Parnitha national Park, Attica, Greece. *iForest: Biogeosciences and Forestry, 11*(6), 757–764. https://doi.org/10.3832/ifor2744-011

Daws, M. I., Mullins, C. E., Burslem, D. F. R. P., Paton, S. R., & Dalling, J. W. (2002). Topographic position affects the water regime in a semideciduous tropical forest in Panamá. *Plant and Soil, 238*(1), 79–89. https://doi.org/10.1023/A:1014289930621

Farr, T. G., Rosen, P. A., Caro, E., Crippen, R., Duren, R., Hensley, S., Kobrick, M., Paller, M., Rodriguez, E., Roth, L., Seal, D., Shaffer, S., Shimada, J., Umland, J., Werner, M., Oskin, M., Burbank, D., & Alsdorf, D. (2007). The Shuttle radar topography mission. *Reviews of Geophysics, 45*(2). https://doi.org/10.1029/2005RG000183

Fox, D. M., Maselli, F., & Carrega, P. (2008). Using SPOT images and field sampling to map burn severity and vegetation factors affecting post forest fire erosion risk. *Catena, 75*(3), 326–335. https://doi.org/10.1016/j.catena.2008.08.001

Fragou, S., Kalogeropoulos, K., Stathopoulos, N., Louka, P., Srivastava, P. K., Karpouzas, S., P. Kalivas, D., & P. Petropoulos, G. (2020). Quantifying land cover changes in a Mediterranean environment using Landsat TM and support vector machines. *Forests, 11*(7). https://doi.org/10.3390/f11070750

Fulé, P. Z., Ribas, M., Gutiérrez, E., Vallejo, R., & Kaye, M. W. (2008). Forest structure and fire history in an old *Pinus nigra* forest, eastern Spain. *Forest Ecology and Management, 255*(3), 1234–1242. https://doi.org/10.1016/j.foreco.2007.10.046

Ganatsas, P., Daskalakou, E., & Paitaridou, D. (2012). First results on early post-fire succession in an *Abies cephalonica* forest (Parnitha National Park, Greece). *iForest: Biogeosciences and Forestry, 5*(1), 6. https://doi.org/10.3832/ifor0600-008

García-Llamas, P., Suárez-Seoane, S., Fernández-Guisuraga, J. M., Fernández-García, V., Fernández-Manso, A., Quintano, C., Taboada, A., Marcos, E., & Calvo, L. (2019). Evaluation and comparison of Landsat 8, Sentinel-2 and Deimos-1 remote sensing indices for assessing burn severity in Mediterranean fire-prone ecosystems. *International Journal of Applied Earth Observation and Geoinformation, 80*, 137–144. https://doi.org/10.1016/j.jag.2019.04.006

Gouveia, C., DaCamara, C. C., & Trigo, R. M. (2010). Post-fire vegetation recovery in Portugal based on spot/vegetation data. *Natural Hazards and Earth System Sciences, 10*(4), 673–684. https://doi.org/10.5194/nhess-10-673-2010

Grissino Mayer, H. D., & Swetnam, T. W. (2000). Century scale climate forcing of fire regimes in the American Southwest. *The Holocene, 10*(2), 213–220. https://doi.org/10.1191/095968300668451235

Hope, A., Tague, C., & Clark, R. (2007). Characterizing post-fire vegetation recovery of California chaparral using TM/ETM+ time-series data. *International Journal of Remote Sensing, 28*(6), 1339–1354. https://doi.org/10.1080/01431160600908924

IPCC. (2022). *Climate change 2022: Impacts, adaptation and vulnerability. Contribution of working group II to the sixth assessment report of the intergovernmental panel on climate change.* Cambridge University Press. https://doi.org/10.1017/9781009325844.002

Ireland, G., & Petropoulos, G. P. (2015). Exploring the relationships between post-fire vegetation regeneration dynamics, topography and burn severity: A case study from the Montane Cordillera Ecozones of western Canada. *Applied Geography, 56*, 232–248. https://doi.org/10.1016/j.apgeog.2014.11.016

Jensen, J. R. (2015). *Introductory digital image processing. A remote sensing perspective* (4th ed.). Pearson Education, Inc.

Jensen, J. R. (2000). *Remote sensing of the environment: An earth resource perspective*. Saddle River, NJ: Prentice Hall.

Joint Research Centre (European Commission), Costa, H., Rigo, D. de, Libertà, G., Houston Durrant, T., & San-Miguel-Ayanz, J. (2020). *European wildfire danger and vulnerability in a changing climate: Towards integrating risk dimensions: JRC PESETA IV project: Task 9 forest fires*. Publications Office of the European Union. https://data.europa.eu/doi/10.2760/46951.

Jucker, T., Bongalov, B., Burslem, D. F. R. P., Nilus, R., Dalponte, M., Lewis, S. L., Phillips, O. L., Qie, L., & Coomes, D. A. (2018). Topography shapes the structure, composition and function of tropical forest landscapes. *Ecology Letters, 21*(7), 989—1000. https://doi.org/10.1111/ele.12964

Kalivas, D. P., Petropoulos, G. P., Athanasiou, I. M., & Kollias, V. J. (2013). An intercomparison of burnt area estimates derived from key operational products: The Greek wildland fires of 2005—2007. *Nonlinear Processes in Geophysics, 20*(3), 397—409. https://doi.org/10.5194/npg-20-397-2013

Karali, A., Varotsos, K. V., Giannakopoulos, C., Nastos, P. P., & Hatzaki, M. (2023). Seasonal fire danger forecasts for supporting fire prevention management in an eastern Mediterranean environment: The case of Attica, Greece. *Natural Hazards and Earth System Sciences, 23*(2), 429—445. https://doi.org/10.5194/nhess-23-429-2023. Scopus.

Katagis, T., Gitas, I. Z., Toukiloglou, P., Veraverbeke, S., Goossens, R., Katagis, T., Gitas, I. Z., Toukiloglou, P., Veraverbeke, S., & Goossens, R. (2014). Trend analysis of medium- and coarse-resolution time series image data for burned area mapping in a Mediterranean ecosystem. *International Journal of Wildland Fire, 23*(5), 668—677. https://doi.org/10.1071/WF12055

López García, A. R. (2020). Study of the severity by the fire of 2012 and regeneration of vegetation in La Primavera Forest, Mexico, using LANDSAT 7 images. *Revista Cartográfica, 2020*(101), 35—50. https://doi.org/10.35424/rcarto.i101.420. Scopus.

Lillesand, T., Kiefer, R. W., & Chipman, J. (2015). Remote sensing and image interpretation. John Wiley & Sons.

Liu, C., Chen, R., & He, B. (2023). Integrating machine learning and a spatial contextual algorithm to detect wildfire from Himawari-8 data in southwest China. *Forests, 14*(5). https://doi.org/10.3390/f14050919

Llorens, R., Sobrino, J. A., Fernández, C., Fernández-Alonso, J. M., & Vega, J. A. (2021). A methodology to estimate forest fires burned areas and burn severity degrees using Sentinel-2 data. Application to the October 2017 fires in the Iberian Peninsula. *International Journal of Applied Earth Observation and Geoinformation, 95*, 102243. https://doi.org/10.1016/j.jag.2020.102243

Louhaichi, M., Toshpulot, R., Moyo, H. P., & Belgacem, A. O. (2021). Effect of slope aspect on vegetation characteristics in mountain rangelands of Tajikistan: Considerations for future ecological management and restoration. *African Journal of Range and Forage Science*, 1—9. https://doi.org/10.2989/10220119.2021.1951840

Milne, A. K. (1986). The use of remote sensing in mapping and monitoring vegetational change associated with bushfire events in Eastern Australia. *Geocarto International, 1*(1), 25—32. https://doi.org/10.1080/10106048609354022

Moeslund, J. E., Arge, L., Bøcher, P. K., Dalgaard, T., & Svenning, J.-C. (2013). Topography as a driver of local terrestrial vascular plant diversity patterns. *Nordic Journal of Botany, 31*(2), 129—144. https://doi.org/10.1111/j.1756-1051.2013.00082.x

Moreira, F., Ascoli, D., Safford, H., Adams, M. A., Moreno, J. M., Pereira, J. M. C., Catry, F. X., Armesto, J., Bond, W., González, M. E., Curt, T., Koutsias, N., McCaw, L., Price, O., Pausas, J. G., Rigolot, E., Stephens, S., Tavsanoglu, C., Vallejo, V. R., ... Fernandes, P. M. (2020). Wildfire management in Mediterranean-type regions: Paradigm change needed. *Environmental Research Letters, 15*(1), 011001. https://doi.org/10.1088/1748-9326/ab541e

Moreira, F., Catry, F., Duarte, I., Acácio, V., & Silva, J. S. (2009). A conceptual model of sprouting responses in relation to fire damage: An example with cork oak (Quercus suber L.) trees in Southern Portugal. In A. G. Van der Valk (Ed.), *Forest ecology: Recent advances in plant ecology* (pp. 77—85). Springer Netherlands. https://doi.org/10.1007/978-90-481-2795-5_7

Nioti, F., Xystrakis, F., Koutsias, N., & Dimopoulos, P. (2015). A remote sensing and GIS approach to study the long-term vegetation recovery of a fire-affected pine forest in southern Greece. *Remote Sensing, 7*(6). https://doi.org/10.3390/rs70607712

Ozelkan, E., Ormeci, C., & Karaman, M. (2011). *Determination of the forest fire potential by using remote sensing and geographical information system, case study-Bodrum/Turkey.* https://bit.ly/3eGklum.

Pérez, C. C., Olthoff, A. E., Hernández-Trejo, H., & Rullán-Silva, C. D. (2022). Evaluating the best spectral indices for burned areas in the tropical Pantanos de Centla Biosphere Reserve, Southeastern Mexico. *Remote Sensing Applications: Society and Environment, 25.* https://doi.org/10.1016/j.rsase.2021.100664. Scopus.

Pandey, P. C., Koutsias, N., Petropoulos, G. P., Srivastava, P. K., & Ben Dor, E. (2021). Land use/land cover in view of earth observation: Data sources, input dimensions, and classifiers—a review of the state of the art. *Geocarto International, 36*(9), 957—988. https://doi.org/10.1080/10106049.2019.1629647

Pausas, J. G., & Vallejo, V. R. (1999). The role of fire in European Mediterranean ecosystems. In E. Chuvieco (Ed.), *Remote sensing of large wildfires: In the European Mediterranean Basin* (pp. 3—16). Springer. https://doi.org/10.1007/978-3-642-60164-4_2

Petropoulos, G. P., Griffiths, H. M., & Kalivas, D. P. (2014). Quantifying spatial and temporal vegetation recovery dynamics following a wildfire event in a Mediterranean landscape using EO data and GIS. *Applied Geography, 50,* 120—131. https://doi.org/10.1016/j.apgeog.2014.02.006

Petropoulos, G. P., & Kalaitzidis, C. (2011). *Multispectral vegetation indices in remote sensing: An overview.* Nova-publishers: Ecological Modeling.

Rouse, J. W., Haas, R. H., Schell, J. A., & Deering, D. W. (1973). Monitoring vegetation systems in the great plains with ERTS. In *3rd ERTS symposium, NASA SP-351, 309—317.*

Rovithakis, A., Grillakis, M. G., Seiradakis, K. D., Giannakopoulos, C., Karali, A., Field, R., Lazaridis, M., & Voulgarakis, A. (2022). Future climate change impact on wildfire danger over the Mediterranean: The case of Greece. *Environmental Research Letters, 17*(4). https://doi.org/10.1088/1748-9326/ac5f94. Scopus.

Said, Y. A., Petropoulos, G. P., & Srivastava, P. K. (2015). Assessing the influence of atmospheric and topographic correction and inclusion of SWIR bands in burned scars detection from high-resolution EO imagery: A case study using ASTER. *Natural Hazards, 78*(3), 1609—1628. https://doi.org/10.1007/s11069-015-1792-9

Sheykhmousa, M., Mahdianpari, M., Ghanbari, H., Mohammadimanesh, F., Ghamisi, P., & Homayouni, S. (2020). Support vector machine versus random forest for remote sensing image classification: A meta-analysis and systematic review. *IEEE Journal of Selected Topics in Applied Earth Observations and Remote Sensing, 13,* 6308—6325. https://doi.org/10.1109/JSTARS.2020.3026724

Smith-Ramírez, C., Castillo-Mandujano, J., Becerra, P., Sandoval, N., Fuentes, R., Allende, R., & Paz Acuña, M. (2022). Combining remote sensing and field data to assess recovery of the Chilean Mediterranean vegetation after fire: Effect of time elapsed and burn severity. *Forest Ecology and Management, 503,* 119800. https://doi.org/10.1016/j.foreco.2021.119800

Tselka, I., Detsikas, S. E., Petropoulos, G. P., & Demertzi, I. I. (2023). Chapter 7 - Google Earth Engine and machine learning classifiers for obtaining burnt area cartography: A case study from a Mediterranean setting. In N. Stathopoulos, A. Tsatsaris, & K. Kalogeropoulos (Eds.), *Geoinformatics for geosciences* (pp. 131—148). Elsevier. https://doi.org/10.1016/B978-0-323-98983-1.00008-9

Veraverbeke, S., Lhermitte, S., Verstraeten, W. W., & Goossens, R. (2010). The temporal dimension of differenced Normalized Burn Ratio (dNBR) fire/burn severity studies: The case of the large 2007 Peloponnese wildfires in Greece. *Remote Sensing of Environment, 114*(11), 2548—2563. https://doi.org/10.1016/j.rse.2010.05.029

Viana-Soto, A., Aguado, I., & Martínez, S. (2017). Assessment of post-fire vegetation recovery using fire severity and geographical data in the Mediterranean region (Spain). *Environments, 4*(4). https://doi.org/10.3390/environments4040090

Wang, C., Wang, A., Guo, D., Li, H., & Zang, S. (2022). Off-peak NDVI correction to reconstruct Landsat time series for post-fire recovery in high-latitude forests. *International Journal of Applied Earth Observation and Geoinformation, 107,* 102704. https://doi.org/10.1016/j.jag.2022.102704

Wittenberg, L., Malkinson, D., Beeri, O., Halutzy, A., & Tesler, N. (2007). Spatial and temporal patterns of vegetation recovery following sequences of forest fires in a Mediterranean landscape, Mt. Carmel Israel. *Catena, 71*(1), 76—83. https://doi.org/10.1016/j.catena.2006.10.007

Yang, L., Driscol, J., Sarigai, S., Wu, Q., Chen, H., & Lippitt, C. D. (2022). Google Earth Engine and artificial intelligence (AI): A comprehensive review. *Remote Sensing, 14*(14). https://doi.org/10.3390/rs14143253

CHAPTER 5

Understanding and monitoring the dynamics of Arctic permafrost regions under climate change using Earth Observation and cloud computing: The contribution of EO-PERSIST project

George P. Petropoulos[1], Vassilia Karathanassi[2], Kleanthis Karamvasis[2], Aikaterini Dermosinoglou[1] and Spyridon E. Detsikas[1]

[1]Department of Geography, Harokopio University of Athens, Athens, Greece; [2]School of Rural, Surveying and Geoinformatics Engineering, National Technical University of Athens, Athens, Greece

1. Climate change and impacts

In recent times, there has been a discernible increase in both scientific and public attention toward the Arctic region, prompted by a heightened awareness of the potential impacts of climate change. According to the United Nations (2023), climate change is a pressing global issue characterized by long-term shifts in temperature patterns, precipitation levels, and other climatic parameters. It is primarily driven by human activities, such as the burning of fossil fuels (Johnsson et al., 2018; Wood & Roelich, 2019´; Soeder, 2020), deforestation (Longobardi et al., 2016; Wolff et al., 2021; Li et al., 2022), and industrial processes (Niñerola et al., 2020; Wadanambi et al., 2020; Sovacool et al., 2021), which release greenhouse gases (GHGs) into the atmosphere. These gases capture heat, leading to a gradual rise in global temperatures, commonly referred to as global warming. According to the IPCC report of 2023, the GHG release has led to a rise in global surface temperature by 1.1°C above the 1850−1900 baseline during 2011−2020. The global emissions of GHG have continued to escalate from 2010 to 2019, with disparities in historical and ongoing contributions stemming from unsustainable energy usage, land use and changes, diverse lifestyles, and consumption and production patterns. These factors exhibit variations across regions, within and between countries, and among individuals and lead to serious consequences on planet's ecosystems and human societies.

Such consequences include rising sea levels, which pose a threat to coastal communities, while extreme weather events like hurricanes, prolonged droughts, heatwaves, and extreme floods, which are becoming more frequent and severe (Bell et al., 2018; Srivastava et al., 2021; Tsatsaris et al., 2021). Shifts in precipitation patterns can disrupt

Geographical Information Science
ISBN 978-0-443-13605-4
https://doi.org/10.1016/B978-0-443-13605-4.00020-5

agriculture and water resources, leading to food and water scarcity in vulnerable regions (Fitton et al., 2019; Liu et al., 2023). Additionally, the melting of polar ice caps and glaciers contributes to rising sea levels and threatens the habitats of numerous species (Amjad et al., 2022). The main strategies for the mitigation of climate change entail reducing human-caused emissions of GHGs, the prominent factor of global warming. Additionally, it is crucial to adapt to the impacts of climate change, using the least number of resources to effectively address climate-related risks (Zhao et al., 2023). Both these approaches require the discernment and quantification of substantial land changes that have occurred to enhance the comprehension of climate change and provide support to policymakers in their endeavors to effectively address and mitigate its impacts.

2. Arctic and permafrost regions and climate change

The Arctic region emerges as a hotspot of climate change, warming three times faster than the rest of the planet (Rantanen et al., 2022; Hu et al., 2023). Climate change is increasingly affecting components of the terrestrial cryosphere and hydrology and its adverse impacts on the active layer freeze-thaw cycle and permafrost conditions is well documented (Wang et al., 2021). Global warming has resulted in significant effects to the local and regional Arctic natural environments, infrastructures, as well as the climate (Hjort et al., 2022). Permafrost changes have important ecological and economic impacts globally (Schuur & Mack, M. C, (2018); Hu et al., 2021). Given the significant concentration of the oil and gas industry, as well as pipeline infrastructure, in the Arctic today, there is a heightened concern regarding the monitoring of permafrost dynamics. Permafrost exhibits significant variability in temperature, extent, and depth. This diversity arises from distinctions in climate patterns, vegetation, and soil characteristics. Building in areas with permafrost necessitates the use of inventive construction methods specifically crafted to isolate frozen ground from the thermal influence of buildings (Suter et al., 2019). This also includes the critical assessment of socioeconomic impacts and associated costs to local communities and businesses, both in terms of preventive measures and postdisaster mitigation efforts. For instance, the collapse of a fuel tank at a power station on May 29, 2020, as documented by Rajendran et al. (2021), exemplifies the potential risks involved. Besides, as permafrost conditions retreat, microbial decomposition of soil organic carbon is accelerating, even more GHGs are unleashed which exacerbating further climate change (Hugelius et al., 2014). In the northern hemisphere, permafrost regions at depths ranging from 0 to 3 m house approximately half of the global soil carbon pool (Feng et al., 2020). Given their capacity to potentially release significant quantities of previously sequestered soil carbon into the atmosphere under warming conditions, these regions emerge as a pivotal determinant in shaping the trajectory of future climate change.

Acquiring a comprehensive understanding of the physical parameters influencing the Arctic climate and assessing the potential socioeconomic impacts of climate change in these environments represents a paramount research priority nowadays. In this context, it is deemed a matter of critical scientific significance necessitating the indispensable development of methodologies and tools to facilitate the advancement of our comprehension in this direction. The latter has been highlighted as one of the most complex and emerging issues to be addressed today (EU Council, 2017). As it is extremely challenging to obtain long-term continuous ground-based observations due to the harsh weather conditions in the Arctic, observations from space using Earth Observation (EO) satellites are critically required to investigate the inner connections between Arctic dynamics and climate change impacts to the society and economy (Hu et al., 2021; Wang et al., 2021). The EU is firmly committed to establishing a strong link between its Arctic endeavors and overarching climate policy, exemplified by the European Green Deal and its emphasis on the blue economy dimension. By recognizing the key global role of permafrost changes in the sustainable development and climate, Chuffart and Raspotnik (2019) has funded several EO and cloud-based platforms (e.g., Permafrost CCI, Globpermafrost, APGC) for supporting Arctic related research. However, information provided via the above-mentioned initiatives is currently dispersed through several web platforms and open datasets provided by these platforms are available in various formats, making their use difficult and in some cases impractical.

3. EO data for climate change available for the Arctic

The ongoing advancement of remote sensing (RS) technologies, including cutting-edge satellites and innovative methods for large-scale data processing, is crucial for enhancing the precision and reliability of climate change research on RS data. For the Arctic region, the contribution of RS data is extremely important due to its remote location and the difficulty of in situ observations.

The surface of the Arctic region can be effectively monitored using a range of optical satellites, with varying degrees of spatial resolution. These include MODIS with a 1 km resolution, Landsat with a 30 m resolution, and Sentinel with a 10 m resolution. Additionally, very high-resolution satellites like PlanetScope (3 m resolution), WorldView (0.5–1.5 m resolution), IKONOS (1–4 m resolution), and Pleiades (0.5 m resolution) provide even more detailed observations. In the context of monitoring ice layers, synthetic aperture radar (SAR) data are commonly employed, distinguished by its capability to acquire data irrespective of prevailing meteorological conditions, particularly in the presence of cloud cover. This sensor type enables us to detect changes in Land Use or Land Cover (LULC), variations of surface temperature dynamics, and quantification of changes in ice volume. Additionally, there is a plethora of global freely available geospatial datasets for the Arctic region. Those have been categorized based on the respective field to which it pertains, as described in the remainder of this section below.

3.1 Climate weather and atmospheric datasets

Climate and atmospheric geospatial data are essential reservoirs for examining and observing the complex processes impacting Earth's weather patterns. Derived from satellite observations and measurements on the ground, these datasets provide critical details on atmospheric dynamics, enabling thorough investigations into climate phenomena and enhancing our understanding of atmospheric patterns and fluctuations. The *Global Land Data Assimilation System (GLDAS)* offers global coverage of key land surface conditions, including soil moisture, temperature, and precipitation. ERA-5 provides hourly estimates of numerous atmospheric, land, and oceanic climate variables, contributing to a nuanced comprehension of climate dynamics. The *Climatic Research Unit Timeseries (CRU TS)* presents monthly time-series data encompassing precipitation, daily temperatures, and cloud cover across Earth's land areas. *WorldClim* delivers high-resolution global weather and climate data, facilitating analyses of current conditions and climate projections. *TerraClimate* provides monthly climate and water balance data globally. Satellite datasets, such as *MODIS Aqua and Terra*, *Sentinel-3*, *Sentinel-5 (TROPOMI)*, *Sentinel-6*, *SMOS*, and *SMAP*, contribute critical information on surface characteristics, ocean and land temperatures, atmospheric trace gases, and soil moisture. Additionally, datasets like *CAMS NearReal-Time Global Forecasts* offer atmospheric composition data for near-real-time global forecasting. The available datasets described above are presented in Table 5.1.

3.2 Land datasets

Geospatial datasets related to land, further categorized into Terrain and LULC datasets, are fundamental resources for analyzing Earth's surface characteristics. Terrain datasets like Digital Elevation Models (DEMs) provide detailed elevation information, allowing for comprehensive analyses of terrain features, slope, and landscape characteristics. Available DEMs for the Arctic regions are presented in Table 5.2.

LULC datasets constitute crucial geospatial resources for discerning the spatial distribution and utilization of land resources within specific geographic regions. These datasets systematically classify and illustrate diverse land cover types and land use patterns, sourced from satellite imagery and ground-based surveys. Most of these data are available for noncommercial use and cover spatial resolution from 2.5 to 500 m (Table 5.3).

From the above it becomes evident that a wealth of geospatial data is available for the Arctic region. The utilization of those datasets is contingent upon the specific requirements of individual tasks, incorporating considerations of their spatial and temporal resolution, and adherence to policies set by data providers. The spatial analysis of these datasets encompasses a range of resolutions, spanning from low to medium and high,

Table 5.1 Available geospatial datasets for the Arctic region, climate-weather-atmosphere category.

Dataset	Spatial resolution	Temporal resolution	Link[1]
Global Land Data Assimilation System (GLDAS)	0.25 degrees (28 km)	3 hours	https://ldas.gsfc.nasa.gov/gldas
ERA-5	30 km	1940—present	https://www.ecmwf.int/en/forecasts/datasets/reanalysis-datasets/era5
CRU TS (Climatic Research Unit Timeseries)	0.5° × 0.5 degrees (∼50 km)	1901—2015	https://data.ceda.ac.uk/badc/cru/data
WorldClim	1 km	1970—2000	https://www.worldclim.org/
TerraClimate	4 km	1958—2020	https://www.climatologylab.org/terraclimate.html
MODIS Aqua and Terra	500 m	Daily, day and night 1999—present	https://modis.gsfc.nasa.gov/about/
Sentinel-3	1 km	2015—present	https://sentinels.copernicus.eu/web/sentinel/missions/sentinel-3
Sentinel-5 (TROPOMI)	7 × 3.5 km	2017—present	https://sentinels.copernicus.eu/web/sentinel/missions/sentinel-5p
Sentinel-6	∼35 km	2020—present	https://sentinels.copernicus.eu/web/sentinel/missions/sentinel-6/overview
CAMS Near Real-Time (NRT) Global Forecasts	0.4° × 0.4 degrees (∼40 km)	2015—present	https://ads.atmosphere.copernicus.eu/cdsapp#!/dataset/cams-global-atmospheric-composition-forecasts?tab=overview
SMOS products	30—50 km	Daily	https://earth.esa.int/eogateway/catalog/smos-science-products
SMAP products	36 km	2—3 days	https://smap.jpl.nasa.gov/data/

[1]Last date accessed on 20/11/2023.

Table 5.2 Available geospatial datasets for the Arctic region, land category (DEMs).

Datasets (digital elevation models)	Spatial resolution (m)	Spatial extent	Link[2]
SRTM	90	Global	https://www.earthdata.nasa.gov/sensors/srtm
ASTER GDEM	30	Global	https://asterweb.jpl.nasa.gov/gdem.asp
MERIT DEM	90	Global	https://hydro.iis.u-tokyo.ac.jp/%7Eyamadai/MERIT_DEM/
ALOS DSM	30	Global	https://www.eorc.jaxa.jp/ALOS/en/dataset/aw3d30/aw3d30_e.htm
NASADEM	30	Global	https://www.earthdata.nasa.gov/esds/competitive-programs/measures/nasadem
EUCOPERNICUS DEM	25	EU	https://land.copernicus.eu/en/products/products-that-are-no-longer-disseminated-on-the-clms-website
TanDEM-X	12	Global	https://data.europa.eu/data/datasets/5eecdf4c-de57-4624-99e9-60086b032aea?locale=en
Arctic DEM	2	Global	https://www.pgc.umn.edu/data/arcticdem/

[2]Last date accessed on 20/11/2023.

enabling the selection of data types adapted to the needs of each research endeavor. Most of them are available on particular cloud platforms, typically requiring user subscription for data acquisition.

4. Introducing EO-PERSIST (https://eo-persist.eu/)

4.1 The framework

The main focus of the proposed project revolves around pioneering research and innovation in the comprehensive accessibility, management, and utilization of EO data suitable for permafrost socioeconomic studies and research. This distinctive approach involves consolidating the available EO data into a unified cloud-based system/platform. The amalgamation and integration of processing workflows, data repositories, and distribution centers within a unified cloud architecture promises not only a more harmonized and resource-efficient generation and accessibility of products but also an enhanced and integrated utilization of the pertinent data and resultant products. The attainment of this objective will involve the following key strategies:

Table 5.3 Available geospatial datasets for the Arctic region, land category (land use/land cover, surface).

Dataset	Spatial resolution (m)	Temporal resolution	Extent	Data policy	Link[3]
MODIS Land Cover Type (NASA)	500	Annual	Global	Free and open access	https://modis.gsfc.nasa.gov/data/dataprod/mod12.php
CORINE (Copernicus)	100	4 years, 1990—2018	Global	Free and open access	https://land.copernicus.eu/en/products/corine-land-cover
Globland30 (National Geomatics Center of China)	30	2000—2010	Global	Free and open access	http://www.globeland30.org/GLC30Download/index.aspx
Dynamic World (ESA)	10	Annual	Global	Free and open access for noncommercial use	https://dynamicworld.app/
World Cover (ESA)	10	Annual	Global	Free and open access for noncommercial use	https://esa-worldcover.org/en
JRC Monthly Water History, Water Recurrence, Surface Water Mapping Layers (ECD JRC/Google)	30	1984—2022	Global	Free and open access for noncommercial use	https://data.jrc.ec.europa.eu/dataset/jrc-gswe-global-surface-water-explorer-v1

Continued

Table 5.3 Available geospatial datasets for the Arctic region, land category (land use/land cover, surface).—cont'd

Dataset	Spatial resolution (m)	Temporal resolution	Extent	Data policy	Link[3]
European Settlement Map (ECD JRC/ Google)	2.5, 10 and 100	1984–2022	EU	Free and open access for noncommercial use	https://land.copernicus.eu/en/products/ products-that-are-no-longer-disseminated-on-the-clms-website
Urban Atlas (Copernicus)	10	2006, 2012, 2018	EU	Free and open access for noncommercial use	https://land.copernicus.eu/en/products/ urban-atlas
Global PALSAR-2/ PALSAR Forest/ Non-Forest Map (JAXA)	25	2017–2020	Global	Free and open access for noncommercial use	https://www.eorc.jaxa.jp/ALOS/en/ dataset/fnf_e.htm
European Impervious Surface Maps (Copernicus)	10	2006, 2009, 2012, 2015, 2018	EU	Free and open access for noncommercial use	https://land.copernicus.eu/en/products/ high-resolution-layer-imperviousness

[3]Last date accessed on 20/11/2023.

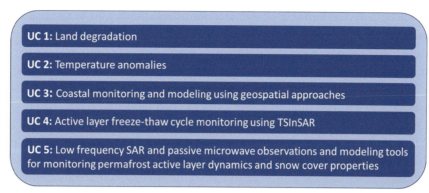

UC 1: Land degradation

UC 2: Temperature anomalies

UC 3: Coastal monitoring and modeling using geospatial approaches

UC 4: Active layer freeze-thaw cycle monitoring using TSInSAR

UC 5: Low frequency SAR and passive microwave observations and modeling tools for monitoring permafrost active layer dynamics and snow cover properties

Figure 5.1 The five use cases which will also serve as Key Performance Indicators (KPIs) of the system.

To assess the viability and effectiveness of the system, experimental analyses will be conducted through five distinct Use Cases (UCs). These UCs will concurrently function as key performance indicators (KPIs) for the system. Specifically, two UCs will be dedicated to the development of innovative algorithms and techniques for permafrost studies, while the remaining three will engage in RS and Geographic Information Systems (GIS)/big data modeling. These latter UCs will lay the foundation for formulating socioeconomic indicators that capture the impact of permafrost changes on various sectors, including industry and local communities. The geographical emphasis will be on European Arctic areas (Fig. 5.1).

4.2 Project aims and innovation aspects

The aim of EO-PERSIST is to establish a dynamic collaborative research and innovation environment through staff exchanges, knowledge sharing, and know-how transfer. This initiative seeks to harness current services, datasets, and emerging technologies to achieve (a) the development of a consistently updated ecosystem featuring EO-based datasets tailored for permafrost applications, (b) methodological advancements in permafrost research by capitalizing on the vast volume of RS datasets, and (c) the generation of indicators directly linked to socioeconomic impacts arising from permafrost dynamics.

Specifically, EO-PERSIST objectives are centered on creating a research and innovation collaboration network facilitated by staff exchanges for the purpose of:

- Exploitation of the cloud processing resources and big data EO archives to construct EO-PERSIST system.
- Design and implementation of innovative RS algorithms (such as TSInSAR methods, convolutional neural networks (CNNs), and super-resolution methods) to facilitate the evaluation of permafrost changes.
- Application of GIS-based models (such as multivariate analysis and gap filling) to large geospatial datasets for monitoring permafrost dynamics.

- Formulation and development of socioeconomic indicators for the sustainable development of the Arctic region, considering factors like land degradation, temperature anomalies, and coastal erosion resulting from permafrost dynamics and changes.
- Exchange of multidisciplinary knowledge (including Earth's processes modeling, RS, GIS, and cloud computing) among partners, fostering the integration of diverse scientific cultures to address scientific inquiries in the permafrost domain.
- Advancement of the skills and careers of staff through promotion and exchange. Reintegration of seconded staff into their respective companies/institutes with a built-in return mechanism to further leverage their newly acquired and enhanced skills and knowledge.
- Facilitation of the transition of complementary research conducted by partners to the market. This involves creating value-added products aligned with (a) a sustainable Arctic EO-based system, (b) enhanced algorithms for permafrost studies, and (c) socioeconomic indicators derived from RS outputs and GIS models, contributing positively to decision-making processes.
- Assurance of project sustainability and long-term business value.

The innovative core of the EO-PERSIST system revolves around the integration of diverse platforms into a unified hub, providing end users with access to extensive collections of EO datasets from multiple sources. This system is designed to merge a big data repository with formidable cloud processing capabilities, offering users a streamlined entry point to a comprehensive computing environment and resources. In essence, users will have the convenience of discovering, processing, and utilizing all datasets from the EO-PERSIST repository within a single, easily accessible location. Key innovative aspects of the EO-PERSIST cloud system include:

- *Conversion of data formats*: The system will employ a mechanism to convert traditional data formats into cloud-optimized formats, enhancing efficiency and compatibility.
- *Continuous data repository updates*: The data repository will undergo continuous updates with the integration of new EO products. This entails ongoing services for data ingestion, processing, and delivery.
- *JupyterHub accessibility*: Users will have access to JupyterHub, providing an ideal solution for those who prefer using different programming languages (such as R, Python2, Python3, Mathics Kernel, or Julia 1.0.3) for data processing. This feature adds a versatile dimension to the system's capabilities, accommodating diverse user preferences and needs.

Additionally, innovative aspects of each experimental Use Case (UC) can be summarized as follows (Figs. 5.2 and 5.3):

4.3 Project beneficiaries

The beneficiaries of EO-PERSIST project are research institutes, universities, and companies from seven countries located in Europe (Fig. 5.1). The project necessitates the

UC 1: Evaluating the influence of permafrost thaw on land degradation and ecosystems.

UC 2: Employing Deep Learning techniques to model climate downscaling for enhancing image data resolution.

UC 3: Investigating novel geospatial data analysis methods for mapping changes in coastal areas and quantifying the socioeconomic impact of climate change on the Arctic region.

UC 4: Harnessing phase information from distributed scatterers to enhance spatial continuity, facilitating detailed recognition and estimation of ground deformation patterns.

UC 5: Unifying algorithms for retrieving soil Freezing/Thawing (F/T) and Snow Water Equivalent (SWE) from Synthetic Aperture Radar (SAR) data, applicable to both L- and C-bands, and develop methods for using physical snow models to support retrievals from active and passive microwave sensors.

Figure 5.2 Summary of EO-PERSIST's innovative aspects, per use case.

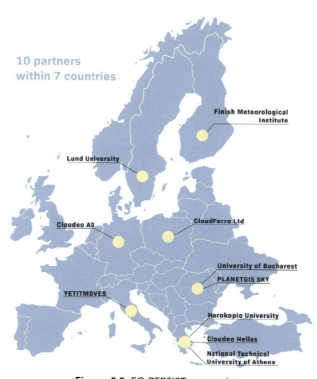

Figure 5.3 EO-PERSIST consortium.

integration of various inter/multidisciplinary forms of knowledge to achieve its objectives. The primary categories of knowledge implicated in the project encompass variables, processes, and models associated with permafrost, RS, system design, and cloud computing, GIS, geospatial socioeconomic modeling, and Global Navigation Satellite System (GNSS). The synthesis of all these knowledge types is imperative for the comprehensive design, implementation, calibration, validation, and demonstration of the EO-PERSIST system. The *National and Technological University of Athens* (https://www.ntua.gr/el/), *University of Bucharest* (https://unibuc.ro/), and *Harokopio University of Athens* (https://www.hua.gr/index.php/en/)—the RS group—have been performing pioneering research in RS, *YETITMOVES* (https://www.yetitmoves.it/) with expertise in GNSS, *CloudFerro* (https://cloudferro.com/), and *CloudEO* (https://www.cloudeo.group/) are key players in cloud environments, *Harokopio University of Athens*—the socioeconomic group—has considerable expertise in geospatial socioeconomics, *PlanetGIS SKY* (https://gis-sky.com/), *CloudEO* Hellas will share their expertise in GIS modeling and *Lund University* (https://www.lunduniversity.lu.se/) and *Finnish Meteorological Institute* (https://en.ilmatieteenlaitos.fi/) have considerable expertise in EPM including permafrost regions, while joint research corresponds to international mobility, based on secondments of research and innovation staff.

4.4 The structure of EO-PERSIST

The research and innovation within EO-PERSIST are structured into five key phases to ensure the quality and credibility of the conducted research and innovation endeavors across the consortium. This framework is specifically crafted to facilitate knowledge transfer and know-how among consortium members, fostering the transformation of inventive concepts into innovative products and services. The overarching goal is to amplify the project's impact by optimizing the entire process from idea generation to tangible outcomes.

The first step encompasses *Communities of Practices (CoP), User Requirements, and EO-PERSIST design*. During this step, a comprehensive list of user requirements and supported datasets within the platform will be established and consistently updated throughout the project's progression. Initial steps involve gathering insights from end-users, including the Advisory Board of EO-PERSIST, through various channels such as questionnaires, interviews, focus groups, workshops, and a combination of these methods. Additionally, external stakeholders, specifically invited by the project, will contribute to the establishment of a CoP group. Concurrently, a detailed analysis will be undertaken, exploring technical, regulatory, and financial aspects critical for the development of the integrated EO-PERSIST system. Based on the requirements identified through user consultations, the consortium will systematically document all available geographic and pertinent data and services, encompassing areas such as hydrology, atmosphere, and meteorology, with a focus on permafrost. A survey will be conducted for each UC, examining existing data and services by referencing analogous projects from the present or past. The inventorying process will cover data and services derived from both RS techniques and on-site measurements, addressing historical as well as real-time data. The inventorying will be meticulous and precise, leveraging high-quality GIS applications and products, along with other database management software, to ensure accuracy and efficiency.

The second step points to the *Socioeconomic Impacts Assessment Related to Thawing Permafrost*. In this phase, socioeconomic indicators will be constructed, using RS and geospatial models relevant to monitoring land degradation (UC1), temperature (UC2), and coastlines (UC3).

The case study on **land degradation** focuses on the environmental impact of permafrost thawing, aiming to assess the chain reaction in geomorphological processes and terrestrial ecosystems. Three specific applications will be explored:

- **Assessing the relationship between permafrost thawing and landslide occurrences**: This aspect of the study involves a comprehensive approach to understanding the interplay between permafrost thawing and landslide occurrences. The detection of landslides is a multistep process, leveraging InSAR analysis and Sentinel-1 imagery. Furthermore, point cloud data obtained from persistent scatterer (PS) analyses undergo evaluation using an artificial intelligence algorithm, allowing for the identification and mapping of space-time anomalies. The integration of these anomalies with geomorphological and geological datasets contributes to the development of an advanced AI model designed specifically for landslide mapping. Additionally, a long short-term memory (LSTM) artificial recurrent neural network is being trained to estimate landslide occurrences under the influence of global climate changes, considering the latest IPCC scenarios.

- **Mapping areas with possible thermokarst induced by permafrost thawing**: In this application, which falls within the realm of geomorphology, the primary objective is to map areas that are susceptible to or already experiencing thermokarst—a phenomenon arising from permafrost thawing. The mapping process involves the use of Sentinel-2 images, applying super-resolution algorithms to enhance resolution to 5 m. Improved Sentinel-2 images are then employed to create a detailed dataset featuring collapsed pingos, sinkholes, and pits, serving as a robust training and validation dataset for the detection model. The development of the detection model incorporates the latest CNN models, with a specific emphasis on testing object detection, object instance segmentation, and pixel classification models for accurate mapping of thermokarst in the Arctic region.

- **Mapping the transformation of terrestrial ecosystems into aquatic ones**: This application is related to the transformation of terrestrial ecosystems into aquatic ecosystems. The methodology involves the use of Sentinel-1 and Sentinel-2 imagery, both of which prove valuable for discerning shifts from nonwater to water within ecosystems. The initial stage of the application encompasses a comprehensive analysis of the advantages and disadvantages associated with these two types of imagery for mapping changes in terrestrial ecosystems. To further enhance understanding, the seasonal cycle from frozen soil/water to water undergoes scrutiny through space-time land cover change detection and monitoring methodologies. This analysis aims to provide insights into the dynamic changes occurring within terrestrial ecosystems. As an outcome, a detailed map with ecosystem changes in the Arctic region is planned to be released during the project's implementation.

The case study on **temperature anomalies** is centered on using thermal RS for industrial monitoring and risk assessment, leveraging historical EO data. These applications are crucial for industry as EO data analysis enables hazard evaluation and the prevention of extensive damage. The study will specifically address two main categories within industrial risk applications, utilizing a methodological approach for a comprehensive exploration.

The first category involves site *monitoring, risk assessment for chemical gases, and plume detection and identification*. In this category, the focus is on monitoring and assessing sites for chemical gases in the event of a spill or accident in a chemical or oil refining plant. Emergency response authorities need to evaluate the plant and its surroundings. The second category focuses on *monitoring/risk assessment for gas, hot water, and high voltage transportation infrastructure leak detection*. In these applications, gas identification is not necessary as the transportation infrastructure is predetermined. The focus is on two key areas: (a) natural gas transportation pipelines and (b) hot water pipelines from central heating plants.

The **coastal monitoring and modeling** case study integrates EO datasets with advanced geospatial modeling approaches, including multivariate models and raster gap-filling approaches. The primary objective is to map coastline changes in Arctic permafrost areas. Advanced algorithms, such as the ArcGIS API for Python, the ArcGIS Notebook Server, and machine learning (ML) methods will be employed to extract shorelines within a GIS environment. Long-term analysis of coastline erosion or accretion rates will be conducted using GIS techniques. A time-series analysis of EO data will reveal coastal dynamics, and the CHAOS modeling system will assess the impact of Arctic permafrost changes on atmospheric conditions.

The third step encompasses the *Scientific Advances for Permafrost Monitoring*. In this phase, the consortium will develop innovative algorithmic approaches within the permafrost domain. The first endeavor involves the development of an innovative approach based on big SAR data to provide information related to the active layer freeze-thaw cycle (UC4). The primary goal of this UC is to design a TSInSAR algorithm specifically for permafrost regions, with the capability to effectively utilize interferometric measurements of distributed scatterers and leverage spatiotemporal information during the unwrapping procedure.

The second entails developing methodologies that harness low-frequency SAR observations to monitor permafrost active layer dynamics and facilitate snow modeling (UC5). The objective of this UC is to develop and assess the capabilities of L-band SAR imagery for retrieving both soil freeze/thaw (F/T) state and snow water equivalent (SWE) with the eventual aim of exploiting the impact of these retrievals on permafrost active layer dynamics. The primary sources of RS data will be ALOS2 and Sentinel-1 observations. A time series of ALOS2 imagery is accessible for various test sites through an ESA-JAXA collaboration involving FMI, and all Sentinel-1 data are publicly distributed.

The fourth step is the *Implementation of EO-PERSIST System*. This phase includes all necessary procedures for designing and implementing the proposed structure of the platform. The EO-PERSIST system will be built upon the existing CloudFerro public cloud infrastructure—CREODIAS. This involves leveraging its EO data repository and utilizing cloud resources, including computing power and storage. CREODIAS stands out as a highly scalable processing cloud, equipped with a local storage capacity exceeding 26 PetaBytes, housing diverse Copernicus data, such as imagery from Sentinel satellite missions, Landsat satellites, Copernicus services, and various datasets. In this step, all partners will closely collaborate and exchange knowledge and expertise to transition the designed system into an operational state.

The fifth and final step focuses on the *Validation of the EO-PERSIST System*. The developed algorithmic approaches will be calibrated and validated exploiting EO-PERSIST cloud-based resources. This process will specifically target regions where high-quality information, such as in situ measurements, is readily available. Complementary knowledge of the socioeconomic activities will be incorporated for enhanced accuracy. Subsequently, utilizing the calibrated and validated EO-PERSIST approaches, an uncertainty assessment will be conducted over permafrost regions in Europe. For each socioeconomic indicator within EO-PERSIST, an uncertainty layer, leveraging all available multisource datasets, will be generated. The engagement of other identified users and stakeholders will be a key component, facilitating platform utilization and collecting valuable feedback on the added value of the service.

5. Conclusions and future outlook

EO-PERSIST assembles a consortium comprising distinguished academic and industry experts across diverse disciplines to address the imperative of monitoring permafrost conditions in the Arctic region. Through the introduction of novel methodologies and innovative tools, leveraging contemporary geoinformation technologies, the project will have a broad applicability and influence on scientific, societal, and economic realms. The avant-garde EO-PERSIST web platform, characterized by technological advancements, including state-of-the-art cloud computing, and access to a spectrum of EO datasets from advanced sensors. The integration of cloud computing is pivotal in providing scalable resources for the efficient processing of the vast EO datasets from. This technological synergy promises to yield unprecedented insights into permafrost studies in the Arctic.

The comprehensive EO dataset made available through EO-PERSIST presents unique opportunities for the exploration of geoinformation technologies in permafrost studies and related applications and also underscores the importance of a cloud-based platform in enhancing accessibility and scalability. The cloud infrastructure ensures the seamless availability of data, fostering collaboration and facilitating real-time processing. Notably, this involves investigating the interplay between climate dynamics and anthropogenic activities and their historical implications for permafrost areas across varying geographical scales.

The embedded case scenarios within EO-PERSIST serve as exemplars, illustrating the platform's efficacy in addressing research inquiries and evaluating socioeconomic impacts associated with permafrost studies. The outcomes from these inquiries have the potential to influence policies and decision-making in permafrost regions, contributing to the refinement of mitigation strategies. The system will be flexible and facilitate communication to improve situational awareness relevant to permafrost. As such, it will provide timely information relevant to permafrost to local and government authorities as well as stakeholders.

The potential influence of comprehending the application of geoinformation technology in permafrost studies in the Arctic extends beyond the realm of science, intersecting with various fields to which EO-PERSIST also makes substantive contributions. As big data is overwhelming all scientific disciplines, the imperative arises for tools capable of collecting, analyzing, and responding to this data in real-time. Noteworthy applications encompass domains such as advanced security cameras, smart-city infrastructure, and autonomous vehicles among others. The cloud-based platform and methodologies envisioned within EO-PERSIST, particularly those related to data processing and modeling, possess significant adaptability for application across diverse disciplines. This adaptability extends to scenarios where combining complementary information from various environmental parameters derived from EO data can enhance output outcomes. In the era of "Big Data," such a conceptual framework is essential, and our project aligns with this imperative. EO-PERSIST holds significant potential for a distinctive contribution by fusing EO data, opening avenues for innovative approaches and capabilities development in various fields, including the cryosphere, ecology, and the study of natural hazards and climate change impacts. For instance, an improved understanding of permafrost processes could lead to a better understanding of the general observation of an acceleration of the water cycle under climate change. The importance of this lies in the substantial and diverse potential impact of climate change on society. Addressing the study of climate impacts on society is considered as a topic of key priority to be addressed today.

Acknowledgments

The present research study has been financially supported by the project "EO-PERSIST", funded by European Union's Horizon Europe Research and Innovation program HORIZON-MSCA-2021-SE-01-01 under grant agreement N. 101086386.

References

Amjad, I., Noor, R., Khalid, A., & Riaz, S. (2022). Polar ice caps melting—A hallmark to vanishing ice along with a global climate change- and addressing solutions to it. *Journal of Environmental Science and Public Health, 06*(01). https://doi.org/10.26502/jesph.96120161

Bell, J. E., Brown, C. L., Conlon, K., Herring, S., Kunkel, K. E., Lawrimore, J., Luber, G., Schreck, C., Smith, A., & Uejio, C. (2018). Changes in extreme events and the potential impacts on human health. *Journal of the Air and Waste Management Association, 68*(4), 265—287. https://doi.org/10.1080/10962247.2017.1401017

Chuffart, R., & Raspotnik, A. (2019). The EU and its Arctic spirit: Solving Arctic climate change from home? *European View, 18*(2), 156—162.

EU Council. (2017). *Council of the European Union (EU) Council conclusions on "Space solutions for a sustainable Arctic", 21/112019, 21 13996/19. 179; EU Council Conclusions on A Space Strategy for Europe, 30/05/2017, 981 and 28/02/2022 IPCC report.*

Feng, J., Wang, C., Lei, J., Yang, Y., Yan, Q., Zhou, X., Tao, X., Ning, D., Yuan, M. M., Qin, Y., Shi, Z. J., Guo, X., He, Z., Van Nostrand, J. D., Wu, L., Bracho-Garillo, R. G., Penton, C. R., Cole, J. R., Konstantinidis, K. T., … Zhou, J. (2020). Warming-induced permafrost thaw exacerbates tundra soil carbon decomposition mediated by microbial community. *Microbiome, 8*(1). https://doi.org/10.1186/s40168-019-0778-3

Fitton, N., Alexander, P., Arnell, N., Bajzelj, B., Calvin, K., Doelman, J., Gerber, J., Havlik, P., Hasegawa, T., Herrero, M., Krisztin, T., van Meijl, H., Powell, T., Sands, R., Stehfest, E., West, P., & Smith, P. (2019). The vulnerabilities of agricultural land and food production to future water scarcity. *Global Environmental Change, 58*, 101944. https://doi.org/10.1016/j.gloenvcha.2019.101944

Hjort, J., Streletskiy, D., Doré, G., Wu, Q., Bjella, K., & Luoto, M. (2022). Impacts of permafrost degradation on infrastructure. *Nature Reviews Earth & Environment, 3*(1), 24—38. https://doi.org/10.1038/s43017-021-00247-8

Hu, J., Wu, J., Petropoulos, G. P., Bao, Y., Liu, J., Lu, Q., Wang, F., Zhang, H., & Liu, H. (2023). Temperature and relative humidity profile retrieval from Fengyun-3D/VASS in the Arctic region using neural networks. *Remote Sensing, 15*(6), 1648. https://doi.org/10.3390/rs15061648

Hu, J., Bao, Y., Liu, J., Liu, H., Petropoulos, G. P., Katsafados, P., Zhu, L., & Cai, X. (2021). Temperature and relative humidity profile retrieval from Fengyun-3D/HIRAS in the Arctic Region. *Remote Sensing*, (13), 1884. https://doi.org/10.3390/rs13101884

Hugelius, G., Strauss, J., Zubrzycki, S., Harden, J. W., Schuur, E. A. G., Ping, C. L., Schirrmeister, L., Grosse, G., Michaelson, G. J., Koven, C. D., O'Donnell, J. A., Elberling, B., Mishra, U., Camill, P., Yu, Z., Palmtag, J., & Kuhry, P. (2014). Estimated stocks of circumpolar permafrost carbon with quantified uncertainty ranges and identified data gaps. *Biogeosciences, 11*(23), 6573—6593. https://doi.org/10.5194/bg-11-6573-2014

Johnsson, F., Kjärstad, J., & Rootzén, J. (2018). The threat to climate change mitigation posed by the abundance of fossil fuels. *Climate Policy, 19*(2), 258—274. https://doi.org/10.1080/14693062.2018.1483885

Li, Y., Brando, P. M., Morton, D. C., Lawrence, D. M., Yang, H., & Randerson, J. T. (2022). Deforestation-induced climate change reduces carbon storage in remaining tropical forests. *Nature Communications, 13*(1). https://doi.org/10.1038/s41467-022-29601-0

Liu, J., Fu, Z., & Liu, W. (2023). Impacts of precipitation variations on agricultural water scarcity under historical and future climate change. *Journal of Hydrology, 617*, 128999. https://doi.org/10.1016/j.jhydrol.2022.128999

Longobardi, P., Montenegro, A., Beltrami, H., & Eby, M. (2016). Deforestation induced climate change: Effects of spatial scale. *PLoS One, 11*(4), e0153357. https://doi.org/10.1371/journal.pone.0153357

Niñerola, A., Ferrer-Rullan, R., & Vidal-Suñé, A. (2020). Climate change mitigation: Application of management production philosophies for energy saving in industrial processes. *Sustainability, 12*(2), 717. https://doi.org/10.3390/su12020717

Rajendran, S., Sadooni, F., Al-Kuwari, H. A. S., Oleg, A. M., Govil, H., Nasir, S., & Vethamony, P. (2021). Monitoring oil spill in Norilsk, Russia using satellite data. *Scientific Reports.* https://doi.org/10.1038/s41598-021-83260-7

Rantanen, M., Karpechko, A. Y., Lipponen, A., Nordling, K., Hyvärinen, O., Ruosteenoja, K., Vihma, T., & Laaksonen, A. (2022). The Arctic has warmed nearly four times faster than the globe since 1979. *Communications Earth & Environment.* https://doi.org/10.1038/s43247-022-00498-3

Schuur, E. A., & Mack, M. C. (2018). Ecological response to permafrost thaw and consequences for local and global ecosystem services. *Annual Review of Ecology, Evolution, and Systematics, 49*(1), 279–301. https://doi.org/10.1146/annurev-ecolsys-121415-032349

Soeder, D. J. (2020). Fossil fuels and climate change. *Fracking and the Environment*, 155–185. https://doi.org/10.1007/978-3-030-59121-2_9

Sovacool, B. K., Griffiths, S., Kim, J., & Bazilian, M. (2021). Climate change and industrial F-gases: A critical and systematic review of developments, sociotechnical systems and policy options for reducing synthetic greenhouse gas emissions. *Renewable and Sustainable Energy Reviews, 141*, 110759. https://doi.org/10.1016/j.rser.2021.110759

Srivastava, P. K. R. K. Pradhan, Petropoulos, G. P., Pandey, V., Gupta, M., Yaduvanshi, A., Jaafar, W., Mall, R. K., & Sahai, A. K. (2021). Long-term trend analysis of precipitation and extreme events over Kosi river basin in India. *Water, 13*, 1695–1703. https://doi.org/10.3390/w13121695. MDPI.

Suter, L., Streletskiy, D., & Shiklomanov, N. (2019). Assessment of the cost of climate change impacts on critical infrastructure in the circumpolar Arctic. *Polar Geography, 42*(4), 267–286. https://doi.org/10.1080/1088937x.2019.1686082

Tsatsaris, A., Kalogeropoulos, K., Stathopoulos, N., Louka, P., Tsanakas, K., Tsesmelis, D. E., Krassanakis, V., Petropoulos, G. P., Pappas, V., & Chalkias, C. (2021). Geoinformation technologies in support of environmental hazards monitoring under climate change: An extensive review. *ISPRS International Journal of Geo-Information, 10*(94), 1–32. https://doi.org/10.3390/ijgi10020094

Wadanambi, R. T., Wandana, L. S., Chathumini, K. K. G. L., N.P., Preethika, D. D. P., Udara, S. P. R., & Arachchige Faculty of Technology, University of Sri Jayewardenepura. (2020). The effects of industrialization on climate change. *Journal of Research Technology and Engineering, 1*(4), 86–94. https://www.jrte.org/wp-content/uploads/2020/10/The-Effects-Of-Industrialization-On-Climate-Change-1-1.pdf.

Wang, X., Yang, B., Bao, Y., Petropoulos, G. P., Liu, H., & Hu, B. (2021). Seasonal trends in clouds and radiation over the Arctic areas from satellite observations during 1982 to 2019. *Remote Sensing, 13*, 3201–3219. MDPI.

Wolff, N. H., Zeppetello, L. R. V., Parsons, L. A., Aggraeni, I., Battisti, D. S., Ebi, K. L., Game, E. T., Kroeger, T., Masuda, Y. J., & Spector, J. T. (2021). The effect of deforestation and climate change on all-cause mortality and unsafe work conditions due to heat exposure in Berau, Indonesia: A modelling study. *The Lancet Planetary Health, 5*(12), e882–e892. https://doi.org/10.1016/s2542-5196(21)00279-5

Wood, N., & Roelich, K. (2019). Tensions, capabilities, and justice in climate change mitigation of fossil fuels. *Energy Research and Social Science, 52*, 114–122. https://doi.org/10.1016/j.erss.2019.02.014

Zhao, S., Liu, M., Tao, M., Zhou, W., Lu, X., Xiong, Y., Li, F., & Wang, Q. (2023). The role of satellite remote sensing in mitigating and adapting to global climate change. *Science of the Total Environment, 904*, 166820. https://doi.org/10.1016/j.scitotenv.2023.166820

SECTION 2

Agricultural environment

CHAPTER 6

Development of algorithms based on the integration of vegetation indices and meteorological data for the identification of low productivity agricultural areas

M. Lanfredi[1,2], R. Coluzzi[1,2], M. D'Emilio[1], V. Imbrenda[1,2], L. Pace[3], C. Samela[1], T. Simoniello[1,2], L. Salvati[4] and J. Mughini Gras[5]

[1]Institute of Methodologies for Environmental Analysis, National Research Council of Italy (IMAA-CNR), Tito Scalo, Italy; [2]NBFC, National Biodiversity Future Center, Palermo, Italy; [3]Department of Earth and Geoenvironmental Sciences, University of Bari "Aldo Moro", Bari, Italy; [4]Department of Methods and Models for Economics, Territory and Finance (MEMOTEF), University of Rome "La Sapienza", Rome, Italy; [5]Octopus S.R.L., Rome, Italy

1. Introduction

The effectiveness of the response to climate change at both regional and local scale crucially depends on the level of preparation and the ability to define potential impacts and territorial priorities. Currently, there is a vast scientific literature that predicts widespread warming and precipitation decreases in most of Mediterranean areas with an increase in frequency and intensity of extreme events (Baronetti et al., 2020; Coluzzi et al., 2020; Greco et al., 2018; Francipane et al., 2021). Furthermore, the effects of climate change are associated with an unstoppable trend of land take (Bianchini et al., 2021; Nickayin et al., 2021; Salvati, Morelli, et al., 2013; Salvati & Carlucci, 2016; Alphan & Güvensoy, 2016; Santarsiero et al., 2022; Smiraglia et al., 2023), especially in the coastal areas of the Mediterranean basin which have always been suited to agricultural activities (Alrteimeiet al., 2022; Zalidis et al., 2022) and are characterized by an extreme fragmentation of land property (Janus et al., 2023) which puts at risk the economic and ecological sustainability of rural districts (Salvati & Carlucci, 2011; Salvati, Tombolini, et al., 2013; Biasi et al., 2015; Smiraglia et al., 2015; Tomao et al., 2017).

In a proactive vision of climate risk management in agricultural areas, it becomes essential to develop effective strategies to evaluate the proneness of exposed areas to be negatively influenced by (climate and land use) changes (Pachauri et al., 2014). The problem is extremely complex and involves also socioeconomic variables, while the vegetation-climate dynamics are the fundamental core on which all the other contributing parameters are based. Especially in a scenario of rapid climate change, the development of adequate strategies for monitoring the relevant parameters to this dynamics is

Geographical Information Science
ISBN 978-0-443-13605-4
https://doi.org/10.1016/B978-0-443-13605-4.00011-4

essential to understand the processes underway before they become irreversible, such as soil degradation (Chuai et al., 2018; Karatayev et al., 2022). Besides solar radiation, the main abiotic factors that influence photosynthetic rates and biomass production are water, temperature, carbon dioxide concentration, and nutrients. Globally, there is an overall relationship of equilibrium between the amount of biomass produced per unit of area and time (NPP, net primary productivity), temperature, and precipitation. The equilibrium conditions are strongly influenced by nutrient limitation, thus estimating NPP is crucial for understanding vegetation dynamics.

In this research field, remote sensing provides cost-effective and efficient tools for estimating NPP proxies, from site scale to regional or global scale (Zheng et al., 2020) since 70s with the advent of Landsat satellite collection which has facilitated an effective environmental remote monitoring (Wulder et al., 2022). Time series of optical imagery can be used to study relevant parameters (photosynthetic activity, water content, chlorophyll, etc.) paying attention to cloud cover and other forms of disturbance (Simoniello et al., 2004; Coluzzi et al., 2018). Specifically, spectral indices derived from hyperspectral, multispectral, and UAV (unmanned aerial vehicle) data reflect peculiar features of soil and vegetation in urban, natural, and agricultural ecosystems (e.g., Pignatti et al., 2015; D'Emilio et al., 2018; Lynch et al., 2020; Simoniello et al., 2022; Candiago et al., 2015). The normalized difference vegetation index (NDVI) is widely used as a proxy variable for photosynthetic activity and NPP (Herrmann et al., 2005; Rouse et al., 1974).

Low NDVI values compared to the average for a given land cover are associated with stressed and/or sparse vegetation. These areas are particularly exposed to the negative impact of possible climate change, both in terms of average values and intensity and frequency of extreme meteorological events. In both cases, the protective action of vegetation cover is reduced, and soils are directly exposed to detrimental phenomena such as erosion and leaching (Santarsiero et al., 2022; Samela et al., 2022; Osman, 2014).

In the perspective of reaching the 2030 Sustainable Development Goals promoting the sustainable management of ecosystems (Sachs, 2012), the mitigation of soil degradation (erosion, compaction, salinization, etc.) is a priority, given the negative effects on agricultural areas, on water quality, and on the sediment production, whose transport can cause many problems also over considerable distances (Bindraban et al., 2012; Borrelli et al., 2018; Issaka & Ashraf, 2017; Santarsiero et al., 2023). This is also an objective to achieve at European level, through EU (European Union) sustainable agricultural policies, such as the Farm to Fork Strategy which is the core of the wider European Green Deal (Delgado et al., 2022). Finally, the possible triggering of land degradation phenomena can cause irreversible effects and thus represents a direct danger (Ferrara et al., 2020; Imbrenda, Coluzzi, et al., 2022; Nickayin et al., 2022) with significant economic damages (Sallustio et al., 2018; Carlucci et al., 2022).

This study focuses on the joint use of meteoclimatic data and satellite-derived NDVI for monitoring vulnerability to climate change. The exploratory analyses illustrated here concern the Metapontum Plain (Basilicata-Southern Italy), which has long been considered a district of considerable importance in the agri-food sector at a national level (Conto et al., 2009; Margiotta et al., 2015). The analysis carried out concerns the phenological year 2019—20 framed in the context of the 20-year period 2000—2020. The core of our analysis is focused on areas classified as arable land according to the European Copernicus Coastal Zones 2018 dataset (available at: https://land.copernicus.eu/en/products/coastal-zones) and characterized by low NDVI values compared to the statistical behavior of the study area in all seasons. Starting with arable land is due to the fact that these areas represent an important agricultural class within the Metapontum Plain and are among the vegetation covers most exposed to erosion processes because of the high portions of fields often remaining bare in the postharvest period (Cerda et al., 2018). The critical areas that we intend to identify are those already showing low photosynthetic activity in the growth period and in which the soils remain mostly uncovered after harvesting. The classification of these areas based on the aridity levels of the local climate adds fundamental insights on the potential ability of lands to face the hottest and driest period of the year.

2. Materials and methods

2.1 Study area

The study site is the Metapontum Plain (Southern Italy), a flat area located in the south-eastern side of Basilicata and overlooking the Ionian Sea (Fig. 6.1). This area represents an agribusiness district of national relevance, specialized primarily in fruit and vegetable production, where several tourist-recreational activities also coexist along the 40 km of Basilicata's Ionian coast (Trivisani et al., 2017; Viccaro et al., 2018). This area is also characterized by a high naturalistic level by virtue of natural forests and coniferous afforestation, located along the coast. Particularly, since 50s, the majority of afforestation has been planted behind the dunes as windbreak barriers to protect agricultural zones during the reclamation works managed by the Land Reclamation Consortium (Margiotta et al., 2015). Forest areas are often home to protected areas (sites belonging to the Natura 2000 network which have become Special Areas of Conservation—SAC in 2013, see Imbrenda, Lanfredi, et al., 2022). Changes in land use/cover that have occurred over the last 2 decades, mainly due to the increasing exploitation of coastal areas for touristic purposes, have consisted mainly in the building of recreational and tourist complexes favoring tourism flows and agricultural intensification (Imbrenda et al., 2018). These land transformations associated with the presence of a prevalent semiarid climate, and the phenomena of shoreline regression, make this area vulnerable to land and water degradation (Ciancia et al., 2018; Muzzillo et al., 2021). Recent analyses of meteorological data (Scalcione et al., 2020) have identified potential signs of ongoing climate change.

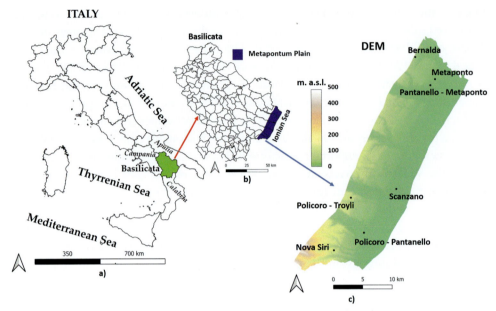

Figure 6.1 (a) Italy subdivided in 20 administrative units, that is, "regions" according to Nomenclature of Territorial Units for Statistics (NUTS2) level; (b) Basilicata and Metapontum Plain (Southern Italy); and (c) Digital Elevation Model (with a spatial resolution of 10 m, see Tarquini et al., 2012) and the seven meteorological stations of the ALSIA network falling in the study area.

2.2 Data collection

For this work, data from meteorological stations were used for rainfall and temperature, while remote data were adopted to estimate photosynthetic activity. The land cover map used is also a product of satellite data processing.

In particular, the meteoclimatic parameters were derived from the network of 45 ground stations of the ALSIA (Agency for Development and Innovation in Agriculture of Basilicata) network. The complete dataset (2000–2020) was used to estimate the aridity index at a regional scale; the seven stations within the Metapontum Plain were used to reconstruct rainfall and temperature trends in the phenological analysis of the year 2020.

At the same time, NDVI was estimated from Sentinel-2 satellite belonging to the Sentinel constellation of the European Copernicus Project by ESA (European Space Agency, see e.g., Gascon et al., 2014).

Sentinel-2 is a European wide-swath, high-resolution, multispectral imaging mission. It samples the electromagnetic spectrum in 13 spectral bands: four bands have a spatial resolution of 10 m, six bands have a spatial resolution of 20 m, while three bands have a spatial resolution of 60 m. Essentially, Sentinel-2 including two twin satellites

(Sentinel-2A and Sentinel-2B) is in the same vein of SPOT and Landsat missions by providing analogous types of data and contributing to current multispectral observations. Sentinel-2 data can be used to support a large assortment of services and applications: land monitoring, agriculture and forest management, disaster control, risk mapping, and other humanitarian concerns.

Specifically, 12 monthly NDVI Sentinel-2 Level-2A images were acquired (one for each month from December 2019 to November 2020) (https://scihub.copernicus.eu/), chosen from those images without clouds in the area of interest.

Finally, the land cover map adopted is the Copernicus Coastal Zones 2018 product which is dedicated to coastal ecosystems as the most fragile environments under threat from human activity. This map provides for the EEA39 European countries a land use/cover dataset with a high level of labeling (71 distinct thematic classes) for coastal areas (from coastline to an inland depth of 10 km, see https://land.copernicus.eu/local/coastal-zones). These data have a minimum mapping unit (MMU) of 0.5 ha and a minimum mapping width (MMW) of 10 m, allowing for informed decisions about the management of these vulnerable areas in ecological and economic terms. For the purposes of the work, the 71 classes populating the map were appropriately aggregated into 10 classes (Fig. 6.2).

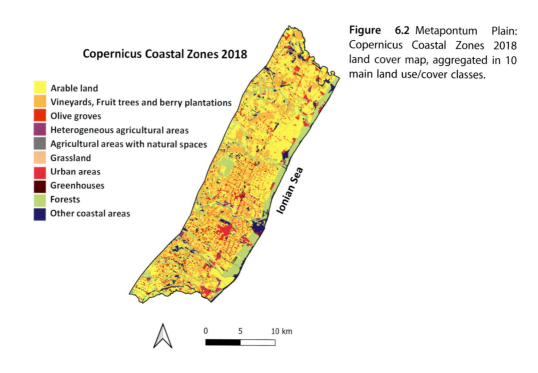

Copernicus Coastal Zones 2018

- Arable land
- Vineyards, Fruit trees and berry plantations
- Olive groves
- Heterogeneous agricultural areas
- Agricultural areas with natural spaces
- Grassland
- Urban areas
- Greenhouses
- Forests
- Other coastal areas

Ionian Sea

0   5   10 km

Figure 6.2 Metapontum Plain: Copernicus Coastal Zones 2018 land cover map, aggregated in 10 main land use/cover classes.

3. Methodology

The rationale behind our study is a combined use of bioclimatic and spectral indexes to detect less productive areas in different climatic conditions. For the sake of simplicity, among the land use classes conveniently grouped, we tested our procedure on the arable land class, as one of the most important land uses in the whole Metapontum Plain.

Through remote data, we derived NDVI to be integrated along the seasons (ΣNDVI) to capture stressed areas, that is, agricultural zones prone to potential harmful impacts of climate change and improper management, especially if the local climate is dry and warm. Thus, these critical areas were classified into different climatic zones on the basis of the bioclimatic index.

Starting from this idea, we choose a bioclimatic index able to synthetize the local climate of the Metapontum Plain: the Aridity Index (AI) as defined by De Martonne (1926), whose equation is:

$$AI = Pa / (Ta + 10) \tag{6.1}$$

where Pa is the total annual rainfall (in mm) and Ta is the average annual temperature (in degrees Celsius).

Although it is one of the oldest aridity indices, De Martonne AI is currently used as a numerical indicator of the level of aridity in a given location and allows for a climatic classification in relation to water availability by effectively identifying dry/wet conditions of the different regions characterizing the examined areas (Hrnjak et al., 2014).

This index was estimated starting from the 2000—2020 data of the ALSIA meteorological network by computing the annual averages over the entire period and then interpolating the values to obtain an index map at 10 m resolution (same spatial resolution of DEM) over the examined area. The interpolation was carried out using altitude regression and ordinary Kriging (for further details see Lanfredi et al., 2015).

As regards the photosynthetic activity to monitor crop productivity, the study was based on the NDVI, the most known metric used as proxy for vegetation vigor and biomass density. The NDVI equation, described below, is based on the knowledge that healthy vegetation reflects more light in the near-infrared (NIR) band and absorbs more light in the red (RED) band than in other wavelengths:

$$NDVI = (\rho_{NIR} - \rho_{RED}) / (\rho_{NIR} + \rho_{RED}) \tag{6.2}$$

where ρ_{NIR} and ρ_{RED} are the reflectances in the NIR and RED wavelengths, respectively. For each acquired image, NDVI was calculated using the bands eight (NIR channel) and four (RED channel) of Sentinel-2 having a spatial resolution of 10 m.

For the evaluation of the phenological status for the year 2020, NDVI maps were averaged into four seasonal maps grouping triplets of months: December-January-February, March-April-May, June-July-August, and September-October-November describing the photosynthesis activity over time.

The statistical distributions of the NDVI values in different seasons were then analyzed using boxplots and all the areas characterized by values lower than the 25th percentile were identified. These critical areas were finally included in the climate zones defined by the AI and the related annual curves of the average NDVI values were placed in correspondence with the homologous curves of temperature and precipitation.

4. Results

The Metapontum Plain falls completely within Mediterranean Italy where the driest and hottest periods of the year are contextual (Lanfredi et al., 2020). Climate change could make these areas drier, further increasing vulnerability to desertification. The estimated values of the De Martonne index identify four climatic zones (Fig. 6.3) in which the degree of aridity increases proceeding from the south-west to the north-east. Areas exhibiting a Mediterranean climate are dominant (more than half of the study area, see Table 6.1), the semiarid and subhumid zones have a similar extent (about 20% of the examined area), while the humid areas represent a climatic niche within the Metapontum Plain (just over 4%).

The analysis of NDVI was focused on identifying areas characterized by low values compared to the statistics for the arable land class extracted from the Copernicus Coastal Zones database (Fig. 6.4). Due to the large presence of agricultural areas with permanent and annual crops, the phenological top is reached in spring, except from coastal forests

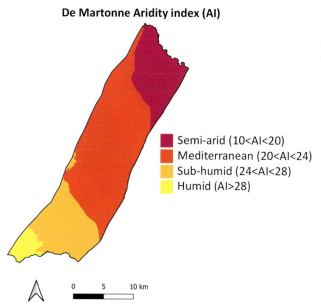

De Martonne Aridity index (AI)

Semi-arid (10<AI<20)
Mediterranean (20<AI<24)
Sub-humid (24<AI<28)
Humid (AI>28)

0   5   10 km

Figure 6.3 De Martonne aridity index computed on the study area and reclassified according to Croitoru et al. (2013).

Table 6.1 Different climate zones characterizing Metapontum Plain.

Climatic zones according to De Martonne AI	Extent (km^2)	Extent (%)
Semiarid	78,86	19,54
Mediterranean	220,87	54,73
Subhumid	85,92	21,29
Humid	17,91	4,44

According to the De Martonne Aridity Index.

that show the peak in summer. Harvest, fire, and drought contribute to make summer the season with the lowest NDVI values (Fares et al., 2017; Marino, 2023; Yildirim et al., 2021).

From the statistical point of view, low values of the distributions (less than the 25th percentile) indicate low rates of photosynthetic activity which in summer correspond to the spectral behavior of bare soils (Fig. 6.5a). The areas that remain in this region throughout the year, defined as critical areas (Fig. 6.5b and c), show low productivity and their soil layers are particularly vulnerable in summer.

The annual production per unit area was estimated as the integral of the NDVI over all seasons. This value for areas identified as critical does not reach 50% of the mean production estimated for the class arable land. Table 6.2 shows the distribution of critical

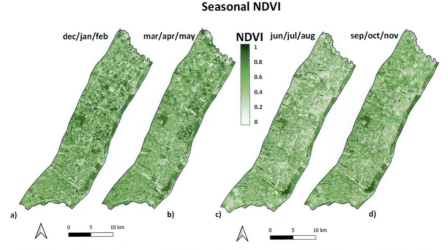

Figure 6.4 Mean NDVI values computed for the four meteorological seasons: (a) winter (Dec 2019–Feb 2020), (b) spring (Mar 2020–May 2020), (c) summer (Jun 2020–Aug 2020), (d) autumn (Sep 2020–Nov 2020) producing a temporal sequence revealing the photosynthesis rate over time.

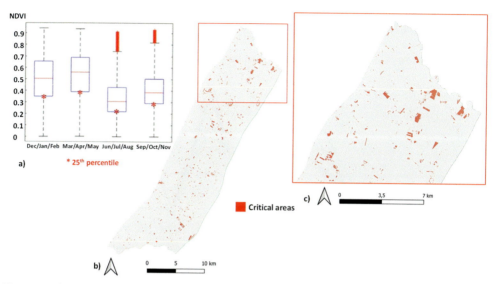

Figure 6.5 (a) Boxplot of the NDVI values for the four seasons of the year 2020. The value corresponding to the 25th percentile is highlighted; (b) critical arable land in the Metapontum Plain (red colored); (c) zoom on the critical arable lands of the semiarid zone (north-eastern of the study area).

areas in different local climates according to the De Martonne Aridity Index. In subhumid, Mediterranean, and semiarid climates, the percentage of critical arable land is around 5% and it halves when moving to a humid climate, suggesting the existence of a direct link between critical areas and climate.

The analysis of rainfall and maximum temperatures recorded during 2020 in the ALSIA meteorological stations located within the study area shows a hot and dry winter with heavier rainfalls in spring (Fig. 6.6). In summer, there are maximum temperature peaks up to 40° and, if we exclude the month of July, little rainy periods, that are in substantial agreement with what is expected for the peculiar climatic context.

Table 6.2 Critical arable lands of the Metapontum Plain split in different climate zones.

Climatic zones according to De Martonne AI	Critical arable land (ha)	Arable land per climatic zone (ha)	Critical areas per climatic zones (%)
Semiarid	201,85	3755,34	5,38
Mediterranean	402,06	8694,15	4,62
Subhumid	162,01	2843,54	5,70
Humid	13,76	996,46	1,38

According to the De Martonne Aridity Index.

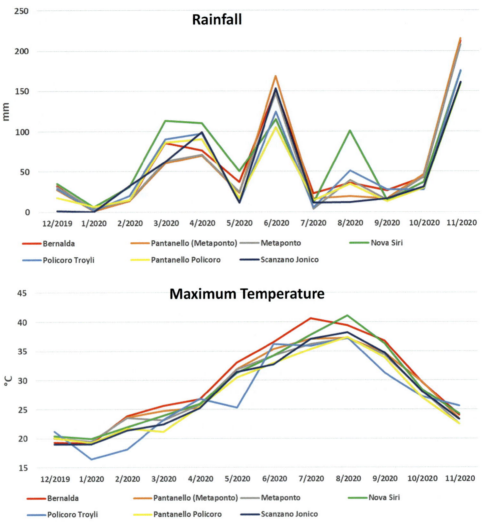

Figure 6.6 Plots of rainfall and maximum temperature values recorded by the seven ALSIA stations falling in the study area.

5. Conclusions

The identification of areas characterized by low NDVI values compared to the statistics of the class composed by arable land has made it possible to identify critical zones falling within subhumid, mediterranean and semiarid climates where productivity is low and the soil is not appropriately protected during the hottest period of the year (July–August). The analysis carried out represents the exploratory phase of a useful study to develop a climate vulnerability monitoring strategy that can be easily extended to other agricultural/natural classes. The next step will be to repeat the analysis for the arable land in the years following 2020. The areas identified here that remained critical also in subsequent years will be considered vulnerable. This will stimulate a more in-depth meteorological study that will take into account also supporting information (e.g., pedology, socioeconomic variables) to better frame the problem of climate vulnerability in the study area. The main gap of this analysis is limiting the investigation to only one land use class (arable land). In the light of this, additional analyses of other land cover types will help to better understand the overall behavior of the examined district.

On the contrary, a strong point of the methodology is its flexibility, being able to make use of a large variety of remote data at different spatial/temporal resolutions. Sentinel's free-of-charge data, their spatial (10 m) and temporal resolution (an image every 2/3 days), contribute to indicate it as an optimal sensor solution for this kind of study.

Furthermore, the developed methodology allows both analyses with sensors spatially coarser (MODIS at 250 m) for investigations at regional/national scale, and analyses relying on commercial sensors with suitable spatial resolutions (PlatetScope at 3 m) for local studies focusing on restricted areas of great (economic/landscape) value.

Overall, on the basis of the results of this exploratory analysis, local authorities and land managers can activate appropriate strategies to prevent that potential or early degradation could turn into actual degradation.

Acknowledgment

This research is inserted within the **ODESSA** (On DEmand Services for Smart Agriculture) project (financed by the European Regional Development Fund Operational Programme 2014–20).

References

Alphan, H., & Güvensoy, L. (2016). Detecting coastal urbanization and land use change in southern Turkey. *Journal of Environmental Engineering and Landscape Management, 24*(2), 97–107.

Alrteimei, H. A., Ash'aari, Z. H., & Muharram, F. M. (2022). Last decade assessment of the impacts of regional climate change on crop yield variations in the Mediterranean region. *Agriculture, 12*(11), 1787.

Baronetti, A., González-Hidalgo, J. C., Vicente-Serrano, S. M., Acquaotta, F., & Fratianni, S. (2020). A weekly spatio-temporal distribution of drought events over the Po Plain (North Italy) in the last five decades. *International Journal of Climatology, 40*(10), 4463–4476.

Bianchini, L., Egidi, G., Alhuseen, A., Sateriano, A., Cividino, S., Clemente, M., & Imbrenda, V. (2021). Toward a dualistic growth? Population increase and land-use change in Rome, Italy. *Land, 10*(7), 749.

Biasi, R., Brunori, E., Smiraglia, D., & Salvati, L. (2015). Linking traditional tree-crop landscapes and agrobiodiversity in Central Italy using a database of typical and traditional products: A multiple risk assessment through a data mining analysis. *Biodiversity and Conservation, 24*, 3009–3031.

Bindraban, P. S., van der Velde, M., Ye, L., Van den Berg, M., Materechera, S., Kiba, D. I., … Van Lynden, G. (2012). Assessing the impact of soil degradation on food production. *Current Opinion in Environmental Sustainability, 4*(5), 478–488.

Borrelli, P., Van Oost, K., Meusburger, K., Alewell, C., Lugato, E., & Panagos, P. (2018). A step towards a holistic assessment of soil degradation in Europe: Coupling on-site erosion with sediment transfer and carbon fluxes. *Environmental Research, 161*, 291–298.

Candiago, S., Remondino, F., De Giglio, M., Dubbini, M., & Gattelli, M. (2015). Evaluating multispectral images and vegetation indices for precision farming applications from UAV images. *Remote Sensing, 7*(4), 4026–4047.

Carlucci, M., Salvia, R., Quaranta, G., Salvati, L., & Imbrenda, V. (2022). Official statistics, spatio-temporal dynamics and local-scale monitoring: Toward integrated environmental-economic accounting for land degradation. *Letters in Spatial and Resource Sciences, 15*(3), 469–491.

Cerda, A., Rodrigo-Comino, J., Novara, A., Brevik, E.,C., Vaezi, A.,R., Pulido, M., … Keesstra, S. D. (2018). Long-term impact of rainfed agricultural land abandonment on soil erosion in the Western Mediterranean basin. *Progress in Physical Geography: Earth and Environment, 42*(2), 202–219.

Chuai, X., Qi, X., Zhang, X., Li, J., Yuan, Y., Guo, X., … Lv, X. (2018). Land degradation monitoring using terrestrial ecosystem carbon sinks/sources and their response to climate change in China. *Land Degradation and Development, 29*(10), 3489–3502.

Ciancia, E., Coviello, I., Di Polito, C., Lacava, T., Pergola, N., Satriano, V., & Tramutoli, V. (2018). Investigating the chlorophyll-a variability in the Gulf of Taranto (North-western Ionian Sea) by a multi-temporal analysis of MODIS-aqua level 3/level 2 data. *Continental Shelf Research, 155*, 34–44.

Coluzzi, R., Fascetti, S., Imbrenda, V., Italiano, S. S. P., Ripullone, F., & Lanfredi, M. (2020). Exploring the use of sentinel-2 data to monitor heterogeneous effects of contextual drought and heatwaves on Mediterranean forests. *Land, 9*(9), 325.

Coluzzi, R., Imbrenda, V., Lanfredi, M., & Simoniello, T. (2018). A first assessment of the Sentinel-2 Level 1-C cloud mask product to support informed surface analyses. *Remote Sensing of Environment, 217*, 426–443.

Conto, F., La Sala, P., & Papapietro, P. (2009). *"The Metapontum agro-food district of quality": A case study of knowledge, innovation and improvement of human capital in territorial rural development (No. 697-2016-47763)* (pp. 367–377).

Croitoru, A. E., Piticar, A., Imbroane, A. M., & Burada, D. C. (2013). Spatiotemporal distribution of aridity indices based on temperature and precipitation in the extra-Carpathian regions of Romania. *Theoretical and Applied Climatology, 112*, 597–607.

D'Emilio, M., Coluzzi, R., Macchiato, M., Imbrenda, V., Ragosta, M., Sabia, S., & Simoniello, T. (2018). Satellite data and soil magnetic susceptibility measurements for heavy metals monitoring: Findings from Agri Valley (Southern Italy). *Environmental Earth Sciences, 77*, 1–7.

De Martonne. (1926). Aerisme, et índices d'aridite. *Comptes Rendus de l'Académie des Sciences, 182*, 1395–1398.

Delgado, L., Garino, C., Moreno, F. J., Zagon, J., & Broll, H. (2022). Sustainable food systems: EU regulatory framework and contribution of insects to the farm-to-fork strategy. *Food Reviews International*, 1–22.

Fares, S., Bajocco, S., Salvati, L., Camarretta, N., Dupuy, J. L., Xanthopoulos, G., … Corona, P. (2017). Characterizing potential wildland fire fuel in live vegetation in the Mediterranean region. *Annals of Forest Science, 74*(1), 1–14.

Ferrara, A., Kosmas, C., Salvati, L., Padula, A., Mancino, G., & Nolè, A. (2020). Updating the MEDALUS-ESA framework for worldwide land degradation and desertification assessment. *Land Degradation and Development, 31*(12), 1593–1607.

Francipane, A., Pumo, D., Sinagra, M., La Loggia, G., & Noto, L. V. (2021). A paradigm of extreme rainfall pluvial floods in complex urban areas: The flood event of 15 July 2020 in Palermo (Italy). *Natural Hazards and Earth System Sciences, 21*(8), 2563–2580.

Gascon, F., Cadau, E., Colin, O., Hoersch, B., Isola, C., Fernández, B. L., & Martimort, P. (September 2014). Copernicus sentinel-2 mission: Products, algorithms and Cal/Val. *In Earth Observing Systems XIX, 9218*, 455–463 (SPIE).

Greco, S., Infusino, M., De Donato, C., Coluzzi, R., Imbrenda, V., Lanfredi, M., … Scalercio, S. (2018). Late spring frost in Mediterranean beech forests: Extended crown dieback and short-term effects on moth communities. *Forests, 9*(7), 388.

Herrmann, S. M., Anyamba, A., & Tucker, C. J. (2005). Recent trends in vegetation dynamics in the African Sahel and their relationship to climate. *Global Environmental Change, 15*(4), 394–404.

Hrnjak, I., Lukić, T., Gavrilov, M. B., Marković, S. B., Unkašević, M., & Tošić, I. (2014). Aridity in Vojvodina, Serbia. *Theoretical and Applied Climatology, 115*, 323–332.

Imbrenda, V., Coluzzi, R., Di Stefano, V., Egidi, G., Salvati, L., Samela, C., … Lanfredi, M. (2022). Modeling spatio-temporal divergence in land vulnerability to desertification with local regressions. *Sustainability, 14*(17), 10906.

Imbrenda, V., Coluzzi, R., Lanfredi, M., Loperte, A., Satriani, A., & Simoniello, T. (2018). Analysis of landscape evolution in a vulnerable coastal area under natural and human pressure. *Geomatics, Natural Hazards and Risk, 9*(1), 1249–1279.

Imbrenda, V., Lanfredi, M., Coluzzi, R., & Simoniello, T. (2022). A smart procedure for assessing the health status of terrestrial habitats in protected areas: The case of the natural 2000 ecological network in Basilicata (Southern Italy). *Remote Sensing, 14*(11), 2699.

Issaka, S., & Ashraf, M. A. (2017). Impact of soil erosion and degradation on water quality: A review. *Geology, Ecology, and Landscapes, 1*(1), 1–11.

Janus, J., Ertunç, E., Muchová, Z., & Tomić, H. (2023). Impact of selected land fragmentation parameters and spatial rural settlement patterns on the competitiveness of agriculture: Examples of selected European and Asian countries. *Habitat International, 140*, 102911.

Karatayev, M., Clarke, M., Salnikov, V., Bekseitova, R., & Nizamova, M. (2022). Monitoring climate change, drought conditions and wheat production in Eurasia: The case study of Kazakhstan. *Heliyon, 8*(1).

Lanfredi, M., Coluzzi, R., Imbrenda, V., Macchiato, M., & Simoniello, T. (2020). Analyzing space–time coherence in precipitation seasonality across different European climates. *Remote Sensing, 12*(1), 171.

Lanfredi, M., Coppola, R., D'Emilio, M., Imbrenda, V., Macchiato, M., & Simoniello, T. (2015). A geostatistics-assisted approach to the deterministic approximation of climate data. *Environmental Modelling and Software, 66*, 69–77.

Lynch, P., Blesius, L., & Hines, E. (2020). Classification of urban area using multispectral indices for urban planning. *Remote Sensing, 12*(15), 2503.

Margiotta, S., Manera, C., Sivolella, C., & Fabrizio, D. (2015). Evolution of the Metaponto district, southern Italy: From land reform to new sustainable scenarios. *Landscape Research, 40*(2), 174–191.

Marino, S. (2023). Understanding the spatio-temporal behavior of crop yield, yield components and weed pressure using time series Sentinel-2-data in an organic farming system. *European Journal of Agronomy, 145*, 126785.

Muzzillo, R., Zuffianò, L. E., Rizzo, E., Canora, F., Capozzoli, L., Giampaolo, V., … Polemio, M. (2021). Seawater intrusion proneness and geophysical investigations in the Metaponto coastal plain (Basilicata, Italy). *Water, 13*(1), 53.

Nickayin, S. S., Coluzzi, R., Marucci, A., Bianchini, L., Salvati, L., Cudlin, P., & Imbrenda, V. (2022). Desertification risk fuels spatial polarization in 'affected' and 'unaffected' landscapes in Italy. *Scientific Reports, 12*(1), 747.

Nickayin, S. S., Salvati, L., Coluzzi, R., Lanfredi, M., Halbac-Cotoara-Zamfir, R., Salvia, R., … Gaburova, L. (2021). What happens in the city when long-term urban expansion and (un) sustainable fringe development occur: The case study of Rome. *ISPRS International Journal of Geo-Information, 10*(4), 231.

Osman, K. T. (2014). *Soil degradation, conservation and remediation* (Vol 820). Dordrecht: Springer Netherlands.

Pachauri, R. K., Allen, M. R., Barros, V. R., Broome, J., Cramer, W., Christ, R., … van Ypserle, J. P. (2014). *Climate change 2014: Synthesis report. Contribution of working groups I, II and III to the fifth assessment report of the intergovernmental panel on climate change* (p. 151). IPCC.

Pignatti, S., Acito, N., Amato, U., Casa, R., Castaldi, F., Coluzzi, R., … Cuomo, V. (July 2015). Environmental products overview of the Italian hyperspectral prisma mission: The SAP4PRISMA project. In *2015 IEEE international geoscience and remote sensing symposium (IGARSS)* (pp. 3997–4000). IEEE.

Rouse, J. W., Haas, R. H., Schell, J. A., & Deering, D. W. (1974). Monitoring vegetation systems in the great plains with ERTS. *NASA Special Publication, 351*(1), 309.

Sachs, J. D. (2012). From millennium development goals to sustainable development goals. *The Lancet, 379*(9832), 2206–2211.

Sallustio, L., Pettenella, D., Merlini, P., Romano, R., Salvati, L., Marchetti, M., & Corona, P. (2018). Assessing the economic marginality of agricultural lands in Italy to support land use planning. *Land Use Policy, 76*, 526–534.

Salvati, L., & Carlucci, M. (2016). Patterns of sprawl: The socioeconomic and territorial profile of dispersed urban areas in Italy. *Regional Studies, 50*(8), 1346–1359.

Salvati, L., & Carlucci, M. (2011). The economic and environmental performances of rural districts in Italy: Are competitiveness and sustainability compatible targets? *Ecological Economics, 70*(12), 2446–2453.

Salvati, L., Morelli, V. G., Rontos, K., & Sabbi, A. (2013a). Latent exurban development: City expansion along the rural-to-urban gradient in growing and declining regions of southern Europe. *Urban Geography, 34*(3), 376–394.

Salvati, L., Tombolini, I., Perini, L., & Ferrara, A. (2013b). Landscape changes and environmental quality: The evolution of land vulnerability and potential resilience to degradation in Italy. *Regional Environmental Change, 13*, 1223–1233.

Samela, C., Imbrenda, V., Coluzzi, R., Pace, L., Simoniello, T., & Lanfredi, M. (2022). Multi-decadal assessment of soil loss in a Mediterranean region characterized by contrasting local climates. *Land, 11*(7), 1010.

Santarsiero, V., Nolè, G., Lanorte, A., Tucci, B., Cillis, G., & Murgante, B. (2022). Remote sensing and spatial analysis for land-take assessment in Basilicata Region (Southern Italy). *Remote Sensing, 14*(7), 1692.

Santarsiero, V., Lanorte, A., Nolè, G., Cillis, G., Tucci, B., & Murgante, B. (2023). Analysis of the effect of soil erosion in abandoned agricultural areas: The case of NE area of Basilicata region (Southern Italy). *Land, 12*(3), 645.

Scalcione, E., Dichio, B., & Fabrizio, G. (2020). Cambiamenti climatici, aumenta la frequenza degli eventi estremi. *Agrifoglio, n.100.*

Simoniello, T., Coluzzi, R., D'Emilio, M., Imbrenda, V., Salvati, L., Sinisi, R., & Summa, V. (2022). Going conservative or conventional? Investigating Farm management strategies in between economic and environmental sustainability in southern Italy. *Agronomy, 12*(3), 597.

Simoniello, T., Cuomo, V., Lanfredi, M., Lasaponara, R., & Macchiato, M. (2004). On the relevance of accurate correction and validation procedures in the analysis of AVHRR-NDVI time series for long-term monitoring. *Journal of Geophysical Research: Atmospheres, 109*(D20).

Smiraglia, D., Cavalli, A., Giuliani, C., & Assennato, F. (2023). The increasing coastal urbanization in the Mediterranean environment: The state of the art in Italy. *Land, 12*(5), 1017.

Smiraglia, D., Ceccarelli, T., Bajocco, S., Perini, L., & Salvati, L. (2015). Unraveling landscape complexity: Land use/land cover changes and landscape pattern dynamics (1954–2008) in contrasting peri-urban and agro-forest regions of northern Italy. *Environmental Management, 56*, 916–932.

Tarquini, S., Vinci, S., Favalli, M., Doumaz, F., Fornaciai, A., & Nannipieri, L. (2012). Release of a 10-m-resolution DEM for the Italian territory: Comparison with global-coverage DEMs and anaglyph-mode exploration via the web. *Computers and Geosciences, 38*(1), 168–170.

Tomao, A., Quatrini, V., Corona, P., Ferrara, A., Lafortezza, R., & Salvati, L. (2017). Resilient landscapes in Mediterranean urban areas: Understanding factors influencing forest trends. *Environmental Research, 156*, 1–9.

Trivisani, A., Simeoni, U., Corbau, C., & Rodella, I. (2017). La percezione dell'offerta turistico-balneare delle spiagge della Costa del Metapontino (Basilicata). *Studi Costieri, 25*, 67–76.

Viccaro, M., Rocchi, B., Cozzi, M., & Romano, S. (2018). SAM multipliers and subsystems: Structural analysis of the Basilicata's agri-food sector. *Bio-Based and Applied Economics, 7*(1), 19–38.

Wulder, M. A., Roy, D. P., Radeloff, V. C., Loveland, T. R., Anderson, M. C., Johnson, D. M., … Cook, B. D. (2022). Fifty years of Landsat science and impacts. *Remote Sensing of Environment, 280*, 113195.

Yildirim, T., Zhou, Y., Flynn, K. C., Gowda, P. H., Ma, S., & Moriasi, D. N. (2021). Evaluating the sensitivity of vegetation and water indices to monitor drought for three Mediterranean crops. *Agronomy Journal, 113*(1), 123–134.

Zalidis, G., Stamatiadis, S., Takavakoglou, V., Eskridge, K., & Misopolinos, N. (2002). Impacts of agricultural practices on soil and water quality in the Mediterranean region and proposed assessment methodology. *Agriculture, Ecosystems and Environment, 88*(2), 137−146.

Zheng, Z., Zhu, W., & Zhang, Y. (2020). Seasonally and spatially varied controls of climatic factors on net primary productivity in alpine grasslands on the Tibetan Plateau. *Global Ecology and Conservation, 21*, e00814.

CHAPTER 7

A comprehensive framework for multiscale soil erosion modeling: A case study of Pag Island, Croatia

Fran Domazetović, Ante Šiljeg and Ivan Marić
University of Zadar, Department of Geography, Center for Geospatial Technologies, Zadar, Croatia

1. Introduction

Soil erosion is among the most complex geomorphic processes, mainly due to its spatial and temporal variability, as well as the multitude of various natural and anthropogenic factors that influence its occurrence and intensity (Morgan, 2009; Poesen, 2018). Scale of soil erosion processes can vary from tiny soil particles eroded by water or wind to the formation of massive gullies and badlands (García-Ruiz et al., 2015). To successfully manage and mitigate soil erosion, a thorough understanding of its various aspects and negative environmental impacts is essential (Panagos & Borrelli, 2017). In order to understand the full complexity of soil erosion, it is necessary to integrate a comprehensive interdisciplinary multiscale framework, which would allow modeling and management of different aspects and scales of soil erosion process.

Main goal of this study is to develop and apply a comprehensive multiscale modeling framework that will allow modeling of different aspects of soil erosion process within chosen study area. Multiscale modeling represents a comprehensive research approach based on the application of various geospatial technologies to study a specific process or phenomenon at two or more research levels (Ferrer et al., 2017; Schmidt, 2000). Each research level within multiscale framework has its own specific research objectives, as well as different spatial extent and resolution of used models (spatial, economic, spectral, and temporal) (Brasington et al., 2012; Ferrer et al., 2017). Multiscale methodological framework has already been successfully applied in few geomorphological studies, including the management of tufa-forming watercourses (Šiljeg et al., 2020), study of interaction between slope and geomorphological and hydrological processes (García-Ruiz et al., 2010), landslide susceptibility modeling (Modugno et al., 2022), and assessment of geomorphic processes in semiarid Badlands (Ferrer et al., 2017). However, according to our knowledge, application of such multiscale approaches for soil erosion modeling is still very rare, mainly due to the complexity and diversity of soil erosion processes.

Geographical Information Science
ISBN 978-0-443-13605-4
https://doi.org/10.1016/B978-0-443-13605-4.00008-4

The basis for developed multiscale soil erosion modeling framework lies in the integration of various geospatial technologies. These technologies allow modeling of complex processes and forms at varying levels of detail (LoDs) depending on the research purpose (Šiljeg et al., 2020). In many previous scientific studies on soil erosion modeling, the research objective usually addressed only one specific scale (e.g., soil erosion susceptibility modeling, spatial-temporal changes quantification, etc.). In contrast, the developed multiscale soil erosion modeling framework enables a comprehensive and systematic examination of various aspects of soil erosion processes while achieving the highest LoD within each research level. Selecting the optimal LoD is crucial for successfully studying desired aspects of soil erosion processes and achieving research objectives (Boulton et al., 2018; Dai et al., 2019; Lu et al., 2017). Therefore, within the developed multiscale framework, special emphasis was given to the selection of optimal LoD and geospatial technology for certain research level.

1.1 Research levels of the multiscale soil erosion modeling framework

Multiscale soil erosion modeling framework was implemented within Pag Island (Croatia), which is known for widespread soil erosion processes (Domazetović et al., 2019a). Developed framework can be divided into three separate research levels: macro-level, meso-level, and micro-level.

At the macro-level, soil erosion modeling is carried on a regional scale, where the region represents a homogenous area with similar characteristics of soil erosion processes (Schmidt, 2000). Macro-level of research covers whole Pag Island, fifth largest island in Croatia (284,29 km^2), located in the middle of the Croatian part of the Adriatic (Fig. 7.1a), which is known by indented coastline and rugged karstic landscapes (Fig. 7.1b). Due to the specific bare karst landscape and harsh climatic conditions, almost 30% of island's total area are susceptible to different soil erosion processes (Domazetović et al., 2019a, 2019b), which is evidenced by the presence of large number of gullies and other erosional forms (Domazetović et al., 2022; Šiljeg et al., 2021). The extensive size of the island and the presence of diverse soil erosion processes led to the restriction of soil erosion modeling to large-scale and lower LoD.

Within the meso-level of multiscale soil erosion modeling framework, the process of soil erosion is studied within specific units, such as certain catchment areas, which share the same or similar characteristics of soil erosion processes (Schmidt, 2000). The Santiš peninsula, characterized by several active karstic gullies, represents the first part of the meso-level of research (Fig. 7.1c-1). Located in the southeastern part of Pag Island, this remote peninsula is known for the presence of soil deposits, within which five karstic gullies have formed. The second part of the meso-level of research focuses on the Santiš gully, the largest and most active gully within the Santiš peninsula (Fig. 7.1c-2). Due to the high diversity of soil erosion processes (sheet erosion, rill erosion, gully erosion), soil

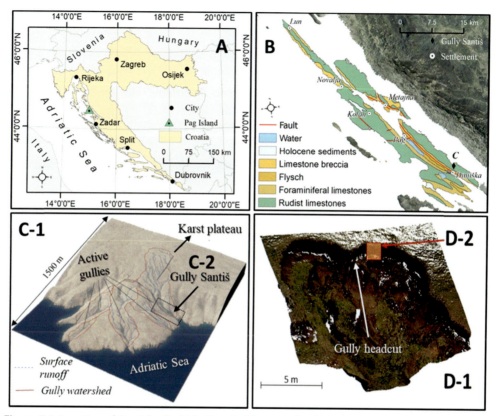

Figure 7.1 Location of Pag island within Croatia (a); macro-level covering Pag Island (b); meso-level covering Santiš peninsula (c-1) and gully Santiš (c-2); micro-level covering most active part of the main gully headcut of gully Santiš (d-1 and d-2).

erosion modeling within gully Santiš was carried at a higher scale and LoD compared to the rest of the peninsula.

Micro-level of multiscale soil erosion modeling framework is typically significantly smaller in area compared to the meso-level and it mostly encompasses individual erosion features within larger erosional units (Schmidt, 2000). Within this research, first part of micro-level covers the chosen part of the main gully headcut of gully Santiš (200 m²) (Fig. 7.1d-1). Main gully headcut of gully Santiš is 92.5 m wide, with steep, at some sections even overhanging walls, whose depth varies from few centimeters up to several meters. Micro-level of research is focused only on most active part of the gully headcut, where most intensive and complex STCs were observed within prior field research (Domazetović et al., 2022; Šiljeg et al., 2021). Second part of micro-level covers one small part of the steep wall of the main gully headcut, where intensive crusting and cracking of soil surface was observed (Fig. 7.1d-2). This area was modeled with sub-centimeter LoD.

1.2 Specific objectives of multiscale soil erosion modeling framework

Although main goal of this study was to develop a comprehensive multiscale modeling framework for facilization of soil erosion modeling, following specific objectives can be distinguished for each research level (Fig. 7.2):

- Macro-level: *Identification of soil erosion hotspots through the high-resolution soil erosion susceptibility modeling*
- Meso-level: *VHR detection and quantification of soil erosion-induced STCs within selected gullies*
- Micro-level: *Monitoring of complex (3D) changes in morphology of gully headcut at subcentimeter scale*

2. Methodology

2.1 Integration of geospatial technologies within the multiscale soil erosion modeling framework

A comprehensive framework was developed for multiscale soil erosion modeling, covering three research levels that encompass different methodological procedures. Due to the complexity and heterogeneity of soil erosion processes, as well as the limitations of certain geospatial technologies, it was impossible to apply an identical methodological approach to all three levels of research. Therefore, within the proposed comprehensive methodological framework, each level has its own methodological framework, adjusted to achieve the maximal accuracy and LoD. Furthermore, each research level integrates different geospatial technologies, allowing the fulfillment of specific research objectives for that level, with maximum possible accuracy and LoD for the created models and carried analysis. Comprehensive analysis of available literature was conducted to establish the optimal geospatial technologies for each of the three research levels. In the end, the following geospatial technologies were integrated into the multiscale framework:

Very-high resolution satellite imagery represents submeter spatial resolution images collected by advanced commercial satellite platforms (Worldview, Geoeye, Ikonos, etc.) (Shean et al., 2016). In contrast to open-source satellite data, commercial satellite imagery is characterized by significantly higher spatial resolution of the collected imagery (e.g., WV-3—30 cm), while maintaining high spectral and temporal resolution. Thanks to a short average revisiting time, such images allow detailed monitoring of specific aspects of soil erosion processes over larger areas, with very high temporal, spectral, and spatial resolutions (Aguilar et al., 2013; Barbarella et al., 2017). Within multiscale framework, VHR satellite imagery is applied for creation of high-resolution models of macro-level of research.

Aerophotogrammetric survey represents a systematic aerial photogrammetric survey, which is carried out using a high-resolution photogrammetric multispectral (MS)

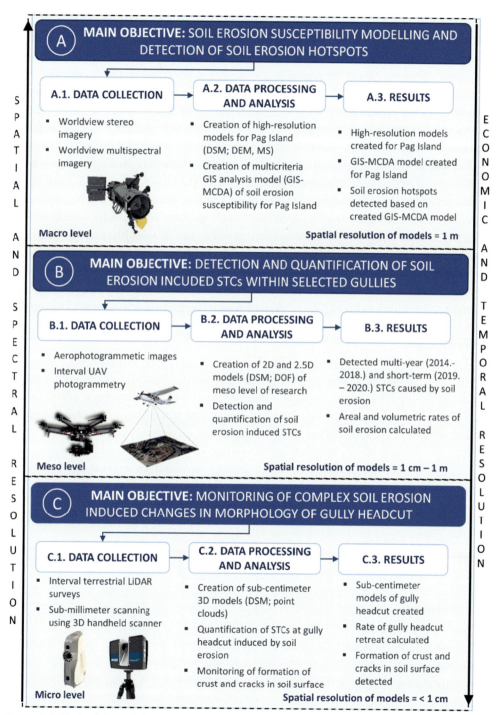

Figure 7.2 Methodological framework for multiscale soil erosion modeling applied at Pag Island, Croatia (a) macro-level; (b) meso-level; (c) micro-level.

camera installed on an aircraft. Systematic aerophotogrammetric surveys are carried in the Republic of Croatia every few years (SGA, 2018), thus allowing detailed and accurate detection of STCs, with multiyear temporal resolution. Similar data have been successfully used to monitor long-term trends in soil erosion processes in other countries (e.g., Spain (Fernández et al., 2017), Italy (Aucelli et al., 2016), etc.), while in Croatia, the full scientific potential of this data is yet to be fully explored and utilized. Within multiscale framework, aerophotogrammetric imagery is applied for the creation of VHR models of Santiš peninsula, within the meso-level of research.

Repeat UAV photogrammetry is a specific branch of aerophotogrammetry based on the use of specialized unmanned aerial vehicles (UAVs) for repeated aerial image acquisition necessary for various measurements and research purposes. The application of repeat UAV photogrammetry, combined with modern software for processing the collected images using structure-from-motion (SfM) and 3D-multiview algorithms, allows for the creation of digital elevation models (DEMs) with subdecimeter precision (Ouédraogo et al., 2014; Peter et al., 2014). Therefore, repeat UAV photogrammetry in soil erosion-related studies is often used for the detection and quantification of different soil erosion-induced STCs. Within multiscale framework, repeat UAV photogrammetry is applied for the creation of VHR models of gully Santiš, within the meso-level of research.

Terrestrial LiDAR scanning (TLS) is one of the most advanced techniques for topographic modeling, widely used in various geomorphological research with a particular focus on detecting, quantifying, and monitoring of STCs (Kromer et al., 2017). Over the years, TLS has been successfully applied to monitor spatial-temporal changes caused by soil erosion (e.g., monitoring headcut migration in gullies (Goodwin et al., 2017; Rengers & Tucker, 2015), calculating the volume of eroded material (Goodwin et al., 2017; Taylor et al., 2018), etc.). Repeat TLS surveys are used for interval data collection, using a local coordinate system, thus achieving centimeter-level accuracy in data positioning (Zhang et al., 2016). Within multiscale framework interval, TLS surveys are applied for the creation of VHR models of chosen part of gully headcut, within the micro-level of research.

3D handheld scanning is a noncontact data acquisition method that combines structured light technology, phase measurement technology, and computer vision technology to gather detailed 3D geometry and texture of a specific shape or surface (Wang et al., 2018, 2019). In comparison to other mentioned technologies, 3D handheld scanning allows collection of data with submillimeter accuracy and resolution, but its application is limited to smaller areas, covering up to several m^2. Therefore, within multiscale framework, 3D handheld scanning is applied for the creation of VHR model of test surface (1 m^2) of soil sediments, forming the part of gully headcut, within the micro-level of research.

Detailed graphical representation of the integration of these geospatial technologies within the methodological framework for multiscale soil erosion modeling is given in Fig. 7.2.

2.2 Macro-scale

Due to the large area of the macro-level of research, the use of geospatial technologies for field data collection and creation of very-high resolution models was not feasible. Instead, available Worldview commercial satellite images were used within this level, enabling the creation of noninterval models with a spatial resolution of 1 m. Created models were then used for the creation of first high-resolution soil erosion susceptibility model in Croatia that allowed the identification of the parts of Pag Island that are more prone to soil erosion occurrence. Methodological framework of the macro-scale can be divided into:

2.2.1 Creation of high resolution models of Pag Island

First part of the methodology at macro-scale covered the creation of high-resolution models of Pag Island, from available Worldview MS and stereo imagery. The area of Pag Island has been lacking high-resolution models, as the existing models had only medium spatial resolution, limiting their usefulness for soil erosion modeling (Domazetović et al., 2019a). Therefore, Worldview stereo imagery was used for the creation of high-resolution digital surface model (DSM) and DEM, while Worldview MS imagery was used for the creation of high-resolution orthorectified MS image.

2.2.1.1 Creation of a high-resolution DSM and DEM of Pag Island

High-resolution DSM of Pag Island was created from the provided WV stereo images in the OrthoEngine 2018 extension of Geomatica 2018 software. Due to the elongated and irregular shape of Pag Island, as well as its large size, it was not possible to create a DSM of the whole island from a single stereo-pair of Worldview imagery. Instead several stereo-pairs had to be used, which complicated the whole process. A total of 44 ground control points (GCPs) were utilized to enhance the accuracy and georeferencing of the generated DSM (Aguilar et al., 2013; Goldbergs et al., 2019.). These GCPs were collected during the fieldwork using the Stonex S10 RTK GNSS at specific locations distributed across all parts of the island. As a result, DSM with 1 m spatial resolution was created, covering whole Pag Island, after which created model was converted to DEM using the automatic and manual filtration of all features except the terrain (e.g., stonewalls, urban objects, vegetation, etc.). After the filtration, a final DEM of Pag Island with a spatial resolution of 1 m was created. Accuracy of created high-resolution DEM was validated using the 273 check points (CPs) distributed evenly within the island and collected with RTK GNSS.

2.2.1.2 Creation of a high-resolution multispectral image of Pag Island

High-resolution MS image of Pag Island was created from Worldview MS images, with eight different spectral bands (*Coastal, Blue, Green, Yellow, Red, Red Edge, NIR1, and NIR2*). Again it was necessary to combine several MS images to cover the whole island. Before the integration of available MS images to a single MS image, all images were pan-sharpened, using the PANSHARP2 algorithm from Geomatica 2018 software. As a result of pansharpening, MS images with spatial resolution of 0.5 m were created and integrated into a single image, covering whole island. Due to the size of created dataset, spatial resolution of created MS had to be downsampled to 1 m. Finally, downsampled image was orthorectified using the created high-resolution DEM. After the orthorectification, a final high-resolution MS (8 band) image of Pag Island was created.

2.2.2 Soil erosion susceptibility modeling using the multicriteria GIS analysis approach

2.2.2.1 Determination of predisposing factors for high-resolution soil erosion susceptibility modeling

Created high-resolution models of Pag Island (MS and DEM) were then used as a basis for a soil erosion susceptibility modeling. Soil erosion susceptibility modeling was based on *multicriteria GIS decision analysis* (GIS-MCDA), which is one of the most commonly used methods for susceptibility modeling that has applications in many different scientific fields (Malczewski & Rinner, 2015, p. 331). Application of GIS-MCDA for soil erosion susceptibility modeling covered six steps, which were automated and simplified using the developed GAMA method (Domazetović et al., 2019b). The initial step of GIS-MCDA involved setting the objective to develop a high-resolution soil erosion susceptibility model. Within the second step, selection and grouping of predisposing and limiting criteria was performed. In total, 17 different soil erosion predisposing criteria were chosen and grouped in following three criteria groups: *morphometric criteria, hydrological criteria*, and *other criteria*. Morphometric criteria group consisted of following seven morphometric criteria that were derived from created high-resolution DEM: *slope* (SLO), *aspect* (ASP), *profile curvature* (PROF), *planar curvature* (PLAN), *mass-balance index* (MBI), *topographic position index* (TPI), and *terrain roughness index* (TRI). Hydrologic criteria group consisted of following seven hydrologic criteria that were derived from created high-resolution DEM: *topographic wetness index* (TWI), *stream power index* (SPI), *length-slope factor* (LSF), *vertical distance to channels* (VDC), *drainage density* (DD) and *convergence index* (ConI). Other criteria group consisted of three criteria derived from created high-resolution MS, which included *land cover* (LC), *normalized difference vegetation index* (NDVI), and *normalized difference water index* (NDWI). Additionally, one constraining criteria, known as Boolean criteria, was also included for high-resolution soil erosion susceptibility modeling. Boolean criteria included all areas where the occurrence of soil erosion-related processes is not possible. In case of Pag Island, Boolean criteria included water areas (e.g., lakes) and urban areas and roads. All 17 predisposing criteria, as well as one Boolean criteria, were created in ArcGIS 10.1 and SAGA GIS softwares.

Due to the easier comparability, all 17 predisposing criteria had to be standardized. The criteria standardization was performed using the *decision maker standardization* (DMS) method, integrated as a DMS tool within the GAMA method (Domazetović et al., 2019b). Utilizing the DMS tool, all chosen predisposing criteria were standardized on a numerical scale, with values from 1 to 5. Classes with the lowest impact on soil erosion susceptibility were assigned a score of 1, while classes with the highest impact received a score of 5.

2.2.2.2 Determination of criteria weight coefficients

The weight coefficients for individual predisposing criteria were determined using the analytical hierarchy process (AHP) (Saaty, 2008), which has been demonstrated in previous research as a dependable method for assigning of weight coefficients to a significant number of criteria (Domazetović et al., 2019a). Weight coefficients were assigned using the AHP procedure to each predisposing criteria. Consistency index (CI) was under 0.1, thus confirming the reliability of assigned weight coefficients.

2.2.2.3 Creation of high-resolution soil erosion susceptibility model

Final step of soil erosion susceptibility model creation covered (1) aggregation of predisposing criteria with their weight coefficients and (2) aggregation of three criteria groups with their respective weight coefficients. This resulted in the creation of a high-resolution susceptibility model. The accuracy of created model was validated using two reference datasets. The first dataset consisted of polygonal data representing the spatial extent of 10 reference gullies, which were previously used as a reference in previous research (Domazetović et al., 2019a). The second reference dataset included 16-point reference locations that were collected during fieldwork within areas with observed presence of soil erosion. Reference datasets were used for the creation of *receiver operating characteristics curves* (ROC curves) and calculation of *area under curve* (AUC), which are one of the most commonly used methods for the validation of the accuracy of GIS-MCDA models (Arabameri et al., 2018, 2020; Domazetović et al., 2019a). Final high-resolution soil erosion susceptibility model was chosen based on calculated AUC values, where most accurate had the highest value of AUC. Created high-resolution soil erosion susceptibility model was classified in to following five classes of soil erosion susceptibility, using the natural jenks classification: very low susceptibility, low susceptibility, medium susceptibility, high susceptibility, and very high susceptibility.

2.3 Meso-scale

Research activities carried within the meso-level of multiscale soil erosion modeling framework can be divided into two parts. First part of meso-level covering the whole

Santiš peninsula is based on the detection of multiyear STCs, where historical aerial photographs (HAPs) provided by the State Geodetic Administration of the Republic of Croatia were used for the creation of VHR DSMs. Second part of meso-level covers the application of repeat UAV photogrammetry for the detection and quantification of short-term STCs, caused by soil erosion within the gully Santiš. Methodological framework of the meso-scale can be divided into:

2.3.1 Detection of multiyear soil erosion-induced STCs within Santiš peninsula
2.3.1.1 Available historical aerial photographs for Santiš peninsula

Basis for the detection and quantification of multiyear soil erosion-induced STCs were HAPs provided by the State Geodetic Administration of the Republic of Croatia. Provided aerial photographs were created as part of systematic aerial surveys of the entire territory of the Republic of Croatia, which the State Geodetic Administration conducts periodically every few years (SGA, 2018). Although the first aerial surveys of certain parts of Croatia began in the early 1930s, the first systematic aerial photogrammetric surveys were carried out only after the Second World War (Biočić, 2014). Even though the first historical aerial images of the Santiš peninsula date back to 1959, providing an opportunity to detect soil erosion-induced STCs within very long timeframe (>60 years), the quality of these early images is very low. Consequently, the detection of multiyear soil erosion-induced STCs relied solely on the most recent aerial images, which had adequate quality for photogrammetric processing and the creation of very high-resolution (VHR) models. These aerial images include images collected in 2014 and 2018 by advanced UltraCam photogrammetric cameras, mounted on the bottom of airplane. Characteristics of the aerial photographs available for photogrammetric processing are given in Table 7.1.

2.3.1.2 Creation of VHR models of Santiš peninsula from available HAPs

Available HAPs were processed in Agisoft Metashape 1.5.1. software, using the identical methodological approach for the processing of images from both 2014 and 2018. Before

Table 7.1 Characteristics of the HAPs used to create the interval VHR models.

ID	Year	Camera	No of images	Focal length (mm)	Image width (m)	Image length (m)	Flight height (m)	GSD (cm)
SGA-1	2014	UltraCam Condor M1 f80	6	79.8	6080	3974	4598	29.96
SGA-2	2018	UltraCam Eagle M3 f100	5	100.5	4866	7572	7189.9	28.62

the start of processing, it was necessary to adjust the calibration parameters for the Ultra-Cam photogrammetric cameras, that were used for the collection of aerial photographs. To enhance the absolute georeferencing and accuracy of VHR models, GCPs and CPs were incorporated. As HAPs were collected several years prior to this study, specific distinctive features present in the collected images and easily recognizable in the field were used to represent both GCPs and CPs. Precise coordinates of GCPs and CPs were collected in the field, using the Stonex 10 RTK GNSS. In total, 8 GCPs and 5 CPs were added, distributed evenly within the entire spatial extent of processed HAPs. As a result of photogrammetric processing interval DSM and digital orthophoto (DOP) with spatial resolution of 0.3 m were created. Validation of accuracy of the interval models was carried out by overlapping them with a reference dataset comprising 22 points collected in the field using Stonex 10 RTK GNSS.

2.3.1.3 Detection of multiyear soil erosion-induced STCs within Santiš peninsula

Created interval VHR models allowed the detection and quantification of soil erosion-induced STCs within the chosen multiyear period (2014—2018.), which was performed using the *Geomorphic Change Detection* 7 add-on for ArcGIS 10.1 software. GCD7 add-on allows the detection of STCs while taking into account the uncertainty of interval models used for change detection (Wheaton et al., 2010; Williams, 2012). Change detection within GCD7 is based on the creation of a *DEM of difference* (DoD) from two interval DSMs. DoD of the Santiš peninsula was generated using initial VHR DSMs from 2014 and subsequent VHR DSM from 2018, with uncertainty determined based on the validated accuracy of those two models.

2.3.2 Detection of short-term soil erosion-induced STCs within gully Santiš
2.3.2.1 Repeat UAV photogrammetric surveys

The detection of short-term soil erosion-induced STCs within the Santiš gully relied on VHR models created through repeat UAV photogrammetry. Repeated UAV photogrammetric surveys were based on repeat aerophotogrammetry system (RAPS), an advanced aerial photogrammetric system developed for research purposes within the Laboratory for Geospatial Technologies (GAL) of the University of Zadar (Šiljeg et al., 2021). Prior to the UAV surveys, local coordinate system was created within the research area, with seven GCPs and six CPs. Due to the extensive impact of soil erosion on soil deposits, it was necessary to position GCPs and CPs outside the areas where STCs could potentially occur during the observation period. Consequently, all GCPs and CPs were carefully situated on a stable carbonate substrate, distinct from the surrounding unconsolidated soil deposits. All seven GCPs and six CP were marked as a chessboard, consisting of two squares with an area of 25 cm^2, while their precise coordinates were collected with Trimble R12i RTK GNSS using the 300 epochs. Identical coordinates for GCPs and CPs were used for both repeat UAV photogrammetric surveys.

UAV photogrammetric surveys were planned ant automated using the Universal Ground Control Software (UgCS), a commercial application designed for planning of UAV flight operations. This software enables users to define various flight operation features, such as flight profiles, flight height, lateral and frontal image overlap, and more. Due to the very-high terrain roughness of gully Santiš, with numerous complex morphological features (e.g., overhangs, steep walls, deep rills, etc.), it was necessary to adjust the parameters of automated UAV flights. Therefore, during the UAV flight planning, following parameters were set, in order to achieve maximal LoD and accuracy of resulting photogrammetric models: front overlap = 75%; side overlap = 75%; flight height = 26.2 m; flight speed = 4.70 m/s; flight mission type = double grid. Additionally, an additional terrain following option was incorporated (*above ground level—AGL*), enabling RAPS to follow the morphology of the terrain and maintain a consistent flight height throughout the entire mission. Ground sampling distance (GSD) of photogrammetric UAV flight missions, planned in UgCS, was set to 0.44 cm, which allowed the creation of VHR models with subcentimeter spatial resolution. The identical automated light missions, generated using the UgCS application, were utilized for both UAV photogrammetric surveys carried with RAPS.

The initial UAV photogrammetric survey (UAV-A) of the gully Santiš took place on December 17, 2019, from 12:00 p.m. to 12:30 p.m. Subsequently, the second UAV photogrammetric survey (UAV-B) occurred on December 17, 2020, from 12:10 p.m. to 12:40 p.m. Throughout both surveys, the weather conditions were highly favorable, with no wind and with permanent cloud cover eliminating the potential shadows. A total of 1534 aerial images were collected over the gully Santiš. An overview of the characteristics of carried repeat UAV photogrammetric surveys is given in Table 7.2.

2.3.2.2 Creation of VHR models from collected UAV imagery

Aerial images collected during the repeat UAV photogrammetric surveys (UAV-A and UAV-B) were processed in Agisoft Metashape 1.5.1. software, using the identical methodological approach (Šiljeg et al., 2021). Prior to the processing, image quality estimation was performed for all collected images, while all images with quality under 0.7 were removed. Since UAV photogrammetric surveys were carried by RAPS, which utilizes

Table 7.2 Characteristics of carried repeat UAV photogrammetric surveys.

Survey ID	Date	UAV system	Flight duration (min)	GSD (cm)	Flight height (AGL) (m)	No of photos	Weather
UAV-A	December 17, 2019	RAPS	23:34	0.44	26.2	1534	Cloudy
UAV-B	December 17, 2020	RAPS	23:39	0.44	26.2	1534	Cloudy

professional grade DSLR camera, there were only few such images. In order to enhance the absolute georeferencing and accuracy of VHR models, coordinates of seven GCPs and six CPs marked and collected in the field were incorporated in photogrammetric project. Furthermore, to achieve maximal LoD of created models, all processing steps were processed using the highest possible parameters (Align photos: Accuracy = highest, Key point limit = 0, Tie point limit = 0; Built dense cloud: Quality = Ultra high, Depth filtering = Aggressive). Accuracy of created VHR models was validated using the six CPs that were integrated within the photogrammetric projects. As a result of photogrammetric processing, VHR DSM and DOP with spatial resolution of 0.44 cm were created.

2.3.2.3 Detection of short-term soil erosion-induced STCs within gully Santiš

Created VHR models allowed the detection and quantification of short-term soil erosion–induced STCs that have occurred in gully Santiš within the chosen 1-year period (2019−2020.). Detection and quantification was performed using the *Geomorphic Change Detection 7* add-on for ArcGIS 10.1 software (Neverman et al., 2016). Due to the subcentimeter spatial resolution of the created initial VHR DSMs from 2019 (UAV-A) and subsequent VHR DSM from 2020 (UAV-B), it was possible to create the DoD of the gully Santiš with spatial resolution of 1 cm. Uncertainty of created DoD was determined based on the validated accuracy of those two models, which was determined in Agisoft Metashape using the six CPs.

2.4 Micro-scale

Research activities carried within the micro-level of multiscale soil erosion modeling framework can be also divided into two parts, with different research objectives and methodological approaches. First part of micro-level is covering most active part of the main gully headcut (200 m^2) of the gully Santiš, where main research objective was to detect the complex gradual retreat of the headcut. Within this research area interval terrestrial LiDAR surveys were used for the creation of subcentimeter VHR DSMs. Second part of micro-level covers even smaller part of this gully headcut (1 m^2), where 3D handheld scanner was applied for the detection of soil crusting and cracking at the millimeter level. 3D handheld scanner was used for scanning of almost vertical part of the gully headcut, from very short distance, thus allowing achievement of highest LoD. Methodological framework of the micro-scale can be divided into:

2.4.1 Monitoring of short-term complex (3D) changes at gully headcut
2.4.1.1 Interval terrestrial LiDAR surveys

Faro Focus M70 terrestrial LiDAR scanner was employed to perform interval LiDAR scanning of the gully Santiš. With a range of up to 70 m and high speed and density of collected points ($\sim$488,000 points/s), this scanner is perfect for the collection of point clouds over complex morphological features. However, terrestrial LiDAR surveying over such complex morphology can be challenging (Domazetović et al., 2022; Goodwin et al., 2017; Neugirg et al., 2016a, 2016b; Perroy et al., 2010). Therefore, within the developed multiscale soil erosion modeling framework, a new systematic framework for the optimization of multitemporal terrestrial LiDAR surveys over a complex gully morphology was also developed. Detailed methodology and achieved results related to the development and application of this systematic framework are published in Domazetović et al. (2022). Developed systematic framework not only greatly facilitated the implementation of multitemporal terrestrial LiDAR surveys but also enabled the achievement of very high accuracy and comprehensive coverage of complex gully morphology. The main emphasis in the developed systematic framework is given to the thorough planning of interval TLS surveys, especially in terms of choosing optimal locations for the placement of scanner and reference targets. Optimal locations for the placement of scanner and reference targets within the developed systematic framework were chosen based on carried visibility analysis. As a result of visibility analysis, local coordinate system with eight optimal scanner positions and seven optimal reference targets positions was built that allowed the achievement of very-high coverage (>95%) over the complex gully morphology.

Based on created local coordinate system, two terrestrial LiDAR surveys were carried over the gully Santiš with Faro Focus M70 scanner. The initial terrestrial LiDAR survey (TLS-A) over gully Santiš was carried on December 17, 2019, from 2:23 p.m. to 5:34 p.m. The entire gully Santiš was scanned from seven optimal scanning positions, resulting in a total of seven individual scans. The second terrestrial LiDAR survey (TLS-B) took place on December 4, 2020, where gully headcut was scanned from only five scanning positions directly related to the chosen section of gully headcut. Thus, the overall duration of the TLS surveying was significantly reduced while still allowing the collection of all necessary data over the chosen section of the gully headcut. The second scanning began at 3:15 p.m. and concluded at 6:11 p.m., resulting in a total of five scans.

2.4.1.2 Creation of interval subcentimeter resolution models of gully headcut

The processing and registration of raw scans collected during the two carried interval terrestrial LiDAR surveys were processed using the Faro SCENE 2019.2 software. This enabled the creation of very-detailed point cloud of the whole gully Santiš. To ensure the comparability of models created from the collected interval TLS surveys, an identical methodology for processing the collected interval laser scans was applied. After the processing of individual TLS scans, automatic detection of reference targets was applied, based on the detection of target spheres with active radii of 69.5 cm. Following the automated detection of targets (spheres) followed automatic registration of processed individual scans, using the *target-based registration* approach. Target-based automatic registration was executed using the forced correspondence by target names. Subsequently, precise target (sphere) coordinates, gathered during the field surveying with Trimble

R12i RTK–GNSS, were integrated into the project. This step facilitated the final absolute georeferencing and automatic registration of the generated LiDAR cluster. The mean TLS survey error was then calculated for each survey, taking into account the mean vertical and horizontal errors, along with the mean distance error. The last phase of the methodology employed for processing of multitemporal terrestrial LiDAR surveys involved the automated generation of point clouds from the registered cluster. As a result, two interval point clouds were created, first for initial TLS survey from 2019 (TLS-A) and second for final TLS survey from 2020 (TLS-B).

2.4.1.3 Detection and quantification of complex (3D) changes at gully headcut

Detection of one-year soil erosion STCs was implemented only in one part of created subcentimeter models, which covers most active part of main gully headcut. The detection of soil erosion–induced STCs was carried out using the Multiscale Model to Model Cloud Comparison (M3C2) algorithm within the CloudCompare software (v2.6.1.). The M3C2 algorithm, being a point cloud-based method, enables the detection of complex 3D soil erosion–induced STCs, effectively overcoming the limitations of 2.5D methods, such as DoDs (Cândido et al., 2020; James et al., 2017). The user-defined parameters of M3C2 algorithm were configured as follows: the radius of the normal (D/2) was set to 10 cm, the cylinder radius (d/2) was set to 20 cm, and the maximum depth was set to 1.5 m. The preferred orientation was set to $+ Z$, and the registration error was set to 1.33 cm. This registration error was calculated as the average of the mean TLS survey errors from the two TLS surveys processed in Faro Scene. As a result of M3C2 application, point cloud of difference was created, which represented all STCs that have occurred between December 2019 and December 2020.

2.4.2 Detection of subcentimeter changes in soil surface
2.4.2.1 Scanning of complex gully headcut face with 3D handheld scanner

Even though TLS enabled the creation of subcentimeter models of the main gully headcut, the LoD of these models was still insufficient for the detection of the smallest soil erosion-related STCs, such as crusting and cracking of the soil surface. Therefore, Artec Eva handheld 3D scanner was used for scanning of small, but complex part of the wall of the main gully headcut. The Artec Eva handheld 3D scanner allows highly accurate and fast capture of 3D data, with a maximum resolution of 0.5 mm (Marić et al., 2020, 2022). Due to its high accuracy and LoD, Artec Eva is suitable for the creation of detailed 3D models of even the smallest soil erosion–induced STCs. Artec Eva handheld 3D scanner was applied for scanning of a chosen section of main gully headcut, where intensive traces of soil surface cracking and crusting were noticed during the carried fieldwork. This small chosen section was almost completely vertical and has a complex surface morphology, shaped by gradual cracking and mass collapse of parts of the soil sediment from the vertical wall.

Operating range of Artec Eva handheld 3D scanner is from 40 to 100 cm, due to which scanning had to be performed from close distance from the headcut. Chosen part of the gully headcut was scanned initially on June 19, 2020, and then afterward on June 26, 2020. Both surveys were carried during the prolonged dry period, which has caused gradual drying, cracking, and crusting of soil sediments. As chosen section of the gully headcut wall was relatively small (1 m²), scanning lasted only about 5 minutes per survey. While scanning, Artec Eva immediately streams the collected data to the connected laptop, allowing real-time visualization of the scans in the Artec Studio software. This was crucial to achieve optimal scanning coverage over the complex morphology of the chosen section, as the scanner had to be directed from multiple angles to scan deep cracks in soil surface.

2.4.2.2 Creation of subcentimeter resolution model of gully headcut face

Data collected with Artec Eva were processed in Artec Studio 15 software, which allows rapid processing of data collected by Artec handheld 3D scanners. First step of the processing covered alignment and registration of individual scans in order to create a solid 3D model. For this purpose, both fine and global registration algorithms were applied. After the registration followed sharp fusion of the registered scans, which created a single 3D model of scanned surface of the gully headcut wall. Final step included texturing of the created 3D model with color information stored by the initial collected scans. As a result of processing in Artec Studio interval, 3D mesh models of gully headcut wall were created with subcentimeter LoD.

2.4.2.3 Detection of crusting and cracking in soil surface

After generating a subcentimeter model of the selected section of the gully headcut, the mesh models had to be transformed into a DSM to facilitate the detection of cracks on the soil surface. This was performed in CloudCompare (v2.6.1.) software, using the two step process. Firstly, created subcentimeter mesh models had to be converted to point cloud, and then created point cloud was rasterized in order to create the DOP and DSM with spatial resolution of 1 mm.

Created DSMs and DOPs were then used for manual vectorization of cracks in the surface of the scanned soil, as well as to detection of STCs related to gradual crusting. Manual vectorization of existing cracks was performed in ArcGIS 10.1 using the created subcentimeter DSMs and DOPs as a base layer.

3. Results

3.1 Macro-level of research

3.1.1 High-resolution models of Pag Island

As a result of processing of stereo Worldview imagery, a high-resolution DEM of Pag Island with spatial resolution of 1 m was created. Average RMSE value for the created high-resolution DEM was 0.59 m. Such LoD and accuracy of created high-resolution

DEM are significantly better than existing DEMs of the Pag Island, which were used in previous research (Domazetović et al., 2019a). Thus, created DEM served as perfect basis for the generation of morphometric and hydrological predisposing criteria for the soil erosion susceptibility modeling.

Second created high-resolution model of Pag Island is an orthorectified MS image with spatial resolution of 1 m of whole Pag Island. This 8-band MS images allowed the creation of spectral indices, such as NDVI and NDSI, as well as the generation of high-resolution land cover of Pag Island that were also used as predisposing criteria for the soil erosion susceptibility modeling.

3.1.2 Soil erosion susceptibility model

A high-resolution soil erosion susceptibility model with spatial resolution of 1 m was created for Pag Island (Fig. 7.3a), based on aggregation of 17 predisposing criteria and

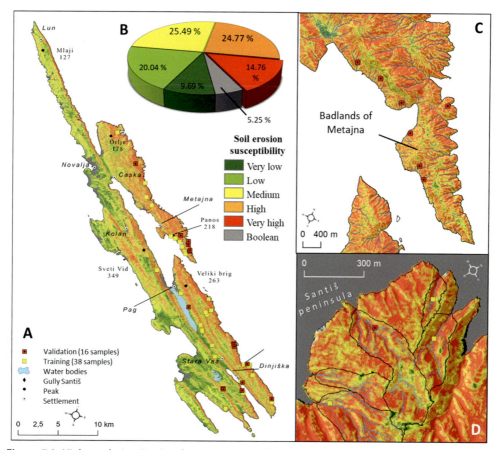

Figure 7.3 High-resolution (1 m) soil erosion susceptibility model of Pag island (a); representation of susceptibility classes within Pag island (b); detected soil erosion hotspots at badlands of Metajna (c) and Šantiš peninsula (d).

one Boolean criteria with their corresponding weight coefficients. The validation of the high-resolution model using two reference datasets revealed that the AUC values are above 0.85 for both reference datasets. Thus, accuracy of created high-resolution suscep-tibility model can be considered as excellent.

Final created high-resolution soil erosion susceptibility model has allowed the detec-tion of soil erosion "hotspots" within the Pag Island (Fig. 7.3c,d). It should be noted that areas that have very high susceptibility to soil erosion are covering approximately 15% of the total area of Pag Island (42 km^2). Moreover, combined area of high and very-high soil erosion susceptible areas covers almost 40% of total area of Pag Island (Fig. 7.3b). How-ever, it should be noted that distribution of these high and very-high susceptible areas within Pag Island is very heterogenic, where it is possible to single out several hotspots of soil erosion (Fig. 7.3). These soil erosion hotspots are mainly concentrated on steeper slopes of northeastern, eastern, and southeastern parts of the Island, where the vegetation cover is poorly developed due to the influence of the strong northeastern "bura" wind. Within these areas, several soil erosion hotspots can be singled out, such as Badlands of Metajna (Fig. 7.3c), or active gullies of Santiš peninsula that have been chosen as a study area for more detailed soil erosion modeling (Fig. 7.3d).

3.2 Meso-level of research

3.2.1 Multiyear soil erosion induced STCs detected within Santiš peninsula

Creation of VHR DoD of Santiš peninsula with spatial resolution of 0.3 m has allowed the detection of multiyear soil erosion-induced STCs. Validation of accuracy of created VHR DSMs, which were used for the creation of DoD, has shown that uncertainty of initial DSM (2014) was 0.36 m, while uncertainty of second DSM (2018) was 0.31 m. These values of model uncertainty were used as a minimal level of detection (LOD$_{min}$) for the detection of STCs in created VHR DoD. Such values of LOD$_{min}$ are proportional or even lower (better) than the values used in similar research studies (Fernández et al., 2017; Stark et al., 2020).

Within the studied multiyear period (2014–2018), soil erosion-induced STCs were detected within three of five existing gullies of Santiš peninsula (Fig. 7.4). STCs within two smallest gullies haven't been detected, probably due to the low intensity of soil erosion processes, as STCs were under the set LOD$_{min}$ of created LoD. Detected soil erosion occurred over a total area of 164.34 m^2 during the observed period, resulting in the total erosion of 201.28 m^3 of soil deposits (Table 7.3). On the other hand, soil accumulation occurred on a small area of only 4.59 m^2, within which only 5.01 m^3 of soil was accumulated. Given that the amount of eroded material significantly exceeds the amount of accumulated material, the remaining eroded material was probably trans-ported to the sea by surface runoff.

Most of the detected soil erosion is concentrated within existing gully headcuts, thus confirming their gradual retreat within the observed multiyear period. It should be noted

Figure 7.4 Distribution of the multiyear erosion within four gully headcuts of Santiš peninsula.

Table 7.3 Total detected STCs caused by soil erosion within Santiš peninsula (2014–2018).

Process	Aerial changes		Volumetric changes		Average vertical change
	m^2	%	m^3	%	m
Erosion	164.34	97.28	201.28	97.57	1.22
Accumulation	4.59	2.72	5.01	2.43	1.08
Total change	168.93	100	206.29	100	–

that the distribution of STCs within gully headcuts is unequal. Two headcuts of gully Santiš (GH–1 and GH–2) correspond to the over 85% of all detected multiyear STCs (Figs. 7.4 and 7.5). Therefore, gully Santiš was chosen for the detection of short–term gully evolution, using the repeat UAV photogrammetry and higher LoD of created VHR models.

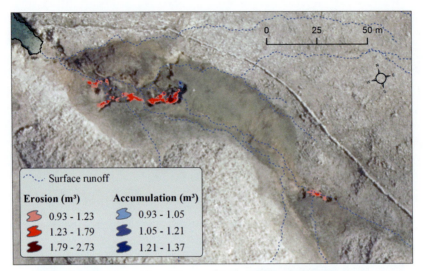

Figure 7.5 Detected multiyear volumetric STCs within gully Santiš.

3.2.2 Short-term soil erosion-induced STCs detected within gully Santiš

Application of repeat UAV photogrammetry has allowed the creation of VHR interval models of gully Santiš with a 1 cm spatial resolution. Uncertainty of created interval models determined based on six CPs was 2.3 cm for initial DSM (UAV-A) and 2.5 cm for second DSM (UAV-B). Therefore, VHR DoD created from these models allowed accurate and detailed detection of all short-term soil erosion-induced STCs that have occurred within the one-year period (2019–2020).

STCs were detected within relatively small portion of gully Santiš (255.99 m^2- 1.84% of total study area), with most STCs concentrated around gully headcuts (GH1, GH2, GH3). While soil erosion was detected within the area of 158.51 m^2, with 13.46 m^3 of eroded material, accumulation was detected within the smaller area (97.48 m^2), with 11.55 m^3 of accumulated material. Net volume is negative, which suggests that some part of eroded material (1.91 m^3) has been transported to the Adriatic Sea.

LoD of created DoD was high enough for the detection of various subprocesses of soil erosion. Most of the material was eroded by gradual retreat of two existing gully headcuts (GH1 and GH3) and mass collapse of part of the carbonate sandstone, caused by the selective erosion (Fig. 7.6). Furthermore, initiation of the formation of new gully headcut (GH2) has been observed within the previously noneroded areas of the gully (Fig. 7.6). Although spatial distribution of detected STCs corresponds to earlier results of multiyear detection (2014–2018), intensity of soil erosion in this studied one-year period is significantly lower. This can be related to the temporal variations in soil erosion occurrence

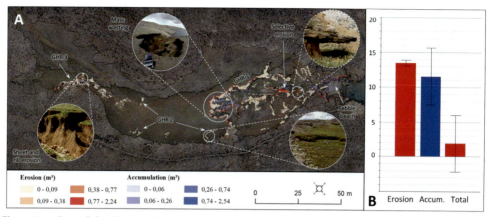

Figure 7.6 Spatial distribution of short-term STCs (a) and total detected volumetric STCs (b) within the gully Santiš.

and intensity, as well as to the difference in LoD and accuracy of models used for the creation of DoDs.

3.3 Micro-level of research

3.3.1 Detected short-term complex (3D) changes at gully headcut

Detection of complex (3D) changes in gully headcut was based on interval point clouds created through the carried interval TLS surveys. Characteristics of created point clouds can be seen in Table 7.4.

High LoD of created points clouds has allowed the detection of mass collapse of parts of the headcut, as well as even smaller soil erosion processes, such as the formation of rills, or sheet wash of material from steep parts of gully headcut (Fig. 7.7). Calculated M3C2 distances indicate that erosion was dominant within most of chosen part of gully headcut, with negative M3C2 distances accounting for 72.31% of all measured distances (Fig. 7.7). Calculated average gradual retreat rate for chosen section of the gully headcut was 4.93 cm/year, with maximum values ranging up to 23.58 cm/year. The main advantage

Table 7.4 Characteristics of created interval TLS point clouds.

TLS survey (year)	Point cloud characteristics			
	Total No of points	Average point density (No/m^2)	SD (RMSE)	Mean TLS survey error (cm)
TLS-A (2019)	26,650,295	92,411,45	0.4 (0.7) cm	1.29
TLS-B (2020)	25,875,206	95,179,63	1.5 (1.8) cm	1.37

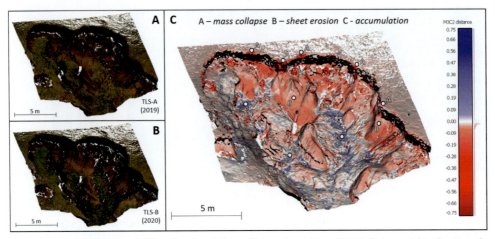

Figure 7.7 Created interval TLS point clouds (a and b) and resulting M3C2 distance (c) indicating the soil erosion-related processes at the gully headcut.

of models created within this level of research is that it was possible to detect and identify small-scale soil erosion-related STCs that were too small for detection within the meso-level of research. However, it should be noted that certain subcentimeter changes in the soil surface, such as soil crusting and cracking, are still hardly detectable with this LoD. Detailed results of this part of micro-level are published in Domazetović et al. (2022).

3.3.2 Detected subcentimeter changes in soil surface

Application of handheld 3D scanning has allowed the detection of even the smallest, sub-centimeter scale processes related to the soil erosion. As field survey was carried in June 2020, during the prolonged period of drought, scanned soil surface of the main gully head-cut was affected by soil crusting that has caused the formation of numerous millimeter-sized cracks in the soil surface (Fig. 7.8). Soil crusting refers to the process in which the surface layer of soil hardens, forming a crust ranging in thickness from several millimeters to several centimeters (Boardman & Poesen, 2007). This compacted and cracked surface layer prevents effective water infiltration, resulting in increased surface runoff that can swiftly remove the fragmented crust (Morgan, 2009). This soil crust can be extremely frag-ile, making it susceptible to even minor natural or human-induced forces, such as rainfall or wind, which can lead to erosion of the large parts of the formed crust. Within the studied short-term period, collapse of the dry and cracked parts of gully headcut was detected, thus confirming the influence of soil crusting and cracking on short-term erosion of material from gully headcut (Fig. 7.8). While this process is widespread, affecting all bare parts of soil sediments, its intensity is too low for detection with less-detailed geospatial technolo-gies, such as TLS or UAV surveys. Therefore, handheld 3D scanning has allowed whole new level of soil erosion research, with unprecedented LoD.

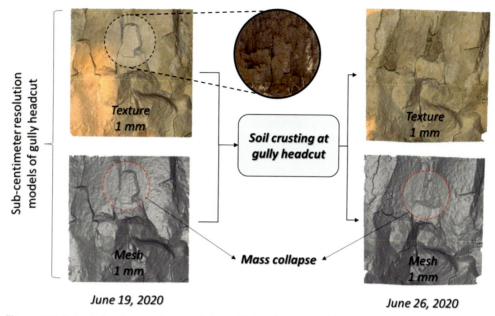

Figure 7.8 Detected collapse of parts of the gully headcut caused by subcentimeter scale cracks in soil surface during the prolonged drought in June 2020.

4. Discussion

Application of developed multiscale soil erosion modeling framework on example of Pag Island demonstrated that selection of optimal LoD and geospatial technology are crucial for the achievement of desired research objectives. Based on carried study, it was possible to distinguish six different LoDs of multiscale soil erosion modeling (Table 7.5). Scale and minimum mapping unit were recalculated from the spatial resolution according to Tobler's rule (Tobler, 1988).

The lowest level of detail (LoD1) is suitable for coarse spatial data with a resolution above 10 m, where the minimum mapping unit is greater than 100 m^2. This LoD is appropriate for studying only the spatial extent of the largest erosion forms, such as badlands. The spatial resolution of LoD1 is not sufficient for more complex analyses, like the detection and quantification of soil erosion-induced STCs, but it is adequate for various regional analyses, such as creating models of soil erosion susceptibility (e.g., Domazetović et al., 2019a). Given that it represents a very low LoD, different open-source data (e.g., DEM –Aster, SRTM, EUDEM, Alos Palsar; MS –Landsat, Sentinel) are sufficient for the creation of the necessary models. Due to its coarse resolution and availability of more detailed and accurate data, this LoD was not used within the application of developed multiscale soil erosion modeling framework on example of Pag Island.

Table 7.5 Proposed six LoDs for multiscale soil erosion modeling.

Level of detail (LoD)	Spatial resolution	Data scale	Minimal mapping width	Minimal mapping unit (m²)	Soil erosion process	Appropriate geospatial technology
LoD1	>10 m	>1: 20,000	>20 m	>100	Badlands	Open-source data
LoD2	1–10 m	1: 2000–1: 20,000	2–20 m	1–100	Badlands and gully erosion	VHR satellite imagery/aerophotogrammetric imagery
LoD3	10 cm–1 m	1: 200–1: 2000	20 cm–2 m	10^{-2}–1	Badlands and gully erosion	Aerophotogrammetric imagery/interval UAV photogrammetry
LoD4	1–10 cm	1: 20–1: 200	2–20 cm	10^{-4}–10^{-2}	Badlands, gully and rill erosion	Interval UAV photogrammetry/TLS survey
LoD5	1 mm–1 cm	1: 2–1: 20	2 mm–2 cm	10^{-6}–10^{-4}	Gully, rill and sheet erosion	TLS survey/3D handheld scanning
LoD6	<1 mm	<1: 2	<2 mm	$<10^{-6}$	Rill, sheet and splash erosion	3D handheld scanning

Level of Detail (LoD)	Soil erosion-related research that has applied certain LoD
LoD1	(Aiello et al., 2015; Domazetović et al., 2019a; Sepuru & Dube, 2018; Seutloali et al., 2017)
LoD2	(Aucelli et al., 2016; Campo-Bescó et al., 2013; Conforti & Buttafuoco, 2017; Shruthi et al., 2015)
LoD3	(Fernández et al., 2017; Koci et al., 2017; Liu et al., 2017; Wang et al., 2016)
LoD4	(Cheng et al., 2019; Eltner et al., 2013, 2015; Koci et al., 2017; Liu et al., 2016; Meinen & Robinson, 2020; Neugirg et al., 2016; Perroy et al., 2010; Šiljeg et al., 2021)
LoD5	(Domazetović et al., 2022; Greenbaum et al., 2020; Li et al., 2020)
LoD6	(Wang et al., 2018, 2019b)

The LoD2 encompasses spatial models with a resolution ranging from 1 to 10 m, with a minimum mapping unit of $1-100$ m^2. The spatial extent of LoD2 models can cover several hundred square kilometers. Similar to LoD1, LoD2 allows the study of large erosion forms, such as badlands or larger individual gullies. Alongside creating models for soil erosion susceptibility, LoD2-level models enable object-based mapping of larger existing soil erosion forms (e.g., Ding et al., 2020; d'Oleire-Oltmanns et al., 2014). Due to their lower LoD, LoD2 models are rarely applied for the detection and quantification of STCs. Data required to create LoD2-level models include various high-resolution commercial satellite images (e.g., Worldview, Ikonos, Quickbird, etc.) and various historical aerial photogrammetric images (e.g., aerial photogrammetric images). Within the developed multiscale soil erosion modeling framework, LoD2 was used within the macro-level of research (Fig. 7.9).

The next level of detail (LoD3) represents a transition from the broader regional character of (LoD1 and LoD2) research to models with centimeter and subcentimeter precision. The spatial extent of LoD3 models can vary depending on the applied geospatial technology, ranging from several square kilometers, up to several tens of square

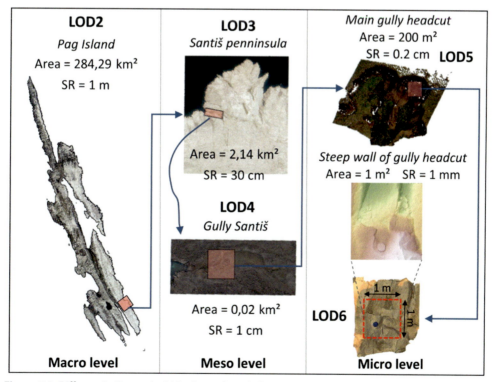

Figure 7.9 Different LoDs used within the multiscale framework for soil erosion modeling on example of Pag Island.

kilometers. LoD3 encompasses spatial models with a resolution ranging from 10 cm to 1 m, with a minimum mapping unit varying from 10^{-2} to 1 m^2. Models at this LoD are suitable for a wide range of applications in soil erosion research. In addition to creating models for soil erosion susceptibility and object-based mapping of existing soil erosion forms, LoD3-level models are suitable for the detection of soil erosion-induced STCs (e.g., Aucelli et al., 2016; Fernández et al., 2017). However, it should be noted that the LoD3 model's level of detail is sufficient for the detection of larger STCs, while the detection of smaller-scale and lower-intensity changes is often not possible. Geospatial technologies suitable for data collection within LoD3 level include HAPs and repeat UAV photogrammetry. Within the developed multiscale soil erosion modeling framework, LoD3 was used within the first part of meso-level of research (Fig. 7.9).

The LoD4 encompasses spatial models with a resolution ranging from 1 to 10 cm, with a minimum mapping unit varying from 10^{-4} to 10^{-2} square meters. Due to the very high LoD and limitations of the required geospatial technologies, the spatial extent of LoD4 models varies from several thousand square meters up to 1 km^2. The primary application of LoD4-level models is the detection and quantification of soil erosion-induced STCs, where these models enable monitoring of very small-scale changes. The development of LoD4 models requires the use of advanced geospatial technologies, such as UAV photogrammetry and/or TLS surveys. Within the developed multiscale soil erosion modeling framework, LoD4 was used within the second part of meso-level of research (Fig. 7.9).

The LoD5 covers spatial models with a resolution ranging from 1 mm to 1 cm, where the minimum mapping unit varies from 10^{-6} to 10^{-4} square meters. Achieving such sub-centimeter resolution models is possible only through the application of the most advanced geospatial technologies, such as interval TLS or 3D handheld scanning. Consequently, the spatial extent of LoD5 models is relatively small, ranging from only several square meters up to several hundred square meters. The very high level of detail in LoD5 models enables the detection and monitoring of complex, small-scale soil erosion processes, such as rill erosion or splash erosion. Therefore, LoD5-level models are suitable exclusively for areas and research where achievement of maximum detail in monitoring of soil erosion processes is required. Within the developed multiscale soil erosion modeling framework, LoD5 was used within the first part of micro-level of research (Fig. 7.9).

The LoD6 encompasses spatial models with a resolution under or equal to 1 mm. Currently, models with such high resolution can only be created through the application of 3D handheld scanning, with the spatial extent of research limited to a maximum of a few square meters. Although LoD6-level models are limited to a very small spatial extent, their LoD surpasses all other LoDs. Therefore, these models can be used for studying of even the smallest erosion forms and processes, such as soil crusting and cracking. Within the developed multiscale soil erosion modeling framework, LoD6 was used within the second part of micro-level of research (Fig. 7.9).

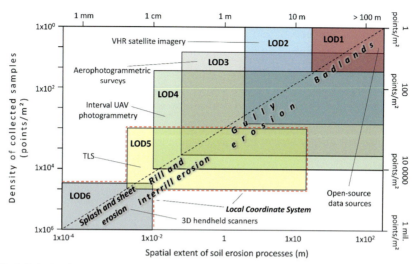

Figure 7.10 Relation between spatial extent of soil erosion processes and different LODs and geospatial technologies within the multiscale soil erosion modeling.

Although developed multiscale framework allows achievement of various LoDs and realization of complex multilayered research objectives, its implementation requires thorough planning. First of all, choosing of optimal geospatial technology and LoD for the achievement of established research objectives should be done prior to the implementation of the multiscale soil erosion modeling framework. While too low LoD can lead to generalized and inaccurate results and conclusions, too high LoD can introduce significant challenges in processing and interpretation of collected data. Fig. 7.10 demonstrates the relationship between optimal LoD and geospatial technology, with the corresponding soil erosion processes that can be surveyed at that level.

5. Conclusion

Developed framework for multiscale soil erosion modeling has allowed comprehensive and systematic examination of various aspects of soil erosion processes while achieving the highest LoD within each research level. Several important conclusions can be highlighted based on carried study:

(1) Although implementation of developed multiscale framework is complex, it has allowed modeling of wide range of soil erosion processes, ranging from millimeter up to kilometer scale.

(2) Applied three research levels have established the better insight into characteristics of soil erosion processes at Pag Island, from detection of soil erosion hotspots, identification different soil erosion processes (gully erosion, rill erosion, sheet erosion, soil crusting, etc.) up to calculation of multiyear and short-term soil erosion intensity.

(3) Established six LoDs should serve as guidelines for planning of soil erosion modeling based on the application of geospatial technologies. Optimal LoD should be determined based on established guidelines prior to the study, taking into account chosen research objectives.

(4) Although developed multiscale framework was applied in this research for modeling of soil erosion, it can be applied for multiscale modeling of any other, similar geomorphic process.

References

Aguilar, M.Á., del Mar Saldaña, M., & Aguilar, F. J. (2013). Generation and quality assessment of stereo-extracted DSM from GeoEye-1 and WorldView-2 imagery. *IEEE Transactions on Geoscience and Remote Sensing, 52*(2), 1259−1271.

Aiello, A., Adamo, M., & Canora, F. (2015). Remote sensing and GIS to assess soil erosion with RUSLE3D and USPED at river basin scale in southern Italy. *Catena, 131*, 174−185.

Arabameri, A., Chen, W., Blaschke, T., Tiefenbacher, J. P., Pradhan, B., & Tien Bui, D. (2020). Gully head-cut distribution modeling using machine learning methods—a case study of NW Iran. *Water, 12*(1), 16.

Arabameri, A., Rezaei, K., Pourghasemi, H. R., Lee, S., & Yamani, M. (2018). GIS-based gully erosion susceptibility mapping: A comparison among three data-driven models and AHP knowledge-based technique. *Environmental Earth Sciences, 77*(17), 628.

Aucelli, P. P., Conforti, M., Della Seta, M., Del Monte, M., D'uva, L., Rosskopf, C. M., & Vergari, F. (2016). Multi-temporal digital photogrammetric analysis for quantitative assessment of soil erosion rates in the Landola catchment of the Upper Orcia Valley (Tuscany, Italy). *Land Degradation and Development, 27*(4), 1075−1092.

Barbarella, M., Fiani, M., & Zollo, C. (2017). Assessment of DEM derived from very high-resolution stereo satellite imagery for geomorphometric analysis. *European Journal of Remote Sensing, 50*(1), 534−549.

Biočić, M. (2014). Data collection with new and accessible measurement methods using aerial vehicles (master thesis). *Faculty of Geodesy, University of Zagreb.*

Boardman, J., & Poesen, J. (Eds.). (2007). *Soil erosion in Europe.* John Wiley & Sons.

Boulton, S. J., & Stokes, M. (2018). Which DEM is best for analyzing fluvial landscape development in mountainous terrains? *Geomorphology, 310*, 168−187.

Brasington, J., Vericat, D., & Rychkov, I. (2012). Modeling river bed morphology, roughness, and surface sedimentology using high resolution terrestrial laser scanning. *Water Resources Research, 48*(11).

Cândido, B. M., James, M., Quinton, J., Lima, W. D., & Silva, M. L. N. (2020). Sediment source and volume of soil erosion in a gully system using UAV photogrammetry. *Revista Brasileira de Ciência do Solo*, 44.

Campo-Bescós, M. A., Flores-Cervantes, J. H., Bras, R. L., Casalí, J., & Giráldez, J. V. (2013). Evaluation of a gully headcut retreat model using multitemporal aerial photographs and digital elevation models. *Journal of Geophysical Research: Earth Surface, 118*(4), 2159−2173.

Cheng, Y. C., Yang, C. J., & Lin, J. C. (2019). Application for terrestrial LiDAR on mudstone erosion caused by typhoons. *Remote Sensing, 11*(20), 2425.

Conforti, M., & Buttafuoco, G. (2017). Assessing space−time variations of denudation processes and related soil loss from 1955 to 2016 in southern Italy (Calabria region). *Environmental Earth Sciences, 76*(13), 1−18.

d'Oleire-Oltmanns, S., Marzolff, I., Tiede, D., & Blaschke, T. (2014). Detection of gully-affected areas by applying object-based image analysis (OBIA) in the region of Taroudannt, Morocco. *Remote Sensing, 6*(9), 8287−8309.

Dai, W., Yang, X., Na, J., Li, J., Brus, D., Xiong, L., … Huang, X. (2019). Effects of DEM resolution on the accuracy of gully maps in loess hilly areas. *Catena, 177*, 114−125.

Ding, H., Liu, K., Chen, X., Xiong, L., Tang, G., Qiu, F., & Strobl, J. (2020). Optimized segmentation based on the weighted aggregation method for loess bank gully mapping. *Remote Sensing, 12*(5), 793.

Domazetović, F., Šiljeg, A., Lončar, N., & Marić, I. (2019). Development of automated multicriteria GIS analysis of gully erosion susceptibility. *Applied Geography, 112*, 102083.

Domazetović, F., Šiljeg, A., Lončar, N., & Marić, I. (2019). GIS automated multicriteria analysis (GAMA) method for susceptibility modelling. *MethodsX, 6*, 2553–2561.

Domazetović, F., Šiljeg, A., Marić, I., & Panđa, L. (2022). A new systematic framework for optimization of multi-temporal terrestrial LiDAR surveys over complex gully morphology. *Remote Sensing, 14*(14), 3366.

Eltner, A., Baumgart, P., Maas, H. G., & Faust, D. (2015). Multi-temporal UAV data for automatic measurement of rill and interrill erosion on loess soil. *Earth Surface Processes and Landforms, 40*(6), 741–755.

Eltner, A., Mulsow, C., & Maas, H. G. (2013). Quantitative measurement of soil erosion from TLS and UAV data. *ISPRS-International Archives of the Photogrammetry, Remote Sensing and Spatial Information Sciences, 1*(2), 119–124.

Fernández, T., Pérez, J. L., Colomo, C., Cardenal, J., Delgado, J., Palenzuela, J. A., … Chacón, J. (2017). Assessment of the evolution of a landslide using digital photogrammetry and LiDAR techniques in the Alpujarras region (Granada, southeastern Spain). *Geosciences, 7*(2), 32.

Ferrer, V., Gómez-Gutiérrez, A., Nadal-Romero, E., Errea, P., & Alonso, E. (2017). *A multiscale approach to assess geomorphological processes in a semiarid badland area (Ebro Depression, Spain).* No. ART-2017-101130.

García-Ruiz, J. M., Beguería, S., Nadal-Romero, E., González-Hidalgo, J. C., Lana-Renault, N., & Sanjuán, Y. (2015). A meta-analysis of soil erosion rates across the world. *Geomorphology, 239*, 160–173.

García-Ruiz, J. M., Lana-Renault, N., Beguería, S., Lasanta, T., Regüés, D., Nadal-Romero, E., … Alatorre, L. C. (2010). From plot to regional scales: Interactions of slope and catchment hydrological and geomorphic processes in the Spanish Pyrenees. *Geomorphology, 120*(3–4), 248–257.

Goldbergs, G., Maier, S. W., Levick, S. R., & Edwards, A. (2019). Limitations of high resolution satellite stereo imagery for estimating canopy height in Australian tropical savannas. *International Journal of Applied Earth Observation and Geoinformation, 75*, 83–95.

Goodwin, N. R., Armston, J. D., Muir, J., & Stiller, I. (2017). Monitoring gully change: A comparison of airborne and terrestrial laser scanning using a case study from Aratula, Queensland. *Geomorphology, 282*, 195–208.

Greenbaum, N., Mushkin, A., Porat, N., & Amit, R. (2020). Runoff generation, rill erosion and time-scales for hyper-arid abandoned alluvial surfaces, the Negev desert, Israel. *Geomorphology, 358*, 107101.

James, M. R., Robson, S., & Smith, M. W. (2017b). 3-D uncertainty-based topographic change detection with structure-from-motion photogrammetry: Precision maps for ground control and directly georeferenced surveys. *Earth Surface Processes and Landforms, 42*(12), 1769–1788.

Koci, J., Jarihani, B., Leon, J. X., Sidle, R. C., Wilkinson, S. N., & Bartley, R. (2017). Assessment of UAV and ground-based structure from motion with multi-view stereo photogrammetry in a gullied savanna catchment. *ISPRS International Journal of Geo-Information, 6*(11), 328.

Kromer, R. A., Abellán, A., Hutchinson, D. J., Lato, M., Chanut, M. A., Dubois, L., & Jaboyedoff, M. (2017). Automated terrestrial laser scanning with near-real-time change detection—monitoring of the Séchilienne landslide. *Earth Surface Dynamics, 5*(2), 293–310.

Li, L., Nearing, M. A., Nichols, M. H., Polyakov, V. O., & Cavanaugh, M. L. (2020). Using terrestrial LiDAR to measure water erosion on stony plots under simulated rainfall. *Earth Surface Processes and Landforms, 45*(2), 484–495.

Liu, K., Ding, H., Tang, G., Na, J., Huang, X., Xue, Z., … Li, F. (2016). Detection of catchment-scale gully-affected areas using unmanned aerial vehicle (UAV) on the Chinese Loess Plateau. *ISPRS International Journal of Geo-Information, 5*(12), 238.

Liu, K., Ding, H., Tang, G., Zhu, A. X., Yang, X., Jiang, S., & Cao, J. (2017). An object-based approach for two-level gully feature mapping using high-resolution DEM and imagery: A case study on hilly loess plateau region, China. *Chinese Geographical Science, 27*(3), 415–430.

Lu, X., Li, Y., Washington-Allen, R. A., Li, Y., Li, H., & Hu, Q. (2017). The effect of grid size on the quantification of erosion, deposition, and rill network. *International Soil and Water Conservation Research, 5*(3), 241–251.

Malczewski, J., & Rinner, C. (2015). *Multicriteria decision analysis in geographic information science.* New York: Springer.

Marić, I., Šiljeg, A., & Domazetović, F. (2022). Precision assessment of Artec Space Spider 3D handheld scanner for quantifying Tufa formation dynamics on small limestone plates (PLs). *Proceedings of the 8th International Conference on Geographical Information Systems Theory, Applications and Management (GISTAM 2022)*, 67–74. https://doi.org/10.5220/0010886900003185

Marić, I., Šiljeg, A., Domazetović, F., & Cukrov, N. (2020). A framework for using handheld 3D surface scanners in quantifying the volumetric tufa growth. *Geomorphometry, 2020*, 22–26.

Meinen, B. U., & Robinson, D. T. (2020). Where did the soil go? quantifying one year of soil erosion on a steep tile-drained agricultural field. *Science of the Total Environment, 729*, 138320.

Modugno, S., Johnson, S. C. M., Borrelli, P., Alam, E., Bezak, N., & Balzter, H. (2022). Analysis of human exposure to landslides with a GIS multiscale approach. *Natural Hazards, 112*(1), 387–412.

Morgan, R. P. C. (2009). *Soil erosion and conservation*. John Wiley & Sons.

Neugirg, F., Stark, M., Kaiser, A., Vlacilova, M., Della Seta, M., Vergari, F., … Haas, F. (2016). Erosion processes in calanchi in the Upper Orcia Valley, Southern Tuscany, Italy based on multitemporal high-resolution terrestrial LiDAR and UAV surveys. *Geomorphology, 269*, 8–22.

Neverman, A. J., Fuller, I. C., & Procter, J. N. (2016). Application of geomorphic change detection (GCD) to quantify morphological budgeting error in a New Zealand Gravel-Bed river: A case study from the Makaroro river, Hawke's bay. *Journal of Hydrology (New Zealand)*, 45–63.

Ouédraogo, M. M., Degré, A., Debouche, C., & Lisein, J. (2014). The evaluation of unmanned aerial system-based photogrammetry and terrestrial laser scanning to generate DEMs of agricultural watersheds. *Geomorphology, 214*, 339–355.

Panagos, P., & Borrelli, P. (2017). Soil erosion in Europe: Current status, challenges and future developments. *All That Soil Erosion: The Global Task to Conserve our Soil Resources*, 20–21.

Perroy, R. L., Bookhagen, B., Asner, G. P., & Chadwick, O. A. (2010). Comparison of gully erosion estimates using airborne and ground-based LiDAR on Santa Cruz Island, California. *Geomorphology, 118*(3–4), 288–300.

Peter, K. D., d'Oleire-Oltmanns, S., Ries, J. B., Marzolff, I., & Hssaine, A. A. (2014). Soil erosion in gully catchments affected by land-levelling measures in the Souss basin, Morocco, analysed by rainfall simulation and UAV remote sensing data. *Catena, 113*, 24–40.

Poesen, J. (2018). Soil erosion in the Anthropocene: Research needs. *Earth Surface Processes and Landforms, 43*(1), 64–84.

Rengers, F. K., & Tucker, G. E. (2015). The evolution of gully headcut morphology: A case study using terrestrial laser scanning and hydrological monitoring. *Earth Surface Processes and Landforms, 40*(10), 1304–1317.

Šiljeg, A., Domazetović, F., Marić, I., Lončar, N., & Panda, L. (2021). New method for automated quantification of vertical spatio-temporal changes within gully cross-sections based on very-high-resolution models. *Remote Sensing, 13*(2), 321.

Šiljeg, A., Marić, I., Cukrov, N., Domazetović, F., & Roland, V. (2020). A multiscale framework for sustainable management of tufa-forming watercourses: A case study of national park "krka", Croatia. *Water, 12*(11), 3096.

Saaty, T. L. (2008). Decision making with the analytic hierarchy process. *International Journal of Services Sciences, 1*(1), 83–98.

Schmidt, J. (Ed.). (2000). *Soil erosion: Application of physically based models*. Springer Science & Business Media.

Sepuru, T. K., & Dube, T. (2018). An appraisal on the progress of remote sensing applications in soil erosion mapping and monitoring. *Remote Sensing Applications: Society and Environment, 9*, 1–9. https://doi.org/10.1016/j.rsase.2017.10.005

Seutloali, K. E., Dube, T., & Mutanga, O. (2017). Assessing and mapping the severity of soil erosion using the 30-m Landsat multispectral satellite data in the former South African homelands of Transkei. *Physics and Chemistry of the Earth, Parts A/B/C, 100*, 296–304.

Shean, D. E., Alexandrov, O., Moratto, Z. M., Smith, B. E., Joughin, I. R., Porter, C., & Morin, P. (2016). An automated, open-source pipeline for mass production of digital elevation models (DEMs) from very-high-resolution commercial stereo satellite imagery. *ISPRS Journal of Photogrammetry and Remote Sensing, 116*, 101–117.

Shruthi, R. B., Kerle, N., Jetten, V., Abdellah, L., & Machmach, I. (2015). Quantifying temporal changes in gully erosion areas with object oriented analysis. *Catena, 128*, 262–277.

Stark, M., Neugirg, F., Kaiser, A., Della Seta, M., Schmidt, J., Becht, M., & Haas, F. (2020). Calanchi badlands reconstructions and long-term change detection analysis from historical aerial and UAS image processing. *Journal of Geomorphology*, 1–24. https://doi.org/10.1127/jgeomorphology/2020/065

State Geodetic Administration. (2018). *Data catalog (version 1.11)*. Retrieved from https://dgu.gov.hr/UserDocsImages/dokumenti/Pristup%20informacijama/Zakoni%20i%20ostali%20propisi/Ostalo/Katalog_podataka_DGU_2018_v11.pdf.

Taylor, R. J., Massey, C., Fuller, I. C., Marden, M., Archibald, G., & Ries, W. (2018). Quantifying sediment connectivity in an actively eroding gully complex, Waipaoa catchment, New Zealand. *Geomorphology, 307*, 24–37.

Tobler, Waldo (1988). Resolution, resampling, and all that. In H. Mounsey, & R. Tomlinson (Eds.), *Building data bases for global science* (pp. 129–137). London: Taylor and Francis.

Wang, P., Jin, X., & Li, S. (2018). Application of handheld 3D scanner in quantitative study of slope soil erosion. *Earth and Environmental Science, 170*(2), 022178.

Wang, P., Jin, X., Li, S., Wang, C., & Zhao, H. (2019). Digital modeling of slope micro-geomorphology based on Artec Eva 3D scanning technology. *IOP Conference Series: Earth and Environmental Science, 252*(5), 052116.

Wang, R., Zhang, S., Pu, L., Yang, J., Yang, C., Chen, J., … Sang, X. (2016). Gully erosion mapping and monitoring at multiple scales based on multi-source remote sensing data of the Sancha River catchment, Northeast China. *ISPRS International Journal of Geo-Information, 5*(11), 200.

Wheaton, J. M., Brasington, J., Darby, S. E., & Sear, D. A. (2010). Accounting for uncertainty in DEMs from repeat topographic surveys: Improved sediment budgets. *Earth Surface Processes and Landforms: The Journal of the British Geomorphological Research Group, 35*(2), 136–156.

Williams, R. (2012). DEMs of difference. *Geomorphological Techniques, 2*(3.2).

Zhang, W., Chen, Y., Wang, H., Chen, M., Wang, X., & Yan, G. (2016). Efficient registration of terrestrial LiDAR scans using a coarse-to-fine strategy for forestry applications. *Agricultural and Forest Meteorology, 225*, 8–23.

CHAPTER 8

Estimation of soybean productivity in a crop livestock integration system from orbital imagens and simplified triangle method

Gustavo Rodrigues Pereira[1], Daniela Fernanda da Silva Fuzzo[2], João Alberto Fischer Filho[2], Bruno Enrique Fuzzo[3] and Gleyce Kelly Dantas Araújo Figueiredo[1]

[1]Faculty of Agricultural Engineering, UNICAMP - Campinas, Brazil; [2]Department of Agricultural and Biological Sciences, University of the State of Minas Gerais, Frutal Unit, Brazil; [3]Master's Program in Environmental Sciences, University of the State of Minas Gerais, Fruit Unit, Brazil

1. Introduction

Brazil is the second largest producer of soybeans in the world, behind only the USA (Embrapa-Soja, 2020). Data from the 2016/2017 harvest indicate occupation and production of 33.8 million hectares and a total production of 113.92 million tons, respectively, consequently generating an average productivity of 3.362 ton/ha according to IBGE.

In view of the importance of agricultural activity in the country, we must also pay attention to technologies that expand production monitoring strategies, given that due to the large territorial extension of Brazil, we have different climate dynamics that directly impact the performance of crop development. Also considering the possible variations in water availability between years and the variation in other meteorological variables call for a more systematic approach in crop monitoring. In this context, we can use remote sensing (RS) techniques as a tool for monitoring agriculture.

Thinking of spectral image acquisition instruments, orbital satellites provide a solution with several sources and a range of products that have their applications for agriculture. Free initiatives for acquiring satellite images such as the Landsat program are of great relevance for agricultural monitoring, as can be seen in works such as Sanches et al. (2005) and Roy and Yan (2020) which demonstrate the viability of the study of time series of agricultural crops. Landsat 8, which brings a better spectral resolution for the bands provided by the sensor onboard the satellite compared to the previous ones, has been used frequently for agricultural monitoring studies. We have some works that show the applicability of the images generated by Landsat 8, such as the work of Amaro (2019) and Santos (2015).

Geographical Information Science
ISBN 978-0-443-13605-4
https://doi.org/10.1016/B978-0-443-13605-4.22001-8

RS also brings the opportunity to generate products from the acquired images, such as indices related to the behavior of vegetation reflectance, for example, the Normalized Difference Vegetation IndexNDVI, which can help in the behavioral analysis of crop development.

Within the scope of agricultural monitoring, productivity estimation is widely and frequently studied in order to obtain an estimation model that is accurate and capable of generating data in advance (Sarmiento et al., 2020). Agrometeorological-spectral models are one of the most accurate ways to generate a reliable product, combining both the local meteorological characteristics and the biophysical characteristics of the plant. However, meteorological variables may make it difficult to apply these models, since Brazil does not have a dense network of meteorological stations to the point of obtaining reliable data for the entire earth's surface (Monteiro et al., 2018).

As an alternative, Silva-Fuzzo et al. (2019) proposed an adaptation of the Doorenbos and Kassam (1979) productivity estimation model by using evapotranspiration data from the simplified triangle method (Silva-Fuzzo & Rocha, 2016). This methodology was performed using thermal images from the MODIS sensor and vegetation index data from the same sensor. Using the simplified triangle method, it is possible to extract moisture and energy flux data through the relationship between surface temperature and fraction of vegetation cover (Silva-Fuzzo et al., 2019). From this, the data are plotted in a scatter plot and the result is points forming a triangle, and then it is possible to extract soil moisture and evapotranspiration data.

However, the major bottleneck in relation to this application is the spatial resolution of the MODIS sensor for surface temperature (1 km) that the authors used, as an alternative, Amaro (2019) who aimed to overcome the small scale monitoring deficiencies of MODIS, bringing the soil moisture estimate for a farm in the municipality of Caiuá in the state of São Paulo using thermal images from the Landsat 8 satellite. However, these data have not yet been tested for the estimation of productivity on small scales.

In this context, this work aims to apply the agrometeorological-spectral model using the simplified triangle method on a reduced spatial scale (local scale farm) to estimate soybean productivity in the 2019/2020 crop year for an area with a system of crop-livestock integration ILP.

2. Triangle model and agrometeorological estimation of agricultural productivity

The study of climate-productivity interactions is developed using models that seek to quantify the effects of climate variations on plant behavior (Robertson, 1983).

Agrometeorological models were the first to be developed and are widely used today. In addition to providing data to feed crop forecasting systems, they allow identifying water stress throughout the crop cycle, whose impacts on productivity can be detected and evaluated (Aparecido et al., 2017, Doorenbos & Kassam, 1979; Rudorff & Batista, 1990).

However, when using precipitation data in a station to infer the water balance of production, we cannot always represent this balance accurately, because when dealing with a localized unit for collecting precipitation, and this does not always portray the spatial scenario we have in the field. Alternatively, it is possible to use meteorological data from meteorological satellites or global atmospheric models.

The TRMM (Tropical Rainfall Measuring Mission) satellite is an alternative to conventional meteorological stations, being used in models such as Jie et al. (2016) and RS models such as Silva-Fuzzo and Rocha (2016).

Thinking about methods that use a more simplified form of data acquisition, we can also consider models such as Doorenbos and Kassam (1979), as proposed by Rao et al. (1988) who take into account the values of evapotranspiration present in each pixel to estimate the productivity of soybean cultivation at different stages of development. More broadly, we can infer that agrometeorological models consider only the influence of climatic factors on crop productivity, functioning as an efficiency meter. The methods used to establish the plant climate relationship range from simple correlation to complex models, that is, production functions that can consider different parameters involved in the production system (Picini et al., 1999).

However, a caveat should be made regarding agrometeorological models, also made by Frizzone et al. (2005), that their applications are often specific to a location, and often do not consider all the factors involved in the development of the crop with the dynamics of the climate and the associated natural resources.

To overcome this type of specificity, we can use agrometeorological-spectral models, where agrometeorological variables express the influence of meteorological conditions on plant growth and development. Spectral variables express, in addition to meteorological conditions, the influence of management practices, cultivars, spatial location, and other elements not included in the agrometeorological component, in determining the final yield (Pereira et al., 2009).

In an attempt to understand the dynamics between soil moisture (Mo) and energy atmosphere fluxes, Gillies and Carlson (1995) and Guilles et al. (1997) described a method that encompasses spectral bands (visible and thermal) to generate estimates of fractional vegetation, soil moisture, and evapotranspiration. The method can be illustrated in a scatter plot where the VI and Ts value of each pixel given a satellite image is related. The dispersion of pixels generates a characteristic shape of a trapezoid or triangle, where the right edge represents the pixels with the highest surface temperature, often identified as "dry edge" or "hot edge" while the left edge represents the "hot edge" "cold edge" or "wet edge," the variations on the abscissa axis represent the combined effect of relief and vegetation cover variations, which at these points have an exposed soil, unlike the top of the trapezoid which can be represented by vegetation cover pixels total.

Points on the scatter plot with the same IV value represent different cooling intensities caused by evapotranspiration, bearing in mind that once the plant is supplied with water, it will release water vapor for thermal control, since the plant suffers a tension of water it will close its stomata causing a minor cooling. The difference of a data dispersion in the form of a trapezoid instead of a triangle is given by changing the value of Ts to the maximum value of VI which indicates a change in the water content of the soil.

From the concepts of the triangle method, the vegetation and temperature values can be scaled in order to reduce the need for atmospheric correction and the dependence on environmental conditions, being able to isolate the cloud pixels Silva-Fuzzo et al. (2019).

3. Simplified triangle method

From this we can generate a new metric that represents a significant physical quantity directly linked to the agricultural production values and minimizes the need for atmospheric corrections, of extreme importance when working with temporal data series of a certain region, the generated value is defined with "Fr" or fractioned vegetation (Eq. 8.1).

$$Fr = \left\{ \frac{(NDVI - NDVIo)}{(NDVIS - NDVIo)} \right\}^2 \tag{8.1}$$

where: NDVIo is the NDVI value corresponding to bare soil and NDVIs is the NDVI value corresponding to maximum vegetation cover (Gillies & Carlson, 1995).

The Ts values are presented in the form of minimum temperature (Tmin) and maximum temperature (Tmax), radiant temperature scale (T*), which varies from 0 (Tmin), the temperature that belongs to a dense group of vegetation in well-watered soils at 1.0 (Tmax) temperature dry soil or bare soil. Thus, we define T (Eq. 8.2).

$$T* = \left\{ \frac{Tir - Tmin}{Tmax - Tmin} \right\} \tag{8.2}$$

where: Tir is the radiant surface temperature, Tmax is the Ts value for bare soil, and Tmin is the corresponding Ts value for maximum vegetation cover (Carlson, 2007). From the two scaled metrics, we can generate two values to compose the dispersion graph, soil moisture (Mo), and evapotranspiration (total EF) represented by Eqs. (8.3) and (8.4):

$$Mo = 1 - T*(pixel)/T*borda\ quente \tag{8.3}$$

$$EF\ total = EF\ solo*(1 - Fr) + Fr\ (pixel)*EF\ veg \tag{8.4}$$

EF veg is the vegetation value for potential evapotranspiration (assumed by the equation to be equal to 1), and T* is represented by the surface temperature along the hot edge. So EF soil = Mo and EF veg = 1.

4. Material and methods

4.1 Study of area

The study was carried out in an area of 200 ha on the Campina farm, located in the western region of the state of São Paulo, in the municipality of Caiuá (Fig. 8.1). The

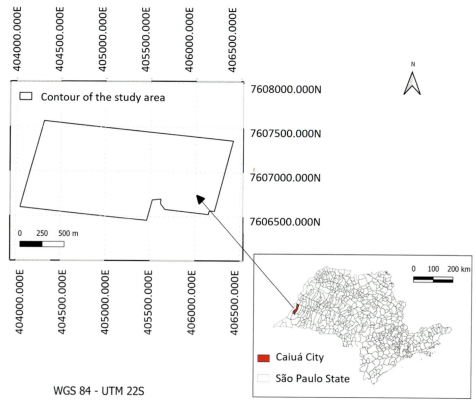

Figure 8.1 Location of the study area in the municipality of Caiuá -SP.

cultivation of soy in this area is part of a crop-livestock integration system, starting the second cycle with the planting of soy in December 2019 with harvest in April 2020 and was followed by the planting of pasture in a consortium system between grass brachiaria (Urochloa ruziziensis) and millet (Pennisetum glaucum).

According to Koppen and Geiger (1928), the region has a climate classification of Aw, rainy tropical with a cold and dry winter. Precipitation data were taken from an automatic conventional meteorological station installed close to the plots.

5. Obtaining productivity in the field and acquiring data

The productivity values were obtained by field collection during the harvest period in the study area where they were collected from the file extracted from the harvester after the completion of the operation in CSV (comma-separated values) format, then the data were converted to the GEOTIFF (Geostationary Earth Orbit Tagged Image File Format) format to facilitate data manipulation.

Images from the Landsat 8 satellite from the OLI (Operational Terra Imager) and TIRS (Thermal Infrared Sensor) sensors were used, with the atmosphere correction

Table 8.1 Date of the images used for the acquisition of total EF values and the respective stages of development for each date.

Data	Development phase
01/20	Full flourishing
03/08	Full granation
03/24	Maturation

and other processing within the Google Earth Engine GEE platform (Gorelick et al., 2017). The OLI images were used to calculate the NDVI vegetation index and the TIRS images to generate the triangle method. One image was selected per soybean phenological stage, however, due to the large amount of clouds in the region, it was not possible to use an image in the vegetative development stage of the crop (Table 8.1).

6. Application of the simplified triangle method

To apply the simplified triangle method, we calculated the NDVI and LST values with the normalization formula from bands 4 and 5 of the OLI sensor, as we can see in Eqs. (8.5) and (8.6).

$$\mathrm{NDVI} = \frac{\mathrm{B5} - \mathrm{B4}}{\mathrm{B5} + \mathrm{B4}} \tag{8.5}$$

where B4: reflectance values of band 4 (red); B5: reflectance values of band 5 (near infrared);

$$\mathrm{LST(^{\circ}C)} = \mathrm{LSTK} - 273.15 \tag{8.6}$$

where, LST: surface temperature in Celsius; LSTK: surface temperature in Kelvin: Next, we can convert the values of LST and NDVI to T* (Radiant temperature) and the fractional vegetation cover "Fr" from Eqs. (8.7) and (8.8).

$$\mathrm{Fr} = \left\{ \frac{(\mathrm{NDVI} - \mathrm{NDVIo})}{(\mathrm{NDVIs} - \mathrm{NDVIo})} \right\}^2 \tag{8.7}$$

where NDVI: NDVI values at each pixel; NDVIo: NDVI value for exposed soil; NDVIs: NDVI values for total land cover;

$$T^* = \left\{ \frac{(\mathrm{Tir} - \mathrm{Tmin})}{(\mathrm{Tmax} - \mathrm{Tmin})} \right\} \tag{8.8}$$

where Tir: radiant surface temperature; Tmin: Minimum surface temperature; Tmax: maximum surface temperature; Finally, we apply the simplified triangle method by geometrically analyzing the distribution graph of points by Eqs. (8.9) and (8.10), using the interpretations of the results described, which produces the availability of soil surface moisture (Mo) and total evapotranspiration (total EF), whose values will be used to adapt the equation made by Silva-Fuzzo and Rocha (2016).

$$Mo = 1 - T^*(pixel) / T^*borda\ quente \qquad (8.9)$$

swhere Mo: soil surface moisture; T^*: radiant temperature of each pixel; T^*hot edge: radiant temperature of the hot edge in the distribution of points in the triangle method.

$$EF\ total = EFsoil^*(1 - Fr) + Fr\ (pixel)^*EF\ veg \qquad (8.10)$$

where, EF total: total evapotranspiration: EF soil: soil evapotranspiration; EF veg: vegetation evapotranspiration;

According to the author, these formulas are based on the assumption that soil surface evaporation and soil moisture availability (Mo) are at a maximum (Mo = 1) along the cold edge of the pixel envelope and are zero (Mo = 0) along the hot edge. It is assumed that plant-only transpiration is always potential unless the leaves are withered. Transpiration and evaporation are combined depending on fractional vegetation cover to produce total evapotranspiration expressed as EF. At the end of applying the equations, we will have the total EF images for the mentioned dates, also in GEOTIFF matrix format for later analysis of the productivity estimate.

7. Productivity estimate

To estimate soybean productivity, we used the agrometeorological model by Doorenbos and Kassam (1979) (Eq. 8.11) which makes use of the ratio between the estimated productivity and the potential productivity of the study area, where we can consider the values collected in the field as the potential productivity values, we can also relate this to the total evapotranspiration values generated from the simplified triangle method (Eqs. 8.2 and 8.10) and with the differential of not being necessary to collect data from a fixed station from which the evapotranspiration values are taken in the model to be described by means of sensors installed in the study area, which shows the potential applicability of RS techniques that precisely dispense with the use of this type of equipment (Silva-Fuzzo and Rocha, 2016)

$$\frac{Ya}{Yp} = \prod_{i=1}^{4} \left[1 - kyi\left(1 - \frac{ETr}{ETp} \right)_i \right] \qquad (8.11)$$

where Ya: estimated productivity; Yp: potential productivity; kyi: penalty coefficient due to water deficiency in each stage of development; ETr: current evapotranspiration; ETp: potential evapotranspiration;

Next, we have the adjustment of the total evapotranspiration parameter generated by the simplified triangle method (Table 8.2).

Table 8.2 Yield adjustment factors for the different soybean development stages.

ky	Phenological stage
0.2	Vegetative development
0.4	Full flowering
0.8	Full development
0.2	Maturation/harvest

Source: Doorenbos and Kassam (1979) and Campbell et al. (1986).

$$\frac{Ya}{Yp} = \prod_{i=1}^{4} [1 - kyi(1 - EF)_i]$$ (8.12)

where Ya: estimated productivity; Yp: potential productivity; kyi: penalty coefficient due to water deficiency in each development phase; EF: total evapotranspiration from the simplified triangle method.

And for potential productivity (Yp), productivity data were acquired from the IEA (Institute of Agricultural Economics) agreement, where the average productivity was 2.5 ton/ha for the year 2017 (last available census). A factor of 10% was also added to the average Yp result to eliminate any environmental effect that could interfere with potential productivity (Silva-Fuzzo & Rocha, 2016). When consulting potential productivity values for soybeans in Caiuá in the 2016/2017 harvests (IBGE), we have a potential productivity of 2.5 ton/ha.

8. Results evaluation

For the statistical evaluations, the Willmott concordance index (d) (Eq. 8.13) and the root mean square error (RMSE) (Eq. 8.14) were used. The "d" index, proposed by Willmott et al. (1985) evaluates the fit of the model in relation to the observed data, indicating the degree of agreement or accuracy between the estimated and observed values, and the closer to 1, the better the accuracy of the model's performance in predicting the dependent variable.

$$d = 1 - \left([\Sigma(Y_e - Y)^2] \right) / \left(\Sigma(|Y_e - \bar{Y}| + |Y - \bar{Y}|)^2 \right)$$ (8.13)

where: Ye = estimated values; Y= observed values; $\bar{Y}$ = total average values. The RMSE is one of the most used error measures to assess the goodness of fit of a model, it represents the root mean squared error of the difference between the prediction and the actual value, being a measure analogous to the standard deviation.

$$RMSE = \sqrt{(1/n) \sum (Y - Y_e)^2}$$ (8.14)

Table 8.3 Classification of Pearson's correlation values.

p	Correlation
0—0.3	Negligible
0.3—0.5	Weak
0.5—0.7	Moderate
0.7—0.9	Strong
>0.9	Very strong

Source: Mukaka (2012).

where: n = number of observations; Ye = estimated values; Y = observed values. In addition to these coefficients, we also used the statistical coefficient R^2 (coefficient of determination) which, according to (Wang et al., 2023), R^2 tells us what percentage is eliminated from the forecast error in the variable when we use the least squares regression on the variable. We also used Pearson's correlation coefficient (r) to assess how much our prediction is related to the dispersion of the actual data, presented in Eq. (8.10).

According to Mukaka (2012), we have certain ranges of values that define the classification of the correlation (Table 8.3) and which will be used to indicate the statistical performance of the model in question and follow the general research flowchart (Fig. 8.2).

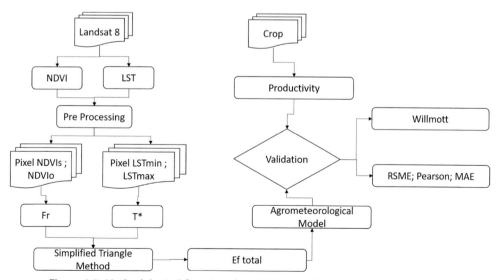

Figure 8.2 Methodological flowchart for carrying out the course completion work.

9. Results and discussion

9.1 Application of the simplified triangle method

In Fig. 8.3 we can observe the typical triangular profiles that are characteristic of the simplified triangle method that has visual analysis where the red lines indicate the hot (dry) edge of the triangle and the blue line indicates the cold (wet) edge. We can understand that each pixel of the images represents a value of humidity (Mo) and fractional evapotranspiration. In addition, the distribution of points illustrates (Fig. 8.3) the development phases described in Table 8.3, where for flowering dispersion we have a greater concentration of points close to the hot edge because we still do not have a fully developed plant, which can generate an increase in radiant temperature in the study area.

As for the phase of filling the pods with grains, we have the beginning of the reproductive activity phase with greater intensity, which explains the points close to the radiant temperature axis. Przezdziecki et al. (2023) Calculations were conducted over the forest area in Tuczno Poland. To calculate the temporal TVDI model 4 Landsat 8 OLI/TIRS scenes were used, and calculation was performed in Python 3 using open source packages. The average moisture conditions of each chosen scene were validated using field data, namely evapotranspiration determined from an eddy covariance.

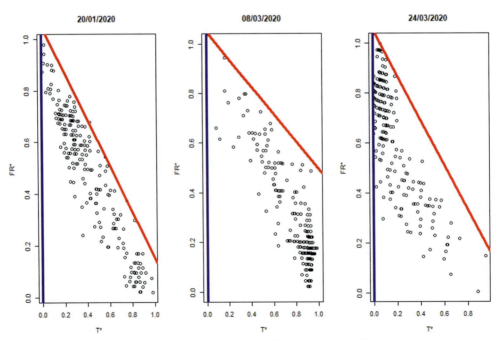

Figure 8.3 Scatter plot of vegetation fractionated by radiant temperature for the images on the dates of flowering, grain filling, and full maturation, respectively.

It is also clear the high vegetative level of soybeans in the full maturation stage, from the dispersion graph, we can observe the points close to the unit value of fractional vegetation. Even though it was not possible to obtain an image of the beginning of soybean vegetative development, a grouping close to the radiant temperature axis could be observed, due to the greater exposure of the soil to incident radiation from the atmosphere.

In March, when the beginning of maturation occurs, in fields 3 and 4 have a similar pattern, with higher NDVI values, different from fields 1 and 2, because of the difference between the selected soybean varieties (Fig. 8.4). In April, after the soybean harvest and subsequent pasture sowing, we verified the behavior of exposed soil in all fields by the predominance of pixels at the bottom of the graphs, which may vary due to the amount of dry material remaining from the soybean (Fig. 8.4).

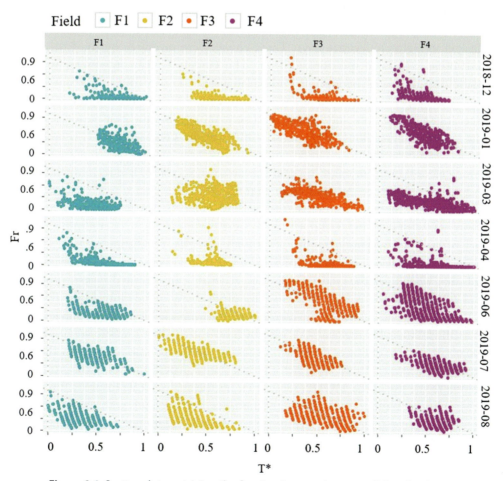

Figure 8.4 Scatter plot containing the fractional vegetation cover (Fr) and radiant.

10. Productivity estimate

It is worth mentioning that each of the dispersion data represents a development phase for soybeans, where due to the massive presence of rain clouds in the study area that fill all four plots in the months of December and January that represent the initial phase of development of soybean (ky1), the coefficients used for each dispersion data previously presented. When analyzing the productivity of the equation used by the author in the calculation of the productivity estimate, each term ky generates a coefficient of mitigation of the potential productivity of the crop, where, by not applying the coefficient at the beginning stage of soybean development, as well as the coefficient for full maturation that was applied in the work, we will have a greater penalty in the quality of the model since the ky coefficient is inversely related to the factor to be applied to potential productivity.

According to Correa et al. (2022), developed tests with different algorithms using a set of 991 spectral curves referring to a healthy soybean plant and under attack by pests, collected in eight consecutive days. These curves were measured by the EMBRAPA team, using a portable spectroradiometer. The Google Collabs interpreter was used and the algorithms were inscribed in Python language, using libraries such as Skit Sklearn. Among the algorithms used, there are random forest, decision tree, support vector machine, logistic regression, and extra-tree. The extra tree has the best performance (F1 score = 80.40%; precision = 81%; recall = 80%) in the proposed task. It is concluded that it is possible to process reflectance spectroscopy measurements with machine learning algorithms to monitor insect attack on soybean plants. It is recommended that the applied approach be tested in other cultures.

When applying the equation adapted by Silva-Fuzzo (2019) and using the proposed metrics, an R^2 of 0.59 and RMSE of 38.8% were obtained (Fig. 8.4), confirming consistency of the model explored by the author with a Pearson correlation of 0.79, a performance by the Willmott coefficient of 52% of the error associated with the productivity prediction when using the model in question, generating an estimated average productivity of 2.01 ton/ha, that is, close to the average productivity of the municipality (Fig. 8.5).

We can make an analysis of the estimate, considering the local scale of application of the method, unlike the regional scale used by Silva-Fuzzo (2016), where we first had the issue of the absence of one of the images necessary for the application of the method that without doubts was a determining factor for the statistical performance of the estimate, in addition to the use of the MODIS sensor that fits the analysis scale made by the author.

Moreira and Nuñez (2023) harmonized images, free of clouds with one before sowing, during plowing, and one image after harvesting. Random points were obtained for each of the six productivity classes and vegetation index values were assigned for each date and class. The data matrix was processed on the Google Collaboratory platform using the Random Forest classifier from the Scikit-learn package. Evaluating all parameters

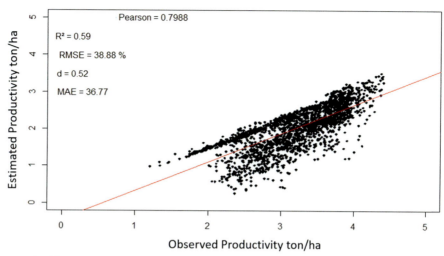

Figure 8.5 Yield estimation using agrometeorological-spectral model in relation to the observed data of the studied área.

allowed by Random Forest, the best score (0.6825) to estimate soybean productivity was obtained using the gini criteria, 85% of samples and 120 estimators, using all recurrent images of the harvest period 21/22 and images before sowing and after harvesting. Traditionally, coarse-resolution satellite data acquired at daily intervals have been used for monitoring. Recently, the harmonized Landsat 8 and Sentinel 2 (HLS) data increased the temporal frequency of the data. Here we investigated if the increased data frequency provided adequate observations to characterize highly dynamic grassland processes (Zhou et al., 2019).

According to Manabem et al. (2018), projects such as the "Integrated crop livestock forest Network" and the "Livestock Rally" have estimated the ICL areas for Brazil on a state or regional basis. However, it remains necessary to create methods for spatial identification of ICL areas. The classification of ICL areas occurred in three phases. Phase 1 corresponded to the classification of land use from 2008 to 2016. In Phase 2, the ICLa areas were identified. Finally, Phase 3 corresponded to the ICLm identification. The framework showed overall accuracies of 86% and 92% for ICL areas. ICLm accounted for 87% of the ICL areas. Considering only agricultural areas or only pasture areas, ICL systems represented 5% and 15%, respectively.

Still, with regard to the results obtained by the study on which this work is based, it was possible to generate values very close to those observed in the field, in addition to showing the statistical superiority in relation to the models using surface station data, where it is clear the potential use of the model as an easy to implement tool, in the absence of auxiliary data or surface information for monitoring and forecasting agricultural yields.

11. Conclusion

It was possible to consistently generate productivity predictions, while still acquiring evapotranspiration data in a simple way using the simplified triangle method. It is worth commenting on the performance of the model that depends on the values of EF to generate the estimate, we could have obtained an even higher performance considering soil reflectance conditions without the presence of clouds for the initial stage of soybean development that directly influences the productivity correction factor (kyi) and consequently the estimate generated by the model, which is evident in the difference between the estimated and observed productivity. A simple way to understand the management of ICLS areas, enabling the identification of the different managements in the fields, with crops or mixed-pasture in the area, throughout the months.

In general terms, we can conclude that the model adapted for small properties offers a plausible alternative for calibrating and estimating productivity in place of agrometeorological data from stations. We believe that more work like this should be developed in areas with changes in land use from agricultural areas and áreas with large-scale deforestation, associated with orbital images with good spectral resolution.

Acknowledgments

To the Research Productivity Scholarship Program (PQ/UEMG and the FAPEMIG Project) for granting scholarships to the second and third authors.

References

Amaro, R.P. (2019). *Soil moisture estimation using the simplified triangle method and satellite images* (Doctoral dissertation, [sn]). https://repositorio.unicamp.br/Busca/Download?codigoArquivo=552734.

Aparecido, L. E. O., Rolim, G. S., Lamparelli, R. A. C., Souza, P. S., & Santos, E. R. (2017). Agrometeorological models for forecasting coffee yield. *Agronomy Journal, 109*(1), 249–258.

Campbell, D. G., Daly, D. C., Prancec, G. T., & Maciel, U. N. (1986). *Quantitative ecological inventory of terra firme and várzea tropical forest on the Rio Xingu* (pp. 369–393). Brittonia: Brazilian Amazon.

Carlson, T. (2007). An overview of the "Triangle Method" for estimating surface evapotranspiration and soil moisture from satellite imagery. *Sensors, 7*, 1612–1629.

Correa, D. V., Ramos, A. P. M., Osco, L. P., & Jorge, L. A. C. (2022). Machine learning to identify soybean plants under insect attack using hyperspectral data. *Colloquium Exactarum, 14*, 146–153.

Doorenbos, J., & Kassam, A. H. (1979). Yield response to water. Rome, FAO. *(Irrigation and Drainage Paper, 33)*, 197.

EMBRAPA SOJA. (2020). Soja. Disponível em. Acesso em 22 abr https://www.embrapa.br/soja/cultivos/soja1.

Frizzone, J. A., Da Silva Cardoso, S., & Rezende, R. (2005). Productivity and quality of melon fruits grown in a protected environment with application of carbon dioxide and potassium via irrigation water. *Acta Scientiarum Agronomy, 27*(4), 707–717.

Gillies, R. R., & Carlson, T. N. (1995). Thermal remote sensing of surface soil water content with partial vegetation cover for incorporation into climate models. *Journal of Applied Meteorology, 34*, 745–756.

Gorelick, N., Hancher, M., Dixon, M., Ilyuschenko, S., Thau, D., & Moore, R. (2017). Google Earth Engine: Planetary-scale geospatial analysis for everyone. *Remote Sensing of Environment, 202*, 18–27.

Guilles, R. R., Krustas, W. P., & Humes, K. S. (1997). A verification of the 'triangle' method for obtaining surface soil water content and energy fluxes from remote measurements of the Normalized Difference Vegetation Index (NDVI) and surface. *International Journal of Remote Sensing, 18*(15), 3145—3166.

Jie, Z. H. E. N. G., Li, L. V., Wen-Lan, F. E. N. G., & Kun, T. U. (2016). Spatial downscaling simulation of monthly precipitation based on TRMM 3B43 data in the Western Sichuan Plateau. *Chinese Journal of Agrometeorology, 37*(02), 245.

Koppen, W., Geiger, R., & der Erde, Klimate (1928). *Gotha: Verlag justus perthes. Wall-Map 150 cm × 200 cm.*

Manabem, V., Melo, M., & Rocha, J. (2018). Framework for mapping integrated croplivestock systemsin Mato Grosso, Brazil. *Remote Sensing, 10*(9), 1322.

Monteiro, L. A., Sentelhas, P. C., & Pedra, G. U. (2018). Assessment of NASA/POWER satellite-based weather system for Brazilian conditions and its impact on sugarcane yield simulation. *International Journal of Climatology, 38*(3), 1571—1581.

Moreira, V. M., & Nuñez, D. N. C. (2023). Estimate of soybean crop productivity in the 2021/22 season: Vegetation indices and Machine Learning. *Brazilian Journal of Science, 2*(1), 7—15.

Mukaka, Mavuto M. (2012). A guide to appropriate use of correlation coefficient in medical research. *Malawi Medical Journal, 24*(3), 69—71.

Pereira, D. D. R., Yanagi, S. D. N. M., Mello, C. R. D., Silva, A. M. D., & Silva, L. A. D. (2009). Performance of reference evapotranspiration estimation methods for the Serra da Mantiqueira region MG. *Ciência Rural, 39*, 2488—2493.

Picini, A. G., Caamargo, M. B. P. D., Ortolani, A. A., Fazuoli, L. C., & Gallo, P. B. (1999). Desenvolvimento e teste de modelos agrometeorológicos para a estimativa de produtividade do cafeeiro. *Bragantia, 58*, 157—170.

Przezdziecki, K., Zawadzki, J. J., Urbaniak, M., Ziemblinska, K., & Miatkows, Z. (2023). Using temporal variability of land surface temperature and normalized vegetation index to estimate soil moisture condition on forest areas by means of remote sensing. *Ecological Indicators, 148*, 110088.

Rao, N. H., Sarma, P. B. S., & Chander, S. (1988). A simple dated water-production function for use in irrigated agriculture. *Agricultural Water Management, 13*, 25—32.

Robertson, G. W. (1983). Guidelines on crop-weather models. In *(World climate application programme, 50)* (p. 115). Geneve: World Meteorological Organization.

Roy, D. P., & Yan, L. (2020). Robust Landsat-based crop time series modelling. *Remote Sensing of Environment, 238*, 110810.

Rudorff, B. F. T., & Batista, G. T. (1990). Yield estimation of sugarcane based on agrometeorological-spectral models. *Remote Sensing of Environment, 33*(3), 183—192.

Sanches, I. D., Epiphanio, J. N., & Formaggio, A. R. (2005). Agricultural crops in multitemporal images from the Landsat satellite. *Agriculture in São Paulo, 52*(1), 83—96.

Santos, J. E. O., Nicolete, D. A. P., Filgueiras, R., Leda, V. C., & Zimback, C. R. L. (2015). Imagens do Landsat-8 no mapeamento de superfícies em área irrigada. *Irriga, 1*(2), 30—36.

Sarmiento, C. M., Coltri, P. P., Alves, M. D. C., & Carvalho, L. G. D. (2020). A spectral agrometeorological model for estimating soybean grain productivity in Mato Grosso, Brazil. *Engenharia Agrícola, 40*, 405—412.

Silva-Fuzzo, D. F., & Rocha, J. V. (2016). Simplified triangle method for estimating evaporative fraction over soybean crops. *Journal of Applied Remote Sensing, 10*(4), 046027.

Silva-Fuzzo, D. F., Carlson, T. N., Nextarios, N. K., & Petropoulos, G. P. (2019). Coupling remote sensing with a water balance model for soybean yield predictions over large areas. *Earth Science Informatics, 13*, 1—15.

Wang, D., Hou, B., Yan, T., Shen, T., Shen, C., & Peng, Z. (2023). New statistical learning perspective for design of a physically interpretable prototypical neural network for machine condition monitoring. *Mechanical Systems and Signal Processing, 188*, 110041.

Willmott, C. J., Ackleson, S. G., & Davis, J. J. (1985). Statistics for the evaluation and comparison of models. *Journal of Geography Research, 90*, 8995—9005.

Zhou, Q., Rover, J., Brown, J., Howard, D., Wu, Z., Gallant, A., Rundquist, B., & Burke, M. (2019). Monitoring landscape dynamics in central U.S. grasslands with Harmonized Landsat-8 and Sentinel-2 time series data. *Remote Sensing, 11*(3), 328.

CHAPTER 9

Desertification risk in agricultural areas under current and future climate conditions—a case study

Nektarios N. Kourgialas

Water Recourses-Irrigation & Environmental Geoinformatics Laboratory, Institute of Olive Tree, Subtropical Crops and Viticulture, Hellenic Agricultural Organization (ELGO Dimitra), Chania, Greece

1. Introduction

Land degradation can be considered as the "reduction or loss in the biological or economic productive capacity of the land caused by human activities and often magnified by the impacts of climate change." A typical land degradation type in arid, semiarid, and dry subhumid environments is desertification. In these ecosystems, water is the core limiting component of land uses. Also, the balance of these environmental systems is very sensitive and a slight modification in climate or in land use management practices can accelerate the desertification risk. Thus, desertification can been described as the highest environmental challenge of our time and climate change is making it worse (Kosmas et al., 1999, pp. 31–47; Morianou et al., 2021).

A number of research studies show that large regions of the Mediterranean area can be documented as prone to desertification (Drake & Vafeidis, 2004; Kourgialas, 2021a; Morianou et al., 2021; Reynolds et al., 2007; Sommer et al., 2011). Especially in agricultural Mediterranean areas, desertification is primarily a result of the intensity of climatic conditions and human activities (Geist & Lambin, 2004; Kepner et al., 2006; Symeonakis et al., 2016). Overgrazing, long-term conservation tillage and deforestation are three of the key reasons for agricultural land degradation. Latest studies have proposed that climate-water effects could significantly affect the agricultural production and increase the desertification risk for many agricultural lands (Kourgialas, 2021b; Weaver et al., 2009). Specifically, many climate models indicate rising temperatures in the Mediterranean, while precipitation is projected to decrease. These tendencies in combination with inherent low levels of organic matter (organic matter increases soil fertility and soil-water holding capacity) in many Mediterranean areas could pointedly affect the risk of desertification in agricultural areas of the Mediterranean leading to reallocation of agricultural land and/or the abandonment of agricultural parcels (Mickley et al., 2004; Zanis et al., 2009; Morianou et al., 2018; Kourgialas, 2021c).

During the past decades, several models and methods have been suggested to evaluate the desertification risk (Kosmas et al., 1999, pp. 31–47; Babaev & Kharin, 1999,

Geographical Information Science
ISBN 978-0-443-13605-4
https://doi.org/10.1016/B978-0-443-13605-4.00022-9

pp. 59—75; Desertlinks, 2004). Kosmas et al. (1999, pp. 31—47) for the Mediterranean region proposing the indicator-based approach Environmentally Sensitive Area Index (ESAI), as the most common approach to assess desertification risk. The environmental sensitivity area (ESA) is a flexible and relatively simple model system that uses and captures a wide variation of parameters. In this system, four key factors of climate, soil, vegetation, and land management are created and combined in a synthetic way to generate the ESAI (Kosmas et al., 1999, pp. 31—47). Up to now, the ESA methodology has been applied to some case studies in the Mediterranean region such as Italy, Spain, Algeria, Egypt, and Greece (Bakr et al., 2012; Basso et al., 2000; Boudjemline & Semar, 2018; Contador et al., 2009; Kosmas et al., 1999, pp. 31—47; Ladisa et al., 2012; Morianou et al., 2018, 2021; Salvati et al., 2015; Salvati and Zitti., 2009; Sorriso-Valvo, 2005; Symeonakis et al., 2016).

The majority of the aforementioned studies assess the desertification risk for large areas incorporating in their methodology (classification scheme) different types of land use without focusing on agricultural regions and the effects of future climatic conditions on them. Thus, this type of result may not be realistic and appropriate in the case of investigating the desertification risk at a specific sector such as the agricultural. The proposed study attempts to fill this gap incorporating in the assessment of agricultural desertification risk, at current and future climate conditions, only the specific characteristics of the studied rural areas. Specifically, in GIS environment, this study investigates and compares the agricultural sensitive areas to desertification under current and a future realistic climate scenario (the RCP 4.5 weak climate change mitigation scenario) using as an example the rural areas of the Mediterranean island of Crete.

2. Methodology

2.1 Study area

Crete is one of the most significant agricultural regions of the Mediterranean zone (Eurostat, 2020). The study area of Crete covers a total area of about 8.265 km^2. On the island of Crete, there are important agricultural lands mainly dominated by olive trees, vineyards, citrus trees, avocado trees, and greenhouse vegetation cultivations. The fact that the island of Crete includes important agricultural lands, which in many cases face a significant risk of abandonment either due to reduced water availability and low soil productivity or due to the negative effects of climate change, makes the study of desertification risk particularly important both under present conditions and in the conditions of the near to medium future. According to the Corine Land Cover maps (2018), the agricultural areas of the island cover 339,166 ha, from which 191,728 ha covering olive groves, 72,683 ha covering land principally occupied by agriculture with significant areas of natural vegetation, 37,940 ha are complex cultivation patterns, while 24,314 ha are vineyards (Fig. 9.1). Regarding soils, in the agricultural areas of Crete, fine soils

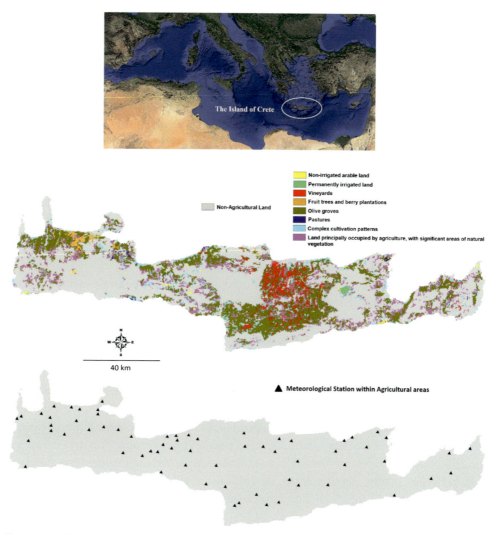

Figure 9.1 The agricultural areas on the island of Crete and the meteorological network within these areas.

(35% < clay <60%) and medium soils (18% < clay <35% and ≥15% sand, or 18% – < clay and 15% < sand) are dominant, while coarse soils (18% < clay and >65% sand) occur mainly in the western part of the island (Panagos et al., 2014).

The agricultural areas on the island of Crete located manly at lowlands (<200 m) include olive, citrus, avocado, vineyards, and vegetation cultivations as well as at semi-mountainous areas (200—800 m) covering mainly olive trees, vineyards, and some areas of chestnut trees. The climate of the agricultural areas in Crete is subhumid Mediterranean with humid and moderately cold winters, and dry and warm summers. The annual precipitation ranges from 300 to 700 mm/year in the lowlands and from 700 to 1000 mm/year in the semimountainous agricultural regions. Noteworthy rainfall variances are recorded between the western and eastern agricultural areas of the island during a year (Kourgialas et al., 2018). Due to the local climate conditions at the western part of the island (higher rainfall amounts than in the eastern part) olive, citrus, and avocado trees are grown. On the other hand, the agricultural parcels at the eastern part of the island are characterized by drier climate conditions and the main cultivations are olive trees and vineyards. In this study, the meteorological data from a meteorological network of 65 stations well distributed within the whole agricultural area of Crete were used. These stations can adequately cover the climatological conditions of the whole rural environment of the study area, a fact that is very crucial for the validity of the results concerning the aim of this study which is the estimation of desertification risk within agricultural areas (Fig. 9.1).

2.2 ESAI method

There are many different parameters that could affect the degree of desertification at a specific agricultural area (Kosmas et al., 1999, pp. 31−47). According to the ESAI methodology, 15 different parameters (layers/maps) are included in four main sensitivity to desertification risk groups which are: climate, soil, vegetation management, and land management. Analytically, the climate sensitivity group includes layers/maps related to rainfall, aridity, and slope aspect. The vegetation sensitivity group includes layers/maps related to fire risk, erosion protection, drought resistance, and plant cover. The soil sensitivity group includes layers/maps related to parent material, soil texture, soil depth, slope gradient, organic content, and water erosion. Finally, the management sensitivity group includes the land use intensity and protection policies layers/maps. The data for generating the above maps were collected from several sources, and their values were standardized between 1 (low sensitivity to desertification risk) and 2 (high sensitivity to desertification risk) (Table 9.1). The four desertification risk groups were assessed using the following equation (Eq. 9.1):

$$\text{Sensitivity}_x = \left(\text{layer}_1 \times \text{layer}_2 \times \text{layer}_3 \times \ldots \text{layer}_n\right)^{1/n} \tag{9.1}$$

where: Sensitivity_x denotes the computed value of each sensitivity and n represents the number of layers/maps used to compute each sensitivity group map.

The climate sensitivity group map was estimated from three climate indicators: rainfall, aridity, and slope aspect. Yearly precipitation was classified in three classes considering that annual rainfall of 300 mm is a critical threshold for plant growth (Sepehr et al., 2007). The aridity index was also incorporated in this study. The aridity index is the ratio of the annual rainfall divided by the annual potential evapotranspiration (P/PET) (Allen et al., 2005). Slope aspect was generated using a digital elevation model (EU-DEM v1.1) (Table 9.1). Also, in this study, the climate change scenario RCP 4.5

Table 9.1 Factors and layers to define ESAI—classification scheme/score and data sources.

Main groups	Parameter	Classes	Index	Data/source used
Climate sensitivity	Rainfall (mm)	>650	1	Data obtained from meteorological stations within the agricultural areas of the island
		300—650	1.5	
		<300	2	
	Aridity	>0.65	1	PET: FAO-56 Penman—Monteith equation. Metrological data from the studied meteorological stations
		0.5—0.65	1.5	
		<0.5	2	
	Aspect	N, NE, NW	1	Digital Elevation Model (EU-DEM v1.1) at 25 m resolution scale
		S, SE, SW	2	
Vegetation sensitivity	Drought resistance	Heterogeneous agricultural areas	1	Corine Land Cover (CLC) for the year 2018. Source: https://land.copernicus.eu/pan-european/corine-land-cover
		Olive tress	1.2	
		Orchards and vines, almonds	1.4	
		Perennial grasslands, pastures, and shrubland	1.7	
		Annual crops (annual grassland, cereals, and maize), horticulture, and very low vegetated	2	
	Erosion protection	Heterogeneous agricultural areas	1	
		Olives, perennial grasslands, pastures, shrubs	1.3	
		Agro-forestry areas	1.6	
		Almonds and orchards	1.8	
		Vines, horticulture, annual crops, very low vegetated	2	

Continued

Table 9.1 Factors and layers to define ESAI—classification scheme/score and data sources.—cont'd

Main groups	Parameter	Classes	Index	Data/source used
	Fire risk	Orchards, vines, olives, irrigated annual crops, and horticulture	1	
		Perennial and annual grasslands, pastures, cereals very low vegetated and shrublands	1.3	
		Mediterranean heterogeneous agricultural areas	1.6	
		Agro-forestry areas	2	
	Plant cover (%)	>40	1	
		10–40	1.8	
		<10	2	
Soil sensitivity	Parent material	Shale, schist, basic, ultrabasic, conglomerates, unconsolidated, clays, and marl	1	Geological maps from the Institute of Geology and Mineral Exploration (I.G.M.E.)
		Limestone, marble, granite, rhyolite, ignimbrite, gneiss, siltstone dolomite	1.7	
		Marl and pyroclastics	2	
	Texture	LSCL, SL, LS, and CL	1	European Soil Data Centre (ESDAC) http://esdac.jrc.ec. europa.eu/content/soilmap-greece-edafologikos-xartis-ellados
		SC, SiL, and SiCL	1.2	
		Si, C, and SiC	1.6	
		S	2	
	Soil depth (cm)	>75	1	
		30–75	1.3	
		15–30	1.6	
		<15	2	
	Soil organic matter (%)	<0.5	2	European Soil Data Centre (ESDAC) https://esdac.jrc.ec. europa.eu/content/octop-
		0.5–1	1.7	
		1–2	1.5	
		2–3	1.2	

Category	Factor	Class	Value	Source
		>3	1	topsoil-organic-carbon-content-europe
	Water erosion (tn/ha/yr)	<1	1	G2 model produced by Panagos et al., 2014; https://esdac.jrc.ec.europa.eu/content/g2-soil-erosion-model-data-crete-greece-and-strymonas-greecebulgaria-ishmi-erzeni-albania
		1–5	1.3	
		5–20	1.7	
		>2	2	
	Slope gradient	<6	1	Digital Elevation Model (EU-DEM v1.1) at 25 m resolution scale.
		6–18	1.2	
		18–35	1.5	
		>35	2	
Management sensitivity	Intensity of land use	Land principally occupied by agriculture, with significant areas of natural vegetation	1	Corine Land Cover (CLC) 2018. Source: https://land.copernicus.eu/pan-european/corine-land-cover
		Irrigated and nonirrigated vineyards, fruit trees, olive groves, nonirrigated arable land, etc	1.5	
		Greenhouses, open field herbaceous, horticultures with spring-summer-autumn cycle	2	
	Protection policies	Water supply areas, parks, and archeological areas	1	Sites of Community Importance and Special Protected Areas—EU Directive 92/43. Sources: http://www.ypeka.gr/?tabid=504
		Woodlands, seminatural/agricultural areas at coastal areas	1.5	
		Agricultural areas not subject to restrictions	2	

After Sepehr et al. (2007), Morianou et al. (2021).

was combined into the rainfall and aridity-related layers/maps of climate sensitivity group in order to examine their effects on the final desertification risk map. Regarding the vegetation sensitivity group map, land use cover is a very critical factor affecting desert-ification risk. So, desertification is associated to the presence of natural/agricultural vege-tation and can be reflected in relation to fire risk and ability to recover, erosion protection offered to the soil, drought resistance, and percentage plant cover. The soil sensitivity group map was created combining the soil data maps of texture, slope, parent material, soil depth, organic matter, and water soil erosion. Finally, the combination of intensity of land use as well as the protection policies maps produces the final management sensitivity group map. More information about the classification processes and the data sources of the above-mentioned parameters can be found in Table 9.1.

2.3 Climate change scenario

Regarding the studied climate change scenario, RCP 4.5 is in accordance with the lowest-emission scenarios (B1) proposed in the IPCC AR4. It is an equilibrium scenario where total radiative forcing is stabilized around 2050 by employment of a range of stra-tegies and technologies for reducing greenhouse gas emissions. This scenario can be char-acterized as a weak climate change mitigation scenario. The MiniCAM modeling team at the Pacific Northwest National Laboratory's Joint Global Change Research Institute (JGCRI) developed the RCP 4.5 (Moss et al., 2010).

For the process of spatial downscaling that can improve the resolution of climate data and therefore produce a better representation of the climate conditions, the data of regional climate models (RCMs) that were implemented in the present study were pro-duced by the RCA4_ MOHC (Met Office Hadley Centre) developed by the Swedish Meteorological and Hydrological Institute (SMHI) driven by the Global Circulation Model (GCM) named "Hadley Centre Global Environmental Model, version 2 Earth System" (Collins et al., 2013; Strandberg et al., 2015). This dataset (RCA4_MOHC) was selected since, according to LIFE ADAPT2CLIMA (2017), it was found to be one of the datasets that adequately reproduced the climatic conditions for the island of Crete. RCA4_MOHC meteorological data were downscaled from 0.11 degrees to 100 m using a bilinear approach (Marke et al., 2011; Herrmann et al., 2016). The period used in the context of the present study is 50 years (2025–70), which is suggested as demonstrative of the near to medium future period.

2.4 Final desertification sensitivity map

According to Sepehr et al. (2007) after the combination of the aforementioned specific parameters of each of the four sensitivity group maps, the quantitative results of each sensitivity group are classified into low, moderate, and high quality classes. Specifically, according to the classification suggested by Sepehr et al. (2007): Climate sensitivity group map was divided into three sensitivity classes: low with value equal to 1, moderate with range of value equal to 1.1—1.5, high with range of value equal to 1.5—2. Vegetation sensitivity group map was divided into three sensitivity classes: low with range of value lower that 1.13, moderate with range of value equal to 1.13—1.38, high with range of value greater that 1.38. Soil sensitivity group map was divided into three sensitivity classes: low with range of value lower that 1.13, moderate with range of value equal to 1.13—1.45, high with range of value greater than 1.45. Finally, regarding management sensitivity group map, it was divided into three sensitivity classes: low with range of value between 1 and 1.3, moderate with range of value equal to 1.3—1.5, high with range of value greater than 1.5.

The final sensitivity to desertification map for the agricultural areas is assessed in a linear way grouping of the above-mentioned four sensitivity group of maps Eq. (9.2):

$$ESAI = (Sensitivity_1 \times Sensitivity_2 \times Sensitivity_3 \times \ldots Sensitivity_n)^{1/n} \qquad (9.2)$$

The ESAI assigns equal weights to layer n, within each sensitivity group map, as well as equal weights to each sensitivity group map when computing the overall ESAI map.

According to Kosmas et al. (1999, pp. 31—47), four general types of environmentally sensitive areas (ESAs) to desertification can be noted as: *Not-Affected* [These areas depict values of the final sensitivity to desertification map lower than 1.17], *Potential* [This type includes regions that may potentially be at risk of desertification. These areas depict values of the final sensitivity to desertification map between 1.17 and 1.22], *Fragile* [This type includes regions in which any change in the balance of natural and human agro-activity is likely to lead to desertification. Fragile class can be divided into three subclasses (F1 with values of the final sensitivity to desertification map between 1.23 and 1.26, F2 with values between 1.26 and 1.32, and F3 with values from 1.33 to 1.37)], and *Critical* [This type includes regions that already are in a high level of degradation, also presenting a threat to the environment of the surrounding areas. Critical class can be divided into three subclasses (C1 includes values of the final sensitivity to desertification map between 1.38 and 1.41, C2 with values between 1.42 and 1.53, and C3 with values above 1.53)].

3. Results and discussion

The results for the four sensitivity groups/maps are depicted in Fig. 9.2—9.5. Concerning the climate sensitivity map at present conditions (Fig. 9.2a), the greater part of the agricultural areas in Crete was noted to have low to moderate climatic sensitivity. However, there are some agricultural areas in the southern and western Crete as well as extented agricultural parts of the eastern island that were depicted as high sensitivity (Fig. 9.2a). This fact could be in line that eastern agricultural areas obtain a smaller amount of precipitation than the western part of the island (Kourgialas et al., 2018). Fig. 9.2b depicts the climate sensitivity map under the average mitigation climate scenario RCP 4.5. In terms of climate, under climate scenation RCP 4.5, extented agricultural regions especially in

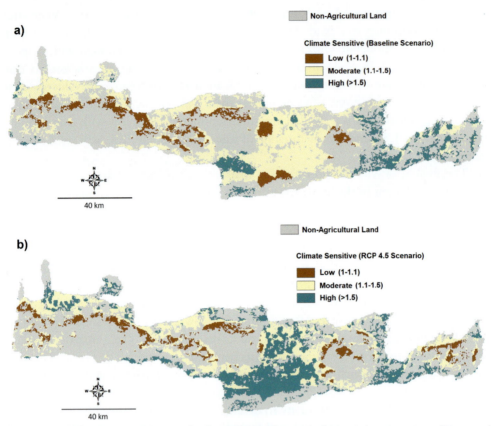

Figure 9.2 (a) Climate sensitivity map for the agricultural areas in Crete under current conditions and (b) climate sensitivity map under the scenario RCP4.5.

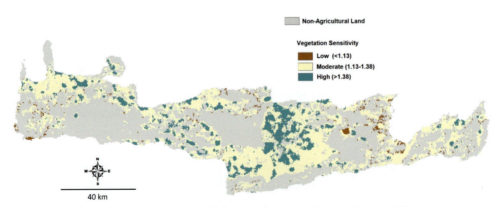

Figure 9.3 Vegetation sensitivity map for the agricultural areas of Crete.

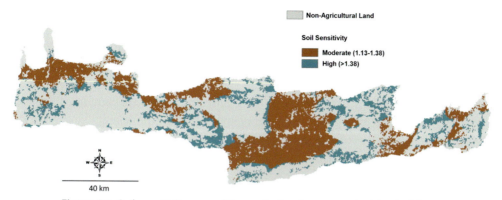

Figure 9.4 Soil sensitivity map of the agricultural areas on the island of Crete.

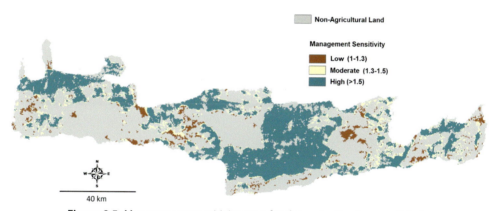

Figure 9.5 Management sensitivity map for the agricultural areas of Crete.

the central part of the island appear to enhance their sensitivity to desertification compared to the current situation. The higher sensitivity of many agricultural areas that was noted under the climate scenario RCP 4.5 could be considered reasonable as the RCM used, under the scenario RCP 4.5, gives in average trend, at specific agricultural areas, less rain compared to current conditions for the period 2025—70.

Concerning the vegetation sensitivity map, Fig. 9.3 indicates that the largest part of the agricultural areas in Crete is illustrated by moderate vegetation sensitivity except for some inland agricultural areas, which are noted to be in high vegetation sensitivity class. Most of these areas are principally occupied by agriculture with significant areas of natural vegetation which are subject to high erosion risk due to overgrazing.

Fig. 9.4 depicts the results of soil sensitivity map for the agricultural areas in Crete. The majority of the agricultural soils in Crete has moderate soil sensitivity. Areas with the most sensitive soils are found in sloppy areas and in areas with fine soil texture and soil depth less than 30 cm. In addition, the most sensitive soils exist in agricultural areas with poor soil organic content.

In terms of management sensitivity, Fig. 9.5 shows that almost the whole of agricultural areas of the island and the major part of the agricultural sector in the central part of Crete belong to high sensitive class. In these areas, human agricultural activity is more intense. In addition, many of the low sensitivity agricultural parts are considered environmentally protected or managed by environmental protection policies.

The final map of desertification risk for the agricultural areas in Crete, using the ESA approach, under current climate conditions is shown in Fig. 9.6a. According to the findings, the substantial majority of the agricultural areas on the island of Crete is considered as critical or fragile for desertification. Specifically, according to Table 9.2 in the case of baseline scenario, the most environmentally sensitive agricultural areas (critical areas C1, C2, C3) in Crete are located mainly in the central-southern and eastern part of the island. Analytically, 57% of the total agricultural area in Crete is considered under critical

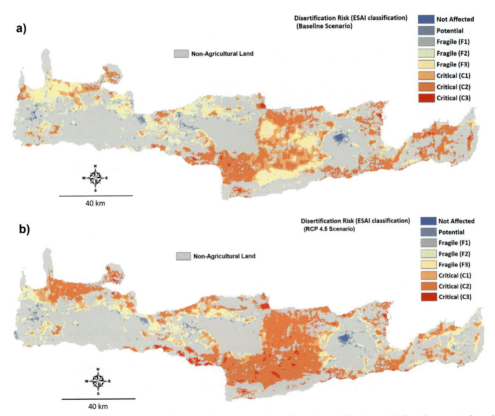

Figure 9.6 Map of environmentally sensitive areas related to desertification risk for the agricultural areas of Crete under (a) current conditions and (b) climate scenario RCP 4.5.

Table 9.2 Environmental sensitivity to desertification agricultural areas under Baseline and RC 4.5 scenarios using ESAI classification.

ESAI classification	Agricultural areas sensitivity to desertification (ha)	Agricultural areas sensitivity to desertification (ha)
	Baseline scenario	*RC4.5 scenario*
Not affected	1554	1501
Potential	4559	4898
Fragile (F1)	9430	8231
Fragile (F2)	39,993	28,224
Fragile (F3)	89,449	48,191
Critical (C1)	72,919	52,994
Critical (C2)	117,641	182,968
Critical (C3)	3621	12,159

sensitivity risk, 40% under fragile sensitivity risk, and only 2% can be characterized as potential or not affected. Rural areas located in Critical C3 zone are estimated to be 3621 ha, which corresponds to a small proportion of 1.1% of the total agricultural land. In Critical C3 zone, the dominated cultivation that is under desertification risk is the olive grove with a percentage of 67%. Moreover, the rural areas located in the Critical C2 zone are 11,7641 ha, reaching a significant percentage of the total agricultural area of the island equal to 34.7%. Regarding the Critical C2 zone, the most sensitive cultivations are olive groves and vineyards occupying the percentages of 55% and 18% of the total area of this zone, respectively.

Regarding the RCP 4.5 climate scenario, the environmental sensitivity to desertification map is depicted in Fig. 9.6b. This figure illustrates that the critical to desertification agricultural areas are significantly extended compared to the baseline scenario. Specifically, under RCP 4.5 climate scenario, 73% of the total rural area in Crete is considered under critical sensitivity risk, 25% under fragile sensitivity risk, and only 1.8% can be characterized as potential or not affected. In addition, based on Table 9.2, the rural areas located in Critical C3 zone are estimated to be 12,159 ha, which corresponds to an increasing proportion compared to the baseline scenario reaching 3.6% of the total agricultural land. In Critical C3 zone, the dominated cultivation that is under desertification risk is olive grove with a percentage of 45% followed by heterogeneous agricultural areas including land principally occupied by agriculture with significant areas of natural vegetation of 30% and complex cultivation patterns with a percentage of 18%. The rural areas located in the Critical C2 zone are 182,968 ha, reaching a high percentage of the total agricultural area on the island equal to 53.9%. Regarding the Critical C2 zone, the most sensitive cultivation areas are olive groves and heterogeneous agricultural areas occupying the percentages of 59% and 15.8% of the total area of this zone, respectively.

The findings of employing the RCP 4.5 climate change scenario suggest that in the near to moderate future, a greater extent of the agricultural regions of Crete will be categorized as critical to desertification in comparison to fragile and potential types. Specifically, under the RCP 4.5 (average mitigation scenario), the critical areas will increase by 16% compared to baseline scenario. There is a local variability of desertification risk between baseline and RCP 4.5 climate scenario, a fact that is closely related to the potential increase or decrease of precipitation at local rural areas in the near to medium future.

These outcomes should raise the awareness of local authorities taking measures to prevent and mitigate desertification risk in terms of policy management and protection of the land. Additionally, the generated ESAI maps could be used as a valued decision-making tool for the local authorities and farmers for taking the appropriate mitigation measures at areas prone to desertification risk. These measures could include soil and crop management improvement approaches and/or projects for investigating, especially at C1 and C2 subclasses of critical areas, the possibility of switching crops to more adapted to drought climate ones such as the carob tree.

4. Conclusions

The present chapter presents a GIS-based ESAI method for the estimation of the desertification risk in the agricultural areas under current and future climate conditions, using as an example the island of Crete. Servicing the aim of this study, the results can be considered more realistic and useful than other similar studies which incorporate in their methodology (classification scheme) different types of land use without focusing on the agricultural land uses. Summarizing the findings, for the current climate conditions, a substantial part of the agricultural areas on the island of Crete is considered as critical or fragile sensitive to desertification. Potential causes could be intense human agro-activities, improper agronomic practices related to soil and crop management, overgrazing and deforestation, water shortage, and the inherent low quality of soils in some agricultural areas (for instance, low levels of organic matter). Generally, the agricultural areas at the eastern and southern parts of Crete seem to be more sensitive to desertification primarily due to the dry climate conditions which seem to increase in the near to medium future due to the increasing temperature and decreasing precipitation. In line to the above, the proposed ESAI methodology especially for the realistic RC4.5 climate change scenario (period 2025–70) suggests that the desertification risk in agricultural areas will greatly increase, a fact that demands attention and actions to be implemented by policy makers or authorities at national/prefecture level and individual farmers or groups of farmers at river basin and parcel level.

References

Allen, R. G., Pereira, L. S., Smith, M., Raes, D., & Wright, J. L. (2005). FAO-56 dual crop coefficient method for estimating evaporation from soil and application extensions. *Journal of Irrigation and Drainage Engineering, 131*(1), 2—13.

Babaev, A., & Kharin, N. (1999). *The monitoring and forecast of desertification processes. Desert problems and desertification in Central Asia* (pp. 59—75). Springer.

Bakr, N., Weindorf, D. C., Bahnassy, M. H., & El-Badawi, M. M. (2012). Multi-temporal assessment of land sensitivity to desertification in a fragile agro-ecosystem: Environmental indicators. *Ecological Indicators, 15*(1), 271—280.

Basso, F., Bove, E., Dumontet, S., Ferrara, A., Pisante, M., Quaranta, G., & Taberner, M. (2000). Evaluating environmental sensitivity at the basin scale through the use of geographic information systems and remotely sensed data: An example covering the Agri basin (Southern Italy). *Catena, 40*(1), 19—35.

Boudjemline, F., & Semar, A. (2018). Assessment and mapping of desertification sensitivity with MEDALUS model and GIS—case study: Basin of Hodna, Algeria. *Journal of Water and Land Development, 36*(1), 17—26.

Collins, M., Knutti, R., Arblaster, J., Dufresne, J.-L., Fichefet, T., Friedlingstein, P., Gao, X., Gutowski, W. J., Johns, T., & Krinner, G. (2013). Long-term climate change: Projections, commitments and irreversibility. In *Climate change 2013-the physical science basis: Contribution of working group I to the fifth assessment report of the intergovernmental panel on climate change* (pp. 1029—1136). Cambridge University Press.

Contador, J. L., Schnabel, S., Gutierrez, A. G., & Fernandez, M. P. (2009). Mapping sensitivity to land degradation in Extremadura. SW Spain. *Land Degradation and Development, 20*(2), 129—144.

Desertlinks. (2004). *Desertification Indicator System for Mediterranean Europe*. Available from: http://www.desire-his.eu/en/assessment-with-indicators/related-sitesthematicmenu-277/204-desertifi.

Drake, N., & Vafeidis, A. (2004). A review of European Union funded research into the monitoring and mapping of Mediterranean desertification. *Advances in Environmental Monitoring and Modelling, 1*(4), 1—51.

Eurostat. (2020). *Agriculture, forestry and fishery statistics—2020 edition*. https://ec.europa.eu/eurostat/en/web/products-statistical-books/-/ks-fk-20-001. (Accessed 15 March 2022).

Geist, H. J., & Lambin, E. F. (2004). Dynamic causal patterns of desertification. *BioScience, 54*(9), 817—829.

Herrmann, F., Kunkel, R., Ostermann, U., Vereecken, H., & Wendland, F. (2016). Projected impact of climate change on irrigation needs and groundwater resources in the metropolitan area of Hamburg (Germany). *Environmental Earth Sciences, 75*(14), 1104.

Kepner, W. G., Rubio, J. L., Mouat, D. A., & Pedrazzini, F. (2006). Desertification in the Mediterranean region. In *A security issue: Proceedings of the NATO Mediterranean dialogue workshop, held in Valencia, Spain, 2—5 December 2003*. Springer Science and Business Media.

Kosmas, C., Ferrara, A., Briasouli, H., & Imeson, A. (1999). *Methodology for mapping environmentally sensitive areas (ESAs) to desertification. The Medalus project Mediterranean desertification and land use. Manual on key indicators of desertification and mapping environmentally sensitive areas to desertification* (pp. 31—47).

Kourgialas, N. N., Anyfanti, I., Karatzas, G. P., & Dokou, Z. (2018). An integrated method for assessing drought prone areas—Water efficiency practices for a climate resilient Mediterranean agriculture. *Science of the Total Environment, 625*, 1290—1300.

Kourgialas, N. N. (2021b). Could advances in geoinformatics, irrigation management and climate adaptive agronomic practices ensure the sustainability of water supply in agriculture? *Water Supply, 21*(6), v—vii.

Kourgialas, N. N. (2021c). A critical review of water resources in Greece: The key role of agricultural adaptation to climate-water effects. *Science of the Total Environment, 775*, 145857.

Kourgialas, N. N. (2021a). Hydroclimatic impact on Mediterranean tree crops area—Mapping hydrological extremes (drought/flood) prone parcels. *Journal of Hydrology, 596*, 125684.

Ladisa, G., Todorovic, M., & Liuzzi, G. T. (2012). A GIS-based approach for desertification risk assessment in Apulia region, SE Italy. *Physics and Chemistry of the Earth, Parts A/B/C, 49*, 103—113.

LIFE ADAPT2CLIMA. (2017). *Deliverable C3: Future Projections on Climatic Indices with Particular Relevance to Agriculture for the Three Islands (Coarse Resolution) and for Each Agricultural Pilot Area (Fine Resolution) Project ADAPT2CLIMA LIFE14 CCA/GR/000928.* Available from: http://adapt2clima.eu/uploads/2017/ADAPT2CLIMA_DEL_C.3_Final_3.pdf. (Accessed 20 May 2023).

Marke, T., Mauser, W., Pfeiffer, A., & Zängl, G. (2011). A pragmatic approach for the downscaling and bias correction of regional climate simulations: Evaluation in hydrological modeling. *Geoscientific Model Development, 4*(3), 759.

Mickley, L. J., Jacob, D. J., Field, B., & Rind, D. (2004). Effects of future climate change on regional air pollution episodes in the United States. *Geophysical Research Letters, 31*(24).

Morianou, G., Kourgialas, N. N., Pisinaras, V., Psarras, G., & Arambatzis, G. (2021). Assessing desertification sensitivity map under climate change and agricultural practices scenarios: The island of Crete case study. *Water Supply, 21*(6), 2916−2934. https://doi.org/10.2166/ws.2021.132

Morianou, G., Kourgialas, N., Psarras, G., & Koubouris, G. (2018). Mapping sensitivity to desertification in Crete (Greece), the risk for agricultural areas. *Journal of Water and Climate Change, 9*(4), 691−702.

Moss, R. H., Edmonds, J. A., Hibbard, K. A., Manning, M. R., Rose, S. K., Van Vuuren, D. P., Carter, T. R., Emori, S., Kainuma, M., & Kram, T. (2010). The next generation of scenarios for climate change research and assessment. *Nature, 463*(7282), 747−756.

Panagos, P., Christos, K., Cristiano, B., & Ioannis, G. (2014). Seasonal monitoring of soil erosion at regional scale: An application of the G2 model in Crete focusing on agricultural land uses. *International Journal of Applied Earth Observation and Geoinformation, 27*, 147−155.

Reynolds, J. F., Smith, D. M. S., Lambin, E. F., Turner, B., Mortimore, M., Batterbury, S. P., Downing, T. E., Dowlatabadi, H., Fernández, R. J., & Herrick, J. E. (2007). Global desertification: Building a science for dryland development. *Science, 316*(5826), 847−851.

Salvati, L., Ferrara, C., & Corona, P. (2015). Indirect validation of the environmental sensitive area index using soil degradation indicators: A country-scale approach. *Ecological Indicators, 57*, 360−365.

Salvati, L., & Zitti, M. (2009). Convergence or divergence in desertification risk? Scale-Based assessment and policy implications in a Mediterranean country. *Journal of Environmental Planning and Management, 52*(7), 957−971.

Sepehr, A., Hassanli, A., Ekhtesasi, M., & Jamali, J. (2007). Quantitative assessment of desertification in south of Iran using MEDALUS method. *Environmental Monitoring and Assessment, 134*(1−3), 243.

Sommer, S., Zucca, C., Grainger, A., Cherlet, M., Zougmore, R., Sokona, Y., Hill, J., Della, Peruta R., Roehrig, J., & Wang, G. (2011). Application of indicator systems for monitoring and assessment of desertification from national to global scales. *Land Degradation and Development, 22*(2), 184−197.

Sorriso-Valvo, M. (2005). To desertification: an application to the Calabrian territory (Italy). *Geomorphological Processes and Human Impacts in River Basins, 299*, 23.

Strandberg, G., Bärring, L., Hansson, U., Jansson, C., Jones, C., Kjellström, E., Kupiainen, M., Nikulin, G., Samuelsson, P., & Ullerstig, A. (2015). *CORDEX scenarios for Europe from the Rossby Centre regional climate model RCA4.* SMHI.

Symeonakis, E., Karathanasis, N., Koukoulas, S., & Panagopoulos, G. (2016). Monitoring sensitivity to land degradation and desertification with the environmentally sensitive area index: The case of lesvos island. *Land Degradation and Development, 27*(6), 1562−1573.

Weaver, C., Liang, X.-Z., Zhu, J., Adams, P., Amar, P., Avise, J., Caughey, M., Chen, J., Cohen, R., & Cooter, E. (2009). A preliminary synthesis of modeled climate change impacts on US regional ozone concentrations. *Bulletin of the American Meteorological Society, 90*(12), 1843−1864.

Zanis, P., Kapsomenakis, I., Philandras, C., Douvis, K., Nikolakis, D., Kanellopoulou, E., Zerefos, C., & Repapis, C. (2009). Analysis of an ensemble of present day and future regional climate simulations for Greece. *International Journal of Climatology: A Journal of the Royal Meteorological Society, 29*(11), 1614−1633.

CHAPTER 10

An evaluation of SMAP soil moisture product using in situ data and Google Earth Engine: A case study from Greece

Spyridon E. Detsikas, Triantafyllia Petsini and George P. Petropoulos
Department of Geography, Harokopio University of Athens, Athens, Greece

1. Introduction

Surface soil moisture (SSM), the relative water content of the upper layer of the soil (Adab et al., 2020), is an important parameter that influences the interactions of the earth–atmosphere system (Srivastava et al., 2013; Vogel et al., 2017). SSM plays a key role in Earth's natural processes regulating global biogeochemical cycles and the global energy balance (Bao et al., 2018; Carlson et al., 2019; Deng, Lamine, Pavlides, Petropoulos, Bao, et al., 2019). Given the importance of SSM, knowledge of its dynamics is essential for understanding a variety of environmental and socioeconomic processes. Such insights can be useful for sustainable water resource management, early warning of natural hazards such as floods and droughts, monitoring natural ecosystems (Choi et al., 2012; Deng, Lamine, Pavlides, Petropoulos, Bao, et al., 2019; Piles et al., 2016), and the impact assessment on vegetation vitality (Fuzzo et al., 2019; Shi et al., 2014). It is also applicable to a variety of hydrological (Suman et al., 2020) and weather forecasting models (Panegrossi et al., 2011) at both regional and global scales. Moreover, the study of SSM at the national wide level can be exploited as a decision-making tool (Deng, Lamine, Pavlides, Petropoulos, Srivastava, et al., 2019). Accurate information on SSM can help farmers in crop yield by making better management in irrigation practices, especially in arid and semiarid regions (Adeyemi et al., 2017). Thus, it is possible to address food and water security challenges that are at risk due to climate change at a global level. Today, proper irrigation management is crucial as most of the water is consumed for irrigation.

Traditionally, soil moisture measurement can be carried out by ground instrumentation (Petropoulos et al., 2013). There are many advantages of in situ instruments for measuring soil moisture distribution. These include easy installation, maintenance and operation, instrument portability, the possibility of measuring at different depths, and relatively direct measurement. Today, soil moisture distribution is considered quite challenging in large-scale areas using traditional methods because of the heterogeneity of landscapes (Gruber et al., 2020). The large spatial and temporal variability of SSM can explain this challenge as soil moisture depends on parameters which are likely to vary

Geographical Information Science
ISBN 978-0-443-13605-4
https://doi.org/10.1016/B978-0-443-13605-4.00018-7

spatially such as soil structure, precipitation, land cover, and topography (Bao et al., 2018; Gupta et al., 2021). As a result, for large-scale areas, point measurement of soil moisture is not useful as the installation of ground networks is quite time-consuming and costly (Liu et al., 2022). However, at local scale, there are some SSM observation networks that have a database of point measurements, such as the International Soil Moisture Network (ISMN) (Dorigo et al., 2021).

In the recent decades, there has been rapid technological development in the domain of Earth Observation (EO). A number of techniques has been developed to estimate SSM over areas of large and diverse geographic scales (Petropoulos, Srivastava, Ferentinos, & Hristopoulos, 2018). The techniques used to retrieve SSM involve optical, thermal, and microwave satellite sensors (Petropoulos, Ireland, & Barrett, 2015). Optical and thermal sensors operate in visible/infrared parts of the electromagnetic radiation spectrum (EMR). Microwave sensors included two main categories, active and passive, are promising because of their direct data acquisition, day, and night due to satellites' revisit period and their low sensitivity to cloud and atmospheric conditions. As for the operation of active sensors, they have better spatial resolution than passive sensors. However, vegetation cover, surface roughness, and topography are some of the factors that affect the retrieval of SSM from active sensors (Li et al., 2021). In contrast, passive sensors have higher accuracy in SSM retrievals due to higher temporal resolution and thus are more suitable for SSM monitoring.

The use of EO technology in SSM retrievals has reached a very satisfactory level of maturity, as evidenced and by the large number of relevant operational products currently available (e.g., see review by Petropoulos, Srivastava, Piles, & Pearson, 2018). Those products are derived from spaceborne systems such as the Advanced Scatterometer with MetOp-A and MetOp-B (C-band scatterometer), the Advanced Microwave Scanning Radiometer2 (AMSR-2) (Bindlish et al., 2018), Windsat/Cariolos, and the FengYun-3C (FY-3C) (Zhang et al., 2020). Two of the most important missions for space-based SSM retrievals are the European Space Agency's (ESA) Soil Moisture and Ocean Salinity (SMOS) mission, and NASA's Soil Moisture Active and Passive (SMAP) mission (Colliander et al., 2021; Kerr et al., 2012; Petropoulos et al., 2014).

Given the large number of satellite operational SM product, it is quite important to carry out studies on the accuracy of the operational products as it achieves data reliability control and improves operational retrieval algorithms (Gruber et al., 2020). In parallel, with the increased volume of the existing operational SSM products, new challenges have been emerged in the context of accessing and processing those datasets. Cloud-based platforms such as the Google Earth Engine (GEE) can help to manage, process, and visualize satellite data more efficiently reducing the volume of data for the user's convenience (Sazib et al., 2018; Greifeneder et al., 2021). SSM datasets can be efficiently processed in terms of cost and time by utilizing cloud computing platforms creating new applications and opportunities for their use.

Several studies have been conducted to evaluate the accuracy of those operational SSM products in different places worldwide with SMAP SM datasets attracting a large of portion of them. Yet, notably studies focusing specifically on soil moisture retrieval from the SMAP satellite are scarce for arid/semiarid regions such as in the Mediterranean environment. For example, Suman et al. (2020) investigated the performance of the SMAP L4 SM product at selected experimental sites for four different continents. The authors reported a high performance of the SMAP product for all the experimental sites included in their study. In another study, Gupta et al. (2021) followed a different way of estimating soil moisture by incorporating soil temperature (ST) using SMAP data for different seasonal periods. In their work, accuracy of the product was obtained by applying linear regression model. At a second time, multiple linear regression was also applied using ST as a third variable to achieve the best accuracy of SM retrieval. A more recent study involving the Tibetan Plateau examined the SMAP L3_36 km and the 9 km product compared to ground data from the Naqu observation network (Deng et al., 2022). The accuracy of those products was equally high with the only difference being that in this study the results are highly dependent on the presence of precipitation which is common in the area and affects the variability of the SSM.

In light of the above, in the current study, the GEE Python Application Programming Interface is utilized to evaluate the SMAP 9 km product across various sites in Central Greece, representing a typical Mediterranean semiarid environment. To achieve this goal, coorbital ground measurements are used together with the SMAP satellite data from year 2020. As part of the agreement examination, the influence of factors that affect the agreement between the compared datasets is examined.

2. Experimental setup

2.1 Study area and in situ measurements

The study area is in Eastern Central Greece, in the prefecture of Boeotia and specifically 110 km northwest of Athens. The current plain has a total area of 215 km^2 and was created by the drainage of Lake Kopaida at the end of the 19th century. The geomorphology of the Kopaida basin has been shaped by the process of karstification, that is, the dissolving action of water on the readily soluble limestone rocks of the area. It is characterized by a flat terrain interrupted by several hills located on the margins of the plain. The area where it is located today is the end of the Boeotian Kifissos River and is a highly arable area with the main crops being cultivated there are wheat, cotton, industrial tomatoes, maize, pulses, and vegetables.

In the present study, an observation network of stations established by Neuropublic (https://www.neuropublic.gr/) in Greece was used for the selection of experimental sites. This network includes field measurements of soil moisture and temperature taken by the Drill and Drop (DnD) instrument at a depth of 5 cm. For the selection of experimental sites, specific criteria based on other studies (Deng, Lamine, Pavlides, Petropoulos, Bao, et al., 2019; Gupta et al., 2021) were considered. These criteria are diversity of crop types, homogeneity of stations within 1 km radius to provide diversity in the study area, large distance of one station from another so that no two stations are in the same pixel and the reliability of results is greater, and dispersion of sites throughout the study area. The area consists of five monitoring stations under two different annual crop types. The geographical location of the selected sites is shown in Fig. 10.1. A summary of the validation sites is described in Table 10.1, where the first column shows the ID of each station, while the second column shows the crop type for each station. The third and fourth columns show the planting and harvesting dates for each crop.

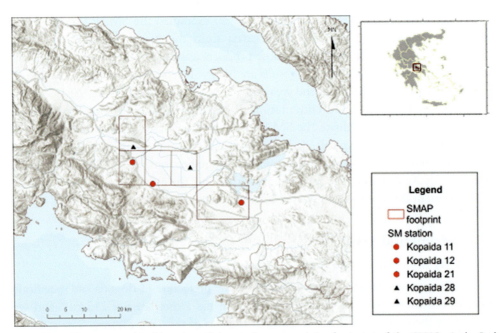

Figure 10.1 The geographical location of the study area and the footprint of the SMAP pixels. *Red* dots represent cotton crops, while the *black* triangle symbol represents tomato crops.

Table 10.1 A description of Neuropublic's SSM monitoring stations utilized in this study area.

Point ID	Crop type	Start date	End date
Kopaida 11	Cotton	June 19, 2020	September 10, 2020
Kopaida 12	Cotton	June 19, 2020	September 10, 2020
Kopaida 21	Cotton	June 19, 2020	September 10, 2020
Kopaida 28	Tomato	June 18, 2020	August 31, 2020
Kopaida 29	Tomato	June 18, 2020	August 31, 2020

2.2 SMAP level 3 SM product (9 km)

The SMAP mission was launched in January 2015 having initial a $\sim$ 36 km spatial that was latter reduced to 9 km, and two−three days temporal resolution (Chan et al., 2018). The SMAP mission has the capability to measure liquid water in the top layer of the soil and can also measure the amount of water between rock materials, organic particles, and minerals if the soil is not covered by water or ice. In this study, the SMAP Level 3 (SPL3SMAP) product was used to retrieve soil moisture estimates using GEE Python API. The SMAP Level 3 (L3) soil moisture product is retrieved from the NSIDC (National Snow and Ice Data Center) (https://nsidc.org/data/spl3smap/versions/3). This product has horizontal (H) and vertical (V) brightness temperatures observed at an incidence angle θ 40 degree (O'Neill et al., 2020). The instrument used is an L–band radiometer of the SMAP active passive system that is designed to measure daily the amount of water in the surface soil on Earth representing the topsoil layer of 0−5 cm. SMAP Level 3 product provides brightness temperature gridded data in both ascending (06:00 p.m. local time) and descending (06:00 a.m. local time) orbits. Together with the SSM predictions are also provided quality assessment flags (O'Neill et al., 2020).

3. Methodology

3.1 Data preprocessing

In SSM validation studies, several preprocessing steps need to be performed on the different datasets prior to direct comparison between the different datasets. These preprocessing steps include filtering of both the spaceborne, and reference datasets based on data quality flags, rescaling of both datasets to the same units, and temporal matching of the datasets. The same principles were therefore applied to our study.

Only the SMAP SSM Soil Moisture product was used to filter the data based on the quality flags of the datasets used. The version of the ground truth datasets used did not have a quality flag. In our case, based on Ayres et al. (2021) and Li et al. (2022), filtering rules are provided by « retrieval_qual_flag_am/pm», where the flag equals "0," the measurement was accepted while when it equals with "1," the measurement was discarded.

Following the implementation of the data quality flags, the in situ data were rescaled to the same units (m^3/m^3) as the SMAP data by a multiplication of 0.01. Finally, the two datasets were time–matched to produce two separate datasets for the 0600- and 1800-hour measurements. The measured soil moisture values were then correlated with the corresponding values from the satellite measurements based on the satellite orbit and retrieval time.

3.2 Statistical assessment

For each collocated data set of in situ and satellite data, descriptive statistics were measured for a more accurate distribution of the data. A series of statistical metrics are used to explore the agreement between ground measurements and the SMAP satellite retrievals based on

Table 10.2 Statistical metrics used in evaluating the accuracy between ground measurements and retrieved SMAP satellite data. "N" represents the in situ observations, "P" represents the "predicted" values, and "O" represents the "observed" values. Subscripts i = 1. The horizontal line represents the mean value.

Name	Description	Mathematical equation
MAE	Mean absolute error	$\text{MAE} = N^{-1} \sum_{i=1}^{N} \lvert P_i - O_i \rvert^2$
R	Correlation coefficient	$R = \dfrac{\sum_{i=1}^{N}(P_i - \overline{P})(O_i - \overline{O})}{\sqrt{\sum_{i=1}^{N}(O_i - \overline{O})^2 (P_i - \overline{O})^2}}$
Bias/MBE	Bias (or mean bias error)	$\text{bias} = \text{MBE} = \frac{1}{N}\sum_{i=1}^{N}(P_i - O_i)$
RMSD	Root mean square difference	$\text{RMSD} = \sqrt{\text{bias}^2 + \text{scatter}^2}$
ubRMSD	Unbiased root mean square difference	$\text{ubRMSD} = \sqrt{\text{RMSD}^2 - \text{bias}^2}$

global validation protocol (Gruber et al., 2020). These statistical metrics are sample size (N), mean absolute error (MAE), correlation coefficient (R), bias (MBE), scatter (MSD), root mean square difference (RMSD), unbiased root mean square difference (ubRMSD), and slope and intercept of the major axis regression linear fit. Their mathematical equations are described in Table 10.2. Further details on each metric are described in Burt et al. (1996), Legates et al. (1999), Silk (1979), and Willmoot (1981). Statistical metrics previously utilized in published studies are used to calculate the comparisons between predicted and measured measurements (Deng et al., 2022; Entekhabi et al., 2010; Gupta et al., 2021; Petropoulos, Ireland, & Srivastava, 2015; Suman et al., 2020).

4. Results

The SMAP L3 data were compared with the corresponding field measurements to evaluate the assessment of the SMAP satellite product. In the presented results, the performance of the model across different datasets and time points is evaluated using various metrics. The main results are presented in Table 10.3 and Fig. 10.2. For each site, a relatively good agreement between the two data sets is identified during both ascending and descending time for all days of comparison. This is evidenced by the RMSD value which ranges for the ascending orbit (06:00 a.m.) between 0.126–0.221 m^3/m^3 and for the descending orbit 18:00 p.m. between 0.120–0.216 m^3/m^3. The highest accuracy is found in the cotton crop for both the morning (MAE = 0.102 m^3/m^3, R = 0.305, MBE = −0.092 m^3/m^3, RMSD = 0.126 m^3/m^3, ubRMSE = 0.087 m^3/m^3) and for the afternoon satellite passtime (MAE = 0.088 m^3/m^3, R = −0.115,

Table 10.3 Results of comparison between SMAP L3 SM and ground data both ascending and descending orbit.

Time	Dataset	N	MAE	R	Bias	RMSD	ubRMSD	Slope	Intercept
0600 a.m.	Kopaida 11	38	0.160	0.467	−0.046	0.179	0.173	0.132	0.151
	Kopaida 12	38	0.102	0.305	−0.092	0.126	0.087	0.174	0.139
	Kopaida 21	38	0.101	−0.237	−0.06	0.136	0.122	−0.11	0.208
	Kopaida 28	33	0.125	0.204	−0.099	0.155	0.119	0.094	0.163
	Kopaida 29	33	0.211	−0.017	−0.21	0.221	0.068	−0.022	0.203
1800 p.m.	Kopaida 11	39	0.170	0.235	−0.042	0.192	0.188	0.057	0.182
	Kopaida 12	37	0.106	0.202	−0.091	0.129	0.091	0.088	0.164
	Kopaida 21	38	0.088	−0.115	−0.057	0.12	0.106	−0.030	0.183
	Kopaida 28	31	0.110	0.250	−0.083	0.137	0.109	0.100	0.166
	Kopaida 29	30	0.209	0.037	−0.209	0.216	0.053	0.030	0.179

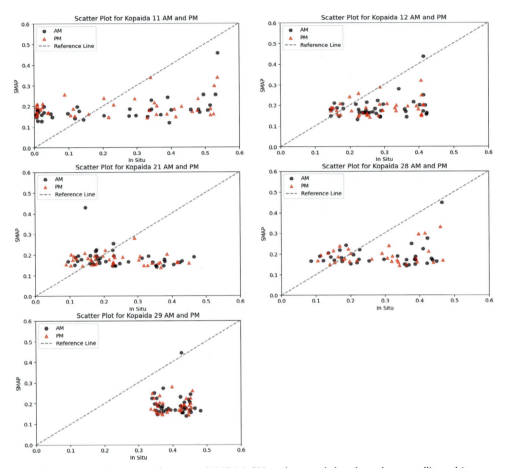

Figure 10.2 Agreement between SMAP L3 SM and ground data based on satellite orbit.

MBE $= -0.057$ m^3/m^3, RMSD $= 0.120$ m^3/m^3, ubRMSD $= 0.106$ m^3/m^3). At the same time, in both analyses, low agreement is obtained for the tomato crop RMSD $= 0.221/0.216$ m^3/m^3, respectively. In overall analysis, a slight underestimation of the satellite data by the field data is observed.

The noticeable agreement between the two datasets during the ascending orbit at 06:00 a.m. and descending orbit at 18:00 p.m. is worth mentioning with the main results being presented in Table 10.4 and Fig. 10.3. This is supported by the RMSD values, which range from 0.126 to 0.221 m^3/m^3 and 0.120 to 0.216 m^3/m^3, respectively, for each crop type. The analysis confirms a solid correlation between soil moisture estimates obtained from both satellite and field measurements. Notably, the cotton crop displays the highest precision, particularly in the morning (MAE $= 0.102$ m^3/m^3, R $= 0.305$,

Table 10.4 Agreement between SMAP L3 SM and ground data both ascending and descending orbit.

	MAE	R	Bias	RMSE	ubRMSE	Slope	Intercept
Tomato (PM)	0.125	0.158	−0.106	0.152	0.152	0.054	0.171
Cotton (PM)	0.170	0.235	−0.042	0.192	0.192	0.057	0.182
Cotton (AM)	0.160	0.467	−0.046	0.179	0.179	0.132	0.151
Tomato (AM)	0.133	0.104	−0.112	0.162	0.162	0.049	0.173

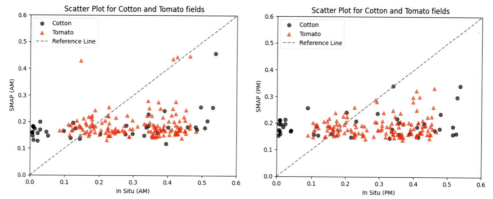

Figure 10.3 SMAP L3 SSM agreement for both the a.m. (*left*) and p.m. (*right*) overpasses for both the tomato and cotton fields.

MBE $= -0.092 \text{ m}^3/\text{m}^3$,   RMSD $= 0.126 \text{ m}^3/\text{m}^3$,   and   ubRMSD $= 0.087 \text{ m}^3/\text{m}^3$). Simultaneously, during the afternoon satellite transit period, the cotton harvest demonstrated exceptional precision (MAE $= 0.088 \text{ m}^3/\text{m}^3$, R $= -0.115$, MBE $= -0.057 \text{ m}^3/\text{m}^3$, RMSD $= 0.120 \text{ m}^3/\text{m}^3$, and ubRMSD $= 0.106 \text{ m}^3/\text{m}^3$). Conversely, both studies highlight reduced accuracy for the tomato crop, denoted by increased RMSD (0.221 and 0.216 m^3/m^3) during the morning and afternoon orbits correspondingly.

5. Discussion

The present's study main objective was to evaluate the soil moisture product SMAP L3 SM using ground data from a network of sites in Kopaida, a typical Mediterranean environment for the summer of 2020. This is an area with intensive agricultural activity and increased water needs due to the crops it produces. Overall, the study gave a reasonable agreement of the product, with the cotton crop showing the best accuracy. However, in all analysis, it is observed that an underestimation of the satellite data by the ground data. However, there are stations where the product showed a low performance. This is likely to be due to a variety of reasons.

One of them is errors caused by the measurement accuracy of the sensors. While the ground truth data are quite useful for estimating satellite products, they are characterized by errors due to problems in the way they are represented and their instrumentation. These problems can be the way and depth of instrument installation, the method of measurement, calibration techniques, and lack of instrument maintenance (Dorigo et al., 2013). The validation of satellite products may depend largely on these errors. Regarding the penetration depth of the instrument, this differs from the field data. Specifically, the L-band of the SMAP has a depth of 3—7 cm and it varies based on SSM, while the in situ sensors are placed at a depth of about 5 cm. This means that the in situ instrumentation is deeper than the L-band (Wang et al., 2021). This reason is likely to account for the bias observed in the results (Tables 10.3 and 10.4).

A possible reason for the underestimation of satellite data compared to ground data and at the same time the dry bias is that it is mainly observed during the summer months as has been shown in other studies (Kang et al., 2016; Oozeer et al., 2020). However, it is not clearly demonstrated by the scientific community that there is a systematic underestimation or overestimation of soil moisture data.

The sample size and spatial distribution of measurements may also explain the low accuracy performance at specific stations (Table 10.3). This is since these crops are classified as annual crops and thus only one growing season is investigated over a three-month period (June—September). The agreement improvement of the satellite product could be achieved by improving point measurements with techniques such as the use of dense networks (Srivastava et al., 2017).

In addition, factors such as topographic complexity and vegetation density play a crucial role in the accuracy of product retrieval (El Hajj et al., 2018; Zhang et al., 2017). The type of land use significantly determines the rate of agreement. In this case, cotton gives higher accuracy than tomato because the microwave signal of the L-band penetrates more easily compared to the canopy of tomato. Thus, RMSD shows a low value, and the R is better compared to the tomato crop. The water content of the vegetation has an equally important role in the accuracy of product recovery as it is affected by microwave radiation. Also, microwave signatures are affected by topographical features (Engman and Chauhan, 1995). However, it should be noted that measurements in the morning hours were slightly better compared to the afternoon hours. Studies show that in the summer months and during the daytime, the canopy layer is more visible compared to the nighttime (Chen et al., 2013).

In summary, the study shows that the performance of the product can be influenced by a variety of factors such as the in situ instrumentation, the depth of penetration of satellite and in situ data, and the type of land cover highlighting the water content of the vegetation and the phenological changes occurring in them. Thus, the results of the present study have a slightly lower agreement compared to the results of other studies (Deng et al., 2022; Gupta et al., 2021; Suman et al., 2020).

6. Conclusions

In conclusion, the comparative assessment of SMAP Level 3 data with field measurements offers valuable insights into the performance and limitations of the satellite product in estimating soil moisture content across agricultural landscapes. The observed congruence between satellite and field data, especially in cotton crops, underscores the reliability of the SMAP satellite product. However, the crop-specific variations highlight the importance of tailoring satellite-based estimations to the unique characteristics of different crops.

The identified underestimation of satellite data relative to field measurements suggests a need for further refinement in calibration or correction methodologies to enhance accuracy. Addressing these inherent biases is crucial for ensuring the reliability of satellite-derived soil moisture estimates in practical applications, such as precision agriculture and water resource management.

Moving forward, future research should focus on refining satellite algorithms to account for crop-specific nuances and improve overall accuracy. Additionally, ongoing efforts to validate and calibrate satellite data against ground measurements will contribute to advancing the reliability of environmental monitoring tools. The findings presented herein contribute to the ongoing dialog on the utilization and enhancement of satellite products, emphasizing the need for nuanced approaches in capturing the complexity of soil moisture dynamics in diverse agricultural settings. Results obtained from this study could yield valuable insights into optimizing irrigation given the high level of agricultural activity in the Kopaida region and the significant importance of water resource management.

Acknowledgments

The authors wish to thank the support of Neuropublic S.A. for providing the in situ measurements. The participation of Spyridon E. Detsikas and George P. Petropoulos was financially supported by the LISTEN-EO project. LISTEN-EO is implemented in the framework of H.F.R.I call "Basic research Financing (Horizontal support of all Sciences)" under the National Recovery and Resilience Plan "Greece 2.0" funded by the European Union—NextGenerationEU (H.F.R.I. Project Number: 15898).

References

Adab, H., Morbidelli, R., Saltalippi, C., Moradian, M., & Ghalhari, G.a. F. (2020). Machine learning to estimate surface soil moisture from remote sensing data. *Water, 12*(11), 3223. https://doi.org/10.3390/w12113223

Adeyemi, O., Grove, I. G., Peets, S., & Norton, T. (2017). Advanced monitoring and management systems for improving sustainability in precision irrigation. *Sustainability, 9*(3), 353. https://doi.org/10.3390/su9030353

Ayres, E., Colliander, A., Cosh, M. H., Roberti, J. A., Simkin, S., & Genazzio, M. A. (2021). Validation of SMAP soil moisture at terrestrial National Ecological Observatory Network (NEON) sites show potential for soil moisture retrieval in forested areas. *IEEE Journal of Selected Topics in Applied Earth Observations and Remote Sensing, 14*, 10903–10918. https://doi.org/10.1109/jstars.2021.3121206

Bao, Y., Lin, L., Wu, S., Deng, K.a. K., & Petropoulos, G. P. (2018). Surface soil moisture retrievals over partially vegetated areas from the synergy of Sentinel-1 and Landsat 8 data using a modified water-cloud model. *International Journal of Applied Earth Observation and Geoinformation, 72*, 76–85. https://doi.org/10.1016/j.jag.2018.05.026

Bindlish, R., Cosh, M. H., Jackson, T. J., Koike, T., Fujii, H., Chan, S. K., Asanuma, J., Berg, A., Bosch, D. D., Caldwell, T. G., Collins, C. H., McNairn, H., Martínez-Fernández, J., Prueger, J. H., Rowlandson, T., Seyfried, M. S., Starks, P. J., Thibeault, M., Van Der Velde, R., ... Coopersmith, E. J. (2018). GCOM-W AMSR2 Soil moisture product validation using core validation sites. *IEEE Journal of Selected Topics in Applied Earth Observations and Remote Sensing, 11*(1), 209–219. https://doi.org/10.1109/jstars.2017.2754293

Burt, J. M., Barber, G. M., & Rigby, D. L. (1996). *Elementary statistics for geographers.* http://ci.nii.ac.jp/ncid/BA06664153.

Carlson, T. N., & Petropoulos, G. P. (2019). A new method for estimating of evapotranspiration and surface soil moisture from optical and thermal infrared measurements: The simplified triangle. *International Journal of Remote Sensing, 40*(20), 7716–7729. https://doi.org/10.1080/01431161.2019.1601288

Chan, S., Bindlish, R., O'Neill, P. E., Jackson, T. J., Njoku, E. G., Dunbar, S., Chaubell, J., Piepmeier, J., Yueh, S., Entekhabi, D., Colliander, A., Chen, F., Cosh, M. H., Caldwell, T. G., Walker, J. P., Berg, A., McNairn, H., Thibeault, M., Martínez-Fernández, J., ... Kerr, Y. H. (2018). Development and assessment of the SMAP enhanced passive soil moisture product. *Remote Sensing of Environment, 204*, 931–941. https://doi.org/10.1016/j.rse.2017.08.025

Chen, Y., Yang, K., Qin, J., Zhao, L., Tang, W., & Han, M. (2013). Evaluation of AMSR-E retrievals and GLDAS simulations against observations of a soil moisture network on the central Tibetan Plateau. *Journal of Geophysical Research Atmospheres, 118*, 4466–4475.

Choi, M., & Hur, Y. (2012). A microwave-optical/infrared disaggregation for improving spatial representation of soil moisture using AMSR-E and MODIS products. *Remote Sensing of Environment, 124*, 259–269. https://doi.org/10.1016/j.rse.2012.05.009

Colliander, A., Reichle, R. H., Crow, W. T., Cosh, M. H., Chen, F., Chan, S., Das, N. N., Bindlish, R., Chaubell, J., Kim, S., Liu, Q., O'Neill, P., Dunbar, S., Dang, L., Kimball, J. S., Jackson, T. J., Jassar, H. K. A., Asanuma, J., Bhattacharya, B. K., ... Yueh, S. (2021). Validation of soil moisture data products from the NASA SMAP mission. *IEEE Journal of Selected Topics in Applied Earth Observations and Remote Sensing, 15*, 364–392. https://doi.org/10.1109/jstars.2021.3124743

Deng, K.a. K., Lamine, S., Pavlides, A., Petropoulos, G. P., Bao, Y., Srivastava, P. K., & Guan, Y. (2019a). Large scale operational soil moisture mapping from passive MW radiometry: SMOS product evaluation in Europe and USA. *International Journal of Applied Earth Observation and Geoinformation, 80*, 206–217. https://doi.org/10.1016/j.jag.2019.04.015

Deng, K.a. K., Lamine, S., Pavlides, A., Petropoulos, G. P., Srivastava, P. K., Bao, Y., Hristopulos, D. T., & Anagnostopoulos, V. (2019b). Operational soil moisture from ASCAT in support of water resources management. *Remote Sensing, 11*(5), 579. https://doi.org/10.3390/rs11050579

Deng, K.a. K., Petropoulos, G. P., Bao, Y., Pavlides, A., Chaibou, A.a. S., & Habtemicheal, B. A. (2022). An examination of the SMAP operational soil moisture products accuracy at the Tibetan Plateau. *Remote Sensing, 14*(24), 6255. https://doi.org/10.3390/rs14246255

Dorigo, W. A., Xaver, A., Vreugdenhil, M., Gruber, A., Hegyiová, A., Sanchis-Dufau, A. D., Zamojski, D., et al. (2013). Global automated quality control of in situ soil moisture data from the international soil moisture network. *Vadose Zone Journal, 12*(3), 0. https://doi.org/10.2136/vzj2012.0097

Dorigo, W., Himmelbauer, I., Aberer, D., Schremmer, L., Petrakovic, I., Zappa, L., Preimesberger, W., Xaver, A., Annor, F. O., Ardö, J., Baldocchi, D. D., Bitelli, M., Blöschl, G., Bogena, H., Brocca, L., Calvet, J., Camarero, J. J., Capello, G., Choi, M., ... Sabia, R. (2021). The international soil moisture network: Serving earth system science for over a decade. *Hydrology and Earth System Sciences, 25*(11), 5749–5804. https://doi.org/10.5194/hess-25-5749-2021

El Hajj, M., Baghdadi, N., Zribi, M., Rodríguez-Fernández, N., Wigneron, J. P., Al-Yaari, A., Al Bitar, A., Albergel, C., & Calvet, J.-C. (2018). Evaluation of SMOS, SMAP, ASCAT and Sentinel-1 soil moisture products at sites in southwestern France. *Remote Sensing, 10*, 569.

Engman, E. T., & Chauhan, N. (1995). Status of microwave soil moisture measurements with remote sensing. *Remote Sensing of Environment, 51*, 189–198.

Entekhabi, D., Njoku, E. G., O'Neill, P. E., Kellogg, K., Crow, W. T., Edelstein, W., Entin, J., Goodman, S. E., Jackson, T. J., Johnson, J. T., Kimball, J. S., Piepmeier, J. R., Koster, R. D., Martin, M. A., McDonald, K. C., Moghaddam, M., Moran, S., Reichle, R. H., Shi, J., … Van Zyl, J. (2010). The soil moisture active passive (SMAP) mission. *Proceedings of the IEEE, 98*(5), 704–716. https://doi.org/10.1109/jproc.2010.2043918

Fuzzo, D. F. S., Carlson, T. N., Kourgialas, N. N., & Petropoulos, G. P. (2019). Coupling remote sensing with a water balance model for soybean yield predictions over large areas. *Earth Science Informatics, 13*(2), 345–359. https://doi.org/10.1007/s12145-019-00424-w

Greifeneder, F., Notarnicola, C., & Wagner, W. (2021). A machine learning-based approach for surface soil moisture estimations with Google earth engine. *Remote Sensing, 13*(11), 2099. https://doi.org/10.3390/rs13112099

Gruber, A. D., De Lannoy, G., Albergel, C., Al-Yaari, A., Brocca, L., Calvet, J., Colliander, A., Cosh, M. H., Crow, W. T., Dorigo, W., Draper, C. S., Hirschi, M., Kerr, Y., Konings, A. G., Lahoz, W., McColl, K. A., Montzka, C., Muñoz-Sabater, J., Peng, J., … Wagner, W. (2020). Validation practices for satellite soil moisture retrievals: What are (the) errors? *Remote Sensing of Environment, 244*, 111806. https://doi.org/10.1016/j.rse.2020.111806

Gupta, D., Srivastava, P., Singh, A., Petropoulos, G., Stathopoulos, N., & Prasad, R. (2021). SMAP soil moisture product assessment over wales, UK. Using observations from the WSMN ground monitoring network. *Sustainability, 13*(11), 6019. https://doi.org/10.3390/su13116019

Kang, C. S., Kanniah, K. D., Kerr, Y., & Varotsos, C. A. (2016). Analysis of in-situ soil moisture data and validation of SMOS soil moisture products at selected agricultural sites over a tropical region. *International Journal of Remote Sensing, 37*(16), 3636–3654. https://doi.org/10.1080/01431161.2016.1201229

Kerr, Y. H., Waldteufel, P., Richaume, P., Wigneron, J., Ferrazzoli, P., Mahmoodi, A., Bitar, A. A., Cabot, F., Gruhier, C., Juglea, S. E., Leroux, D., Mialon, A., & Delwart, S. (2012). The SMOS soil moisture retrieval algorithm. *IEEE Transactions on Geoscience and Remote Sensing, 50*(5), 1384–1403. https://doi.org/10.1109/tgrs.2012.2184548

Legates, D. R., & McCabe, G. J. (1999). Evaluating the use of "goodness-of-fit" measures in hydrologic and hydroclimatic model validation. *Water Resources Research, 35*(1), 233–241. https://doi.org/10.1029/1998wr900018

Li, X., Wigneron, J.-P., Fan, L., Frappart, F., Yueh, S.-H., Colliander, A., Ebtehaij, A., Gao, L., Fernandez-Moran, R., Liu, X., Wang, M., Ma, H., Moisy, C., & Ciais, P. (2022). A new SMAP soil moisture and vegetation optical depth product (SMAP-IB): Algorithm, assessment and inter-comparison. *Remote Sensing of Environment, 271*, 112921. https://doi.org/10.1016/j.rse.2022.112921

Li, Z., Leng, P., Zhou, C., Chen, K., Zhou, F., & Shang, G. (2021). Soil moisture retrieval from remote sensing measurements: Current knowledge and directions for the future. *Earth-Science Reviews, 218*, 103673. https://doi.org/10.1016/j.earscirev.2021.103673

Liu, Y., & Yang, Y. (2022). Advances in the quality of global soil moisture products: A review. *Remote Sensing, 14*(15), 3741. https://doi.org/10.3390/rs14153741

O'Neill, P. E., Chan, S., Njoku, E. G., Jackson, T., Bindlish, R., & Chaubell, J. (2020). *SMAP L3 radiometer global daily 36 km EASE-grid soil moisture, version 7*. Boulder, Colorado USA: NASA National Snow and Ice Data Center Distributed Active Archive Center. https://doi.org/10.5067/HH4SZ2PXSP6A

Oozeer, M. Y., Fletcher, C. D., & Champagne, C. M. (2020). Evaluation of satellite-derived surface soil moisture products over agricultural regions of Canada. *Remote Sensing, 12*(9), 1455. https://doi.org/10.3390/rs12091455

Panegrossi, G., Ferretti, R., Pulvirenti, L., & Pierdicca, N. (2011). Impact of ASAR soil moisture data on the MM5 precipitation forecast for the Tanaro flood event of April 2009. *Natural Hazards and Earth System Sciences, 11*(12), 3135–3149. https://doi.org/10.5194/nhess-11-3135-2011

Petropoulos, G. P., Ireland, G., & Barrett, B. (2015). Surface soil moisture retrievals from remote sensing: Current status, products & future trends. *Physics and Chemistry of the Earth, Parts A/B/C, 83*, 36–56. https://doi.org/10.1016/j.pce.2015.02.009

Petropoulos, G. P., Ireland, G., & Srivastava, P. K. (2015). Evaluation of the soil moisture operational estimates from SMOS in Europe: Results over diverse ecosystems. *IEEE Sensors Journal, 15*(9), 5243—5251. https://doi.org/10.1109/jsen.2015.2427657

Petropoulos, G. P., Ireland, G., Srivastava, P. K., & Ioannou-Katidis, P. (2014). An appraisal of the accuracy of operational soil moisture estimates from SMOS MIRAS using validated in situ observations acquired in a Mediterranean environment. *International Journal of Remote Sensing, 35*(13), 5239—5250. https://doi.org/10.1080/2150704x.2014.933277

Petropoulos, G. P. (2013). *Remote sensing of energy fluxes and soil moisture content.* CRC Press.

Petropoulos, G. P., Srivastava, P. K., Ferentinos, K. P., & Hristopoulos, D. (2018a). Evaluating the capabilities of optical/TIR imaging sensing systems for quantifying soil water content. *Geocarto International, 35*(5), 494—511. https://doi.org/10.1080/10106049.2018.1520926

Petropoulos, G. P., Srivastava, P. K., Piles, M., & Pearson, S. (2018b). Earth observation-based operational estimation of soil moisture and evapotranspiration for agricultural crops in support of sustainable water management. *Sustainability, 10*(2), 181. https://doi.org/10.3390/su10010181

Piles, M., Petropoulos, G. P., Sánchez, N., González-Zamora, Á., & Ireland, G. (2016). Towards improved spatio-temporal resolution soil moisture retrievals from the synergy of SMOS and MSG SEVIRI space-borne observations. *Remote Sensing of Environment, 180*, 403—417. https://doi.org/10.1016/j.rse.2016.02.048

Sazib, N., Mladenova, I. E., & Bolten, J. D. (2018). Leveraging the Google Earth Engine for drought assessment using global soil moisture data. *Remote Sensing, 10*(8), 1265. https://doi.org/10.3390/rs10081265

Shi, Q., & Liang, S. (2014). Surface-sensible and latent heat fluxes over the Tibetan Plateau from ground measurements, reanalysis, and satellite data. *Atmospheric Chemistry and Physics, 14*(11), 5659—5677. https://doi.org/10.5194/acp-14-5659-2014

Silk, J. (1979). *Statistical concepts in geography.* Allen & Unwin Australia.

Srivastava, P. K. (2017). Satellite soil moisture: Review of theory and applications in water resources. *Water Resources Management, 31*, 3161—3176.

Srivastava, P. K., Han, D., Rico-Ramirez, M. A., Al-Shrafany, D., & Islam, T. (2013). *Data fusion techniques for improving soil moisture deficit using SMOS satellite and WRF-NOAH land Surface model.* Water Resources Management. https://doi.org/10.1007/s11269-013-0452-7

Suman, S., Srivastava, P. K., Petropoulos, G. P., Pandey, D. K., & O'Neill, P. (2020). Appraisal of SMAP operational soil moisture product from a global perspective. *Remote Sensing, 12*(12), 1977. https://doi.org/10.3390/rs12121977

Vogel, M. M., Orth, R., Chéruy, F., Hagemann, S., Lorenz, R., Van Den Hurk, B., & Seneviratne, S. I. (2017). Regional amplification of projected changes in extreme temperatures strongly controlled by soil moisture-temperature feedbacks. *Geophysical Research Letters, 44*(3), 1511—1519. https://doi.org/10.1002/2016gl071235

Wang, X., Lü, H., Crow, W. T., Zhu, Y., Wang, Q., Su, J., Zheng, J., & Gou, Q. (2021). Assessment of SMOS and SMAP soil moisture products against new estimates combining physical model, a statistical model, and in-situ observations: A case study over the Huai river basin, China. *Journal of Hydrology, 598*, 126468. https://doi.org/10.1016/j.jhydrol.2021.126468

Willmott, C. J. (1981). On the validation of models. *Physical Geography, 2*(2), 184—194. https://doi.org/10.1080/02723646.1981.10642213

Zhang, C., Cheng, Q., Yang, J., Zhao, J., & Cui, T. J. (2017). Broadband metamaterial for optical transparency and microwave absorption. *Applied Physics Letters, 110*, 143511.

Zhang, S., Weng, F., & Yao, W. (2020). A multivariable approach for estimating soil moisture from microwave radiation imager (MWRI). *Journal of Meteorological Research, 34*(4), 732—747. https://doi.org/10.1007/s13351-020-9203-x

SECTION 3

Coastal/aquatic environment

CHAPTER 11

Charting the changes: Geographic Information System and Remote Sensing study on soil erosion and coastal transformations in Maliakos Gulf, Greece

Nikolaos Stathopoulos[1], Kleomenis Kalogeropoulos[2], Eleni Vasileiou[3], Panagiota Louka[4], Demetrios E. Tsesmelis[5] and Andreas Tsatsaris[2]

[1]Operational Unit "BEYOND Centre for Earth Observation Research and Satellite Remote Sensing", Institute for Astronomy, Astrophysics, Space Applications and Remote Sensing, National Observatory of Athens, Athens, Greece; [2]Department of Surveying and Geoinformatics Engineering, University of West Attica, Athens, Greece; [3]Department of Geological Sciences, School of Mining Engineering and Metallurgy, National Technical University of Athens, Athens, Greece; [4]Department of Natural Resources Development and Agricultural Engineering, Agricultural University of Athens, Athens, Greece; [5]Laboratory of Technology and Policy of Energy and Environment, School of Science and Technology, Hellenic Open University, Patra, Greece

1. Introduction

Soil erosion is a complex and dynamic phenomenon that affects many areas around the world. It is one of the most serious land degradation problems and a major source of environmental degradation, being the biggest environmental problem facing the world after population growth. As an example, the United States is losing land 10 times faster than the rate of natural replenishment, while the corresponding loss for China and India is 30–40 times faster (Pimentel, 2006). Globally, the total land area affected by water-induced erosion is 10,940,000 km^2, of which 7,510,000 km^2 is significantly degraded (Lal, 2003). The annual transport of sediment to the ocean from rivers worldwide has been estimated to be 15–30 billion tons (Milliman & Syvitski, 1992; Walling & Webb, 1996). In addition, soil erosion affects the geomorphological characteristics of a region in terms of soil fertility, agricultural productivity, water quality, water reservoir capacity, and the evolution of coastal areas in the sedimentary environment (Demirci & Karaburun, 2011; Lal, 2001; Xu et al., 2013).

Erosion, from a geological point of view, is a slow and steady process through which soil formation takes place. It results from the removal of soil particles by water or wind and their transport to other locations. It is a natural process, the extent and magnitude of which are determined by a multitude of environmental factors such as climate, soil, topography, inappropriate land use, Renschler inherent soil characteristics, and

Geographical Information Science
ISBN 978-0-443-13605-4
https://doi.org/10.1016/B978-0-443-13605-4.00005-9

207

vegetation cover (Butt et al., 2010; Mutua et al., 2006; Renard et al., 1997; Renschler et al., 1999; Wischmeier & Smith, 1978). However, the influence of human activity through land use diversification, deforestation, agricultural practices, and construction activities can accelerate the rate of soil erosion (Ganasri & Ramesh., 2016; Thomas et al., 2018).

It is helpful to note that semiarid and arid soils, which lack a protective layer of vegetation cover, can erode at a rate that is on average 10 to 50 times greater than the erosion rate of soils in humid climate regions (Miller & Donahue, 1990, p. 768). Forested areas have also been found to protect against soil erosion and excessive sediment transport caused by extreme rainfall events (Bathurst et al., 2011). Effective modeling can provide information on the current erosion situation and prevailing trends and allow an analysis of different scenarios (Ganasri & Ramesh, 2016). In addition, it is an important source of information when making land management decisions by developing alternative land use scenarios and evaluating their outcomes (Fistikoglu & Harmancioglu, 2002). Often quantitative assessment is required (Kothyari, 1996) to extract the extent and magnitude of soil erosion problems in order to develop sound management strategies on a regional basis using field measurements. The main weakness of erosion models is the validation of their results due to the scarce data available to compare model estimates with actual soil losses (Lazzari et al., 2015; Gitas et al., 2009).

The coastal zone including the shoreline is the area separating the land from the marine environment. This area has clear and specific boundaries which do not remain fixed but are constantly changing over time. The rate of change of the coastline reveals the different types of impacts that have occurred over time. Sometimes these are the result of natural processes and sometimes of human activities, since a large proportion of human activities take place near or along the coastal zone (Eurosion, 2004).

Coastal erosion and coastline change is a natural process and is a result of coastal dynamics, in combination with the river action, that is, the solid discharge in the estuary, of the hydrological system into the sea, it contributes to the redistribution of the water balance in the sea. Sediment balance along the coast and it is worth noting that although coastal erosion has always been active in the coastal system, contributing to the formation of various coastal landscapes (shorelines, tidal flats, beaches, deltas, etc.), the human interventions in the coastal zone, in particular urbanization and economic activities (industries, tourism, etc.), have transformed coastal erosion from a natural phenomenon to a problem of increasing intensity (Eurosion, 2004).

The categorization of shoreline change patterns, based on geomorphological rules, helps us to interpret more accurately the shoreline data and provides us with a practical management tools (Del Rio et al., 2013). A shoreline change type is defined as a distinct pattern of shoreline movement identified by a particular change in shape, rhythm, or periodicity. Shoreline change types can occur as two-dimensional (linear) or three-dimensional (along the shoreline) changes that evolve at different temporal and spatial scales (Galgano & Leatherman, 2005). The historical evolution of the shoreline is inextricably linked to the evolution of river basins and their ability to deliver sediment to the shoreline (Carter & Woodroffe, 1994), a factor that must be taken into account to draw reliable conclusions in estimates of shoreline change.

Coastlines are constantly changing through time mainly due to the vertical and horizontal movement of sediments toward the coast (sediment transport) but in particular due to the dynamic nature of sea level during coastal boundary such as waves, tides, groundwater, storm waves, etc. Shoreline changes due to natural pressures are evident in all time scales. Changes due to anthropogenic pressures and man-made pressures occur after decades or even hundreds of years. In many cases, short-term shoreline change can be over a period of several decades, much greater than its long-term change. The rate of shoreline change is a very important parameter used by scientists working on coastal zone dynamics. It is expressed in volume/length/year, (m^3/m/year) but because the rate of change is often used as a synonym for shoreline retreat/advance, it is ultimately expressed in m/year. It is assigned a positive sign for shoreline advance toward the sea (accretion) and a negative sign for shoreline retreat toward the shore (erosion) (Marchand et al., 2011). The choice of method for calculating the rate of change of shorelines depends on the geology of the area under consideration and the purpose of the survey. There is no single method that is used in all cases or is the best (Anderson et al., 2015; Anderson & Frazer, 2014).

Floods are natural phenomena with profound effects, causing a spectrum of consequences that span all areas of life (environmental, social, and economic). Beyond the immediate threat to human settlements, floods wreak havoc on agriculture, infrastructure, and ecosystems. They are capable of reshaping landscapes, eroding soil, changing coastlines due to sediment transport, etc. Climate change has increased the frequency and severity of floods worldwide, highlighting the urgent need for comprehensive mitigation and adaptation measures (Kalogeropoulos et al., 2011; Kalogeropoulos & Chalkias, 2013; Chalkias et al., 2016; Tsanakas et al., 2016; Tsatsaris et al., 2021; Kalogeropoulos et al., 2023).

In this research, primary data were derived from multitemporal Landsat images, forming a 30-year time series obtained from historical archives, which underwent thorough analysis and interpretation. Utilizing unsupervised classification techniques, various land cover types within the specified area were classified, and subsequent coastline delineation was executed. High-resolution orthophotos were also employed to gather information and validate results derived from Landsat data analysis. These orthophotos were integrated into a geographic information system (GIS) context, serving as basemaps for overlaying various data to identify changes. The outcomes demonstrated effective monitoring and evaluation of observed coastline and land cover changes and trends.

2. Data and methodology

2.1 An extensive overview of the study area

The Sperchios river basin is located in Central Greece, encompassing an area of significant geographical and ecological importance (Fig. 11.1). The basin is situated in the heart of the country and covers a diverse landscape characterized by both mountainous and lowland areas. The Sperchios River, the main watercourse of the basin, originates from the Agrafa Mountains and flows southeastward through the fertile Thessaly Plain before reaching the Aegean Sea. The basin's location in Central Greece positions it within close proximity to major cities such as Lamia, Volos, and Larissa, facilitating its socioeconomic significance. The region's geographic features, including the river's flow path and surrounding topography, play a crucial role in influencing hydrological processes, sediment transport, and ecological dynamics within the Sperchios river basin (Stahl et al., 1975). It extends in the northern part of the water district of Eastern Central Greece and is bounded to the west by Tymfristos (2315m), to the north by Orthris, and to the south by Vardousia (2285m), Oiti (2141m), and Kallidromos (1419m), while to the east it is open to the sea and the Maliakos Gulf. Its total surface area is 2116 km^2.

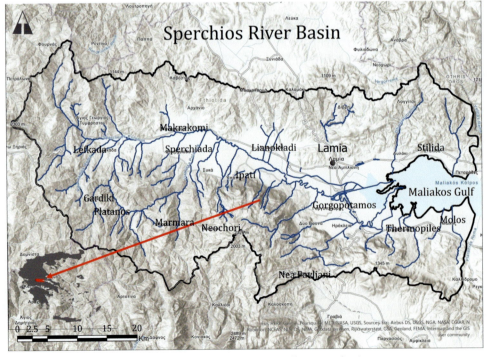

Figure 11.1 The study area—Sperchios river basin.

Around the perimeter, the main cities observed are Lamia, Stylida, Makrakomi, Spherchiada, Ipati, Gorgopotamos, Thermopylae, and Kamena Vourla, which are the main commercial centers of the wider region. Administratively, the Sperchios basin belongs to the prefecture of Fthiotida, where the region of Central Greece is located and has its headquarters in the city of Lamia.

The Sperchios valley is a taffroid neotectonic depression with mainly mountainous features and with a longitudinal axis coinciding with the riverbed (Fig. 11.2). Along this axis, the basin is divided into two parts, with the southern part having higher altitudes than the northern part. This is because, as mentioned above, the central bed of the Sperchios River is a tectonic depression, where the southern part rises while the northern part sinks (Mariolakos, 1976). The elevations of the whole basin range from 0 to 2295m with an average value of 607m. For two-thirds of its length, the valley has steep slopes, which give the river a mountainous-torrential character, with sharp flood peaks and very intense solid discharge. On the other hand, during the last third of its course, the Sperchios River gradually becomes a lowland river and crosses low-lying areas, where it often causes significant flooding.

The morphological topography of the catchment based on the Dikau (1989) system is classified into the following categories (Fig. 11.3):
- Field areas <150 m
- 150−600 m
- Semimountainous areas, with mountains and high hills 600−900 m
- Mountainous areas >900 m

According to the percentages of areas per category, there is a relatively equal distribution between lowland, hilly, semimountainous, and mountainous, with hilly and mountainous areas slightly predominating (HCMR, 2015). Moreover, at altitudes below 100 m, the area is quite large, which is due to the elongation of the central bed of the Sperchios River and the spread of deltaic deposits (Psomiadis, 2010) (Table 11.1).

Figure 11.2 Sperchios river basin and flow axis.

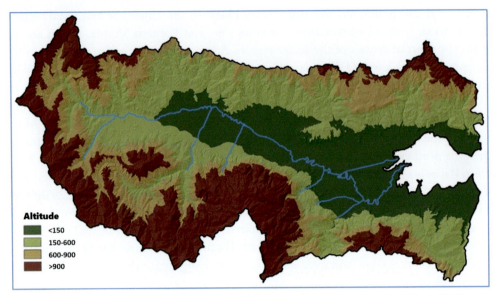

Figure 11.3 The morphological topography of the catchment based on the Dikau (1989).

Table 11.1 Basin's percentages by altitude category (Elke, 2015).

Altitude (m)	Description	Area (%)
<150	Field soil	20
150–600	Sloping ground	31
600–900	Semimountainous	19
>900	Mountainous	30

The slopes of the basin (Fig. 11.4) range from 0 to 88° with an average value of 17°, while the morphological relief of the basin based on the IGU system is classified into the following categories (Demek, 1972):

- <2° Flat to gently sloping (floodplains, terraces, etc.)
- 2–5° Slightly sloping (valley foothills, dune beds, etc.)
- 5–15° Steeply sloping (valley floors, tectonic terraces)
- 15–35° Steep to extremely steep (valley floors of medium mountain ranges).
- 35–55° Steep
- >55° Vertical.

The percentages of the basin areas corresponding to each category are listed in the table below (Table 11.2). It is observed that steep to extremely steep terrain predominates, followed by steeply sloping terrain (HCMR, 2015).

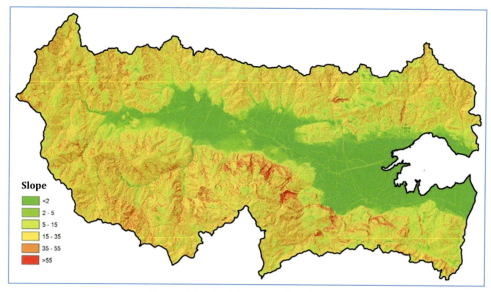

Figure 11.4 Basin's slopes classification.

Table 11.2 Basin's percentages by slope category (HCMR, 2015).

Slope	Description	Area (%)
<2	Level as a slightly inclined plane	19
2–5	Slightly inclined level	9
5–15	Slightly sloping relief	29
15–35	Steep to very steep relief	40
35–55	Steep relief	3
>55	Vertical relief	0

Sperchios river basin covers an area of 2116 km², with an average altitude of approximately 810m, while the river is recharged by many streams of permanent and periodic flow. The high gradients which are present within approximately 2/3 of the total length of the river course form a rather mountainous topography-streamy, with crucial flooding peaks and very intense sediment loads yield. On the contrary, within the last downstream part of its course, the river is transformed gradually into a lowland relief, where cases of severe flooding have been observed and reported. The river discharges in Maliakos Gulf and the basin's extent at the exit is 1830 km², while the average altitude is 626 m. The deltaic alluvial part of the valley covers an area of approximately 200 km² with a highly increasing formation rate during the last 150–200 years; estimated at 130 acres annually. Water balance calculations involved the interpretation of hydrological data, whereas for the area of main scientific focus—adjacent to the coastal zone—the surface water volume is found 1026.8 hm^3, while the evapotranspiration is about 72% of the rainfall. The average value of the river's sediment load yield was estimated by using and cross-referencing various methods. The aim of this research is the monitoring and assessment of coastline changes within the coastal deltaic part of a typical Mediterranean hydrological basin, by coupling remote sensing techniques with available hydrological and climate data. Remotely sensed data are widely used for land cover and/or coastline change detections. The hydrographic network of the Sperchios river basin is extensive and of dendritic type. The form of the hydrographic network is directly related to the hydrogeological structure, tectonics, the physicochemical characteristics of the strata, precipitation, the seasonal distribution of rainfall, the degree of forest cover, etc. (Sotiropoulou, 2012). It is also noted that the hydrographic network of the area has undergone significant interventions in the context of flood protection of the basin (HCMR, 2015).

Sperchios river basin in Greece has, unfortunately, been prone to flooding, posing significant challenges to the local communities and ecosystems. The region experiences periodic floods, often exacerbated by intense rainfall or rapid snowmelt from the surrounding mountainous terrain (Stathopoulos et al., 2017, 2019, 2023).

The highest average monthly flows occur from November to April (64.6–132.6 hm^3), while the lowest occur from July to September (9.4–15.9 hm^3) (Psomiadis, 2010).

The main hydrographic branch that dominates the Sperchios valley is the homonymous river, which is the drainage recipient of all other rivers and torrents that contribute to it and which operates with permanent and seasonal flows. The main riverbed is fed by 63 streams with the permanent and periodic flows. All the meanders and the main bed of the Sperchios River on the left bank of the valley are located along a nearly straight line, with an axial direction of 120 degrees (Kakavas, 1984).

The Sperchios River, with a total course of about 80 km, originates from the eastern sides of Mount Tymphristos (Asproneria springs), runs from west to east, crossing the plain of Lamia and passing south of Makrakomi, Lianokladi, and the bridge of Alamana, it flows into the Malian Gulf, north of the straits of Thermopylae. The bed of the river is fed by streams of permanent and intermittent flow, the main ones being Roustianitis, Vistritsa, Gorgopotamos, Asopos, and Xirias Lamia (Fig. 11.5). The area has a strong relief with steep slopes, giving the river a mountainous and torrential character, with sharp flood peaks and very high solid discharge.

2.2 Remote sensing data

Remotely sensed data are widely used for land cover and/or coastline change detections. In this study, multitemporal Landsat images were the main source of information. A 30-year time series of the Landsat images from past archives were obtained and interpreted. Classification of the various land cover types within the area of interest and the

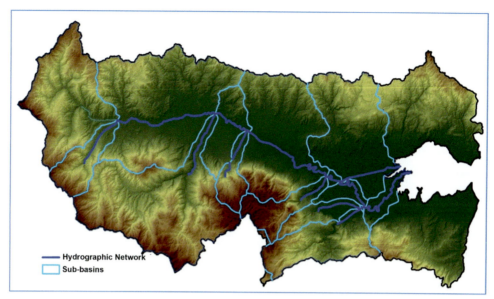

Figure 11.5 Basin's slopes classification.

subsequent delineation of the coastline was performed using unsupervised classification techniques. High-resolution (approximately 0.5 m) orthophotos available through the WMS service of the Greek Cadastral Agency have been also used to acquire information and verify the results obtained from the analysis of Landsat data. The available orthophotos were linked with a GIS system, acting as basemaps on which several data were overlaid in order to identify changes. The results were proved satisfactory in terms of effectively projecting and evaluating relevant observed coastline and land cover changes and trends. The analysis and interpretation of satellite images was performed using a combination of channels 4–5–3 by the professional version of TNT mips editor. Coastline and land cover changes was the main subject of this study. The next two figures present the Corine Land Cover for Sperchios river basin for 2000 (Fig. 11.6) and for 2006 (Fig. 11.7).

3. Results and discussion

3.1 Summary hydrogeological analysis of the area

In the basin of the river Sperchios, successive aquifers are formed in the alluvial deposits. The free aquifer, which is located in the lowland part of the area and specifically in the deposits of the Quaternary period, is of moderate capacity and is characterized as heterogeneous, due to the heterogeneity of the formations in which it is created. In the lower topographic areas located in the southeast and near the village of Anthili, an artesian aquifer develops, due to the nonuniform grain size of the formations and their transition

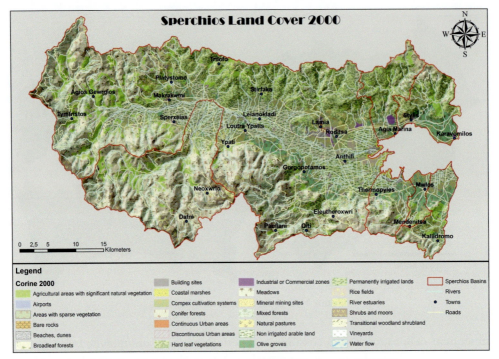

Figure 11.6 Corine land cover of Sperchios river basin (2000).

from anhydrous to finer toward the center of the basin (Karli, 2013). The aquifers formed in the Neogene formations show a large variation in terms of potentiality, which directly depends on the percentage of participation of conglomerates, sandstones, and marly limestones. The quantities escaping to the sea from the systems (a) Yliki-Paralimni, (b) the south–western side of Parnassos, and (c) the alluvial aquifer of Sperchios are estimated at 210 hm^3 (Koutsogiannis, 2003).

The karstic units of Sperchios (strongly tectonized and deconsolidated with increased secondary porosity and permeability) include the central Oiti unit with buffer reserves of 65 hm^3/year, the Oiti-Kallidromou unit with reserves of 21 hm^3/year, and the north-western Orthryos unit with reserves of 50 hm^3/year. The aquifers formed within the limestone formations have rich groundwater reserves, with the result that both the springs formed and the existing boreholes have very high flow rates. Important karstic aquifers are discharged into the sea in the form of coastal springs with high discharge rates. The karst springs are also sometimes the main source of the main surface runoff (Sperchios, Boeotiko Kifissos) and many of them are used for water supply and irrigation of the adjacent lowland areas (Koutsogiannis, 2003). The geological formations occupying the Sperchios basin exhibit different hydrological behavior, which is mainly a function of lithological composition, porosity, and water permeability. With these basic elements,

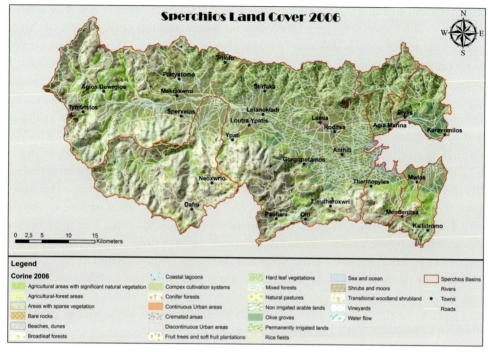

Figure 11.7 Corine land cover of Sperchios river basin (2006).

from a hydrolithological point of view, these formations are described as follows (Kaka-vas, 1984):

• Hydropermeable formations (carbonate rocks, Magoula Formation, conglomerates of Pleistocene lake sediments, anomeric materials of cone caves, river terraces, torrential terraces, torrent ribs, mixed materials of cones and terraces, coarse-grained alluvial formations, modern deposits of the Sperchios river bed)

• Semipermeable to permeable formations (deep deposits of the Sperchios delta, rem-nants of the Mesohellenic groove, various petrological types of the ophiolite complex)

• Semipermeable formations (various quaternary petrological formations, decomposed mantles of phyllite and schistoceratolites)

• Impermeable formations (shale and shale-clay formations, phyllosilicate formations, older deposits of the Sperchios River, hot spring deposits, aeolian deposits).

3.2 Sperchios river basin potential analysis

The climate of the Sperchios basin is characterized as subtropical-Mediterranean, with hot and dry summers and wet and mild winters, while it varies significantly depending on altitude. The lower altitudes (Malia Gulf and the river valley) belong to the

Mediterranean-lowland continental climate zone and are classified as type Csa (climate subdivision of Köppen & Geiger, 1936), that is, "Mediterranean or Mesothermal climate type," with dry and warm summers. Areas with altitudes above 500 m are characterized by a mountainous continental climate with cold winters. The mountainous areas have an average annual temperature of less than 22°C and show a tendency to equal distribution in precipitation during the year and are classified as Cfb (Efthymiou et al., 2005). The determining factor of the climate configuration of the Sperchios river basin is the sea to the east (Maliakos Gulf), as well as the mountain ranges surrounding it (HCMR, 2015). The climate ranges from dry to semihumid. The average temperature is 16.8°C in Lamia. The rainfall distribution at all stations is normal. The analysis of meteorological data showed decrease of rainfall (about 4 mm/year) and runoff (3 mm/year). The annual rain fall in the area is about 893 mm/year. In Lamia meteorological station, in the east coastal part of the area, the average precipitation is about 561 mm. The total amount of evaporation is high in about 72%, the infiltration and the surface run off is 28%. There is strong correlation between water table and discharge (Figs. 11.8 and 11.9).

The next figure (Fig. 11.10) presents the water balance in the Lamia meteorological station.

The delta of the Sperchios River extends about 4 km east of the village of Anthili, southeast of Lamia. It is the fourth largest delta on the Aegean coast and the sixth largest in Greece. The delta area was formerly a sea that gradually receded due to the river's accretions. The deltaic alluvial part of the valley has an area of almost 200 km^2 and is continuously forming at a rate unique in Greece. This rate seems to have increased over the last 150–200 years and is estimated at 130 ha per year (Koutsogiannis, 2003).

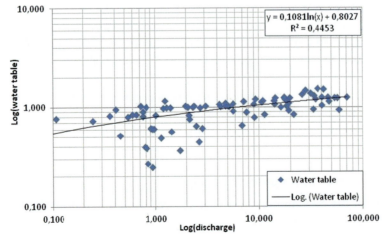

Figure 11.8 Correlation of water table and discharge in Sperchios River (1961–82).

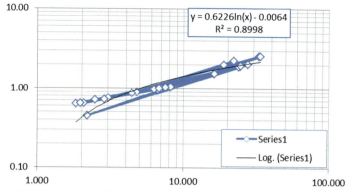

Figure 11.9 Correlation of water table and discharge in Kompotades (1973−80).

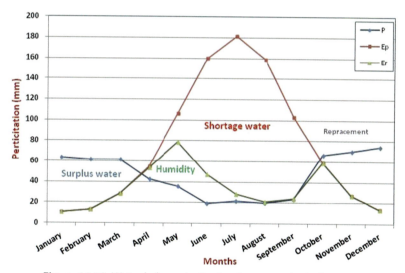

Figure 11.10 Water balance in the Lamia meteorological station.

The deltaic system of the River Sperchios (Fig. 11.11) consists of three distinct branches (HCMR, 2015). These are (1) the southern old Sperchios estuary, which was active until the end of the 19th century in the southern part of the system, (2) the central natural estuary, which has been active since the beginning of the 20th century until today, and (3) the northern new estuary of the river diversion, which is due to the Sperchios diversion works in the area of Anthili (1957−58 until today). The transfer of the delta from the south to the central part is attributed by some to the collapse of the embankments of the old riverbed after a major flood in 1889 (Philippson, 1950), while others link it to land reclamation and flood control works carried out in the area. The central part of the delta is still active today (Pehlivanidou, 2012).

Figure 11.11 Deltaic system of the Sperchios River, (1) old estuary, (2) natural estuary, (3) diversion estuary.

The form of the deltaic plain is mainly determined by the erosive processes of the Sperchios catchment and its hydrographic network, which supply it with large amounts of sediment. This results in the degradation of the mountainous areas on the one hand and the overflow of the river in the delta on the other (Efthymiou et al., 2005). The morphology of the Sperchios estuary is constantly changing due to natural processes of accretion, transport, and deposition of sediments and marine erosion, but also due to human interventions with the construction of roads and railways and gradual changes in land use.

In the lowland areas of the delta, especially that east of the Athens–Lamia highway, which has been created by the transport and deposition of sediment from the Sperchios, there are alternations of sands, sandy clays, silty clays, and clay up to a depth of 306 m, with an estimated total sediment thickness of 2000–2500 m. The influence of the sea also plays a very decisive role in the formation and development of the river delta. Due to the very small gradients (even negative in certain locations) prevailing in the coastal area, the flow of the river water is calm, and in case of flooding, the sea water penetrates the riverbed and on land, flooding large areas (Karli, 2013). The delta of the Sperchios River is one of the few areas in Greece that is estimated to have a serious flooding problem, due to sea level rise (about 1 m in the next 100 years), due to climate changes (HCMR, 2015).

The highly dynamic regime of the deltaic part of the Sperchios River, as described above, causes significant changes in the shoreline. Analysis of these spatiotemporal changes using Landsat satellite imagery over a period of approximately 30 years (Figs. 11.12 and 11.13) shows that the shoreline section in the area of the natural estuary is retreating landward mesostatically, while the shoreline section in the area of the new estuary diversion is advancing seaward. The part of the shoreline between the two estuaries shows a significant seaward advance in the 30-year period (Stathopoulos et al., 2012).

This pattern of shoreline change is explained by the fact that the new diversion bed, which is enclosed and straight, carries a larger volume of soil material from the erosion processes of the basin formations, which is deposited (most of it) in the area of the estuary. In contrast, the natural bed, once the diversion bed is in operation, has a significantly reduced flow rate and therefore less solid transport. In addition, the natural bed (which is not encased), from the point of diversion onwards, has a longer length to its outfall compared to the diversion bed, as well as several meanders, resulting in a reduced water flow velocity and therefore a significant proportion of the transported material being deposited in the

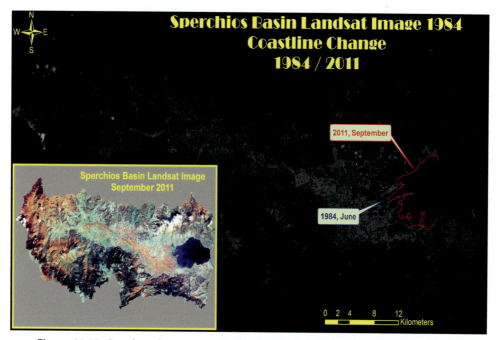

Figure 11.12 Coastline change over the period 1984–2011 (Stathopoulos et al., 2012).

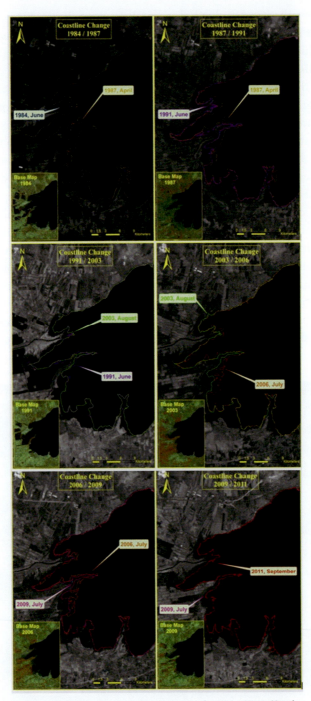

Figure 11.13 Intermediate coastline changes for the period 1984—2011 (Stathopoulos et al., 2012).

path before it even reaches the outfall. The intermediate variations that occur in shorter periods, compared with the general picture of the shoreline evolution pattern obtained from the analysis of the satellite images, are due to the specific and different conditions that prevailed during the periods of variation in the shoreline evolution pattern (e.g., diversion of smaller quantities from the divertor to the diversion bed, intense flooding, technical works in the area, etc.). An example is the period 2003–06, when the shoreline in the area of the natural bed appears to have moved significantly toward the sea.

The preceding analysis confirms the intense erosion processes taking place in the basin, the intense soil loss and solid transport of the eroded material from the hydrographic network, and the high rates of change of the deltaic part of the Sperchios River and the coastline, following the processes of deposition of the soil material and/or sea penetration on land. Finally, the influence of the human factor (artificial diversion bed, cultivation, etc.) on these natural processes and changes in the basin is readily apparent.

3.3 Sediment load empirical estimations

Estimating sediment load in rivers and streams is a fundamental aspect of sediment transport studies and riverine ecosystem management. Various methods have been developed and employed for sediment load estimation, allowing researchers and practitioners to gain insights into sediment dynamics and make informed decisions. Some commonly used methods include the rating curve method, suspended sediment sampling, bedload sampling, and sediment rating curves. The rating curve method establishes a relationship between water discharge and sediment concentration, enabling the estimation of sediment load based on measured or inferred discharge values. Suspended sediment sampling involves the collection of water samples at different locations and depths, followed by laboratory analysis to determine the suspended sediment concentration. Multiplying the concentration by the water discharge yields an estimate of the sediment load (Hauer & Lamberti, 2017).

Bedload sampling techniques, such as sediment samplers and sediment traps, are used to collect sediment samples from the riverbed. These samples are then analyzed to determine bedload sediment concentration. Multiplying the concentration by the water discharge provides an estimation of the bedload sediment load (Reid & Dunne, 1996). Sediment rating curves utilize historical data of sediment and water discharge measurements to establish empirical relationships. These curves enable the estimation of sediment load based on water discharge alone (Lisle, 1989, 1995).

Sediment yield estimations for Sperchios River are achieved mainly from simple empirical models that relate mean annual sediment yield (t/km^2) to catchment properties, including drainage area, topography, and climate and vegetation characteristics. In some cases, catchment area (in km^2) seems to be the only explanatory variable used to predict sediment yield. The relation between sediment yield and catchment area is presented in the next table (Table 11.3).

Table 11.3 Sediment yield estimations empirical models applied for Sperchios river basin.

Literature	Equation	Results in Sperchios
Dendy and Bolton (1976)	$S_y = 674 \times A^{-0.16}$	197.96 tn/km^2
Avendano Salas et al. (1997)	$S_y = 4139 \times A^{-0.43}$	153.79 tn/km^2
Webb and Griffiths (2001)	$Q_y = 193 \times A^{1.04}$	554,745.26 tn/year
Lu et al. (2003)	$S_y = 849.15 \times A^{-0.0785}$	645.51 tn/km^2
Moulder and Syvitski (1996)	$\log(Q_s) = 0.0406 \times \log(A) + 1.279 \times \log(H_{max}) - 3.679$	24.22 tn/year

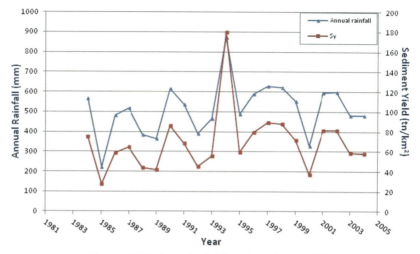

Figure 11.14 Annual rainfall and sentiment yield.

Different equations were applied to estimate the sediment yield of Sperchios river basin. The results of these empirical models are:

Many are the factors, which influence the sediment load in this area: (a) surface, (b) lithology, (c) rainfall, (d) vegetation, and (e) erosion due to climatic conditions. Koutsogiannis (2003) has proposed two additional empirical equations, for estimating the sediment yield, considering all these factors.

$$G = 15 \times \gamma \times e^{3P}$$

and

$$G = 14.4 \times e^{2.9P/1000}$$

The diagram (Fig. 11.14) below presents the fluctuation of annual rainfall and the sediment yield, regarding these two equations.

4. Concluding remarks

Soil erosion is a phenomenon that occurs when the forces of nature wear away the land that meets the sea. It is a natural process that has been happening for thousands of years, shaping our coastlines and creating the beaches, cliffs, and headlands that we see today. The rate of coastline change has been accelerating in recent years, threatening not only our natural habitats and ecosystems but also our coastal communities and infrastructure. The causes of coastal erosion are complex and can vary depending on the location and geology of the area. However, the main drivers of coastal erosion are the power of waves, currents, tides, and storm surges, which can wear away and transport sediments along the coast. Human activities, such as coastal development, construction of ports and harbors, and dredging of coastal waters, can also exacerbate the rate of erosion by disrupting the natural balance of sediment deposition and transport.

The impacts of coastal erosion can be devastating. Coastal communities can lose their homes and businesses to erosion, and entire coastlines can disappear over time. Erosion can also lead to the destruction of habitats and ecosystems that depend on the coastal environment, including coral reefs, salt marshes, and dunes. In addition, erosion can damage infrastructure such as roads, bridges, and buildings, leading to significant economic losses. To address the issue of coastal erosion, a range of management strategies and techniques have been developed. These include coastal defense measures such as sea walls, and beach nourishment, which aim to protect the coastline from the forces of nature. Additionally, natural-based solutions such as the restoration of coastal habitats and the reintroduction of natural processes can help to reduce erosion rates and promote the resilience of coastal ecosystems.

While these solutions can be effective, they also have limitations and can be expensive and challenging to implement. Moreover, the long-term effectiveness of these measures depends on a range of factors, including the location, geology, and natural processes of the area, and the effectiveness of monitoring and maintenance programs. In conclusion, coastal erosion is a complex and pressing issue that affects both our natural environment and human communities.

In this study, the main conclusions were:
- The coastline part in the area of the old river bed, in deltaic part, in September of 2011 has moved toward inland, in comparison with June of 1984.
- The coastline part in the area of the new diverted riverbed, in deltaic part, shows a small change, small accession toward sea, in September of 2011 June of 1984.
- The coastline part, between the two riverbeds depicts a significant accession in September of 2011 compared to June of 1984.
- There are not metric stations in the area, for having a secure estimation for the sediment yield. The estimations are empirical.
- Remote sensing methods can help to evaluate this phenomenon and correlate all the parameters, which participate in this processing.
- All the hydrological parameters (run off, water table, discharge, meteorological data) are necessary for defining the changes in the coastline and the changes in land use along the riverbed of Sperchios River.

Addressing the important issue of coastline change requires a coordinated effort from governments, communities, and industries to develop and implement effective management strategies that protect our coasts, promote resilience, and ensure a sustainable future for generations to come.

References

Anderson, T. R., Frazer, L. N., & Fletcher, C. H. (2015). Long-term shoreline change at Kailua, Hawaii, using regularized single transect. *Journal of Coastal Research, 300,* 464–476. https://doi.org/10.2112/JCOASTRES-D-13-00202.1

Anderson, T. R., & Frazer, L. N. (2014). Toward parsimony in shoreline change prediction (III): B-splines and noise handling. *Journal of Coastal Research, 296,* 729–742. https://doi.org/10.2112/JCOASTRES-D-13-00032.1

Avendano Salas, C., Sanz Montero, E., Rayan, C., & Gomez Montana, J. L. (1997). Sediment yield at Spanish reservoirs and its relationship with the drainage area. In *Proceedings of the 19th symposium of large dams* (pp. 863–874). Florence: International Committee on Large Dams, Paris.

Bathurst, J. C., Birkinshaw, S. J., Cisneros, F., Fallas, J., Iroume, A., Iturraspe, R., Gavino Novillo, M., Urciuolo, A., Alvarado, A., Coello, C., Huber, A., Miranda, M., Ramirez, M., & Sarandon, R. (2011). Forest impact on floods due to extreme rainfall and snowmelt in four Latin American environments 2: Model analysis. *Journal of Hydrology, 400,* 292–304.

Butt, M. J., Waqas, A., Mahmood, R., & Hydrology Research Group. (2010). The combined effect of vegetation and soil erosion in the water resource management. *Water Resources Management, 24*(13), 3701–3714.

Carter, R. W. G., & Woodroffe, C. D. (1994). *Coastal evolution: Late quaternary shoreline morphodynamics.* Cambridge University Press.

Chalkias, C., Stathopoulos, N., Kalogeropoulos, K., & Karymbalis, E. (2016). Applied hydrological modeling with the use of geoinformatics: Theory and practice. In H. Mamun (Ed.), *Empirical modeling and its applications* (pp. 61–86). InTECH. https://doi.org/10.5772/62824

Del Río, L., Gracia, F. J., & Benavente, J. (2013). Shoreline change patterns in sandy coasts. A case study in SW Spain. *Geomorphology, 196,* 252–266.

Demek, J. (1972). *Manual of detailed geomorphological mapping* (p. 334). Prague: Academia.

Demirci, A., & Karaburun, A. (2011). Estimation of soil erosion using RUSLE in a GIS framework: A case study in the Buyukcekmece Lakewatershed, Northwest Turkey. *Environmental Earth Sciences, 66,* 903–913.

Dendy, F. E., & Bolton, G. C. (1976). Sediment yield –runoff drainage area relationships in the United States. *Journal of Soil and Water Conservation, 31,* 264–266.

Dikau, R. (1989). The application of a digital relief model to landform analysis in geomorphology. In *Three dimensional applications in geographical information systems* (pp. 51–77).

Efthimiou, G., Mertzakis, A., Sapountzis, M., & Zakynthinos, G. (2005). *Anthropogenic effects in the Sperchios Delta. Measures to protect, promote and manage natural ecosystems. International exhibition and conference on environmental technology.* Athens: HELECO, 2005 (in Greek).

Eurosion. (2004). *Living with coastal erosion in Europe sediment and space for sustainability.* European Commission.

Fistikoglu, O., & Harmancioglu, N. B. (2002). Integration of GIS with USLE in assessment of soil erosion. *Water Resources Management, 16*(6), 447–467.

Galgano, F., & Leatherman, S. (2005). Modes and patterns of shoreline change. In M. Schwartz (Ed.), *Encyclopedia of coastal science, encyclopedia of earth science series* (pp. 651–656). Springer Netherlands.

Ganasri, B. P., & Ramesh, H. (2016). Assessment of soil erosion by RUSLE model using remote sensing and GIS-A case study of Nethravathi Basin. *Geoscience Frontiers, 7*(6), 953–961.

Gitas, I. Z., Douros, K., Minakou, C., Silleos, G. N., & Karydas, C. G. (2009). Multi-temporal soil erosion risk assessment in N. Chalkidiki using a modified USLE raster model. *EARSeL eProceedings, 8,* 40–52.

Hauer, F. R., & Lamberti, G. (2017). Methods in stream ecology. In *Vol. 1: Ecosystem structure* (3rd ed.). Elsevier, ISBN 9780124165588.

HCMR. (2015). *Development of an integrated watershed management system and the connected coastal and marine zone – technical report with the results of the initial ecological quality assessment.* Development proposals of research organizations - KRIPIS.

Köppen, W., & Geiger, P. (1936). In W. Köppen, & R. Geiger (Eds.), *Das Geographische System der Klimate, Handbuch der Klimatologie* (pp. C-1−C-44). Berlin: Gebruder Borntraeger.

Kakavas, N. (1984). *Hydrological water balance of basin of Sperchios river.* Phd Thesis. Athens: I.G.M.E..

Kalogeropoulos, K., & Chalkias, C. (2013). Modelling the impacts of climate change on surface runoff in small Mediterranean catchments: Empirical evidence from Greece. *Water and Environment Journal, 27*(4), 505−513. https://doi.org/10.1111/j.1747-6593.2012.00369.x

Kalogeropoulos, K., Chalkias, C., Pissias, E., & Karalis, S. (2011). Application of the SWAT model for the investigation of reservoirs creation. In N. Lambrakis, G. Stournaras, & K. Katsanou (Eds.), *Environmental earth sciences: Vol. 2. Advances in the research of aquatic environment* (pp. 71−79). Berlin, Heidelberg: Springer. https://doi.org/10.1007/978-3-642-24076-8_9 (Print ISBN 978-3-642-24075-1, Online ISBN 978-3-642-24076-8).

Kalogeropoulos, K., Tsanakas, K., Stathopoulos, N., Tsesmelis, D. E., & Tsatsaris, A. (2023). Cultural heritage in the light of flood hazard: The case of the "ancient" Olympia, Greece. *Hydrology, 10,* 61. https://doi.org/10.3390/hydrology10030061

Karli, A. (2013). *Hydrogeological and hydrochemical conditions of the aquifers of the Sperchios river basin.* Patras: Post-graduate work. University of Patras (in Greek).

Kothyari, U. C. (1996). Erosion and sediment problems in India. In *Proceedings of the exeter symposium on erosion and sediment yield: Global and regional perspectives July 1996* (pp. 531−540). IAHS Publ. No. 236.

Koutsoyiannis, D. (2003). *Draft program for the management of the country's water resources.* Athens: Ministry of Development - Directorate of Water and Natural Resources (in Greek).

Lal, R. (2001). Soil degradation by erosion. *Land Degradation and Development, 12,* 519−539. https://doi.org/10.1002/ldr.472

Lal, R. (2003). Soil erosion and the global carbon budget. *Environmental International, 29,* 437−450. https://doi.org/10.1016/S0160-4120(02)00192-7

Lazzari, M., Gioia, D., Piccarreta, M., Danese, M., & Lanorte, A. (2015). 2015-sediment yield and erosion rate estimation in the mountain catchments of the Camastra artificial reservoir (southern Italy): A comparison between different empirical methods. *Catena, 127,* 323−339.

Lisle, T. E. (1995). Particle size variations between bed load and bed material in natural gravel bed channels. *Water Resources Research, 31*(4), 1107−1118.

Lisle, T. E. (1989). Sediment transport and resulting deposition in spawning gravels, north coastal California. *Water Resources Research, 25*(6), 1303−1319.

Lu, X. X., Ashmore, P., & Wang, J. (2003). Sediment yield mapping in a large river basin: The upper Yangtze, China. *Environmental Modeling and Software, 18,* 339−353.

Marchand, M., Sanchez-Arcilla, A., Ferreira, M., Gault, J., Jiménez, J. A., Markovic, M., Mulder, J., van Rijn, L., Stănică, A., Sulisz, W., & Sutherland, J. (2011). Concepts and science for coastal erosion management − an introduction to the conscience framework. *Ocean and Coastal Management, 54*(12), 859−866. https://doi.org/10.1016/j.ocecoaman.2011.06.005. ISSN 0964-5691.

Mariolakos, H. (1976). Thoughts and opinions on some problems of the geology and tectonics of the Peloponnese. *Annales Geologioues Des Pays Helleniques, 27,* 215−313 (in Greek).

Miller, R. W., & Donahue, R. L. (1990). *Soils: An introduction to soils and plant growth* (6th ed.). Englewood Cliffs, New Jersey, USA: Prentice Hall.

Milliman, J. D., & Syvitski, J. P. M. (1992). Geomorphic/tectonic control of sediment discharge to the ocean: The importance of small mountainous rivers. *The Journal of Geology, 100,* 525−544. https://doi.org/10.1086/629606

Moulder, T., & Syvitski, J. P. M. (1996). Climatic and morphologic relationships of rivers: Implications of sea level fluctuations on river loads. *The Journal of Geology, 104,* 509−523.

Mutua, B. M., Klik, A., & Loiskandl, W. (2006). Modelling soil erosion and sediment yield at a catchment scale: The case of Masinga catchment, Kenya. *Land Degradation and Development, 17*(5), 557−570.

Pehlivanidou, S. (2012). *Sedimentological and physiogeographical models of the development of Holocene deltaic sequences in the Sperchios river valley. Doctoral thesis.* Thessaloniki: Aristotle University of Thessaloniki (in Greek).

Philippson, A. (1950). *Die griechischen Landschaften. Band I, Teil 1: Thessalien und die Spercheios-Senke* (p. 308). Frankfurt-am-Main: Klostermann, 4 maps.

Pimentel, D. (2006). Soil erosion: A food and environmental threat. *Environment, Development and Sustainability, 8*(1), 119—137.

Psomiadis, E. (2010). *Research of Geomorphological and environmental changes in the hydrological basin of the Sperchios river using new technologies.* Doctoral specialization thesis. Athens: Agricultural University of Athens (in Greek).

Reid, L. M., & Dunne, T. (1996). Sediment production from forest road surfaces. *Water Resources Research, 32*(4), 959—968.

Renard, K. G., Foster, G. R., Weesies, G. A., McCool, D. K., & Yoder, D. C. (1997). *Predicting soil erosion by water—a guide to conservation planning with the Revised Universal Soil Loss Equation (RUSLE).* Washington, DC: United States Department of Agriculture, Agricultural Research Service (USDA-ARS) Handbook No. 703. United States Government Printing Office.

Renschler, C. S., Mannaerts, C., & Diekkrüger, B. (1999). Evaluating spatial and temporal variability in soil erosion risk - rainfall erosivity and soil loss ratios in Andalusia, Spain. *Catena, 34,* 209—225.

Sotiropoulou, K. (2012). *Preparation of flood maps according to directive 2007/60/EU. Application to the Sperchios basin. Postgraduate work.* Athena: National Technical University of Athens (in Greek).

Stahl, W., Aust, H., Dounas, A., & Kakavas, A. (1975). Groundwater investigations, Sperchios basin. In *Central Greece 1970-74. Internal. Report, IX: Stable isotope composition of different ground and surface waters from the Sperchios valley* (p. 40).

Stathopoulos, N., Kalogeropoulos, K., Chalkias, C., Dimitriou, E., Skrimizeas, P., Louka, P., & Papadias, V. (2019). A robust remote sensing — spatial modeling — remote sensing (R-M-R) approach for flood hazard assessment. In H. R. Pourghasemi, & C. Gokceoglu (Eds.), *Spatial modeling in GIS and R for earth and environmental science* (pp. 391—410). Elsevier (Paperback ISBN: 9780128152263).

Stathopoulos, N., Kalogeropoulos, K., Polykretis, C., Skrimizeas, P., Louka, P., Karymbalis, E., & Chalkias, C. (2017). Introducing flood susceptibility index using remote-sensing data and geographic information systems: Empirical analysis in Sperchios river basin, Greece. In George P. Petropoulos, & Tanvir Islam (Eds.), *Remote sensing of hydrometeorological hazards* (pp. 381—400). CRC Press (ISBN 9781498777582 - CAT# K29774).

Stathopoulos, N., Kalogeropoulos, K., Zoka, M., Louka, P., Tsesmelis, D. E., & Tsatsaris, A. (2023). An integrated approach for a flood impact assessment on Land Uses/Cover based on SAR images & spatial analytics. The case of an extreme event in Sperchios river basin, Greece. In Nikolaos Stathopoulos, Andreas Tsatsaris, & Kleomenis Kalogeropoulos (Eds.), *GeoInformatics for geosciences advanced geospatial analysis using RS, GIS & soft computing* (pp. 247—260). Elsevier. Paperback, ISBN 9780323989831. https://doi.org/10.1016/B978-0-323-98983-1.00015-6

Stathopoulos, N., Vasileiou, E., Charou, E., Perrakis, A., Kallioras, A., Rozos, D., & Stefouli, M. (2012). Coupling of remote sensing methods and hydrological data processing for evaluating the changes of Maliakos Gulf coastline (Greece). In *The wider area of Sperchios river basin.* Vienna, Austria: European Geosciences Union General Assembly 2012.

Thomas, J., Joseph, S., & Thrivikramji, K. P. (2018). Assessment of soil erosion in a tropical mountain river basin of the southern Western Ghats, India using RUSLE and GIS. *Geoscience Frontiers, 9,* 893—906.

Tsanakas, K., Gaki-Papanastassiou, K., Kalogeropoulos, K., Chalkias, C., Katsafados, P., & Karymbalis, E. (2016). Investigation of flash flood natural causes of Xirolaki Torrent, Northern Greece based on GIS modeling and geomorphological analysis. *Natural Hazards, 84*(2), 1015—1033. https://doi.org/10.1007/s11069-016-2471-1

Walling, D. E., & Webb, B. W. (1996). Erosion and sediment yield: A global overview. In D. E. Walling, & B. W. Webb (Eds.), *Erosion and sediment yield: Global and regional perspectives proceedings of the exeter symposium, July 1996* (Vol. 236, pp. 3—19). IAHS Publication No.

Webb, R. H., & Griffiths, P. G. (2001). *Sediment delivery by ungaged tributaries of the Colorado River in Grand Canyon. USGS fact sheet 018-01.*

Wischmeier, W. H., & Smith, D. D. (1978). In *Predicting rainfall erosion losses: A guide to conservation planning*. Washington, DC: Agriculture Handbook No. 537. USDA.

Xu, L., Xu, X., & Meng, X. (2013). Risk assessment of soil erosion in different rainfall scenarios by RUSLE model coupled with information diffusion model: A case study of Bohai Rim, China. *Catena, 100*, 74—82.

Further reading

Conscience. (2007). *Concepts and science for coastal erosion management*. European Commission. https://doi.org/10.1016/j.ocecoaman.2011.06.005

CHAPTER 12

Exploring aquatic environments through geographical information science: A comprehensive review and applications

Smrutisikha Mohanty[1], Md. Wasim[1], Prem C. Pandey[1] and Prashant K. Srivastava[2]
[1]Department of Life Sciences, School of Natural Sciences, Shiv Nadar Institution of Eminence (Deemed to be University), Greater Noida, Uttar Pradesh, India; [2]Remote Sensing Laboratory, Institute of Environment and Sustainable Development, Banaras Hindu University, Varanasi, Uttar Pradesh, India

1. Introduction

Environmental monitoring includes tools and strategies for determining, evaluating, and establishing parameters for environmental conditions to quantify the effects of various activities on the environment (Vos et al., 2000). It is a technique used to assess environmental trends, facilitate the formulation and implementation of policies, and produce data for reporting to national and global policy-makers and the general public (Lindenmayer & Likens, 2010; Virapongse et al., 2016). A few nations in Europe and Central Asia have been able to continue their current monitoring efforts over the years. The efficacy of policy tools like emission levies is affected by improper monitoring of solid and hazardous wastes and urban air pollution (Blackman et al., 2018; Matheson, 2022). Furthermore, many of these countries lack consistent national techniques across various monitoring regions and their classification systems are frequently incompatible with international standards (Verburg et al., 2011).

For monitoting purpose, there are two categories of instruments: the first is called "snapshot instruments" that provide snapshot measurements (i.e., a record of conditions at a specific moment), and the other is a continuous record of environmental conditions. The use of these devices or instruments depend upon several factors, such as cost, output, precision, durability, and ease of use, to determine which type of device is best for data acquisition and monitoring requirements. Thus, there are spot snapshot readers that only show the current conditions such as thermometers, hygrometers, and humidity indicator strips, and continuous reading devices such as hygro-thermographs and data loggers that make temperature and humidity monitoring easier. The development of advanced automated monitoring applications along with smart sensors are critical for improving the accuracy of the environmental monitoring process as the human population, energy usage, and industrial activities continue to grow (Ahmad et al., 2016; Bansal et al., 2022;

Geographical Information Science
ISBN 978-0-443-13605-4
https://doi.org/10.1016/B978-0-443-13605-4.00019-9

Bibri, 2018). Various industries, businesses, and individuals must work together to implement the laws and regulations that protect nature (Glicksman et al., 2023; Tao, 2013). Moreover, environmental management not only talks about sustainability and environmental protection but also includes the management of all environmental aspects like wildlife, water, soil, and forest resources (Hull et al., 2003; Wassie, 2020).

The intricacy of environments with adverse events like degradation of natural resources is better understood and studied with the help of environmental monitoring (Artiola et al., 2004; Gavrilescu et al., 2015). It can be done in real time or through samples. Software for real-time monitoring such as Sinay Hub makes it easy and quick to examine important environmental indicators so that decision may be made with accuracy (AbdelAziz et al., 2024; Zahra et al., 2023). Water turbidity and air pollution are two examples of indicators (Golge et al., 2013; Steinwender et al., 2008). In the past, resolving environmental issues has involved taking data out of the sensor collector and storing it on certain platforms. Data transformation is also necessary to enable editing of the gathered data during subsequent analysis, which requires human labor to be finished. However, if the data processing software is integrated with the data collecting software, the efficiency is best achieved. The use of modern AI algorithms in environmental monitoring may provide crucial indications in real time, allowing organizations to comprehend the direct impacts of their activities (Tien, 2017).

The benefit of environmental monitoring is based on raising the standard of living of societies by establishing the relationship between health and the environment (Conrad & Hilchey, 2011). Hence, it is essential to turn environmental monitoring data into knowledge and promptly share actionable insights with the community to keep people aware of their surroundings (Clements et al., 2017). Recently, GIS has been used in all sectors and fields concerning environmental monitoring. Climate change, global warming, and pollution are some of the major environmental problems that can be studied with the help of GIS. Moreover, GIS has potential to assess and monitor aquatic environment such as riverine ecosystem, wetlands, lakes, ocean for different applications, such as habitat mapping, chlorophyll assessment, flood vulnerability, catchment area mapping, ground water potential, glaciars retreat etc. The aim of the present study is to provide comprehensive review of few reserch themes using GIS. Hence, the current study provides an overview of the role of GIS technology in monitoring and assessing aquatic environment highlighting the fresh water and coastal environment.

2. Role of GIS in environmental monitoring

A geographical information system (GIS) can be defined as a tool for gathering, storing, retrieving, processing, and displaying geographic data from the real world for specific applications (Chen et al., 2022). To capture different types of data, GIS has its own method. Numerous scholarly works have addressed the application of GIS and remote sensing to environmental management (Goodchild, 2003; Lovett, 2014). It is feasible to scan and digitize the data that have been printed on the paper directly (Nagy & Wagle, 1979).

Survey data can also be entered into the system using a coordinate system, whereas position data are simply input into the GIS system (Demers, 2008). GIS separates the world into objects and attribute tables. A raster or vector dataset can be used to represent the world spatially (Reddy, 2018). Many tools for spatial analysis and modeling, as well as visualizations for communication and stakeholder participation in the planning process, are available with a GIS (Enan & Xi, 2015). Multiple criteria analysis has been utilized as a decision-making method in a variety of situations (Kiker et al., 2005). A wide range of scientific fields, including computer science, statistics, remote sensing, and cartography, are all included in GIS (Wegmann et al., 2016). Natural environmental variables, societal factors, and human factors are all part of the environmental monitoring system (Vos et al., 2000). Environmental parameters and impact assessments can be determined by viewing and superimposing factors such as slope steepness, aspects, and vegetation (Feoli et al., 2002; Tomczyk, 2011). The use of GIS and remote sensing to monitor the ecological environment provides an effective framework for current environmental protection (Miller & Rogan, 2007; Twumasi & Merem, 2006). GIS provides comprehensive spatial analysis techniques and a range of tools specifically designed for analyzing data related to water resources, including groundwater, flood risk assessment, water quality, and habitat mapping (Tsihrintzis et al., 1996). Its versatility and integration with remote sensing data make it a powerful tool for researchers and practitioners working in aquatic environments. This study will provide few case studies related to the application of GIS in aquatic system and its surrounding environment.

3. GIS in water environment monitoring and risk assessment

Water, which plays a fundamental role in all aspects of our lives, is one of the Earth's most significant sources of energy (Sunitha et al., 2012). The primary sources of consumable water include precipitation, groundwater, and surface water bodies like rivers, ponds, and lakes. With the increasing population, agriculture irrigation, rapid urbanization, human and industrial consumption, pollution from cities, climate changes, and aquifer degradation, these water resources are becoming increasingly strained throughout the world (Boretti & Rosa, 2019). Hence, for planning, monitoring, and management of in situ and geospatial data, an integrated method should be implemented to reduce the stress on water resources (Ray & Abeysingha, 2023). GIS has emerged as a potent tool for developing solutions to the challenges of water resources locally (Hossen et al., 2018) and regionally (Rahmati et al., 2015; Youssef et al., 2011). It helps to understand the spatial distribution of water quality parameters, identify pollution sources, assess the vulnerability of groundwater flow, classify and map the groundwater potential zones (GPZs), and flood risk assessment and mapping (Dash & Sar, 2020; Fenglin et al., 2023; Gaikwad et al., 2020; Megahed, 2020; Oseke et al., 2021; Soleimani et al., 2018; Tarawneh et al., 2019).

3.1 GIS in flood vulnerability and risk assessment

Floods are one of the most frequent and destructive natural calamities that pose a major threat to society because of their profound effect on people's lives, homelessness to millions of people, and socioeconomic conditions (Qi & Altinakar, 2011). Geospatial planning is an effective technique to lower the danger of floods, via promoting decision-making using GIS and satellite remote sensing and examining floods, creating flood hazard risk maps, and analyzing floods in flood-prone regions (Albano et al., 2014; Dewan & Dewan, 2013; Khailani & Perera, 2013; Marchi et al., 2010; Samela et al., 2017; Wondim, 2016). A study was performed using GIS techniques and exploratory regression modeling to map the flood susceptibility (Fig. 12.1) in the Awash river basin, Ethiopia's Afar Region. Eight potential parameters for determining flood susceptibility have been investigated and examined: slope, elevation, rainfall, soil types, land use cover, drainage and lineament density, and topographic compound index in Arc GIS 10.3.1 software. The empirical Bayesian Kriging (EBK), spatial interpolation technique is employed as it automatically adjusts parameters via subsetting and simulations to provide reliable results. In accordance with its susceptibility to risks, each parameter was assigned a weight and was classified into five categories from very high to very low. The result shows that the land use pattern of Awash river basin is the major factor impacting its susceptibility to flood (Fenglin et al., 2023). Another study applied the GIS-based spatial multicriteria decision analysis (MCDA) method for coastal flooding risk assessment (Fig. 12.2) in Bandar Abbas City, Iranian south coast. Three major components, that is, hazard (sea-level rise, tidal range, storm surge, wave and wind set up), social vulnerability (population, age, gender), and exposure (population density) were analyzed to generate a flood susceptibility map (FSM). The analytical hierarchy process (AHP) model was applied to weight these components and create a classification map labeled as very high, high, moderate, low, and very low. Around 14.8% of the flooded areas were found to be at high and very high risk zone, representing the eastern, central, and western parts of the city. The findings of this study revealed that decision-makers can effectively implement risk reduction strategies in the high risk flood zones by utilizing GIS techniques (Hadipour et al., 2020).

3.2 GIS in groundwater potential zones mapping and monitoring

Groundwater is an underground ecosystem, where the natural sources of water are stored and distributed in addition to providing a range of socially and economically significant services to society (Hunt, 2007). The continuous extraction of groundwater will lead to a gradual decline in the water table (Naghibi et al., 2016; Prasad et al., 2020). For any management action to result in a sustainable development strategy for underground natural resources, there is a requirement for a GWP zonation map (Khan & Jhamnani, 2023). An integrated approach of satellite-based remote sensing and GIS with multicriteria

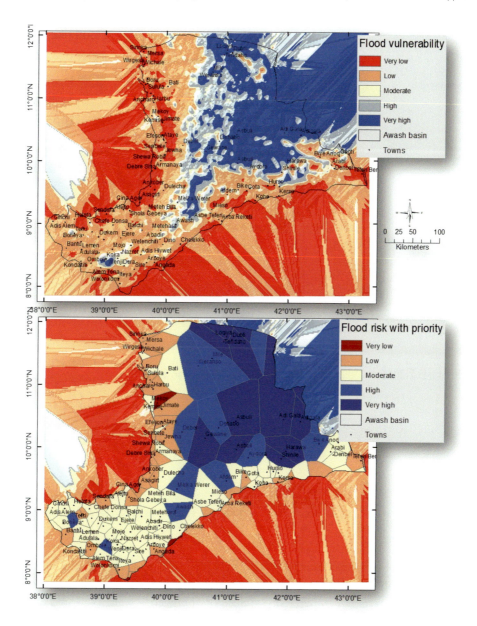

Figure 12.1 Map showing the spatial distribution of flood vulnerability and flood risk of Awash basin, Ethiopia. *(Adapted from Fenglin et al. (2023).)*

decision-making analysis (MCDMA) has been used by several researchers to detect the GPZs (Adimalla & Taloor, 2020). GIS-based MCDM, weighted overlay analysis (WOA), AHP, fuzzy logic, frequency ratio model, and influencing factors techniques are most popular for detecting the GPZ (Das & Pardeshi, 2018; Das et al., 2017; Guru et al., 2017; Magesh et al., 2012; Mohamed & Elmahdy, 2017; Singh et al., 2018).

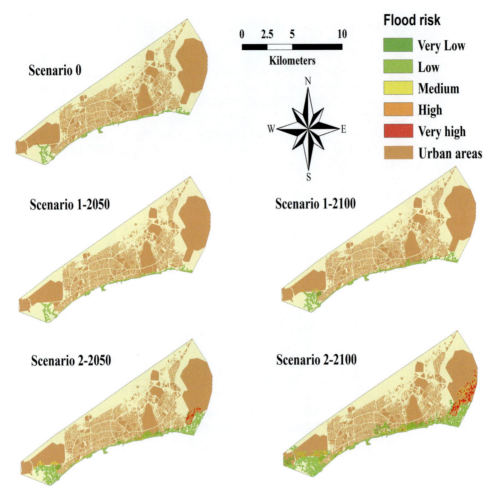

Figure 12.2 Map showing coastal flooding risk under various scenarios in Bandar Abbas city *(Adapted from Hadipour et al. (2020).)*

The advantages of GIS-based method is that they identify the GPZ faster and with more accurate computation results than conventional field methods (Zolekar & Bhagat, 2015).

An integrated remote sensing, GIS, and AHP approaches were used in a study to delineate the GPZs (Fig. 12.3) in the semiarid basin of San Luis Potosi (SLP), Mexico. The thematic layer of seven dominating parameters include slope, topology, drainage density, land use and land cover (LULC), topographic wetness index (TWI), lineament density, and rainfall were generated in raster format. The weight and ranking were assigned to these parameters depending their influence on groundwater potential, and the thematic layer was merged with the raster calculator in Arc GIS software to obtain the GPZ map. The outcome shows that around 68.21% of the region was categorized

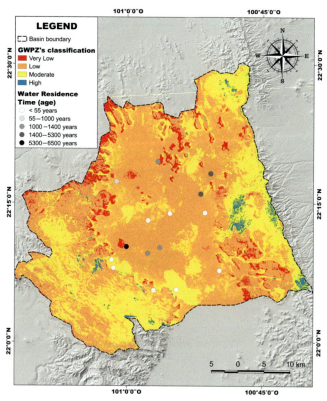

Figure 12.3 Map showing the spatial distribution of groundwater potential zone San Luis Potosi basin, Mexico. *(Adapted from Uc Castillo et al. (2022).)*

as a low groundwater zone and 26.30% as a moderate groundwater zone. Decision-makers can use the important information from this study to prepare the action plan for groundwater management with the help of GIS in future (Uc Castillo et al., 2022). Another relevant study uses a GIS-based multiinfluencing factor (MIF), MCDA, and electrical resistivity survey techniques for the assessment and mapping of GPZ in Raipur City, Chhattisgarh, India. Nine potential parameters namely slope, texture and type of soil, rainfall, lineaments, geology, geomorphology, drainage density, and LULC were assigned a weight based on their influence. Groundwater potential area was delineation using AHP techniques, which were further categorized into five zones, very low, low medium, high, and very high potential. The result of the study shows that the high-potential zones were found in the west and north-eastern regions, and in the central and eastern region low to medium groundwater potential is found (Jhariya et al., 2021). In a similar study, remote sensing techniques and GIS approaches were applied to reveal the spatial distribution of groundwater prospective areas (Fig. 12.4) in the Wadi Al Hamand watershed,

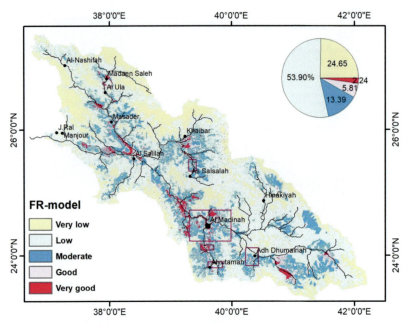

Figure 12.4 Map showing groundwater potential zone, using FR-model and validated field data of Wadi Al Hamand watershed. *(Adapted from Abdekareem et al. (2023).)*

Saudi Arabia. Twelve groundwater controlling factors such as elevation, slope, terrain roughness index, curvature, drainage density (Dd), TWI, distance from the river, lithology, soil, lineaments, NDVI, and rainfall were processed, normalized, and combined with GIS-based frequency ratio (FR) and overlay analysis model. The outcome showed the areas with potential for groundwater, which is divided into five classes: very good, good, moderate, low, and very low, which together account for 2.24%, 5.81%, 13.39%, 53.90%, and 24.65% of the total area (Abdekareem et al., 2023).

3.3 GIS in water quality mapping and monitoring

Water is a vital factor for all aspects of sustainable development, and water resources are an essential tool in meeting humans' needs to increase productivity, economic development, protection of the environment, and ecosystem services (Connor, 2015). Since the past few decades, an increase in anthropogenic activities (domestic sewage and industrial waste discharges, agriculture practices, landscape changes, human settlement, and encroachment) and climate change, the quality of water has decreased dramatically (Duran Zuazo et al., 2009, pp. 27–48; Nagaraju et al., 2016; Usali & Ismail, 2010). The term water quality refers to the suitability of water bodies for the intended purpose and monitoring water quality by collecting the samples of water bodies and analyzing to provide the status of water quality of groundwater and inland water bodies (river, lake, reservoir, wetlands, etc.) (Ray & Abeysingha, 2023; Tyagi et al., 2013). GIS plays a crucial role in determining the spatial variability of ground and surface water quality on local and regional levels (Deepchand et al., 2022; Megahed, 2020; Oseke et al., 2021).

An integrated approach of water quality indices (WQIs) with GIS offers decision-makers with rapid, detailed, and precise information to adopt water pollution and scarcity measures (Singh et al., 2013). Different water quality parameters such as total dissolved solids (TDSs), pH, electric conductivity (EC), alkalinity, salinity, dissolved oxygen (DO), total hardness (TH), fluoride (F$^-$), chloride (Cl$^-$), bromide (Br$^-$), nitrate (NO$_3^-$), phosphate (PO$_4^{3-}$), sulfate (SO$_4^{2-}$), carbonate (CO$_3^{2-}$), and bicarbonate (HCO$_3^-$), are measured and mapped through GIS (Deepchand et al., 2022; Rawat & Singh, 2018; Shil et al., 2019).

A study integrated the WQI with GIS and multivariate statistical techniques for the assessment of the water quality of the Mahanadi River (refer to Fig. 12.5) to analyze the suitability for drinking, agricultural, and industrial uses. Parameters such as temperature, pH, EC, DO, acidity, alkalinity, hardness, TDS, sodium (Na), potassium (K),

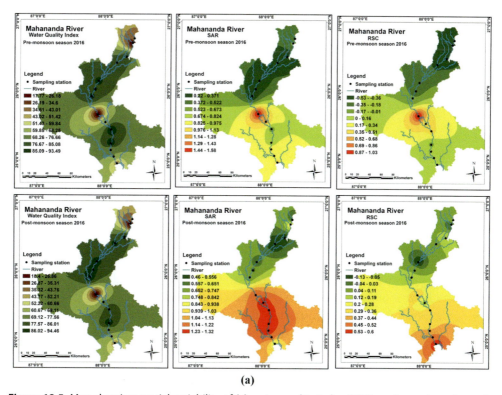

(a)

Figure 12.5 Map showing spatial variability of (a) water quality index (WQI), sodium adsorption ratio (SAR), and residual sodium carbonate (RSC), (b) magnesium hazard (MH), Kelly's index (KI), and permeability index (PI). *(Adapted from Shil et al. (2019).)*

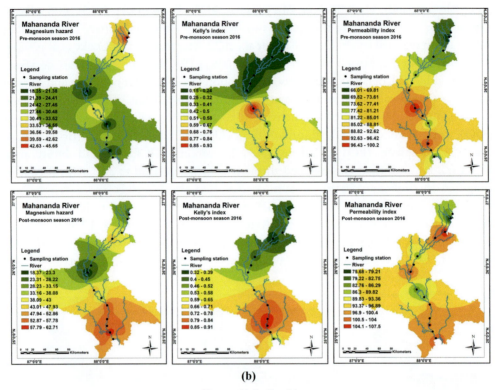

(b)

Figure 12.5 Cont'd

calcium (Ca), magnesium (Mg), fluoride (F$^-$), chloride (Cl$^-$), bromide (Br$^-$), nitrate (NO3$^-$), phosphate (PO$_4^{3-}$), sulfate (SO$_4^{2-}$), carbonate (CO$_3^{2-}$), and bicarbonate (HCO$_3^-$) were collected from fourteens sampling station in premonsoon and postmonsoon season, 2016. GIS-based IDW spatial interpolation technique was used to interpolate these parameters and prepare the spatial map. WQI, agriculture indices such as sodium residual ratio (SAR), sodium percentage (Na%), residual sodium carbonate (RSC), residual sodium bicarbonate (RSBC), Kelly's index, permeability index, potential salinity, and industrial-related indices such as Langelier Saturation Index (LSI), aggressive index, and Ryznar Stability Index (RSI) were used analyze the water quality of the river. The result reveals that the water quality of the river is moderate to good while providing poor to very poor quality at a few sampling stations (Shil et al., 2019). Another study applied the WQI and GIS-based evaluation to observe the decadal groundwater quality of Kanchipuram district of Tamil Nadu, India (refer to Fig. 12.6). Thirteen groundwater-controlling quality parameters were collected and statically analyzed. WQIs thematic maps were created using the IDW interpolation techniques in ArcGIS 10 software. A decadal physicochemical water quality parameter was used to calculate three distinct

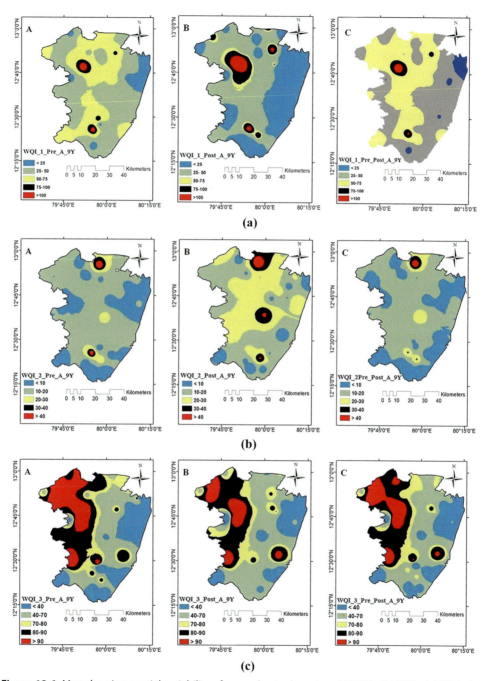

Figure 12.6 Map showing spatial variability of groundwater based on (a) WQI$_1$, (b) WQI$_2$, (c) WQI$_3$ during premonsoon, postmonsoon, and difference of pre-post monsoon. *(Adapted from Rawat & Singh (2018).)*

WQI models, primarily for drinking (WQI1 and WQI2) and irrigation (WQI3) needs. The study demonstrated that the groundwater of the area is impaired by anthropogenic activities which need proper management using both WQI and GIS methods providing useful information for water quality assessment in the future (Rawat & Singh, 2018).

4. GIS in coastal ecosystem management

The coastal ecosystem acts as a supporting system for many forms of life and provides a wide range of essential services for the development of human activities, for example, food and raw resources, tourism, carbon sequestration, storm and wave protection, erosion control, and water purification (Barbier et al., 2011). Globally, the multifunction and dynamic coastal ecosystems face severe stress due to factors such as the rapid growth in population worldwide (projected to reach 8 billion on November 15, 2022), urbanization, industrialization, trade, tourism activities, coastal development, pollution, eutrophication, increase in sea level rise (0.25−0.30 m, in the next 30 years 2020−2050), and climate change (UN DESA, 2022; Blackburn et al., 2019; Bunce et al., 2010; Burkett & Davidson, 2012; Nixon & Fulweiler, 2009; NOAA, 2023a,b; Rangel-Buitrago, 2020). Approximately 20% of the world's population lives within 25 km of the coastline and almost twice the population residing within 100 km of the coastal zone (Ahmed et al., 2021; Cohen et al., 1997). However, as per the estimation by Nicholls and Mimura (1998), about 600 million global populations will occupy the coastal floodplain land by 2100, and, a dramatic increase in coastal urbanization will adverse impact on the coastal landscape (Munizaga et al., 2022). Therefore, monitoring of the coastal ecosystem and changes in the coastal landscape should be done in continuous time intervals (Cloern & Jassby, 2012). Recently, integrated GIS and satellite-based remote sensing approaches have become effective tools for the monitoring, assessment, and mapping of, landscape change for different time periods, coastal regulation, and environmentally sensitive zones (Chettiyam Thodi et al., 2023; Marzouk et al., 2021; Yang, 2008). GIS is widely used for spatiotemporal analysis of shoreline change detection (Baig et al., 2020; Ciritci, 2019), coastal geomorphology mapping (Kaliraj et al., 2017; Vassilopoulos et al., 2008), and habitat mapping (Kaliraj et al., 2019; Wright & Heyman, 2008). GIS-based fuzzy logic (Mullick et al., 2019), AHP (Behera et al., 2019), artificial neural network (ANN) (Peponi et al., 2019), and machine learning (Mohanty et al., 2023) methods are mostly used for coastal erosion and vulnerability mapping.

4.1 GIS in spatiotemporal analysis of coastline change and erosion

The interface between the land and sea is known as the coastline, which varies erratically depending on a combination of factors such as its, dynamic environment setting, morphological characteristics, climate, or geological factors in nature (Mentaschi et al., 2018; Mujabar & Chandrasekar, 2013). Globally, coastal erosion become a major concern that influences practically every nation that has a coastline (Gracia et al., 2018). The term coastal erosion means the landward displacement of the shoreline caused by the forces of waves and currents (Prasad & Kumar, 2014). Shoreline change ongoing long-term process which may take millions of years and occurs as a result of both anthropogenic and natural factors (Gopinath et al., 2023; Mujabar et al., 2013). The phenomenon of coastline erosion and displacement in shoreline occurs due to different natural disasters such as hurricanes, tsunamis, cyclones, landslides, sea level rise, tectonic movement, tides, storms, waves, and coastal flooding (Gayathri et al., 2017; Lawrence, 1994; NOAA, 2023a,b; Oppenheimer et al., 2019; Passeri et al., 2015), and also executed by anthropogenic activities such as sand mining, construction activities along the coastline, offshore dredging, boat wake, and artificial water feature (breakwaters) (Gopinath et al., 2023; Neal, 2010). GIS combined with remote sensing techniques become popular these days for the spatiotemporal analysis of coastline change and vulnerability mapping (Al-Nasrawi et al., 2017; Fatima et al., 2015; Sousa et al., 2018). In the recent past, several studies have been conducted by researchers to detect the shoreline change, erosion, and accretion rate using GIS and remote sensing methods across the nation that has a coastline (Al-Attar & Basheer, 2023; Alesheikh et al., 2007; Kumar et al., 2010; Murray et al., 2023; Santos et al., 2021). A study was conducted to analyze the spatiotemporal change in the shoreline, of Vishakhapatnam coast, Andhra Pradesh, India, using GIS and multitemporal Landsat satellite imageries for the years 1991−2018. Digital shoreline analysis system (DSAS) tool was used to observe the coastal erosion and accretion rate using Arc GIS 10.6 software. The shoreline was delineated through visual digitization using the multitemporal (1991-2018) Landsat satellite and weighted linear regression (WLR), end point rate (EPR), and linear regression rate (LRR) were calculated to observe the shoreline change rate. Erosion rate was further classified into five classes, that is, high and low erosion, no change, and high and low accretion. 5.8 km of coastline region show high erosion followed by low erosion of 46.2 km. However, approximately 34.7 km of coastline region showed no change. Moderate and high accretion was observed within 30.5 and 17.8 km of the coastline region of Vishakhapatnam (refer to Fig. 12.7). The outcome of the study reveals that erosion dominates in Vishakhapatnam taluk, Ankapalli taluk, and Yellamanchili taluk, whereas Bhemunipatnam faces accretion due to the influence of natural and manmade (Baig et al., 2020).

5. Challenges and future recommendations

GIS-based interpolation methods are used to predict the variables at the unsampled site to create spatially distributed surface grids or contour maps (Li & Heap, 2011). However, the quality of the interpolation results may be reduced due to the uneven distribution of sampling data points, which only produce the maximum and minimum values of the interpolated surface at sample points (Li & Heap, 2014; Mitas & Mitasova, 1999;

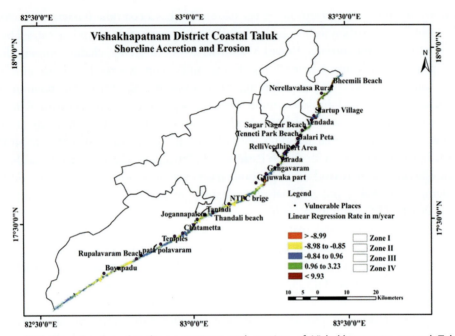

Figure 12.7 Map showing shoreline accretions and erosion of Vishakhapatnam coastal Taluks. *(Adapted from Baig et al. (2020).)*

Zaresefat et al., 2023). Another challenge in GIS monitoring includes the unavailability of precise software for handling spatial statistics and complex datasets (Yeh, 1991). The precision and accuracy of the results may vary due to geographical errors as GIS handles large datasets (Paramasivam, 2019). In the upcoming decades, governments and businesses around the world are increasingly using GIS as a way of mapping and analyzing geographic data, providing user-friendly information in order to better manage their resources and services (Lü et al., 2019; Stepniak and Turek, 2020). In the field of water resource management and hydrological studies, GIS has not yet been integrated with spatial decision support system (SDSS), as there is a continuous variation in the water parameters, becoming a major challenge for researchers. Combining virtual (VR), augmented (AR), and mixed reality (MR) environments with a geographic analysis model, GIS can provide dynamic geographic scenarios for geographic analysis in a combined virtual and realistic way (Çöltekin et al., 2020), further allow us to enter a real virtual geographic environment (VGE) age (Priestnall et al., 2012). In near future, decision support system (DSS) can be used to develop the decision support indices (DSIs), which may help the researchers and environmentalists to prepare a prospective plan for coastal vulnerability assessment and monitoring (Barzehkar et al., 2021). The integration of statistical methods in GIS software will enhance the capability of spatiotemporal analysis at local, regional, and global scale (Machiwal et al., 2018).

6. Conclusion

GIS technology plays a critical role in the management of the environmental system. Large-scale environmental studies find GIS to be an essential tool. It facilitates quick information access, safety, efficiency, and improved decisions for resource management. GIS integrates tools that enable real-time data display and is more appropriate for emergency operations. GIS also facilitates real-time monitoring of aquatic environments, enabling precise mapping of groundwater potential and quality. It aids in assessing flood vulnerability by analyzing topographical data and hydrological patterns. Researchers can create comprehensive models to understand and manage groundwater resources potential through GIS. Therefore, its integration with advanced techniques enhances water quality mapping as well as seasonal changes assessment, offering valuable insights for sustainable aquatic ecosystem management. In this study, the application of GIS is broadly focused on water and coastal environment monitoring and assessment. The increasing usability of GIS has increased their relevance as tools for non-GIS specialists, like senior managers and subject matter experts from other fields of expertise. Further research should be performed to integrate GIS with new technologies such as big data, computer modeling, machine learning, deep learning, and other decision-making algorithms to produce web-based SDSSs.

Acknowledgment

Authors are thankful to SNIOE for the help and support throughout this research work.

References

Abdekareem, M., Abdalla, F., Al-Arifi, N., Bamousa, A. O., & El-Baz, F. (2023). Using remote sensing and GIS-based frequency ratio technique for revealing groundwater prospective areas at Wadi Al Hamdh watershed, Saudi Arabia. *Water, 15*, 1154.

Abdelaziz, N. M., Eldrandaly, K. A., Al-Saeed, S., Gamal, A., & Abdel-Basset, M. (2024). Application of GIS and IoT technology based MCDM for disaster risk management: Methods and case study. *Decision Making: Applications in Management and Engineering, 7*, 1−36.

Adimalla, N., & Taloor, A. K. (2020). Hydrogeochemical investigation of groundwater quality in the hard rock terrain of South India using Geographic Information System (GIS) and groundwater quality index (GWQI) techniques. *Groundwater for Sustainable Development, 10*, 100288.

Ahmad, M. W., Mourshed, M., Mundow, D., Sisinni, M., & Rezgui, Y. (2016). Building energy metering and environmental monitoring—A state-of-the-art review and directions for future research. *Energy and Buildings, 120*, 85−102.

Ahmed, N., Howlader, N., Hoque, M. A. A., & Pradhan, B. (2021). Coastal erosion vulnerability assessment along the eastern coast of Bangladesh using geospatial techniques. *Ocean and Coastal Management, 199*, 105408.

Al-Nasrawi, A. K., Hopley, C. A., Hamylton, S. M., & Jones, B. G. (2017). A spatio-temporal assessment of landcover and coastal changes at Wandandian delta system, Southeastern Australia. *Journal of Marine Science and Engineering, 5*(4), 55.

Al-Attar, I. M., & Basheer, M. A. (2023). Multi-temporal shoreline analysis and future regional perspective for Kuwait coast using remote sensing and GIS techniques. *Heliyon, 9*(9).

Albano, R., Sole, A., Adamowski, J., & Mancusi, L. (2014). A GIS-based model to estimate flood conse-quences and the degree of accessibility and operability of strategic emergency response structures in ur-ban areas. *Natural Hazards and Earth System Sciences, 14,* 2847−2865.

Alesheikh, A. A., Ghorbanali, A., & Nouri, N. (2007). Coastline change detection using remote sensing. *International Journal of Environmental Science and Technology, 4,* 61−66.

Artiola, J. F., Pepper, I. L., & Brusseau, M. L. (2004). *Environmental monitoring and characterization.* Academic Press.

Baig, M. R. I., Ahmad, I. A., Shahfahad, Tayyab, M., & Rahman, A. (2020). Analysis of shoreline changes in Vishakhapatnam coastal tract of Andhra Pradesh, India: An application of digital shoreline analysis system (DSAS). *Annals of GIS, 26*(4), 361−376.

Bansal, S. K., Singh, S., Sarkar, S., Pandey, P. C., Verma, J., Yadav, L., Chandra, L., et al. (2022). Environ-mental impact of sensing devices. In R. K. Sonkar, K. Singh, & R. Sonkawade (Eds.), *Smart Nanostructure Materials and Sensor Technology* (pp. 113−137). Singapore: Springer Nature Singapore. https://doi.org/10.1007/978-981-19-2685-3_6.

Barbier, E. B., Hacker, S. D., Kennedy, C., Koch, E. W., Stier, A. C., & Silliman, B. R. (2011). The value of estuarine and coastal ecosystem services. *Ecological Monographs, 81*(2), 169−193.

Barzehkar, M., Parnell, K. E., Soomere, T., Dragovich, D., & Engström, J. (2021). Decision support tools, systems and indices for sustainable coastal planning and management: A review. *Ocean and Coastal Man-agement, 212,* 105813.

Behera, R., Kar, A., Das, M. R., & Panda, P. P. (2019). GIS-based vulnerability mapping of the coastal stretch from Puri to Konark in Odisha using analytical hierarchy process. *Natural Hazards, 96,* 731−751.

Bibri, S. E. (2018). The IoT for smart sustainable cities of the future: An analytical framework for sensor-based big data applications for environmental sustainability. *Sustainable Cities and Society, 38,* 230−253.

Blackburn, S., Pelling, M., & Marques, C. (2019). Megacities and the coast: Global context and scope for transformation. In *Coasts and estuaries* (pp. 661−669). Elsevier.

Blackman, A., Li, Z., & Liu, A. A. (2018). Efficacy of command-and-control and market-based environ-mental regulation in developing countries. *Annual Review of Resource Economics, 10,* 381−404.

Boretti, A., & Rosa, L. (2019). Reassessing the projections of the world water development report. *NPJ Clean Water, 2,* 15.

Bunce, M., Rosendo, S., & Brown, K. (2010). Perceptions of climate change, multiple stressors and liveli-hoods on marginal African coasts. *Environment, Development and Sustainability, 12,* 407−440.

Burkett, V., & Davidson, M. (2012). *Coastal impacts, adaptation, and vulnerabilities.* Island Press.

Chen, L., Mao, Y., & Zhao, R. (2022). GIS application in environmental monitoring and risk assessment. In *3rd international conference on geology, mapping and remote sensing (ICGMRS)* (pp. 908−917). IEEE.

Chettiyam Thodi, M. F., Gopinath, G., Surendran, U. P., Prem, P., Al-Ansari, N., & Mattar, M. A. (2023). Using RS and GIS techniques to assess and monitor coastal changes of coastal islands in the marine envi-ronment of a humid tropical region. *Water, 15*(21), 3819.

Ciritci, D., & Türk, T. (2019). Automatic detection of shoreline change by geographical information system (GIS) and remote sensing in the Göksu delta, Turkey. *Journal of the Indian Society of Remote Sensing, 47,* 233−243.

Clements, A. L., Griswold, W. G., Rs, A., Johnston, J. E., Herting, M. M., Thorson, J., Collier-Oxandale, A., & Hannigan, M. (2017). Low-cost air quality monitoring tools: From research to practice (a workshop summary). *Sensors, 17,* 2478.

Cloern, J. E., & Jassby, A. D. (2012). Drivers of change in estuarine-coastal ecosystems: Discoveries from four decades of study in San Francisco Bay. *Reviews of Geophysics, 50*(4).

Çöltekin, A., Griffin, A. L., Slingsby, A., Robinson, A. C., Christophe, S., Rautenbach, V., Chen, M., Pettit, C., & Klippel, A. (2020). Geospatial information visualization and extended reality displays. *Manual of Digital Earth,* 229−277.

Cohen, J. E., Small, C., Mellinger, A., Gallup, J., & Sachs, J. (1997). Estimates of coastal populations. *Science, 278*(5341), 1209−1213.

Connor, R. (2015). *The United Nations world water development report 2015: Water for a sustainable world.* UNESCO publishing.

Conrad, C. C., & Hilchey, K. G. (2011). A review of citizen science and community-based environmental monitoring: Issues and opportunities. *Environmental Monitoring and Assessment, 176*, 273–291.

Das, S., Gupta, A., & Ghosh, S. (2017). Exploring groundwater potential zones using MIF technique in semi-arid region: A case study of Hingoli district, Maharashtra. *Spatial Information Research, 25*, 749–756.

Das, S., & Pardeshi, S. D. (2018). Integration of different influencing factors in GIS to delineate groundwater potential areas using IF and FR techniques: A study of Pravara basin, Maharashtra, India. *Applied Water Science, 8*, 1–16.

Dash, P., & Sar, J. (2020). Identification and validation of potential flood hazard area using GIS-based multi-criteria analysis and satellite data-derived water index. *Journal of Flood Risk Management, 13*, e12620.

Deepchand, Khan, N. A., Saxena, P., & Goyal, S. K. (2022). Assessment of supply water quality using GIS tool for selected locations in Delhi—a case study. *Air, Soil and Water Research, 15*.

Demers, M. N. (2008). *Fundamentals of geographic information systems*. John Wiley & Sons.

Dewan, A., & Dewan, A. M. (2013). *Hazards, risk, and vulnerability*. Springer.

Duran Zuazo, V. H., Rodríguez Plegezuelo, C., Flanagan, D. C., Francia Martinez, J. R., & Martinez Raya, A. (2009). *Agricultural runoff: New research trends. Agricultural runoff, coastal engineering and flooding* (pp. 27–48). Indiana, USA: Nova Science Publishers, Inc.

Enan, N. M., & Xi, C. (2015). Geographic information system as a tool of environmental management solution in Rwanda. *East African Journal of Science and Technology, 5*.

Fatima, H., Arsalan, M. H., Khalid, A., Marjan, K., & Kumar, M. (September 2015). Spatio-temporal analysis of shoreline changes along Makran Coast using remote sensing and geographical information system. In *Proceedings of the fourth international conference on space science and technology (ICASE)* (pp. 2–4). Islamabad, Pakistan.

Fenglin, W., Ahmad, I., Zelenakova, M., Fenta, A., Dar, M. A., Teka, A. H., Belew, A. Z., Damtie, M., Berhan, M., & Shafi, S. N. (2023). Exploratory regression modeling for flood susceptibility mapping in the GIS environment. *Scientific Reports, 13*, 247.

Feoli, E., Vuerich, L. G., & Zerihun, W. (2002). Evaluation of environmental degradation in northern Ethiopia using GIS to integrate vegetation, geomorphological, erosion and socio-economic factors. *Agriculture, Ecosystems and Environment, 91*, 313–325.

Gaikwad, S. K., Kadam, A. K., Ramgir, R. R., Kashikar, A. S., Wagh, V. M., Kandekar, A. M., Gaikwad, S. P., Madale, R. B., Pawar, N. J., & Kamble, K. D. (2020). Assessment of the groundwater geochemistry from a part of west coast of India using statistical methods and water quality index. *HydroResearch, 3*, 48–60.

Gavrilescu, M., Demnerová, K., Aamand, J., Agathos, S., & Fava, F. (2015). Emerging pollutants in the environment: Present and future challenges in biomonitoring, ecological risks and bioremediation. *New Biotechnology, 32*, 147–156.

Gayathri, R., Bhaskaran, P. K., & Jose, F. (2017). Coastal inundation research: An overview of the process. *Current Science*, 267–278.

Glicksman, R. L., Buzbee, W. W., Mandelker, D. R., Hammond, E., & Camacho, A. (2023). *Environmental protection: Law and policy*. Aspen Publishing.

Golge, M., Yenilmez, F., & Aksoy, A. (2013). Development of pollution indices for the middle section of the Lower Seyhan basin (Turkey). *Ecological Indicators, 29*, 6–17.

Goodchild, M. F. (2003). Geographic information science and systems for environmental management. *Annual Review of Environment and Resources, 28*, 493–519.

Gopinath, G., Thodi, M. F. C., Surendran, U. P., Prem, P., Parambil, J. N., Alataway, A., ... Mattar, M. A. (2023). Long-term shoreline and islands change detection with digital shoreline analysis using RS data and GIS. *Water, 15*(2), 244.

Gracia, A., Rangel-Buitrago, N., Oakley, J. A., & Williams, A. T. (2018). Use of ecosystems in coastal erosion management. *Ocean and Coastal Management, 156*, 277–289.

Guru, B., Seshan, K., & Bera, S. (2017). Frequency ratio model for groundwater potential mapping and its sustainable management in cold desert, India. *Journal of King Saud University Science, 29*, 333–347.

Hadipour, V., Vafaie, F., & Deilami, K. (2020). Coastal flooding risk assessment using a GIS-based spatial multi-criteria decision analysis approach. *Water, 12*, 2379.

Hossen, H., Ibrahim, M. G., Mahmod, W. E., Negm, A., Nadaoka, K., & Saavedra, O. (2018). Forecasting future changes in Manzala Lake surface area by considering variations in land use and land cover using remote sensing approach. *Arabian Journal of Geosciences, 11*, 1—17.

Hull, R. B., Richert, D., Seekamp, E., Robertson, D., & Buhyoff, G. J. (2003). Understandings of environmental quality: Ambiguities and values held by environmental professionals. *Environmental Management, 31*, 0001—0013.

Hunt, C. E. (2007). *Thirsty planet: Strategies for sustainable water management.* Academic Foundation.

Jhariya, D., Khan, R., Mondal, K., Kumar, T., Indhulekha, K., & Singh, V. K. (2021). Assessment of groundwater potential zone using GIS-based multi-influencing factor (MIF), multi-criteria decision analysis (MCDA) and electrical resistivity survey techniques in Raipur city, Chhattisgarh, India. *AQUA—Water Infrastructure, Ecosystems and Society, 70*, 375—400.

Kaliraj, S., Chandrasekar, N., & Amachandran, K. K. (2019). Coastal habitat vulnerability of Southern India: A multiple parametric approach of GIS based HVI (habitat vulnerability index) model. *Geografia Fisica e Dinamica Quaternaria, 42*, 27—41.

Kaliraj, S., Chandrasekar, N., & Ramachandran, K. K. (2017). Mapping of coastal landforms and volumetric change analysis in the south west coast of Kanyakumari, South India using remote sensing and GIS techniques. *The Egyptian Journal of Remote Sensing and Space Science, 20*(2), 265—282.

Khailani, D. K., & Perera, R. (2013). Mainstreaming disaster resilience attributes in local development plans for the adaptation to climate change induced flooding: A study based on the local plan of Shah Alam city, Malaysia. *Land Use Policy, 30*, 615—627.

Khan, Z. A., & Jhamnani, B. (2023). Identification of groundwater potential zones of Idukki district using remote sensing and GIS-based machine-learning approach. *Water Supply, 23*.

Kiker, G. A., Bridges, T. S., Varghese, A., Seager, T. P., & Linkov, I. (2005). Application of multicriteria decision analysis in environmental decision making. *Integrated Environmental Assessment and Management: An International Journal, 1*, 95—108.

Kumar, T., Mahendra, R., Nayak, S., Radhakrishnan, K., & Sahu, K. (2010). Coastal vulnerability assessment for Orissa State, east coast of India. *Journal of Coastal Research, 263*, 523—534. https://doi.org/10.2112/09-1186.1

Lü, G., Batty, M., Strobl, J., Lin, H., Zhu, A.-X., & Chen, M. (2019). Reflections and speculations on the progress in geographic information systems (GIS): A geographic perspective. *International Journal of Geographical Information Science, 33*, 346—367.

Lawrence, P. L. (1994). Natural hazards of shoreline bluff erosion: A case study of Horizon View, Lake Huron. *Geomorphology, 10*(1—4), 65—81.

Li, J., & Heap, A. D. (2011). A review of comparative studies of spatial interpolation methods in environmental sciences: Performance and impact factors. *Ecological Informatics, 6*, 228—241.

Li, J, & Heap, AD (2014). Spatial interpolation methods applied in the environmental sciences: A review. *Environmental Modelling & Software, 53*, 173—189.

Lindenmayer, D. B., & Likens, G. E. (2010). The science and application of ecological monitoring. *Biological Conservation, 143*, 1317—1328.

Lovett, A. (2014). GIS and environmental management. In *Environmental science for environmental management.* Routledge.

Machiwal, D., Cloutier, V., Güler, C., & Kazakis, N. (2018). A review of GIS-integrated statistical techniques for groundwater quality evaluation and protection. *Environmental Earth Sciences, 77*, 1—30.

Magesh, N. S., Chandrasekar, N., & Soundranayagam, J. P. (2012). Delineation of groundwater potential zones in Theni district, Tamil Nadu, using remote sensing, GIS and MIF techniques. *Geoscience Frontiers, 3*, 189—196.

Marchi, L., Borga, M., Preciso, E., & Gaume, E. (2010). Characterisation of selected extreme flash floods in Europe and implications for flood risk management. *Journal of Hydrology, 394*, 118—133.

Marzouk, M., Attia, K., & Azab, S. (2021). Assessment of coastal vulnerability to climate change impacts using GIS and remote sensing: A case study of Al-alamein new city. *Journal of Cleaner Production, 290*, 125723.

Matheson, T. (2022). Disposal is not free: Fiscal instruments to internalize the environmental costs of solid waste. *International Tax and Public Finance, 29*, 1047—1073.

Megahed, H. A. (2020). GIS-based assessment of groundwater quality and suitability for drinking and irrigation purposes in the outlet and central parts of Wadi El-Assiuti, Assiut Governorate, Egypt. *Bulletin of the National Research Centre, 44*, 1—31.

Mentaschi, L., Vousdoukas, M. I., Pekel, J. F., Voukouvalas, E., & Feyen, L. (2018). Global long-term observations of coastal erosion and accretion. *Scientific Reports, 8*(1), 12876.

Miller, J., & Rogan, J. (2007). Using GIS and remote sensing for ecological mapping and monitoring. *Integration of GIS and remote sensing, 3*, 233.

Mitas, L., & Mitasova, H. (1999). Spatial interpolation. In: Geographical Information Systems: Principles, Techniques, Management and Applications. *1*(2), 481—492.

Mohamed, M. M., & Elmahdy, S. I. (2017). Fuzzy logic and multi-criteria methods for groundwater potentiality mapping at Al fo'ah area, the United Arab Emirates (UAE): An integrated approach. *Geocarto International, 32*, 1120—1138.

Mohanty, B., Sarkar, R., & Saha, S. (2023). Preparing coastal erosion vulnerability index applying deep learning techniques in Odisha state of India. *International Journal of Disaster Risk Reduction, 96*, 103986.

Mujabar, P. S., & Chandrasekar, N. (2013). Shoreline change analysis along the coast between Kanyakumari and Tuticorin of India using remote sensing and GIS. *Arabian Journal of Geosciences, 6*, 647—664.

Mullick, M. R. A., Tanim, A. H., & Islam, S. S. (2019). Coastal vulnerability analysis of Bangladesh coast using fuzzy logic-based geospatial techniques. *Ocean and Coastal Management, 174*, 154—169.

Munizaga, J., García, M., Ureta, F., Novoa, V., Rojas, O., & Rojas, C. (2022). Mapping coastal wetlands using satellite imagery and machine learning in a highly urbanized landscape. *Sustainability, 14*(9), 5700.

Murray, J., Adam, E., Woodborne, S., Miller, D., Xulu, S., & Evans, M. (2023). Monitoring shoreline changes along the southwestern coast of South Africa from 1937 to 2020 using varied remote sensing data and approaches. *Remote Sensing, 15*(2), 317.

Nagaraju, A., Thejaswi, A., & Sreedhar, Y. (2016). Assessment of groundwater quality of Udayagiri area, Nellore district, Andhra Pradesh, South India using multivariate statistical techniques. *Earth Sciences Research Journal, 20*, E1—E7.

Naghibi, S. A., Pourghasemi, H. R., & Dixon, B. (2016). GIS-based groundwater potential mapping using boosted regression tree, classification and regression tree, and random forest machine learning models in Iran. *Environmental Monitoring and Assessment, 188*, 1—27.

Nagy, G., & Wagle, S. (1979). Geographic data processing. *ACM Computing Surveys, 11*, 139—181.

National Ocean and Atmospheric Administration (NOAA). (2023a). https://oceanservice.noaa.gov/hazards/sealevelrise/sealevelrise-tech-report.html. (Accessed 29 November 2023).

National Ocean and Atmospheric Administration (NOAA). (2023b). https://oceanservice.noaa.gov/facts/coastalthreat.html#:~:text=The%20threats%20to%20coastal%20communities,declarations%20is%20related%20to%20flooding. (Accessed 30 November 2023).

Neal, W. J. (2010). Coastal topography, human impact on. *Marine Policy and Economics*, 424.

Nixon, S. W., & Fulweiler, R. W. (2009). Nutrient pollution, eutrophication, and the degradation of coastal marine ecosystems. *Global Loss of Coastal Habitats: Rates, Causes and Consequences*, 23—58.

Oppenheimer, M., Glavovic, B., Hinkel, J., Van de Wal, R., Magnan, A. K., Abd-Elgawad, A., … Sebesvari, Z. (2019). *Sea level rise and implications for low lying islands, coasts and communities.*

Oseke, F. I., Anornu, G. K., Adjei, K. A., & Eduvie, M. O. (2021). Assessment of water quality using GIS techniques and water quality index in reservoirs affected by water diversion. *Water-Energy Nexus, 4*, 25—34.

Paramasivam, C. (2019). *Merits and demerits of GIS and geostatistical techniques* (pp. 17—21). GIS and Geostatistical Techniques for Groundwater Science.

Passeri, D. L., Hagen, S. C., Medeiros, S. C., Bilskie, M. V., Alizad, K., & Wang, D. (2015). The dynamic effects of sea level rise on low-gradient coastal landscapes: A review. *Earth's Future, 3*(6), 159—181.

Peponi, A., Morgado, P., & Trindade, J. (2019). Combining artificial neural networks and GIS fundamentals for coastal erosion prediction modeling. *Sustainability, 11*(4), 975.

Prasad, D. H., & Kumar, N. D. (2014). Coastal erosion studies—a review. *International Journal of Geosciences, 3.*

Prasad, P., Loveson, V. J., Kotha, M., & Yadav, R. (2020). Application of machine learning techniques in groundwater potential mapping along the west coast of India. *GIScience and Remote Sensing, 57*, 735—752.

Priestnall, G., Jarvis, C., Burton, A., Smith, M., & Mount, N. J. (2012). Virtual geographic environments. *Teaching Geographic Information Science and Technology in Higher Education*, 257–288.

Qi, H., & Altinakar, M. S. (2011). A GIS-based decision support system for integrated flood management under uncertainty with two dimensional numerical simulations. *Environmental Modelling and Software, 26*, 817–821.

Rahmati, O., Nazari Samani, A., Mahdavi, M., Pourghasemi, H. R., & Zeinivand, H. (2015). Groundwater potential mapping at Kurdistan region of Iran using analytic hierarchy process and GIS. *Arabian Journal of Geosciences, 8*, 7059–7071.

Rangel-Buitrago, N., Neal, W. J., Bonetti, J., Anfuso, G., & de Jonge, V. N. (2020). Vulnerability assessments as a tool for the coastal and marine hazards management: An overview. *Ocean and Coastal Management, 189*, 105134.

Rawat, K., & Singh, S. K. (2018). Water quality indices and GIS-based evaluation of a decadal groundwater quality. *Geology, Ecology, and Landscapes, 2*, 240–255.

Ray, R. L., & Abeysingha, N. S. (2023). Introductory chapter: Water resources planning, monitoring, conservation, and management. In *River basin management-under a changing climate*. IntechOpen.

Reddy, G. O. (2018). Spatial data management, analysis, and modeling in GIS: Principles and applications. *Geospatial Technologies in Land Resources Mapping, Monitoring and Management*, 127–142.

Samela, C., Troy, T. J., & Manfreda, S. (2017). Geomorphic classifiers for flood-prone areas delineation for data-scarce environments. *Advances in Water Resources, 102*, 13–28.

Santos, C. A. G., do Nascimento, T. V. M., Mishra, M., & da Silva, R. M. (2021). Analysis of long-and short-term shoreline change dynamics: A study case of João Pessoa city in Brazil. *Science of the Total Environment, 769*, 144889.

Shil, S., Singh, U. K., & Mehta, P. (2019). Water quality assessment of a tropical river using water quality index (WQI), multivariate statistical techniques and GIS. *Applied Water Science, 9*, 1–21.

Singh, L. K., Jha, M. K., & Chowdary, V. (2018). Assessing the accuracy of GIS-based multi-criteria decision analysis approaches for mapping groundwater potential. *Ecological Indicators, 91*, 24–37.

Singh, S. K., Srivastava, P. K., Pandey, A. C., & Gautam, S. K. (2013). Integrated assessment of groundwater influenced by a confluence river system: Concurrence with remote sensing and geochemical modelling. *Water Resources Management, 27*, 4291–4313.

Soleimani, H., Abbasnia, A., Yousefi, M., Mohammadi, A. A., & Khorasgani, F. C. (2018). Data on assessment of groundwater quality for drinking and irrigation in rural area Sarpol-e Zahab city, Kermanshah province, Iran. *Data in Brief, 17*, 148–156.

Sousa, W. R. D., Souto, M. V., Matos, S. S., Duarte, C. R., Salgueiro, A. R., & Neto, C. A. D. S. (2018). Creation of a coastal evolution prognostic model using shoreline historical data and techniques of digital image processing in a GIS environment for generating future scenarios. *International Journal of Remote Sensing, 39*(13), 4416–4430.

Stepniak, C., & Turek, T. (2020). Possibilities of using GIS technology for dynamic planning of investment processes in cities. *Procedia Computer Science, 176*, 3225–3234.

Steinwender, A., Gundacker, C., & Wittmann, K. J. (2008). Objective versus subjective assessments of environmental quality of standing and running waters in a large city. *Landscape and Urban Planning, 84*, 116–126.

Sunitha, V., Reddy, B. M., Kumar, M. J., & Reddy, M. R. (2012). GIS based groundwater quality mapping in southeastern part of Anantapur District, Andhra Pradesh, India. *International Journal of Geomatics and Geosciences, 2*, 805–814.

Tao, W. (2013). Interdisciplinary urban GIS for smart cities: Advancements and opportunities. *Geo-spatial Information Science, 16*, 25–34.

Tarawneh, M. S. M., Janardhana, M., & Ahmed, M. M. (2019). Hydrochemical processes and groundwater quality assessment in North eastern region of Jordan valley, Jordan. *HydroResearch, 2*, 129–145.

Tien, J. M. (2017). Internet of things, real-time decision making, and artificial intelligence. *Annals of Data Science, 4*, 149–178.

Tomczyk, A. M. (2011). A GIS assessment and modelling of environmental sensitivity of recreational trails: The case of Gorce National Park, Poland. *Applied Geography, 31*, 339–351.

Tsihrintzis, V. A., Hamid, R., & Fuentes, H. R. (1996). Use of geographic information systems (GIS) in water resources: A review. *Water resources management, 10*, 251–277. https://doi.org/10.1007/BF00508896

Twumasi, Y. A., & Merem, E. C. (2006). GIS and remote sensing applications in the assessment of change within a coastal environment in the Niger Delta region of Nigeria. *International Journal of Environmental Research and Public Health, 3*, 98–106.

Tyagi, S., Sharma, B., Singh, P., & Dobhal, R. (2013). Water quality assessment in terms of water quality index. *American Journal of water resources, 1*, 34–38.

Uc Castillo, J. L., Martínez Cruz, D. A., Ramos Leal, J. A., Tuxpan Vargas, J., Rodríguez Tapia, S. A., & Marín Celestino, A. E. (2022). Delineation of groundwater potential zones (GWPZs) in a semi-arid basin through remote sensing, GIS, and AHP approaches. *Water, 14*, 2138.

United Nations Department of Economic and Social Affairs, Population Division. (2022). *World population prospects 2022: Summary of results*. UN DESA/POP/2022/TR/NO. 3.

Usali, N., & Ismail, M. H. (2010). Use of remote sensing and GIS in monitoring water quality. *Journal of Sustainable Development, 3*, 228.

Vassilopoulos, A., Green, D. R., Gournelos, T. H., Evelpidou, N., Gkavakou, P., & Koussouris, S. (2008). Using GIS to study the coastal geomorphology of the Acheloos river mouth in West Greece. *Journal of Coastal Conservation, 11*, 209–213.

Verburg, P. H., Neumann, K., & Nol, L. (2011). Challenges in using land use and land cover data for global change studies. *Global Change Biology, 17*, 974–989.

Virapongse, A., Brooks, S., Metcalf, E. C., Zedalis, M., Gosz, J., Kliskey, A., & Alessa, L. (2016). A social-ecological systems approach for environmental management. *Journal of Environmental Management, 178*, 83–91.

Vos, P., Meelis, E., & Ter Keurs, W. (2000). A framework for the design of ecological monitoring programs as a tool for environmental and nature management. *Environmental Monitoring and Assessment, 61*, 317–344.

Wassie, S. B. (2020). Natural resource degradation tendencies in Ethiopia: A review. *Environmental Systems Research, 9*, 1–29.

Wegmann, M., Leutner, B., & Dech, S. (2016). *Remote sensing and GIS for ecologists: Using open source software*. Pelagic Publishing Ltd.

Wondim, Y. K. (2016). Flood hazard and risk assessment using GIS and remote sensing in lower Awash sub-basin, Ethiopia. *Journal of Environment and Earth Science, 6*, 69–86.

Wright, D. J., & Heyman, W. D. (2008). Introduction to the special issue: Marine and coastal GIS for geomorphology, habitat mapping, and marine reserves. *Marine Geodesy, 31*(4), 223–230.

Yang, X. (Ed.). (2008). *Remote sensing and geospatial technologies for coastal ecosystem assessment and management*. Springer Science & Business Media.

Yeh, A. G.-O. (1991). The development and applications of geographic information systems for urban and regional planning in the developing countries. *International journal of geographical information system, 5*, 5–27.

Youssef, A., Pradhan, B., & Tarabees, E. (2011). Integrated evaluation of urban development suitability based on remote sensing and GIS techniques: Contribution from the analytic hierarchy process. *Arabian Journal of Geosciences, 4*.

Zahra, S. M., Shahid, M. A., Maqbool, Z., Sabir, R. M., Safdar, M., Majeed, M. D., & Sarwar, A. (2023). *Application of geospatial techniques in agricultural resource management*.

Zaresefat, M., Hosseini, S., & Ahrari Roudi, M. (2023). Addressing nitrate contamination in groundwater: The importance of spatial and temporal understandings and interpolation methods. *Water, 15*(24), 4220.

Zolekar, R. B., & Bhagat, V. S. (2015). Multi-criteria land suitability analysis for agriculture in hilly zone: Remote sensing and GIS approach. *Computers and Electronics in Agriculture, 118*, 300–321.

CHAPTER 13

Assessing coastal vulnerability to climate change-induced hazards in the Eastern Mediterranean: A comparative review of methodological approaches

Dimitrios-Vasileios Batzakis, Efthimios Karymbalis and Konstantinos Tsanakas
Department of Geography, School of Environment, Geography and Applied Economics, Harokopio University of Athens, Athens, Greece

1. Introduction

Coastal zones represent the transitional area where land meets the marine environment. They serve as the interface between aquatic and terrestrial ecosystems and possess inherent environmental worth due to their rich biological diversity, which plays a crucial role in facilitating the delivery of numerous ecosystem services that are fundamental for human well-being (Maes et al., 2013, pp. 1—58; Reid et al., 2005). Coastal areas house highly dynamic and intricate environmental systems, influenced both directly and indirectly by various factors that have contributed to their evolution over time. The present-day coastal landscape is the combined result of interactions among natural processes inherent to the system and external climatic and marine forces (Davidson—Arnott et al., 2019; Masselink and Gehrels 2014, p. 448). Historically, coastal regions have consistently attracted human populations and served as appealing places for human settlements, thanks to their unique natural characteristics that foster the concentration of human activities (Costanza, 1999). The alteration of land cover in these areas is experiencing a dramatic worldwide increase due to the growing and intensifying human use of the land. Consequently, human activities also wield considerable influence in shaping coastal dynamics, frequently imposing extra pressures that can surpass the influence of natural processes (Nicholls et al., 2007).

Over recent decades, there has been a substantial increase in human occupation along coastal regions (World Resources Institute, 2010). It is estimated that approximately 86 million people live within 10 km of the European coastline (European Environment Agency, 2006). Moreover, over the course of the previous century, the population residing along the Mediterranean coast has experienced remarkable growth. Approximately one-third of the total Mediterranean population is concentrated along these coastal areas, with around 120 million inhabitants residing in hydrological basins located in the southern region of the Mediterranean Sea (European Environment Agency, 2015). The Mediterranean coastal zone has a history spanning millennia of more or less intensive

Geographical Information Science
ISBN 978-0-443-13605-4
https://doi.org/10.1016/B978-0-443-13605-4.00013-8

human utilization. Alterations in coastal landscapes primarily stem from shifts in human land use patterns, which, in turn, are a consequence of broader changes in the socioeconomic context. In contemporary times, coastal rural areas are transforming because these locales no longer serve solely as agricultural sites, as they did in the past (Chalkias et al., 2011). Furthermore, many rural areas in coastal zones have evolved into "peri-urban" regions. Demographic trends have accentuated the disparity between populations residing in urban centers and those in the surrounding rural regions. However, starting in the 1980s, numerous Mediterranean cities underwent a swift shift from the traditional "compact growth" model to more "dispersed" ones characterized by extensive expansion of the built-up areas around the city core (Catalán et al., 2008; Salvati et al., 2012). The migration of urban dwellers to coastal rural and peri-urban areas has introduced new challenges related to changes in land use, land cover, and evolving perceptions of rural areas, local needs, and priorities for rural development (Romano & Zullo, 2014). These shifts in land utilization and land cover increase both the exposure and the vulnerability of coastal communities to various natural hazards (Satta et al., 2017). As a result, today, low-lying coastal regions around the Mediterranean face threats from climate change-induced hazards, such as accelerated global mean sea-level rise and extreme storm surge events (Hague et al., 2023; Hoggart et al., 2014; Jervejeva et al., 2014). Moreover, recent analysis indicates an escalating exposure and vulnerability to climate-induced hazards in Mediterranean coasts, particularly due to projected sea-level rise scenarios by the year 2100 (Falciano et al., 2023).

The purpose of this chapter is to present the indices that have been applied in areas of the Eastern Mediterranean to quantify the vulnerability of coastal areas to climate change-related hazards such as long-term sea-level rise and storm surges. Special mention is made of the parameters included in these indices, the methodological approach for their calculation, and the results of their application in various regions.

2. Climate change impacts on coastal areas of the Mediterranean Sea

The accelerated rise in sea levels, attributed to global warming, is a natural hazard with the potential to have a significant impact on coastal regions worldwide in the near future. This global mean sea-level rise results from several factors, including the expansion of seawater due to rising ocean temperatures, the melting of glaciers and ice sheets, and a decrease in liquid water stored on land (Allan & Komar, 2006). In a recent study by Nicholls et al. (2021), it was revealed that over the past 20 years, the global mean sea level has been rising at a pace of 2.5 mm per year. Furthermore, this rate is four times higher in coastal regions experiencing subsidence, resulting in an average relative sea-level increase ranging from 7.8 to 9.9 mm per year. Although the exact future rates of global mean sea-level rise caused by global warming are still uncertain, the latest models and the most severe climate scenarios (referred to as shared socioeconomic pathways or SSP) suggest that by the end of the 21st century, the global mean sea level could potentially rise by as much as 1.10 m (Masson-Delmotte et al., 2021). This upward trend is anticipated to persist in the coming years.

In addition to the long-term rise in sea level, there is another coastal threat posed by extreme storm surge events. These events, also known as meteorological residuals or meteorological tides, are significant components of extreme water levels along coastal areas, alongside waves, and tidal oscillations (Losada et al., 2013; Lowe et al., 2010). Storm surges are driven by wind-induced water movements toward or away from the coast and by variations in water level caused by atmospheric pressure changes, known as the inverse barometric effect. Several studies have reported an increase in both the intensity and frequency of extreme water levels in various coastal regions worldwide (Ullmann & Monbaliu, 2010; Wang et al., 2014; Weisse et al., 2014). Therefore, the expected rise in extreme total water levels due to relative sea-level rise can be further amplified by an increase in the level of extreme storm surges, which can surpass 30% of the relative sea-level rise (Vousdoukas et al., 2016).

Current sea-level trends in the Mediterranean basin have been determined by examining tide-gauge records spanning more than 35 years. For the longest records, a sea-level increase of approximately 1.2–1.5 ± 0.1 mm/year has been calculated (Marcos & Tsimplis, 2008). Decadal sea-level patterns in the Mediterranean sometimes diverge from global trends. Notably, during the 1990s, the Mediterranean experienced an accelerated rise in sea level, reaching up to 5 mm/year, exceeding the global average (largely attributed to higher temperatures). The sea-level trend observed in Alexandria (Egypt) between 1944 and 1989, which is measured at the sole long-term gauge station in the Eastern Mediterranean, stands at 1.9 ± 0.2 mm/year, the highest among the stations in the basin (Marcos & Tsimplis, 2008). Projections indicate that by 2050, the Mediterranean Sea is expected to experience a rise in its overall level, estimated to be between 7 and 12 cm compared to previous decades (Gualdi et al., 2013). This sea-level rise is anticipated to be more pronounced along the coasts of the Eastern and Southern Mediterranean. Furthermore, the latest Assessment Report on "Climate and Environmental Change in the Mediterranean basin" underscores that the Mediterranean region is experiencing a warming trend 20% more rapid than the global average, making it a prominent climate change hotspot on a global scale (Cramer et al., 2020).

Regional projections of storm surge levels in the Mediterranean have been developed in multiple studies (Conte et al., 2013; Jordà et al., 2012; Marcos et al., 2011). These projections indicate a general downward trend in the intensity of extreme storm surges in the Mediterranean Sea over 150 years (1951–2100) across most of the climate change scenarios. This decline is primarily linked to a decrease in the frequency of local peak events and alterations in storm surges' duration and geographical extent. However, certain subregions within the Mediterranean may experience an increase in the magnitudes of sea surface elevation extremes during the 21st century. Notably, the contributions of wind patterns and atmospheric pressure variations to sea-level heights differ significantly among various Mediterranean regions, and seasonal fluctuations in extreme values also exhibit marked variations. The Aegean and Adriatic Seas are characteristic examples, with elevated surge levels primarily attributed to low-pressure systems and favorable wind patterns, respectively (Androulidakis et al., 2015).

Some of the notable adverse consequences stemming from climate change-induced coastal hazards include coastal erosion, resulting in the loss of land of immense economic, social, and environmental importance, as well as damage to critical infrastructure. Moreover, there is the recurring issue of low-lying coastal plains experiencing frequent flooding, ecologically significant wetlands being inundated, and cultural and historical resources facing potential threats. Additionally, this situation implies significant losses in coastal and marine habitats and ecosystems (Brierley & Kingsford, 2009).

According to findings derived from the "Large Scale Integrated Sea-level and Coastal Assessment Tool," a significant portion of the global sandy coastline is currently experiencing erosion (Vousdoukas et al., 2020). Furthermore, half of the world's sandy beaches could be classified as highly threatened by the end of the century if no measures to mitigate greenhouse gas emissions are put into effect. The results of these studies align with earlier analyses that have focused on assessing the historical changes in shorelines globally (Mentaschi et al., 2018) and identifying areas prone to inundation (Kulp & Strauss, 2019). These data underscore the persistent threat faced by coastal regions and their associated environments at a global scale due to the intensification of climate- and marine-related coastal hazardous processes (Kirezci et al., 2020). Recent research conducted in the Mediterranean region indicates that without coastal protection or adaptation strategies and under the worst-case climate conditions, there could be a 48% increase in the world's land area at risk of flooding by 2100, endangering 52% of the global population and 46% of global assets (Kirezci et al., 2020). Therefore, forthcoming climate scenarios are expected to indicate a notable rise in climate-related risks in the Mediterranean region in the coming decades, with many coastal areas vulnerable to the effects of rising sea levels (Antonioli et al., 2020) and storm surges.

3. Revealing the vulnerability of the coastal areas during climate change

Given the above considerations, there is a pressing requirement to comprehend the potential changes in Mediterranean coastal areas due to climate change and to create assessment methods that can gauge the resulting vulnerabilities and risks. The assessment of coastal vulnerability and risk, combined with the sustainable preservation of coastal regions, holds significant importance for integrated coastal management (Nicholls et al., 2021). It is essential to pinpoint "vulnerable" stretches of coastline and evaluate the social vulnerability of coastal communities as a prerequisite for formulating coastal management plans. Numerous methods, primarily in the form of indices, have been suggested to forecast the changes in the coastal area and gauge the susceptibility of coastal communities to the impacts of rising sea levels. The relative vulnerability of various coastal environments to sea-level rise can be assessed by taking into account key variables that influence coastal changes in a specific region. These variables may include coastal geomorphology, slope, shoreline shifting, relative sea-level change rate, tide range, wave height, and other relevant factors. The methods suggested for assessing coastal risk to climate change effects at the regional level involve the consideration of qualitative and quantitative spatial characteristics, which encompass physical, ecological, and socioeconomic factors. There is a notable variation in the number of variables included in previously established coastal risk and vulnerability indices, both in earlier and more recent publications. Lately, studies on coastal hazards have emphasized the necessity of integrating factors beyond physical variables to encompass what is known as social vulnerability (Furlan et al., 2021). Social vulnerability encompasses a range of attributes that influence a community's capacity to react to, manage, recover from, and adapt to environmental hazards. According to an expanding body of research, socioeconomic and demographic factors are the primary determinants of social vulnerability (Flanagan et al., 2011). In this context, efforts have been made to assess risk in coastal areas by considering not only natural factors but also socioeconomic variables (Tate et al., 2010).

Geoinformatics, particularly Geographic Information Systems (GIS) and remote sensing, represents a crucial tool in environmental sciences. Its adoption for assessing the vulnerabilities of coastal areas to climate change-induced hazards is on the rise and gaining significance.

4. Assessing coastal vulnerability through index-based methodologies

Assessing the coastal vulnerability to climate change is a complex and multidisciplinary task that usually requires a combination of scientific approaches. In the recent years, a variety of different methodologies and techniques have been developed to assess the coastal vulnerability to climate change and its related hazards, such as sea-level rise, the increasing intensity of storm surges, coastal erosion, and coastal floods. Various studies conducted worldwide employ various approaches and tools to evaluate coastal areas' susceptibility to climate change impacts. These methodologies work toward rendering the level of sensitivity exhibited by the coastal environment and its components in response to coastal hazards (Rocha et al., 2023). Widely used tools in coastal management and adaptation planning are the coastal vulnerability indices. These indices combine multiple physical and, many times, socioenvironmental and socioeconomic factors/indicators to provide a holistic view of vulnerability (Anfuso et al., 2021; Roukounis & Tsihrintzis, 2022). Through the quantification of the vulnerability, they can provide crucial information to decision-makers for making choices regarding the short-term and long-term adaptation strategies and the mitigation policies of the physical and socioeconomic impact of climate change (Sarkar et al., 2022).

This chapter provides a comprehensive review of methodologies for evaluating coastal vulnerability through indices. This review pertains to the documentation of 34 case studies concerning the assessment of coastal vulnerability. The review encompasses studies that focus on countries in the Eastern Mediterranean and includes 17 studies from Greece, seven from Italy, two from Egypt and Lebanon, and one from Croatia, Slovenia, Malta, Turkey, and Syria (Fig. 13.1). All the methodologies reviewed in this chapter attempt to quantify coastal vulnerability via index-based methodologies.

The most popular index-based methodology is the Coastal Vulnerability Index (CVI) proposed by Gornitz (1991) and modified by Thieler and Hammar-Klose (1999). CVI is specifically designed to calculate and quantify the vulnerability of the coastal areas to the impacts of climate change, particularly the sea-level rise. The original formula (Gornitz, 1991) incorporates for calculations a range of physical parameters, including the coastal relief, the rock type (with respect to the relative resistance to erosion), the coastal landforms, the mean tide range, the maximum wave height, the relative sea-level change, and the shoreline displacement. However, the modified CVI of Thieler and Hammar-Klose (1999), (which considers as parameters the geomorphology, the coastal slope, the relative sea-level rise rate, the shoreline erosion/accretion rate, the mean tide range, and the mean

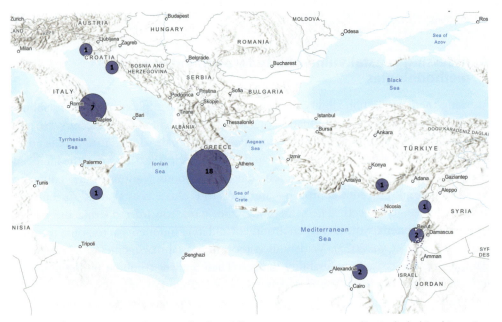

Figure 13.1 Map showing the coastal vulnerability assessment case studies included in this review.

wave height), has become more commonly used. These parameters are ranked into five classes according to their relative influence on the vulnerability (1: very low; 2: low; 3: moderate; 4: high; 5: very high). The ranking and the range of parameters' values may vary across different case studies as they are influenced by the data types employed and the specific characteristics of the local environment (Šimac et al., 2023). The CVI formula is expressed as the square root of the multiplication of the parameters ranking (1−5) divided by the number of parameters involved. The formula provides a single value that represents the integrated vulnerability. Higher CVI values indicate greater vulnerability. The results are classified into five vulnerability categories (very low, low, moderate, high, and very high). This review includes 12 case studies in the Eastern Mediterranean (see Table 13.1) that have applied this exact methodology to assess the vulnerability of the coasts of Greece, Egypt, Syria, and Lebanon (Alexandrakis et al., 2010; El Hage et al., 2011; Faour et al., 2013; Filippaki et al., 2023; Gaki-Papanastassiou et al., 2010; Georgiou and Alexandridis, 2021; Ghoussein et al., 2018; Hereher, 2015; Karymbalis et al., 2012; Komi et al., 2022; Mavromatidi & Karymbalis, 2016; Nasopoulou et al., 2012, p. 10). Moreover, variations of the original formula are used in many other case studies. Modifications of CVI are often driven by the need to provide more accurate assessment by incorporating additional parameters or altering the formula structure to better suit the specific objectives of the case studies or the characteristics of the coastal region being assessed. For example, Pantusa et al. (2022) added more physical variables (emerged beach width, dunes, river discharge, vegetation behind the back beach, coverage of Posidonia Oceanica) to the core parameters, expanding the list of parameters considered in the CVI to 11. Other researchers develop custom CVI formulas. Boumboulis et al. (2021) introduce a new index (CVI_{WF}), applied in the Gulf of Patras, by replacing the parameter of coastal geology and geomorphology with the geotechnical characterization, adding at the same time a higher weight factor into this variable to reflect its relative importance and their detailed work for evaluating it. In other studies, researchers develop more complex and multidimensional formulas for assessing coastal vulnerability that consider a wider array of impacts and receptors. These sophisticated models offer a deeper understanding of the various interacting factors influencing vulnerability. Torresan et al. (2012), applied in North Adriatic Sea an advanced model and considered how different impacts (e.g., sea-level rise, storm surge flooding, coastal erosion) affect various receptors (e.g., beaches, river mouth, wetlands, biological systems) by examining the complex relationships between these impacts and receptors.

Coastal vulnerability assessments often incorporate socioeconomic factors like population density, demographic characteristics, economic activities, land use patterns, etc. These factors play a significant role in determining the Mediterranean communities' exposure and their ability to adapt to climate change impacts (Furlan et al., 2021). By combining physical with socioeconomic factors in coastal vulnerability

Table 13.1 Summary of coastal vulnerability indices, their geographical application and the variables needed to implement them.

Index	Formula	Physical variables considered	Socioeconomic variables considered	Geographical application	Coastline Length	References
Coastal sensitivity index (CSI)	$CSI = \sqrt{\dfrac{a*b*c*d*e*f}{6}}$	Geomorphology, shoreline erosion/accretion rate, coastal slope, relative sea-level, mean significant wave height, mean tide range		Greece (Southern coast of the Gulf of Corinth)	148.00 km	Karymbalis et al. (2012)
Coastal vulnerability index (CVI)	$CVI = \sqrt{\dfrac{a*b*c*d*e*f}{6}}$	Geomorphology, shoreline erosion/accretion rate, coastal slope, relative sea-level, mean significant wave height, mean tide range		Greece (Aegean Sea coastline)	960.50 km	Alexandrakis et al. (2010)
				Greece (Northern coast of Gulf of Corinth)	63.62 km	Nasopoulou et al. (2012)
				Greece (Pieria Prefecture)	32.00 km	Mavromatidi and Karymbalis (2016)
				Greece (Thasos Island)	100.00 km	Georgiou and Alexandridis (2021) Conference Paper
				Greece (Coastal zone of Messolonghi)	101.10 km	Filippaki et al. (2023)
				Greece (Argolikos Gulf)	247.17 km	Gaki-Papanastassiou et al. (2010)
				Greece (Ios island)	110.74 km	Komi et al. (2022)
				Egypt (Mediterranean coast)	1000.00 km	Hereher (2015)
				Syria	183.00 km	Faour et al. (2013)
				Lebanon	375.00 km	El Hage et al. (2011)
				Lebanon (Southern coast)	93.00 km	Ghoussein et al. (2018)

Index	Formula	Variables	Variables	Location	Length	Reference
Integrated coastal vulnerability index (ICVI)	$\mathrm{ICVI.1} = \sqrt{\dfrac{a*b*c*d*e*f*g*h*i}{10}}$ $\mathrm{ICVI.2} = \dfrac{\mathrm{PVI}+\mathrm{SVI}}{2}$	Geomorphology, shoreline erosion/accretion rate, coastal slope, relative sea-level, mean significant wave height, mean tide range	Population density, land use/land cover, road network (distance from the coastline in km), railway network (distance from the coastline in km)	Greece (Southern coast of the Gulf of Corinth)	147.00 km	Ramnalis et al. (2023)
Integrated coastal vulnerability index (ICVI)	$\mathrm{ICVI.1} = \sqrt{\dfrac{a*b*c*d*e*f*g*h*i}{10}}$ $\mathrm{ICVI.2} = \dfrac{\mathrm{PVI}+\mathrm{SVI}}{2}$	Coastal slope, coastline landforms/features, significant wave height, shoreline change rate, sea-level rise, tidal data, coastal elevation	Population, road network, land use/land cover	Italy (Apulian coast)	40.00 km	De Serio et al. (2018)
Flood vulnerability index (FVI), based on the flood intermediate parameter (FIP)	$\mathrm{FIP} = \dfrac{(Ru+\sigma_{RU})+\xi}{B}$ $\mathrm{FIP} = \dfrac{Ru+\xi}{B}$	Storm surge, Berm/beach height, energy content, significant wave height, deepwater wave length, run-up, beach face slope, wave peak period		Greece (Aegean Sea—Thrace and Chania)	150.00 and 30.00 km, respectively	Kokkinos et al. (2014)
Coastal vulnerability index (CVI)	$\mathrm{CVI} = \sqrt{\dfrac{a*b*c*d*e*f*g}{7}}$	Mean absolute sea-level rise, mean significant wave height, mean tidal range, lithology and geomorphology, coastal slope, mean shoreline change, mean coastal vertical velocity		Greece (Rhodes Island)	253.00 km	Vandarakis et al. (2021)

Continued

Table 13.1 Summary of coastal vulnerability indices, their geographical application and the variables needed to implement them.—cont'd

Index	Formula	Physical variables considered	Socioeconomic variables considered	Geographical application	Coastline Length	References
Coastal vulnerability index (CVI_WF)	$CVI_{WF} = \sqrt{\dfrac{a*b*c*d*e*f}{6}}$	Shoreline change rates, significant wave height, tidal range, coastal slope relief, relative sea-level, geotechnical parameters		Greece (Western part of the Southern coastline of Gulf of Corinth)	40.00 km	Depountis et al. (2023)
Coastal vulnerability index (CVI) Social vulnerability index (SVI)	$CVI = \sqrt{\dfrac{a*b*c*d*e*f}{6}}$ $SVI_i = \sum_{A=1}^{6} x'_{A,i}$	Geomorphology, shoreline erosion/accretion rate, coastal slope, relative sea-level, mean significant wave height, mean tide range	Population density, share of women in t.p., share of persons above 65 in t.p., share of children below 5 in t.p., share of foreign-born in t.p., share of low educated in t.p., number of communities	Greece (Peloponnese)	1481.70 km	Tragaki et al. (2018)
Coastal vulnerability index (CVI) Socioeconomic vulnerability index (SocCVI)	$CVI = \sqrt{\dfrac{GEO*CS*RSLR*ERO*H_s}{6}}$ $SocCVI = \sqrt{\dfrac{CC+CF+SE}{3}}$	Geomorphology, shoreline erosion/accretion rate, coastal slope, relative sea-level, mean significant wave height, mean tide range	Settlement, cultural heritage, transport, land use, economic activities	Greece (Elounda Bay, Crete)	N/A	Alexandrakis et al. (2014)
Coastal vulnerability index (CVI)	$CVI = \sqrt{\dfrac{P1*...*P6*P9*...*P14}{12}}$	Seabed sediment thickness, beach sediment, distance from major faults, shoreline evolution, land slope, marine slope, sea-level rise due to climate change, mean		Greece (Eastern part of the Northern coastline of Gulf of Corinth)	5.19 km	Tsaimou et al. (2023)

Coastal vulnerability index	Equation	Variables	Additional variable	Location	Coastline length	Reference
Coastal vulnerability index (CVI$_{WF}$)	$CVI_{WF} = \sqrt{\dfrac{a*b*c*d*e*f}{6}}$	range of the astronomical tide, storm surge, significant wave height, extreme significant wave height, cross–shore profile erosion, beach width, distance from vegetation		Greece (Gulf of Patras)	10.80 km	Boumboulis et al. (2021)
Coastal vulnerability index (CVI)	$CVI = \sqrt[5]{R_{Habitats}*R_{Shorline\ Type}*R_{Relief}*R_{Waves}*R_{Surge}}$	Geotechnical characterization, shoreline erosion/accretion rate, coastal slope, relative sea level, mean significant wave height, mean tide range; Habitat type, local bathymetry and topography, relative wind, storm wave force	Population density	Greece (Lesvos Island)	N/A	Papasarafianou et al. (2023)
Coastal vulnerability index (CVI)	$CVI = \sqrt{\dfrac{a*b*c*d*e}{6}}$	Geological fabric, coastal slope, beach width, significant wave height	Land use	Croatia (Krk Island)	7.72 km	Ružić et al. (2019)
Coastal vulnerability index (CVI)	$CVI = \sqrt{\dfrac{a1*a2*a3*a4*a5*a6*a7}{7}}$	Dune height, barrier type, beach type, relative sea-level change, shoreline erosion/accretion, mean tidal range, mean wave height		Egypt (Northern Nile delta)	275.00 km	Frihy (2017)

Continued

Table 13.1 Summary of coastal vulnerability indices, their geographical application and the variables needed to implement them.—cont'd

Index	Formula	Physical variables considered	Socioeconomic variables considered	Geographical application	Coastline Length	References
Physical vulnerability index (PVI)	$PVI = \sqrt{\frac{a*b*c*d*e*f*g}{7}}$	Coastal elevation, coastal slope, orientation of the coast, seabed slope, coastline covered by (artificial) protection structures, beach width, geomorphological processes		Slovenia	46.60 km	Poklar and Brečko Grubar (2023)
Coastal vulnerability index (CVI)	$CVI = \sqrt{\frac{a*b*c*d*e*f*g*h*i}{10}}$	Geomorphology, coastal slope, Shoreline erosion/accretion rate, emerged beach width, dune width, relative sea-level change, mean significant wave height, mean tide range, width of vegetation behind the beach, Posidonia oceanica		Italy (Apulian coastline)	12.00 km	Pantusa et al. (2018)
Coastal vulnerability index (CVI)	$CVI = \sqrt{\frac{CS*3S*3D*3G*3WH*3TR}{6}}$	Coastal slope, subsidence, displacement, geomorphology, wave height, tidal range		Greece (Western Peloponnese)	53.70 km	Doukakis (2005)

Regional vulnerability assessment (RVA)	$V_{j,k} = W_A \dfrac{\sum_{a=1}^{n} W_{a,k,j} SF_{a,k,j}}{\sum_{a=1}^{n} W_{a,k,j}} + W_B \dfrac{\sum_{b=1}^{m} W_{b,k,j} PF_{b,k,j}}{\sum_{b=1}^{n} W_{b,k,j}} +$ $W_c' = \dfrac{\sum_{c=1}^{p} W_{c,k,j} VF_{c,k,j}}{\sum_{c=1}^{p} W_{c,k,j}}$ $V_{j,k}' = (10-1)\dfrac{V_{j,k} - V_{j,k(\min)}}{V_{j,k(\max)} - V_{j,k(\min)}}$	Elevation, distance from coastline, coastal slope, geomorphology, artificial protections, sediment budget, wetland extension, vegetation cover, mouth typology, dunes	Urban typology, agricultural typology, protection level	Italy (North Adriatic coast)	286.00 km	Torresan et al. (2012)
Coastal vulnerability index	$V = \Sigma(w_i X_i)$ $N(v_i) = ((v_i - V\min)/(V\max - V\min)) * 5$	elevation, dune coverage, shoreline covered by artificial protection structures, recent shoreline change	Land cover	Italy (Ravena Province)	34.00 km	Sekovski et al. (2020)
Multidimensional coastal vulnerability index (CVI)	$SI_{a,t}^p = 100 \dfrac{\sum_{n=1}^{N} \beta_{n,a,t}^p - N_a}{M_{n,a,t}^p - m_{n,a,t}^p}$ $MDim - CV_t^p = \dfrac{\sum_{i=1}^{4} SI_{a,t}^p}{4}$	Extreme sea-levels, shoreline evolution trend, distance from shoreline, elevation, coastal slope, geological coastal type, land roughness, conservation designation, coastal protection structures	Population <5 years, population >65 years, land use patterns, gross domestic product	Italian coasts	8970.00 km	Furlan et al. (2021)

Continued

Table 13.1 Summary of coastal vulnerability indices, their geographical application and the variables needed to implement them.—cont'd

Index	Formula	Physical variables considered	Socioeconomic variables considered	Geographical application	Coastline Length	References
Coastal vulnerability index (CVI)	$CVI = \sqrt{\dfrac{a*b*c*d*e*f*g*h*i*j*k}{11}}$	Geomorphology, shoreline erosion/accretion rates, coastal slope, emerged beach width, dune, river discharge, sea-level change, mean significant wave height, mean tidal range, vegetation behind the back-beach, coverage of Posidonia Oceanica		Italy (Calabrian coastline)	20.00 km	Pantusa et al. (2022)
Coastal vulnerability assessment	$CVA = IRu + IR + ID + E + T$	Wave run-up distance index, maximum beach erosion, potential index for the shoreline, backshore coastal protection structures stability index, multiyear beach erosion rate index, tidal range that is negligible in a microtidal environment		Italy (Kaulonia, Ionian coast of Calabria)	3.50 km	Di Luccio et al. (2018)

Overall vulnerability index	$\text{OVI} = \sqrt{\text{Physical vulnerability} \times \text{Social vulnerability}}$	Pressure and impact factors proposed by EUROSION project	Pressure and impact factors proposed by EUROSION project, land use, transport, utilities	Malta (Gozo Island)	N/A	Rizzo et al. (2020)
Coastal vulnerability assessment	$\text{CVI}_{\text{impact}} = \dfrac{\left(\sum_{1}^{n} PP_n R_n W_n\right) + \left(\sum_{1}^{m} HP_m R_m W_m\right)}{\text{CVI}_{\text{leastvulnerable}}}$ $\text{CVI(SLR)}_n = \dfrac{\sum^{\text{Parameters of Impacts Group}} }{\sum \text{Least Vulnerable Case of the Group}}$	**Physical parameters** Rate of sea-level rise, geomorphology, coastal slope, H1/3, sediment budget, tidal range, proximity to coast, type of aquifer, hydraulic conductivity, depth to groundwater level above sea, water depth at downstream, discharge	**Human influence parameters** Reduction of sediment supply, river flow regulation, engineered frontage, natural protection degradation, coastal protection structures, groundwater consumption, land use pattern	Turkey (Göksu Delta and Amasra)	N/A	Ozyurt et al. (2010)

indices, a holistic vulnerability assessment may be provided. Relevant studies (e.g., De Serio et al., 2018; Ramnalis et al., 2023) combine the original CVI, as a subindex (named Physical Vulnerability Index—PVI), to provide a more complete assessment of vulnerability, through the Integrated Vulnerability Index (ICVI). Rizzo et al. (2020) introduced the Overall Vulnerability Index (OVI), applied on Gozo Island in Malta, by incorporating physical and social indicators through the consideration of physical exposure and social vulnerability. However, Tragaki et al. (2018) compare physical vulnerability with social vulnerability by overlying the results of the original CVI onto the results of the proposed Social Vulnerability Index (SVI), which involves socioeconomic and demographic parameters, to reveal the high-risk coastal municipalities along the coast of Peloponnese.

The coastal vulnerability indices can be applied at various spatial scales. The spatial scale refers to the geographical extent that these methodologies are utilized to evaluate the vulnerability. The scale may be varied from local to regional or national. The choice of spatial scale for CVI application depends on the objectives of the study, the available data, and the specific needs of the decision-makers (Rocha et al., 2023). The regional and national scale involves assessing vulnerability across larger coastal regions, like coastlines of entire states, provinces, or regions. At the national scale, the coastal vulnerability indices assess the vulnerability of an entire coastal nation or a significant portion of its coastline (e.g., Furlan et al., 2021). National-scale assessments are essential for informing national-level policies and strategies to address climate change impacts on coastal areas.

In local-scale applications of the coastal vulnerability indices, the utilization of the physical parameters has particular importance. The accuracy and level of detail associated with these parameters are essential for a comprehensive understanding of the specific vulnerabilities of a given coastal area. The level of detail needed for these parameters is relatively high due to the focus on a specific, often limited between some tens of kilometers, geographic area. High-resolution topography of the coastal area, including the precise elevation and the slope, is crucial for identifying elevation changes and the low-lying areas that are susceptible to sea-level rise and flooding. LiDAR (Light Detection and Ranging) is a common method used to collect high-resolution topography data. Aucelli et al. (2017) use LiDAR data to produce a digital elevation model (DEM) with a 1-meter resolution, for assessing the vertical ground displacements in their study area, in Southern Italy, while Poklar and Brečko Grubar, 2023 use a high-resolution DEM created from LIDAR data, with a 0.5-meter horizontal resolution to obtain the coastal elevation information in Slovenia. Elevation data collected from unmanned aerial vehicles (UAVs) play, also, a significant role in producing high-resolution and site-specific topographic information. Ružić et al. (2019) use a 3.26-cm resolution DEM originating from UAV data. Bathymetry is also a critical component of coastal vulnerability assessment indices. Bathymetric data provide detailed information about the underwater topography, including water depth and the seabed morphology. Depountis et al. (2023) use bathymetric sonal to produce bathymetric data with a high resolution of 2 m in the Gulf of Patras in Greece.

Another important parameter in local-scale applications is the shoreline change rate. The shoreline change rate is calculated as the change in distance divided by the time interval. It's typically expressed in meters per year. Accurate shoreline change data allow for better prediction of future vulnerabilities. It is crucial to understand how natural processes, climate change, or human activities are affecting coastal areas, including potential erosion or accretion. The shoreline change rate can be obtained from aerial or satellite imagery (Depountis et al., 2023). Historical imagery is usually the main source that depicts the coastline changing in time. Tsaimou et al. (2023) to estimate the shoreline evolution analyze aerial historical imagery from 3 years (1945, 1969, and 1992) to identify past coastlines and use UAV images to define the current coastline of specific beaches studied in the Northeastern Gulf of Corinth, Greece. Hereher (2015) estimates the erosion/accretion rates along the entire Mediterranean coastline of Egypt, based on the comparison of shoreline positions in historical Landsat Multi Spectral Scanner (MSS) images from 1972, which had a spatial resolution of 57 m, with more recent Landsat Enhanced Thematic Mapper plus (ETM+) images from 2005, featuring a higher spatial resolution of 28.5 m. In a separate study conducted by Sekovski et al. (2020), the shoreline change rate in Ravenna province in Italy is calculated by analyzing aerial photos of the year 1954 and high-resolution multispectral WorldView-2 satellite imagery of 1.84 m resolution of the year 2021.

The accurate mean tide range is also essential when estimating the coastal vulnerability on a local scale. Tide gauge stations are widely used to collect tide data. Gauges provide continuous measurements of water levels, allowing for the calculation of tidal variations over time. Tide gauge data are usually accurate and well documented, and their data are widely used as an input in the coastal vulnerability indices. In other cases, oceanographic models that simulate tidal variations based on various inputs, including bathymetry, wind patterns, and astronomical forces can be used to estimate tide ranges, especially in data-sparse coastal regions. For instance, Boumboulis et al. (2021) used tide range tide from a tide gauge system and a wave buoy installed in their study area as part of a research program. Filipaki et al. (2023) assess the coastal vulnerability in the Messolonghi coastal area using data obtained from the Wind and Wave Atlas of the Greek Seas (Soukisian et al., 2007). In a similar vein, Vandarakis et al. (2021) relied on results derived from a hydrodynamic–numerical model of tides in the Aegean Sea (Tsimplis, 1994). Similarly, studies utilize data for the mean significant wave height derived from gauge stations.

Additionally, numerical models are frequently employed to perform wave simulations. In this context, numerous studies rely on the SWAN wave model to obtain accurate estimates of wave parameters in coastal areas (e.g., Kokinos et al., 2014; Ružić et al., 2019; 2021).

GIS play an important role in coastal vulnerability assessment applications. The management of geospatial data is an essential aspect of GIS in the context of coastal vulnerability assessments. GIS serve as a helpful tool in collecting, storing, and organizing a multitude of geospatial data associated with the underexamination coastal areas (Parthasarathy & Deka, 2019). Extensive datasets, which include information concerning the physical and socioeconomic parameters would often be challenging to manage effectively without the functions integrated into GIS. Through the data integration GIS process, information derived from different sources and geospatial datasets can be unified creating a coherent database. In the context of index-based coastal vulnerability assessment, data integration involves the processing of various geospatial datasets related to physical and socioeconomic parameters that influence the coastal vulnerability, such as the topography and bathymetry, the geology and geomorphology information, wave height and tide range data, land use, infrastructure, and more (Ružić et al., 2019).

Spatial analysis is a crucial component of GIS in coastal vulnerability assessment. It involves operations such as proximity analysis, spatial statistics, and map overlay, which enable researchers to assess the spatial relationships between various factors influencing coastal vulnerability. In this context, GIS provides valuable analysis functions. Once the geospatial data are integrated, researchers can utilize GIS tools to perform spatial analysis by overlaying different datasets. This operation allows researchers to explore relationships between various factors. In the context of social vulnerability, physical vulnerability can be overlaid with population or demographic data, gaining insights into how sea-level rise affects coastal communities. Such spatial analysis is critical in identifying vulnerable areas and evaluating the potential impacts of climate change on the coastal environment and communities (e.g., Furlan et al., 2021; Torresan et al., 2012; Tragaki et al., 2018).

The spatial distribution of coastal vulnerability is provided by the process of visualization through GIS-produced thematic maps, which are a visual representation of coastal vulnerability across the study area. The most common technique of visualization is the classification of coastal vulnerability into five distinct classes. This classification simplifies the interpretation of the results and facilitates a clear understanding of the vulnerability variation. Typically, these five classes represent a spectrum, ranging from areas with very low vulnerability to those with very high vulnerability. The classification of vulnerability for the individual case studies conducted in the Eastern Mediterranean, which are considered in this review, is presented in Table 13.2.

Table 3.2 Coastal vulnerability scores for various geographical applications in the Eastern Mediterranean.

Geographical application	Authors		Coastline Length (km)	Vulnerability Categories (scores % of the studied coastline)				
				Very Low	Low	Moderate	High	Very high
Greece (Southern coast of the Gulf of Corinth)	Karymbalis et al. (2012)		148.00	14.60	21.60	25.10	8.20	30.50
Greece (Northern coast of Gulf of Corinth)	Nasopoulou et al. (2012)		63.62	0.00	69.71	28.54	1.74	0.00
Greece (Pieria prefecture)	Mavromatidi and Karymbalis (2016)		32.00	6.30	21.90	6.51	21.69	43.60
Greece (Thasos Island)	Georgiou and Alexandridis (2021) Conference Paper		100.00	28.08	28.24	22.11	19.6	1.97
Greece (Coastal zone of Mesolonghi)	Filippaki et al. (2023)		101.10	19.62		2.99	64.79	12.6
Greece (Argolikos Gulf)	Gaki-Papanastassiou et al., 2010		247.17	46.60	9.60	9.10	14.50	4.00
Greece (Ios island)	Komi et al. (2022)		110.74	92.37	1.69	2.64	1.74	1.56
Egypt (Mediterranean coast)	Hereher (2015)		1000.00	40.00	6.00	9.00	42.00	3.00
Syria	Faour et al. (2013)		183.00	1.50	22.00	40.80	16.10	19.50
Lebanon	El Hage et al. (2011)		375.00	33.00	18.00	32.00	12.00	5.00
Lebanon (Southern coast)	Ghoussein et al. (2018)		93.00	0.00	4.00	66.00	30.00	0.00
Greece (Southern coast of the Gulf of Corinth)	Rammalis et al. (2023)	CVI_1	147.00	10.80	28.50	24.70	17.70	18.30
		CVI_2		7.10	26.20	26.80	30.80	9.10
Italy (Apulian coast)	De Serio et al. (2018)	CVI_1	40.00	52.50	–	–	–	3.75
		CVI_2		12.50	–	–	–	15.00
Greece (Rhodes Island)	Vandarakis et al. (2021)		253.00	7.00	32.00	21.00	28.00	22.00
Greece (Peloponnese)	Tragaki et al. (2018)	CVI	1481.70	46.50	18.70	17.50	12.00	5.20
		SVI		11.00	20.50	30.10	19.20	19.20
Greece (Elounda Bay, Crete)	Alexandrakis et al. (2014)	CVI	N/A	0.52	85.55	0.28	7.81	5.84
		SocCVI		0.00	57.21	37.46	5.09	0.22
Greece (Gulf of Patras)	Boumboulis et al. (2021)		10.80	14.56	25.91	20.04	36.48	2.98
Croatia (Krk Island)	Ružić et al. (2019)		7.72	45.01	15.28	15.54	11.20	12.95

Continued

272 Geographical Information Science

Table 13.2 Coastal vulnerability scores for various geographical applications in the Eastern Mediterranean.—cont'd

Geographical application	Coastline Length (km)	Authors	Vulnerability Categories (scores % of the studied coastline)				
			Very Low	Low	Moderate	High	Very high
Egypt (Northern Nile delta)	275.00	Frihy (2017)	-	42.80	19.70	16.30	21.30
Slovenia	46.60	Poklar and Brečko Grubar (2023)	85.90		4.10	7.50	2.50
Italy (Apulian coastline)	12.00	Pantusa et al. (2018)	0.00	12.50	41.66	20.84	25.00
Greece (Western Peloponnese)	53.70	Doukakis (2005)	-	-	59.03	26.00	14.97
Italy (Ravena Province)	34.00	Sekovski et al. (2020)	0.00	8.34	61.10	22.22	8.34
Italy (Calabrian coastline)	20.00	Pantusa et al. (2022)	0.00	49.00	33.30	7.70	10.00
Malta (Gozo Island)	N/A	Rizzo et al. (2020)	0.00	20.40	61.30	7.30	11.00

5. Conclusion

Coastal zones stand as vital interfaces that connect land and sea, providing diverse wildlife and vital ecosystem benefits. Over the course of history, these areas have attracted human settlements due to their advantages. Nowadays, coastal communities are confronted with higher exposure and vulnerability to a range of natural and climate change-induced hazards. The accelerating global mean sea-level rise, primarily due to climate change, is a significant and immediate threat to coastal areas. These challenges demand urgent global attention due to their significant impacts on both coastal communities and ecosystems. The vulnerability of the Mediterranean region to climate change highlights the need for robust assessment methods capable of measuring associated risks and vulnerabilities. These methods are crucial not only for the efficient management of coastal areas but also for the preservation of these invaluable coastal areas in a sustainable manner.

The CVI is a widely used methodology for assessing coastal vulnerability to climate change, particularly sea-level rise. Moreover, modified CVI is more commonly used. Modifications of the CVI formula often aim to provide more accurate assessments by incorporating additional parameters or altering the formula structure. Coastal vulnerability indices can be applied at various spatial scales, ranging from local to regional or national. The choice of scale depends on the study's objectives, available data, and decision-makers' needs. National-scale assessments are essential for informing policies and strategies at the national level. In local-scale applications, precision is crucial, necessitating high-resolution topography data, such as LiDAR or UAV-collected information, to identify low-lying areas susceptible to sea-level rise and flooding. Additionally, accurate measurements of the mean tide range, often collected by tide gauge stations or oceanographic models, are vital for local-scale vulnerability assessments. Shoreline change rate data, typically obtained from historical imagery and satellite imagery, are essential for understanding coastal dynamics. These local-scale data and spatial analysis are essential components of coastal vulnerability assessments. GIS plays a fundamental role in managing, analyzing, and visualizing the complex geospatial data involved. Thematic maps produced through GIS provide visual representations of coastal vulnerability, simplifying result interpretation and highlighting variations in vulnerability across the study area.

This chapter provides a review of index-based methodologies for assessing coastal vulnerability. It encompasses 34 case studies focused on the Eastern Mediterranean, including countries such as Greece, Italy, Egypt, and Lebanon, among others. In these case studies, numerous techniques and methodologies have been used to assess the vulnerability of coastal areas to climate change-induced hazards, such as sea-level rise, the increasing intensity of storm surges, coastal erosion, and coastal floods. These studies employ a variety of tools to precisely measure the crucial parameters, which are essential for the methodologies' result accuracy. Moreover, they attempt to provide a comprehensive overview of the vulnerability in these various coastal areas by calculating the diverse vulnerability levels across the studied coastlines. The different vulnerability levels, that vary from very low to very high, stress the need for region-specific approaches to protect these coastal environments and their communities in a rapidly changing climate.

6. Key takeaways

The coastal vulnerability indices serve as a crucial metric for assessing the susceptibility of coastal areas to environmental changes and hazards by providing a comprehensive picture of coastal vulnerability. This information is vital for researchers, policymakers, and communities to understand and prioritize regions at higher risk and implement protective measures effectively.

In the context of climate change, the coastal vulnerability assessment through indices gains added significance. The various methodologies are instrumental in addressing the direct impacts of the sea-level rise, changing storm patterns, and other climate change-related hazards on coastal regions. By quantifying and qualifying this vulnerability, the indices provide crucial insights into the potential risk faced by coastal areas. This information is essential for the development of adaptive strategies, enabling communities and decision-makers to proactively plan and implement measures that enhance resilience and mitigation to the effects of climate change along the coast.

GIS play an important role in the coastal vulnerability indices effectiveness and allow the integration and analysis of diverse spatial data sets, enabling a more accurate and detailed assessment of coastal vulnerability. It facilitates mapping, visualization, and modeling of complex environmental variables, providing actionable information. GIS technology enhances the precision of the indices by incorporating geospatial data, which is fundamental for understanding the spatial relationships and interactions that influence coastal vulnerability.

References

Alexandrakis, G., Karditsa, A., Poulos, S., Ghionis, G. K. N. A., & Kampanis, N. A. (2010). An assessment of the vulnerability to erosion of the coastal zone due to a potential rise of sea level: The case of the Hellenic Aegean coast. Environmental systems. In Achim Sydow (Ed.), *Encyclopedia of life support systems (EOLSS), developed under the auspices of the UNESCO*. Oxford, UK: Eolss Publishers.

Alexandrakis, G., Petrakis, S., Ghionis, G., Kampanis, N., & Poulos, S. E. (2014). Natural and human induced indicators in coastal vulnerability and risk assessment. In *10th International congress of the hellenic geographical society* (pp. 644—655).

Allan, J. C., & Komar, P. D. (2006). Climate controls on US West Coast erosion processes. *Journal of Coastal Research, 22*(3), 511—529.

Androulidakis, Y., Kombiadou, K., Makris, C., Baltikas, V., & Krestenitis, Y. (2015). Storm surges in the Mediterranean Sea: Variability and trends under future climatic conditions. *Dynamics of Atmospheres and Oceans, 71*, 56—82.

Anfuso, G., Postacchini, M., Di Luccio, D., & Benassai, G. (2021). Coastal sensitivity/vulnerability characterization and adaptation strategies: A review. *Journal of Marine Science and Engineering, 9*(1), 72.

Antonioli, F., De Falco, G., Lo Presti, V., Moretti, L., Scardino, G., Anzidei, M., … Mastronuzzi, G. (2020). Relative sea-level rise and potential submersion risk for 2100 on 16 coastal plains of the Mediterranean Sea. *Water, 12*(8), 2173.

Aucelli, P. P. C., Di Paola, G., Incontri, P., Rizzo, A., Vilardo, G., Benassai, G., … Pappone, G. (2017). Coastal inundation risk assessment due to subsidence and sea level rise in a Mediterranean alluvial plain (Volturno coastal plain—southern Italy). *Estuarine, Coastal and Shelf Science, 198*, 597—609.

Boumboulis, V., Apostolopoulos, D., Depountis, N., & Nikolakopoulos, K. (2021). The importance of geotechnical evaluation and shoreline evolution in coastal vulnerability index calculations. *Journal of Marine Science and Engineering, 9*(4), 423.

Brierley, A. S., & Kingsford, M. J. (2009). Impacts of climate change on marine organisms and ecosystems. *Current Biology, 19*(14), R602—R614.

Catalán, B., Saurí, D., & Serra, P. (2008). Urban sprawl in the Mediterranean?: Patterns of growth and change in the Barcelona Metropolitan region 1993—2000. *Landscape and Urban Planning, 85*(3—4), 174—184.

Chalkias, C., Papadopoulos, A., Ouils, A., Karymbalis, E., & Detsis, V. (2011). Land cover changes in the coastal periurban zone of Corinth, Greece. *Proceedings of the Tenth International Conference on the Mediterranean Coastal Environment, 2*, 913—923.

Conte, D., & Lionello, P. (2013). Characteristics of large positive and negative surges in the Mediterranean Sea and their attenuation in future climate scenarios. *Global and Planetary Change, 111*, 159—173.

Costanza, R. (1999). The ecological, economic, and social importance of the oceans. *Ecological Economics, 31*(2), 199—213.

Cramer, W., Guiot, J., & Marini, K. (2020). Climate and environmental change in the Mediterranean basin—current situation and risks for the future. In *First mediterranean assessment report. MedECC (mediterranean experts on climate and environmental change)*. Marseille, France: Union for the Mediterranean, Plan Bleu, UNEP/MAP.

Davidson-Arnott, R., Bauer, B., & Houser, C. (2019). *Introduction to coastal processes and geomorphology*. Cambridge university press.

De Serio, F., Armenio, E., Mossa, M., & Petrillo, A. F. (2018). How to define priorities in coastal vulnerability assessment. *Geosciences, 8*(11), 415.

Depountis, N., Apostolopoulos, D., Boumpoulis, V., Christodoulou, D., Dimas, A., Fakiris, E., … Sabatakakis, N. (2023). Coastal erosion identification and monitoring in the Patras Gulf (Greece) using multi-discipline approaches. *Journal of Marine Science and Engineering, 11*(3), 654.

Di Luccio, D., Benassai, G., Di Paola, G., Rosskopf, C. M., Mucerino, L., Montella, R., & Contestabile, P. (2018). Monitoring and modelling coastal vulnerability and mitigation proposal for an archaeological site (Kaulonia, Southern Italy). *Sustainability, 10*(6), 2017.

Doukakis, E. (2005). Coastal vulnerability and risk parameters. *European Water, 11*(12), 3—7.

El Hage, M., Faour, G., & Polidori, L. (2011). L'impact de l'elevation du niveau de la mer (2000-2100) sur le littoral libanais une approche par teledetection et cartographie diachronique. *Revue Française de Photogrammétrie et de Télédétection n, 194*(2), 36.

European Environment Agency. (2015). *Mediterranean Sea region briefing—the European environment—state and outlook. EEA report*. Copenhagen, Denmark: European Environment Agency.

European Environment Agency. (2006). *The changing faces of europe's coastal areas. EEA report*. Copenhagen, Denmark: European Environment Agency.

Falciano, A., Anzidei, M., Greco, M., Trivigno, M. L., Vecchio, A., Georgiadis, C., … Doumaz, F. (2023). The SAVEMEDCOASTS-2 webGIS: The online platform for relative sea level rise and storm surge scenarios up to 2100 for the Mediterranean coasts. *Journal of Marine Science and Engineering, 11*(11), 2071.

Faour, G., Fayad, A., & Mhawej, M. (2013). GIS-based approach to the assessment of coastal vulnerability to sea level rise: Case study on the eastern mediterranean. *Journal of Surveying and Mapping Engineering, 1*(3), 41—48.

Filippaki, E., Tsakalos, E., Kazantzaki, M., & Bassiakos, Y. (2023). Forecasting impacts on vulnerable shorelines: Vulnerability assessment along the coastal zone of Messolonghi area—Western Greece. *Climate, 11*(1), 24.

Flanagan, B. E., Gregory, E. W., Hallisey, E. J., Heitgerd, J., & Lewis, B. (2011). A social vulnerability index for disaster management. *Journal of Homeland Security and Emergency Management, 8*, 1–22.

Frihy, O. E. (2017). Evaluation of future land-use planning initiatives to shoreline stability of Egypt's northern Nile delta. *Arabian Journal of Geosciences, 10*, 1–14.

Furlan, E., Dalla Pozza, P., Michetti, M., Torresan, S., Critto, A., & Marcomini, A. (2021). Development of a multi-dimensional coastal vulnerability index: Assessing vulnerability to inundation scenarios in the Italian coast. *Science of the Total Environment, 772*, 144650.

Gaki-Papanastassiou, K., Karymbalis, E., Poulos, S. E., Seni, A., & Zouva, C. (2010). Coastal vulnerability assessment to sea-level rise based on geomorphological and oceanographical parameters: The case of Argolikos Gulf, Peloponnese, Greece. *Journal of Geosciences, 45*, 109–122.

Georgiou, A., & Alexandridis, V. (2021). Geospatial analysis for coastal risk assessment on the island of Thassos, Greece. In *Proceedings of the eighth international conference on environmental management, engineering*. Thessaloniki, Greece: Planning and Economics.

Ghoussein, Y., Mhawej, M., Jaffal, A., Fadel, A., El Hourany, R., & Faour, G. (2018). Vulnerability assessment of the South-Lebanese coast: A GIS-based approach. *Ocean and Coastal Management, 158*, 56–63.

Gornitz, V. (1991). Global coastal hazards from future sea level rise. *Palaeogeography, Palaeoclimatology, Palaeoecology, 89*(4), 379–398.

Gualdi, S., Somot, S., May, W., Castellari, S., Déqué, M., Adani, M., … Xoplaki, E. (2013). Future climate projections. In *Regional assessment of climate change in the Mediterranean: Volume 1: Air, sea and precipitation and water* (pp. 53–118).

Hague, B. S., McGregor, S., Jones, D. A., Reef, R., Jakob, D., & Murphy, B. F. (2023). The global drivers of chronic coastal flood hazards under sea-level rise. *Earth's Future, 11*(8), Article e2023EF003784.

Hereher, M. E. (2015). Coastal vulnerability assessment for Egypt's Mediterranean coast. *Geomatics, Natural Hazards and Risk, 6*(4), 342–355.

Hoggart, S. P. G., Hanley, M. E., Parker, D. J., Simmonds, D. J., Bilton, D. T., Filipova-Marinova, M., … Thompson, R. C. (2014). The consequences of doing nothing: The effects of seawater flooding on coastal zones. *Coastal Engineering, 87*, 169–182.

Jevrejeva, S., Moore, J. C., Grinsted, A., Matthews, A. P., & Spada, G. (2014). Trends and acceleration in global and regional sea levels since 1807. *Global and Planetary Change, 113*, 11–22.

Jordà, G., Gomis, D., Álvarez-Fanjul, E., & Somot, S. (2012). Atmospheric contribution to Mediterranean and nearby Atlantic sea level variability under different climate change scenarios. *Global and Planetary Change, 80*, 198–214.

Karymbalis, E., Chalkias, C., Chalkias, G., Grigoropoulou, E., Manthos, G., & Ferentinou, M. (2012). Assessment of the sensitivity of the southern coast of the Gulf of Corinth (Peloponnese, Greece) to sea-level rise. *Central European Journal of Geosciences, 4*, 561–577.

Kirezci, E., Young, I. R., Ranasinghe, R., Muis, S., Nicholls, R. J., Lincke, D., & Hinkel, J. (2020). Projections of global-scale extreme sea levels and resulting episodic coastal flooding over the 21st Century. *Scientific Reports, 10*(1), 11629.

Kokkinos, D., Prinos, P., & Galiatsatou, P. (September 2014). Assessment of coastal vulnerability for present and future climate conditions in coastal areas of the Aegean Sea. In *11th international conference on hydro-science and engineering: Hydro-engineering for environmental challenges*. Hamburg.

Komi, A., Petropoulos, A., Evelpidou, N., Poulos, S., & Kapsimalis, V. (2022). Coastal vulnerability assessment for future sea level rise and a comparative study of two pocket beaches in seasonal scale, Ios Island, Cyclades, Greece. *Journal of Marine Science and Engineering, 10*(11), 1673.

Kulp, S. A., & Strauss, B. H. (2019). New elevation data triple estimates of global vulnerability to sea-level rise and coastal flooding. *Nature Communications, 10*(1), 1–12.

Losada, I. J., Reguero, B. G., Méndez, F. J., Castanedo, S., Abascal, A. J., & Mínguez, R. (2013). Long-term changes in sea-level components in Latin America and the Caribbean. *Global and Planetary Change, 104*, 34–50.

Lowe, J. A., Woodworth, P. L., Knutson, T., McDonald, R. E., McInnes, K. L., Woth, K., … Weisse, R. (2010). Past and future changes in extreme sea levels and waves. In *Understanding Sea-level rise and variability*. London, UK: Wiley-Blackwell.

Maes, J., Teller, A., Erhard, M., Liquete, C., Braat, L., Berry, P., … Bidoglio, G. (2013). Mapping and assessment of ecosystems and their services. In *An analytical framework for ecosystem assessments under action 5 of the EU biodiversity strategy to 2020*. Luxembourg, Luxembourg: Publications Office of the European Union.

Marcos, M., Jordà, G., Gomis, D., & Pérez, B. (2011). Changes in storm surges in southern Europe from a regional model under climate change scenarios. *Global and Planetary Change, 77*(3–4), 116–128.

Marcos, M., & Tsimplis, M. N. (2008). Coastal sea level trends in Southern Europe. *Geophysical Journal International, 175*(1), 70–82.

Masselink, G., & Gehrels, R. (2014). *Coastal environments and global change*. West Sussex, UK: John Wiley and Sons.

Masson-Delmotte, V., Zhai, P., Pirani, A., Connors, S. L., Péan, C., Berger, S., … Zhou, B. (2021). Climate change 2021: The physical science basis. *Contribution of Working Group I to the Sixth Assessment Report of the Intergovernmental Panel on Climate Change, 2*.

Mavromatidi, A., & Karymbalis, E. (2016). Assessment of susceptibility to sea-level rise in the coastal area of Pieria prefecture. *Bulletin of the Geological Society of Greece, 50*(3), 1721–1729.

Mentaschi, L., Vousdoukas, M. I., Pekel, J. F., Voukouvalas, E., & Feyen, L. (2018). Global long-term observations of coastal erosion and accretion. *Scientific Reports, 8*(1), 12876.

Nasopoulou, I., Poulos, S. E., Karymbalis, E., & Gaki-Papanastassiou, K. (2012). Coastal vulnerability assessment of the northern coast (Antirrio-Eratini) of the western Corinth Gulf to the anticipated sea-level rise. In *Proceedings of the 10th panhellenic symposium of oceanography and fisheries, athens, 7-11/5/2012*. Number of Paper 079.

Nicholls, R. J., Lincke, D., Hinkel, J., Brown, S., Vafeidis, A. T., Meyssignac, B., … Fang, J. (2021). A global analysis of subsidence, relative sea-level change and coastal flood exposure. *Nature Climate Change, 11*(4), 338–342.

Nicholls, R. J., Wong, P. P., Burkett, V., Codignotto, J., Hay, J., McLean, R., … Saito, Y. (2007). Coastal systems and low-lying areas. Climate change 2007. Impacts, adaptation, and vulnerability. In M. L. Parry, O. F. Canziani, J. P. Palutikof, P. J. van der Linden, & C. E. Hanson (Eds.), *Contribution of working group II to the fourth assessment report of the intergovernmental panel on climate change* (pp. 315–356). Cambridge, UK: Cambridge University Press.

Ozyurt, G., Ergin, A., & Baykal, C. (2010). Coastal vulnerability assessment to sea level rise integrated with analytical hierarchy process. *Coastal Engineering Proceedings*, (32), 6.

Pantusa, D., D'Alessandro, F., Frega, F., Francone, A., & Tomasicchio, G. R. (2022). Improvement of a coastal vulnerability index and its application along the Calabria Coastline, Italy. *Scientific Reports, 12*(1), 21959.

Pantusa, D., D'Alessandro, F., Riefolo, L., Principato, F., & Tomasicchio, G. R. (2018). Application of a coastal vulnerability index. A case study along the Apulian Coastline, Italy. *Water, 10*(9), 1218.

Papasarafianou, S., Gkaifyllia, A., Iosifidi, A. E., Sahtouris, S., Wulf, N., Culibrk, A., … Tzoraki, O. (2023). Vulnerability of small rivers coastal part due to floods: The case study of Lesvos north—west coast. *Environmental Sciences Proceedings, 25*(1), 44.

Parthasarathy, K. S. S., & Deka, P. C. (2019). Remote sensing and GIS application in assessment of coastal vulnerability and shoreline changes. *ISH Journal of Hydraulic Engineering, 27*.

Poklar, M., & Brečko Grubar, V. (2023). Assessing coastal vulnerability to sea level rise: The case study of Slovenia. *The Egyptian Journal of Environmental Change, 15*(1), 7–24.

Ramnalis, P., Batzakis, D. V., & Karymbalis, E. (2023). Applying two methodologies of an integrated coastal vulnerability index (ICVI) to future sea-level rise. Case study: Southern coast of the Gulf of Corinth, Greece. *Geoadria, 28*(1), 1–24.

Reid, W. V., Mooney, H. A., Cropper, A., Capistrano, D., Carpenter, S. R., Chopra, K., … Zurek, M. B. (2005). *Ecosystems and human well-being-synthesis: A report of the millennium ecosystem assessment*. Island Press.

Rizzo, A., Vandelli, V., Buhagiar, G., Micallef, A. S., & Soldati, M. (2020). Coastal vulnerability assessment along the north-eastern sector of Gozo Island (Malta, Mediterranean Sea). *Water, 12*(5), 1405.

Rocha, C., Antunes, C., & Catita, C. (2023). Coastal indices to assess sea-level rise impacts-a brief review of the last decade. *Ocean and Coastal Management, 237*, 106536.

Romano, B., & Zullo, F. (2014). The urban transformation of Italy's Adriatic coastal strip: Fifty years of unsustainability. *Land Use Policy, 38*, 26–36.

Roukounis, C. N., & Tsihrintzis, V. A. (2022). Indices of coastal vulnerability to climate change: A review. *Environmental Processes, 9*(2), 29.

Ružić, I., Benac, Č., Jovančević, S. D., & Radišić, M. (2021). The application of UAV for the analysis of geological hazard in Krk Island, Croatia, Mediterranean Sea. *Remote Sensing, 13*(9), 1790.

Ružić, I., Dugonjić Jovančević, S., Benac, Č., & Krvavica, N. (2019). Assessment of the coastal vulnerability index in an area of complex geological conditions on the krk Island, Northeast Adriatic Sea. *Geosciences, 9*(5), 219.

Šimac, Z., Lončar, N., & Faivre, S. (2023). Overview of coastal vulnerability indices with reference to physical characteristics of the Croatian Coast of Istria. *Hydrology, 10*(1), 14.

Salvati, L., Munafò, M., Morelli, V. G., & Sabbi, A. (2012). Low-density settlements and land use changes in a Mediterranean urban region. *Landscape and Urban Planning, 105*(1–2), 43–52.

Sarkar, N., Rizzo, A., Vandelli, V., & Soldati, M. (2022). A literature review of climate-related coastal risks in the Mediterranean, a climate change hotspot. *Sustainability, 14*(23), 15994.

Satta, A., Puddu, M., Venturini, S., & Giupponi, C. (2017). Assessment of coastal risks to climate change related impacts at the regional scale: The case of the Mediterranean region. *International Journal of Disaster Risk Reduction, 24*, 284–296.

Sekovski, I., Del Río, L., & Armaroli, C. (2020). Development of a coastal vulnerability index using analytical hierarchy process and application to Ravenna province (Italy). *Ocean and Coastal Management, 183*, 104982.

Soukissian, T., Hatzinaki, M., Korres, G., Papadopoulos, A., Kallos, G., & Anadranistakis, E. (2007). *Wind and wave atlas of the hellenic seas.* Anavissos: Hellenic Centre for Marine Res. Publ.

Tate, E., Cutter, S. L., & Berry, M. (2010). Integrated multihazard mapping. *Environment and Planning B: Planning and Design, 37*(4), 646–663.

Thieler, E. R., & Hammar-Klose, E. S. (1999). *National assessment of coastal vulnerability to sea-level rise: Preliminary results for the US Atlantic coast (No. 99-593).* US Geological survey.

Torresan, S., Critto, A., Rizzi, J., & Marcomini, A. (2012). Assessment of coastal vulnerability to climate change hazards at the regional scale: The case study of the North Adriatic Sea. *Natural Hazards and Earth System Sciences, 12*(7), 2347–2368.

Tragaki, A., Gallousi, C., & Karymbalis, E. (2018). Coastal hazard vulnerability assessment based on geomorphic, oceanographic and demographic parameters: The case of the Peloponnese (Southern Greece). *Land, 7*(2), 56.

Tsaimou, C. N., Papadimitriou, A., Chalastani, V.I., Sartampakos, P., Chondros, M., & Tsoukala, V. K. (2023). Impact of spatial segmentation on the assessment of coastal vulnerability—insights and practical recommendations. *Journal of Marine Science and Engineering, 11*(9), 1675.

Tsimplis, M. N. (1994). Tidal oscillations in the Aegean and Ionian seas. *Estuarine, Coastal and Shelf Science, 39*(2), 201–208.

Ullmann, A., & Monbaliu, J. (2010). Changes in atmospheric circulation over the North Atlantic and sea-surge variations along the Belgian coast during the twentieth century. *International Journal of Climatology: A Journal of the Royal Meteorological Society, 30*(4), 558–568.

Vandarakis, D., Panagiotopoulos, I. P., Loukaidi, V., Hatiris, G. A., Drakopoulou, P., Kikaki, A., … Kapsimalis, V. (2021). Assessment of the coastal vulnerability to the ongoing sea level rise for the exquisite Rhodes Island (SE Aegean Sea, Greece). *Water, 13*(16), 2169.

Vousdoukas, M. I., Ranasinghe, R., Mentaschi, L., Plomaritis, T. A., Athanasiou, P., Luijendijk, A., & Feyen, L. (2020). Sandy coastlines under threat of erosion. *Nature Climate Change, 10*(3), 260–263.

Vousdoukas, M. I., Voukouvalas, E., Annunziato, A., Giardino, A., & Feyen, L. (2016). Projections of extreme storm surge levels along Europe. *Climate Dynamics, 47*, 3171–3190.

Wang, X. L., Feng, Y., & Swail, V. R. (2014). Changes in global ocean wave heights as projected using multimodel CMIP5 simulations. *Geophysical Research Letters, 41*(3), 1026–1034.

Weisse, R., Bellafiore, D., Menéndez, M., Méndez, F., Nicholls, R. J., Umgiesser, G., & Willems, P. (2014). Changing extreme sea levels along European coasts. *Coastal engineering, 87*, 4–14.

World Resources Institute. (2010). *Decision making in a changing climate.* Washington, DC, USA: United Nations Development Programme World Bank; World Resources Institute.

CHAPTER 14

Advances in geographical information science for monitoring and managing deltaic environments: Processes, parameters, and methodological insights

Konstantinos Tsanakas, Efthimios Karymbalis and Dimitrios-Vasileios Batzakis
Department of Geography, School of Environment, Geography and Applied Economics, Harokopio University of Athens, Athens, Greece

1. Introduction—theoretical framework

River deltas are important geomorphic and sedimentary environments and are of great significance in terms of earth and environmental monitoring. Primarily, river deltas function as natural indicators of terrestrial and marine processes and environmental changes (Bianchi & Allison, 2009; Syvitski et al., 2005). They act as sediment sinks, accumulating vast quantities of sediments transported by rivers from their upstream catchments (Milliman & Syvitski, 1992). The sediments deposited in deltas carry crucial information about the geological, climatic, and anthropogenic factors that influence river basins (Day et al., 2016; Walling, 1999). Monitoring the sediments gives information on land-use changes, erosion patterns, and shifts in hydrological regimes, which are fundamental for understanding and mitigating environmental impacts (Syvitski et al., 2009).

The evolution of deltas depends on the interplay of river discharge, sediment deposition, and the dynamic interaction between riverine and marine forces (Edmonds & Slingerland, 2007). The synergy between riverine forces, favoring sediment accumulation, and marine forces, dispersing sediments, orchestrates the ongoing evolution of deltas. Understanding these interactions is vital for predicting deltaic changes and enhancing our capacity to manage and conserve these landforms. In this context, delta morphodynamic models (DMMs) have played an important role in the simulation of the complex dynamics of river deltas as they integrate variables influencing delta behavior, including sediment transport, river discharge, wave action, tides, and more (Edmonds et al., 2022). Such models contribute to our understanding of delta dynamics, helping researchers, policymakers, and conservationists make informed decisions regarding delta management and sustainable conservation efforts.

Geographical Information Science
ISBN 978-0-443-13605-4
https://doi.org/10.1016/B978-0-443-13605-4.00007-2

River deltas are also vital hubs of biodiversity. The deltaic regions support diverse ecosystems, including wetlands, marshes, and mangrove forests, which host a variety of plant and animal species (Barbier et al., 2011). They are critical breeding and feeding grounds for numerous aquatic species, making deltas key areas for fisheries and aquaculture (Arkema et al., 2013). The monitoring of deltaic ecosystems aids in the preservation of biodiversity, sustainable resource management, and the protection of endangered species (Mora et al., 2013).

The impacts of climate crisis are particularly pronounced in deltaic regions (Barbier et al., 2011). Sea-level rise, a prominent consequence of climate change, poses a significant threat to deltas as it can lead to saltwater intrusion, contaminating freshwater resources and endangering local agriculture and ecosystems (Nicholls et al., 1999, 2016). Furthermore, sea-level rise coupled with climate-induced changes in storm patterns, exacerbate the vulnerability of deltas to erosion and land loss. The combination of storm surges, wave action, and sea-level rise can accelerate the degradation of deltaic land, posing a direct threat to the livelihoods of the people who depend on these regions (Fagherazzi et al., 2013; Hinkel et al., 2014). As a result, climate-induced changes in deltas may lead to the displacement of human populations, necessitating relocation and giving rise to social, economic, and political challenges (Anthoff et al., 2010). To better understand the complex dynamics of deltas in the context of climate change and to inform adaptation and mitigation strategies, the monitoring and study of these regions are of paramount importance.

1.1 The role of Geographic Information Systems and remote sensing

Geographic Information Systems (GIS) combine spatial data, geographic analysis, and visualization techniques. In the context of monitoring and mapping, GIS provides a multidimensional approach to data management, analysis, and interpretation (Nicholls et al., 1999). This technology enables the integration of diverse data sources, such as satellite imagery, topographic maps, and hydrological data, into a unified spatial database (Balasubramanian, 2019). By harmonizing datasets, it is possible to gain a holistic perspective on deltaic systems, encompassing land cover changes, water quality variations, and morphological changes (Gao, 1996). Through GIS, sophisticated models and simulations can be developed to predict and assess the impacts of environmental changes, including sea-level rise, land subsidence, and anthropogenic activities (Palace et al., 2015). These predictive capabilities are instrumental in devising strategies for mitigating risks associated with deltaic vulnerabilities. Moreover, GIS-based decision support systems empower stakeholders to make informed choices regarding land use planning, habitat conservation, and disaster risk reduction within deltaic regions (Tang et al., 2022).

Remote sensing (RS) technologies, such as satellite imagery, LiDAR (Light Detection and Ranging), and unmanned aerial vehicle (UAV) photography, provide important data for the physical attributes and temporal dynamics of river deltas. Satellite imagery offers a bird's-eye view of deltaic landscapes, capturing high-resolution data over extensive areas giving the ability to monitor changes in land cover, water levels, sediment transport, and vegetation health. Such data are also critical for tracking deltaic evolution, including the expansion of urban areas, alterations in agricultural practices, and shifts in hydrological regimes (Turner & Rabalais, 2003). LiDAR technology and interferometry enables precise elevation modeling within deltas. This capability is crucial for assessing land subsidence, a common concern in many deltas. Aerial photography complements remote sensing efforts, offering also historical imagery that aids in long-term deltaic analyses (Pettorelli et al., 2014).

In conclusion, river deltas are dynamic environments with far-reaching significance in Earth and environmental monitoring. Their role as indicators of environmental change, support for biodiversity, and natural protective barriers against disasters underscores the importance of ongoing research and observation in these regions. Understanding their processes is not only essential for their preservation but also for the broader goals of sustainable environmental management and climate adaptation. The effective monitoring and mapping of river deltas hinge upon the integration of cutting-edge technologies, with GIS and remote sensing emerging as indispensable tools in this endeavor.

The primary purpose of this chapter is to provide a comprehensive exploration of river delta monitoring, with a focus on the theoretical foundations, methodological approaches, and practical applications that underpin this crucial field of research. Readers will gain a comprehensive understanding of the challenges and opportunities related to river delta monitoring. This chapter aims to serve as a valuable resource for researchers, practitioners, and policy-makers engaged in the field of environmental conservation and sustainable management.

2. Methodological approaches

Over time, methodologies for river delta monitoring have witnessed significant evolution, shaped by advances in technology, shifts in research priorities, and a growing recognition of the environmental significance of deltas. Early observations of deltas were relatively limited and depended upon rudimentary documentation by explorers and scientists during expeditions. However, with the advent of aerial photography and improved cartography in the mid-20th century, researchers gained the ability to create detailed maps and analyze changes in delta morphology and land use. Nowadays, concurrent advances in remote sensing technology, including high-resolution optical and radar sensors, enhanced our ability to monitor deltas, enabling precise measurements of parameters like vegetation cover, shoreline changes, and water quality. Integration of Geographic Information Systems (GIS) with remote sensing data allowed sophisticated

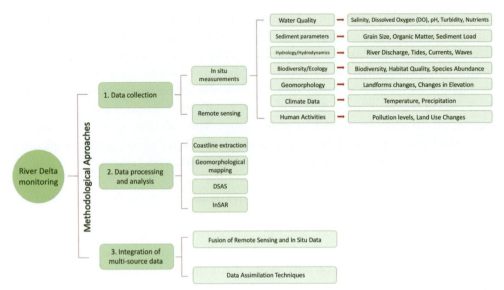

Figure 14.1 Schematic overview of the methodological approaches employed in river delta monitoring.

analysis and modeling of delta processes. Hydrodynamic and sediment transport models also became more accurate and sophisticated, aiding predictions about delta responses to environmental changes and human interventions. In general, river delta monitoring methodological approaches have focused toward understanding deltas' vulnerability to climate change impacts, with experts from various fields collaborating to grasp the complex interactions between physical, ecological, and socioeconomic aspects. Fig. 14.1 gives a graphical overview of the methodological approaches covered in this chapter.

2.1 Data collection

2.1.1 In situ measurements

Numerous studies have highlighted the effectiveness of in situ measurements in monitoring various aspects of deltaic systems. Tessler et al. (2015) have emphasized their role in assessing sediment fluxes in large rivers, shedding light on the sediment transport within deltas. Additionally, several studies have underlined the significance of in situ measurements in understanding the morphology of coastal environments. In any case, a multitude of parameters can be assessed by in situ measurements.

2.1.1.1 Water quality parameters

Salinity: The assessment of salinity is important in decoding the estuarine environment. It reveals the concentration of dissolved salts in the water, a critical factor shaping the delta's hydrodynamics and the composition of its biota. Salinity variations are indicators of the delicate balance between freshwater inflow and seawater intrusion (Smith & Hollibaugh, 1993).

Dissolved oxygen (DO): Monitoring dissolved oxygen (DO) levels is essential for gauging the health and vitality of aquatic life within the delta. Oxygen is a life-sustaining element for aquatic organisms, and its fluctuations can be indicative of ecosystem stressors (Breitburg et al., 2009). DO is commonly measured using dissolved oxygen probes.

pH: The measurement of pH, representing the acidity or alkalinity of water, provides insights into water quality. Altered pH levels can affect the chemical reactions and biological processes within the ecosystem (Kemp et al., 1990). pH can be measured directly using pH meters.

Turbidity: Turbidity denotes the cloudiness or clarity of water, which significantly impacts light penetration and nutrient transport within the delta. High turbidity can obstruct sunlight from reaching submerged vegetation and impact primary productivity. Furthermore, it can transport sediments and contaminants, altering the delta's morphology and ecological balance (Lorenzen, 1967). It can be measured using a nephelometer.

Nutrient concentrations: Monitoring the concentrations of essential nutrients, such as nitrogen and phosphorus, is important for understanding the trophic dynamics of the delta. While nutrients are vital for primary production, excess nutrients can lead to eutrophication, potentially resulting in harmful algal blooms, oxygen depletion, and disruptions in the food web (Elser et al., 2007).

2.1.1.2 Sediment parameters

Grain size: Establishing the grain size distribution is important to understand the sediment transport and deposition processes within a delta. This is often done by using sediment sieving or laser diffraction equipment. Grain size characteristics are also valuable for managing sediment resources and assessing the impact of upstream alterations on delta stability (Folk & Ward, 1957).

Organic matter content: The assessment of organic matter content in sediments reveals the percentage of decaying plant and animal material. Organic matter influences sediment stability, nutrient cycling, and microbial activity. Understanding the organic content aids in comprehending the delta's biogeochemical processes (Middelburg et al., 1997).

Sediment load: Measuring sediment load is important to understand sediment transport patterns and sedimentation rates which provide critical information for studying changes in delta morphology, erosion, and accretion. Sediment transport is closely linked to river discharge, tides, and wave activity (Syvitski et al., 2005). Instruments like sediment traps and sediment samplers can be used to measure sediment load. Rating curves can also be developed to estimate sediment load under various flow conditions.

2.1.1.3 Hydrology/hydrodynamics

River discharge: The volume of water flowing into the delta, often quantified as river discharge, is also a fundamental parameter. Variations in river discharge can lead to shifts in the delta's shape and size. Monitoring river discharge is essential for understanding the hydrological regime of the delta (Syvitski et al., 2005). To measure river discharge, a variety of methods can be employed, including stream gauging stations, acoustic doppler instruments, and flow rate calculations based on river cross-sectional area and velocity measurements.

Tides: Tides have a major role in shaping deltaic environments. Monitoring tidal patterns and their impact on water levels and saltwater intrusion is critical for assessing the delta's hydrodynamics. Tides influence the extent of tidal marshes and the distribution of various species within the delta (Bradshaw et al., 2016). Some common methods and tools for measuring tides include tide gauges, satellite altimetry, acoustic doppler devices, radar and pressure sensors, etc.

Currents: Understanding the speed and direction of water currents is essential, especially in regions with strong river discharge. Currents affect sediment transport, nutrient dispersion, and the distribution of plankton and fish larvae. Assessing these dynamics aids in comprehending the delta's ecological structure (Elbing et al., 2013). They can be measured with current meters, drifters, underwater gliders, and moorings.

Waves: Wave activity is also an important hydrodynamic parameter and its monitoring is critical during extreme weather events like hurricanes, as it helps assess the risk of storm surges and coastal flooding (Poulos et al., 1993). They are most commonly measured with wave buoys, acoustic wave sensors, and wave gliders.

2.1.1.4 Biodiversity and ecology

Biodiversity: Conducting ecological surveys to identify species diversity, including fish, invertebrates, and wetland vegetation, provides information into the delta's ecological richness. Biodiversity assessments help monitor changes in species composition and the health of the delta's ecosystems (Sala et al., 2000).

Habitat quality: The condition of critical habitats, such as mangroves, salt marshes, and submerged aquatic vegetation, influences the delta's overall ecological health. Monitoring habitat quality involves assessing the structural and functional attributes of these habitats (Moilanen et al., 2005).

Species abundance: Counting the number of individuals of specific species in the delta offers a window into the distribution and abundance of key organisms. This parameter is vital for tracking changes in species populations and evaluating the effectiveness of conservation efforts (Bakus et al., 2007).

2.1.1.5 Geomorphology

Landform changes: The ongoing evolution of deltaic landforms, including the expansion or erosion of the delta front, is closely monitored to assess the delta's response to various environmental factors. Changes in landforms impact habitat availability, sediment distribution, and the vulnerability of the delta to sea-level rise (Karalis et al., 2022; Karymbalis et al., 2016, 2018).

Elevation changes: Evaluating land subsidence or elevation alterations due to sediment deposition and compaction is crucial for understanding the delta's resilience and its susceptibility to sea-level rise. Such changes can affect flood risk, habitat suitability, and the delta's adaptability to environmental stressors (Zebker & Villasenor, 1992).

2.1.1.6 Climate data

Temperature: Continuous monitoring of air and water temperature provides essential data for understanding temperature variations within the delta. Temperature impacts metabolic processes in aquatic organisms, biogeochemical cycles, and overall ecosystem health (Oke, 1995).

Precipitation: Recording rainfall patterns and their effects on river discharge and sediment transport is vital for comprehending the delta's hydrological response to climate variability. Precipitation patterns influence river inflow, sediment transport, and habitat availability (Chen et al., 2022).

2.1.1.7 Human activities

Pollution levels: Regular assessments of pollutant presence, contaminants, and heavy metals in the delta's waters and sediments are essential for identifying and mitigating threats to water quality and ecosystem health. Monitoring pollution levels supports efforts to preserve the integrity of deltaic environments (Berg et al., 2007).

Land-use changes: Monitoring human activities, such as urbanization, agriculture, and industrial development, is crucial for understanding their impact on the delta's environment. These activities can influence water quality, sediment dynamics, and habitat alteration. Assessing land-use changes is key to sustainable delta management (Tran et al., 2015).

2.1.2 Remote sensing

The effective utilization of RS techniques is of great importance in ensuring the efficiency when monitoring river deltas. Acquiring proficiency in these methods not only aids in identifying potential deficiencies in the usability of the collected data but also fosters the innovation of novel approaches. It is widely acknowledged that satellite-based observations serve as valuable complements to conventional in situ measurements

(Manfré et al., 2012). Remote sensing stands as an invaluable instrument for generating geospatial resources that support decision-making systems for a range of natural disasters. Over the past 2 decades, there has been a noticeable increase in the availability of RS satellite constellations. A wealth of imagery with diverse spatial, spectral, and radiometric resolutions has enhanced our understanding of natural disasters (Metternicht et al., 2005).

Traditionally, optical remote sensing satellites with moderate and high spatial resolutions have been conventionally employed for the surveillance of coastal environmental changes. Landsat satellites have been consistently active since 1972, providing a continuous stream of high-quality Earth observations and measurements at spatial resolutions of approximately 30 m. The enduring reliability and consistency of Landsat data enable extended analyses to track changes in deltaic regions (Zhu & Woodcock, 2014). Modern approaches, which utilize spatial and temporal data fusion models, have emerged to integrate Landsat imagery with less frequent yet higher spatial resolution observations, along with MODIS and VIIRS data offering lower spatial resolution but greater temporal frequency (Gao, 1996). In this context, Sentinel-2, consisting of the Sentinel-2A and Sentinel-2B sensors, offers a routine global revisit interval of 5 days at spatial resolutions ranging from 10 to 60 m. Virtual satellite constellations can be constructed by amalgamating Landsat, Sentinel-2, and other optical remote sensing satellite data that share similar resolutions, effectively augmenting global revisit rates. For instance, the combined use of Landsat-8, Sentinel-2A, and Sentinel-2B data can achieve a comprehensive revisit period of 2.9 days (Metternicht et al., 2005). This capability contributes to the mapping of dynamic coastal geomorphological changes at a fine temporal resolution, directly benefiting the monitoring of deltaic evolution (Wulder et al., 2015).

Additionally, remote sensing technologies that operate within the microwave spectrum have found extensive application in retrieving information related to the geometric and electromagnetic attributes of observed scenes. Synthetic aperture radar (SAR) represents a highly valuable microwave instrument capable of acquiring high-resolution data from Earth, operating day and night under various weather conditions (Berardino et al., 2002). Sequences of SAR data are harnessed through interferometric SAR techniques (InSAR) (Shanker & Zebker, 2007) to facilitate coastal mapping and monitoring, pinpointing areas where substantial changes have transpired over time, and establishing correlations between these changes and shifts in shorelines, coastal erosion, groundwater utilization, and more. Other methodologies, such as the assessment of the Doppler centroid anomaly (DCA), offer a reliable and precise estimation of sea surface displacements based on a single SAR image, which are interconnected with ocean currents and wind patterns.

Nevertheless, it's crucial to acknowledge certain limitations inherent in InSAR systems, as they can potentially influence the dependability of final outcomes. These limitations include noise introduced by wind (Pierdicca et al., 2013, p. 139–149), heavy precipitation (Berardino et al., 2002), atmospheric disturbances (Klees & Massonnet,

1998), decorrelation phenomena (Just & Bamler, 1994), topographical decorrelation (Lee & Liu, 2001), and other challenges that may arise from improper data processing. A failure to comprehend these limitations adequately can lead to unreliable findings, potentially resulting in skewed and unrealistic analyses.

Case study: Coastal environmental monitoring using remotely sensed data and GIS techniques in the Modern Yellow River delta, China (Zhang, 2011).

The Yellow River is famous for its high sediment levels. It reshaped its course, leading to the vast Yellow River Delta. Human interventions since the 1950s have significantly impacted its course (Milliman & Meade, 1983). The delta's growth is mainly due to sediment deposition, making it one of the world's largest newly added land areas. Dongying City and the Shenli Oil Field are recent developments in the delta. The Yellow River Harbor (YRH) was established in 1985, and a coastal levee was built from YRH to Zhuang 106 (Z106) between 1986 and 1988. In December 2009, China's State Council approved the "Yellow River Delta Economic Zone and efficient eco-development plan," signifying its national importance.

In this study, multitemporal Landsat data from 1987 to 2008 were chosen, as depicted in Table 14.1. In addition to the remotely sensed data, the nautical charts of the study area with various scales, topographical and tidal data, and related literature were collected as the reference materials for remote sensing data analysis.

To better represent the land-ocean boundary, the waterline is utilized. Waterlines are prominent and continuous in remote sensing images, although they can be influenced by coastal factors like beach slope and tidal range. The relatively low mean tidal range (0.73–1.77 m) along the MYR delta coast, coupled with close acquisition times of Landsat data, allowed the authors to discount seasonal sea-level fluctuations. Tidal effects on waterlines were corrected by considering the relationship between beach slope and tidal level. Tidal stage and beach slope data were calculated using tidal prediction tables, daily tidal process lines, relevant literature, and expert insights. These calculations served as indicators for the study's macro-temporal spatial evolution analysis, as the research focuses on understanding broader patterns rather than precise engineering applications.

Reflectance characteristics of tidal flats, transitioning from flat surfaces to shallow seawater, are influenced by factors like particle sizes, moisture content, local slope, turbidity of seawater, and creek presence. NIR and SWIR bands are effective for

Table 14.1 Description of landsat data used in the study (Zhang, 2011).

Sensor (TM)	Acquisition date	Band (µm)	Spatial resolution (m)	Tidal condition	Path/row
Landsat-5	1987.09.12	B3:0.63–0.69	30	Medium tide	121/34
Landsat-5	1996.09.20	B4:0.76–0.90	30	Neap tide	121/34
Landsat-5	2008.10.07	B5:1.55–1.75	30	Neap tide	121/34

detecting sandy coastlines and clear water boundaries. In tidal flat areas, the TIR band is particularly sensitive to waterline location, although its spatial resolution may be insufficient. Alternatively, the combination of TM5, 4, and 3 bands, in conjunction with TM6, produces an RGB composite map that effectively distinguishes water bodies, land, and tidal flat features. This selection minimizes the need for extensive preprocessing.

The MYR delta's complex coastal environment, combined with occasional cloud cover, makes automatic classification challenging. Therefore, manual visual interpretation was chosen, using ERDAS 9.2 and ArcGIS 9.2 software. Extracted coastlines were refined through integrated analysis, considering spatial characteristics, nautical charts, literature, and tidal correction. The final output comprises multitemporal coastline distribution maps.

Coastline distribution maps were used to generate land-water masks. To assess net changes, derivative, color-coded images illustrating land loss and gain during the periods 1987–96 and 1996–2008 were produced. Measurement points were created at 250 m intervals, allowing for a detailed analysis of coastal change while ensuring manageable data for the entire Delta coast. The shortest distances between measurement points and coastlines in 1987 and 2008 were calculated. These distances served as indicators of coastline change distance, highlighting temporal-spatial changes within each coastal section. The difference between these distances provided indicators of acceleration rates of coastline change, helping identify coastal areas that experienced significant changes between the two periods (1987–96 and 1996–2008).

The results from the analysis of the coastal area changes in the MYR delta reveal important temporal-spatial patterns and trends in land accretion and erosion. Historically, the delta experienced rapid land accretion during the mid-20th century, with annual accretion rates exceeding 28 square kilometers per year. However, recent Landsat data from 1987 to 2008 indicate a deceleration in accretion rates (Fig. 14.2).

During 1987–96, land-gain values were around +21.59 square kilometers per year, mainly attributed to reduced water and sediment discharge from the Yellow River after 1987. Land loss due to coastal erosion during this period was approximately −9.67 square kilometers per year. These observations suggest relatively stable development.

In the northern part of the delta (region A), land was eroded at a rate of approximately −6.70 square kilometers per year during 1987–96. In the Yellow River Estuary region (region C), there was an accretion of +19.73 square kilometers per year. The coast experienced a transition from erosion to slow accretion between 1996–2008.

In general, during the entire 1987–2008 period, the northern part of the MYR delta showed an erosion pattern of approximately −1.22 square kilometers per year. The Yellow River Estuary region showed significant accretion at approximately +8.92 square kilometers per year. The area around YRH and GOF, with artificial banks, remained relatively stable, while the coast in the south of the current Yellow River Estuary region showed little change.

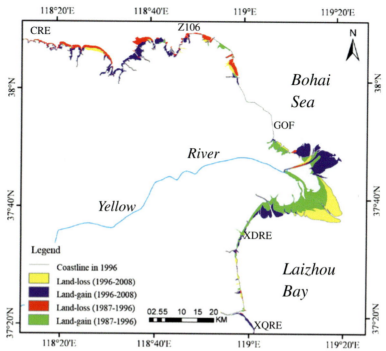

Figure 14.2 Overall changes of temporal—spatial pattern on the yellow river delta coast (Zhang, 2011).

Between the CRE and Z106 in the north, the coast transitioned from rapid erosion to accretion from 1987—2008. The same area between Z106 and CRE saw a shift from erosion to dynamic balance. This pattern of change is associated with human activities, including aquaculture and construction.

Between Z106 and GOF, the coast had a general erosion pattern from 1987 to 2008. The rate of erosion decreased after the construction of coastal levees. In some areas, erosion transitioned to accretion as human interventions altered sediment dynamics.

The coast between the south border of GOF and XDRE experienced rapid accretion from 1987—96, especially in areas near Q8. Since 1996, there has been a shift from accretion to dynamic balance and erosion, which is indicative of human activity and changes in sediment dynamics.

Between XDRE and XQRE, the coast had an overall accretion pattern, with the north part experiencing faster accretion. The south part was more stable. The acceleration rates of change during 1987—96 compared to 1996—2008 showed changing patterns of erosion followed by accretion.

The study examined changes in the coastal environment of China's MYR delta from 1987 to 2008. It found that while the region continued to accrete land, the rate of accretion had slowed. Different coastal areas showed distinct patterns: the north part and artificial banks experienced varying degrees of erosion, the region between CRE and Z106 saw initial erosion followed by accretion, and the area between the south border of GOF and XDRE transitioned from rapid accretion to dynamic balance with increasing erosion. The MYR delta's evolution is shaped by interactions between river and ocean systems and significant human activities, including water and sediment regulation and various coastal developments. The study highlights the interplay between natural processes and human actions in shaping coastal changes.

2.2 Data processing and analysis

2.2.1 Coastline extraction using RS and GIS

Optical RS technology can help efficiently extract and measure coastline changes with high accuracy over large deltaic areas (Zhang, 2011). These changes can be stimulated by natural phenomena, such as sea-level rise, the effects of hurricanes and possibly, in case of large ruptures, the impacts of coastal earthquakes. Coastline changes can also be due to the direct involvement of humans. With RS techniques, the change detection of the coastline's orientation, position, and shape (Lee & Jurkevich, 1990) can retrieve valuable information on the interrelationships between sea-level rise, tidal currents, wave energy, land subsidence, and human intervention. Thus, the mapping of coastline changes is fundamental for coastal management, land use land change investigations, autonomous navigation, coastal erosion study, the planning of protection infrastructures, and an overall assessment of the increased risk of inundation at deltaic areas.

Coastline extraction is related to boundary extraction and image segmentation and belongs to the class of machine vision approaches (Boak & Turner, 2005). Coastline changes have manually been outlined by expert photo interpreters who have visually inspected aerial photographs for many decades. However, these operations were costly and time-consuming. The frequently used automatic interpretation-based methods are mainly based on edge detection, index analysis, threshold segmentation, region growth, neural network, and subpixel (Chen et al., 2016; Sánchez-García et al., 2019). Landsat imagery has efficiently been applied to monitor coastline evolution using dozens of years of data (Sánchez-García et al., 2019). Long-term coastal changes of the Yellow River Delta have been determined based on decades of Landsat imagery. Sediment siltation, erosion, and human activities drive the dramatic deltaic coastline change (Sánchez-García et al., 2019).

Detecting coastal erosion with optical remote sensing imagery strongly depends on the accuracy and frequency of coastline extraction. High-resolution optical remote sensing imaging methods, such as IKONOS, WorldView-2, WorldView-3, QuickBird, and GF-1, have been applied to extract coastline changes and geomorphological changes and map coastal erosion at a fine resolution (Tian et al., 2016). Twenty-nine deltas of the world are in a state of overall erosion, determined by complementary analysis and quantification of coastline changes using satellite images of Landsat, SPOT-5, and SPOT-6 taken from 1972 to 2015 (Besset et al., 2019).

Although many coastline extraction methods using optical remote sensing imagery have been developed over the decades, there is no universally valid coastline extraction technique. All existing algorithms are applied to images of individual satellites or for applications. It is also challenging to choose an algorithm for an application, because there is still no established theory for this purpose and no index for comparing and evaluating the performance of different methods in other specific cases. Complete automatic coastline extraction methods and algorithms using high-resolution optical remote sensing imagery still need to be developed, and time series images of "30 m MODIS" and virtual satellite constellations promise future research directions. Despite the advantages of optical satellite remote sensing, there are also some obvious limitations due to cloud coverage on rainy days, solar illumination, and natural meteorological conditions. Microwave data can circumvent these problems, because they are not significantly affected by these factors. As clarified below, the exploitation of independent information from sets of radar images helps obtain accurate coastline change maps. Some approaches also use coherence information between couples of complex-valued SAR images to discriminate the land and sea boundaries (Dellepiane et al., 2004).

2.2.2 Shoreline change calculation—Digital Shoreline Analysis System

Understanding and quantifying shoreline changes is crucial for assessing the impact of natural forces like erosion and sediment deposition, as well as human interventions. It informs delta management strategies, helps in addressing the challenges posed by sea-level rise and climate change, and aids in mitigating the risks associated with storm surges and flooding. Moreover, it enables the sustainable use of delta resources and the preservation of fragile ecological systems.

The Digital Shoreline Analysis System (DSAS) software, a complement to the Esri ArcGIS desktop, gives the ability to compute dynamic shift rates by analyzing numerous historical coastal boundaries (Kumar Das et al., 2021). It is a tool that simplifies the establishment of measurement sites, executes rate calculations, assembles the essential statistical information for assessing rate reliability, and encompasses a shoreline projection model (Apandi et al., 2022). This model offers the possibility to formulate forecasts for shoreline changes over 10 and/or 20-year periods, along with the option of generating corresponding margins of uncertainty (Abou Samra & Ali, 2021). It incorporates a user-intuitive interface which smoothly steers users through the principal stages of shoreline transition examination.

The DSAS operates with historical shoreline data obtained from sources like aerial imagery and satellite observations, all in digital formats. DSAS establishes a reference baseline along the coast, which serves as the basis for subsequent measurements. Automatic generation of measurement transects, perpendicular to this baseline, allows for precise calculations of the distances between the baseline and shoreline positions at various points in time (Baig et al., 2020). These measurements enable DSAS to compute rates of shoreline change, providing valuable insights into the dynamics of coastal evolution (Moussaid et al., 2015). DSAS is a useful and important tool for researchers, coastal managers, and policymakers in their efforts to understand and address the impacts of coastal processes and human activities on shorelines.

Case study: Long-term spatial and temporal shoreline changes of the Evinos River delta, Gulf of Patras, Western Greece (Karymbalis et al., 2022).

The Evinos River delta, situated on the northern shore of the Gulf of Patras in western Greece, covers an area of approximately 92 square kilometers. It is a vital part of the Messologi wetland, which is protected by the Ramsar Convention, signifying its ecological importance.

The delta formed during the late Holocene due to favorable conditions. High precipitation, erodible geological formations, and steep slopes in the upper reaches of the river contribute to the transportation of substantial amounts of fluvial sediments to the coast (Maroukian & Karymbalis, 2004). The relatively shallow and tectonically subsiding Gulf of Patras, combined with low tidal range and mild wave conditions, has allowed sediments to accumulate at the river mouth.

The coastline of the delta features several abandoned paleo-mouths, indicating the dynamic history of the delta's formation. There have been human interventions in the region, including the construction of a dam and channel straightening, which have significantly altered the natural landscape and dynamics of the delta.

In this study, the authors utilized GIS techniques to analyze coastal changes over a period spanning from 1945 to 2015. They compared coastlines from 1945, 1969, and 2015, using georeferenced orthophoto mosaics for 1945 and contemporary data from 2015. The orthomosaic data from 1945 covered the entire Greek territory and was derived from aerial photographs taken in that year. For 1969, georeferenced topographic diagrams at a 1:5000 scale, based on 1969 aerial photos, were employed.

A baseline, 1 km inland was established parallel to the oldest (1945) coastline of the delta, spanning 26.6 km. Transects were created perpendicular to this baseline every 50 m, with each transect being 1.5 km in length. This allowed for the measurement of the position of the historic coastlines concerning this baseline. A total of 532 transects were analyzed, and the DSAS was used to calculate coastal change rates and shoreline displacement phenomena for the periods of 1945—69 and 1945—2015. The Net Shoreline Movement (NSM) statistical analysis module helped quantify the displacement distances and rates between the 1945 shoreline and subsequent coastlines.

The study also employed medium-resolution Landsat TM satellite images to analyze coastal area variations and land losses or gains associated with shoreline changes over the periods 1993—2002 and 2002—18. The authors examined three low-lying parts of the delta and calculated delta area changes and land gain/loss rates using these satellite images.

Field observations were conducted to correlate the geomorphological features along the Evinos delta coastline with the results of the shoreline displacement assessment. This comprehensive approach provided valuable insights into the long-term changes and dynamics of the Evinos delta's shoreline.

The study of shoreline changes in the Evinos delta, spanning roughly 71 years, revealed several key findings. Erosion has been the dominant process affecting the delta's coastline during this period. From the easternmost part of the delta to the east end of the pre-1959 river mouth, approximately 1.4 km of the coast have eroded, while only 150 m have experienced accretion. Erosion rates were particularly high, reaching 4 m per year between 1945 and 1969 (Fig. 14.3).

The continuous erosion of the eastern delta can be attributed to its earlier abandonment, marked by the presence of palaeomeanders cut off along the shore. Additionally, the orientation of the shoreline, exposure to prevalent winds and waves, and reduced sediment supply due to channel straightening and mouth closure contributed to this phenomenon.

The area around the pre-1959 closed river mouth saw significant retreat, with a shoreline shift of up to 790 m (rates of up to 17 m per year) over the last 47 years. The westward redistribution of sediments from the east abandoned delta and the pre-1959 river mouth created elongated sand spits and barriers parallel to the deltaic coastline (Fig. 14.4).

Furthermore, human activities have played a role in these changes. The construction of the Agios Dimitrios dam upstream has impounded sediment-laden river water, while extensive sand and gravel mining in the delta has disrupted natural sediment transport processes, contributing to shoreline changes.

In summary, this study reveals that about 60% of the Evinos delta experienced erosion between 1945 and 2015. The main factors contributing to these changes include the construction of the Agios Dimitrios dam, sand and gravel mining, an earth dam at the delta's apex, and artificial irrigation canals. Recognizing the connection between human activities and sediment supply is crucial for addressing the delta's increasing vulnerability. This research highlights the effectiveness of using remote sensing and GIS techniques for coastal erosion analysis and management. These methods provide valuable insights for monitoring and mitigating shoreline changes, facilitating more informed decision-making and sustainable management in the region.

2.2.3 Geomorphological mapping

Geomorphological mapping is an important tool for the monitoring of river deltas. By representing deltaic landforms, it can provide crucial insights into delta dynamics, shoreline erosion, sediment deposition, and the impact of factors such as sea-level rise and anthropogenic interventions (Lee, 2001). Historically, geomorphological mapping in river delta monitoring relied heavily on traditional survey methods and remote

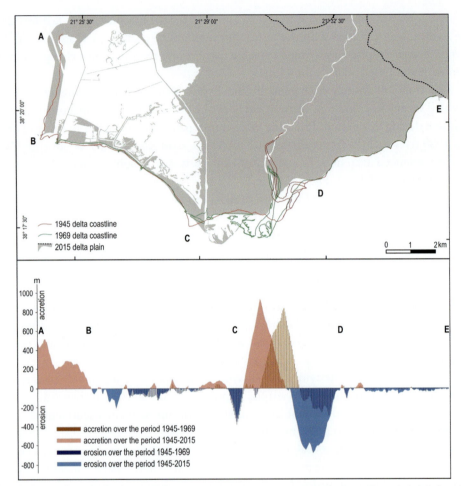

Figure 14.3 (a) Evinos delta coastline changes over the period 1945—2015. (b) Graph of shoreline changes, in terms of absolute length changes, for the Evinos river delta over the periods 1945—69 and 1945—2015. The horizontal axis corresponds to the baseline (the 1945 coastline of the delta) which was divided into 532 segments (50 m each). For each segment, the accretion and/or erosion for the given period was measured perpendicular to the baseline (Karymbalis et al., 2022).

sensing techniques (Dramis et al., 2011). Topographic maps, aerial photographs, and satellite imagery were often used to assess the delta's landforms and track their changes over time (Smith et al., 2011). While these methods were effective to some extent, they had limitations in terms of accuracy, resolution, and the ability to provide up-to-date information (Demek & Embleton, 1978). The static nature of paper maps and the time lag associated with data acquisition and processing were significant drawbacks (Seijmonsbergen, 2013).

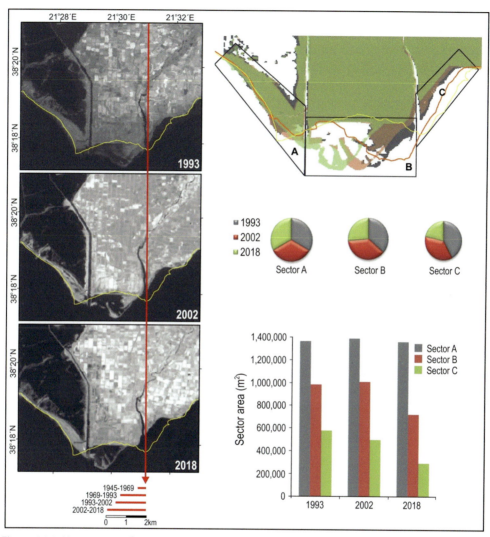

Figure 14.4 Net recent surface area changes for the three sectors of the active part of the Evinos delta estimated using landsat satellite images. The yellow line over the landsat images corresponds to the 1945 coastline (Karymbalis et al., 2022).

In recent years, there have been significant advancements in geomorphological mapping that have revolutionized delta monitoring. One of the most notable developments is the integration of Geographic Information Systems (GIS) (Magliulo & Valente, 2020) with high-resolution satellite imagery and aerial LiDAR data (Bocco et al., 2001). Furthermore, the use of UAVs, commonly known as drones, has become a game-changer in geomorphological mapping (Polyakov et al., 2022). Drones equipped with high-resolution cameras and LiDAR sensors can capture data at a level of detail and precision that was previously unattainable (Camarretta et al., 2020). They can access hard-to-reach areas, capture data at different scales, and provide an immediate overview of the delta's current state. This technology offers advantages not only in terms of accuracy but also in reducing costs and time for data collection. In addition to these technological advancements, remote sensing capabilities have improved considerably, providing a wealth of data sources. SAR systems, for instance, can penetrate cloud cover and monitor changes in land surface deformation, including subsidence and sea-level rise, both of which have critical implications for delta regions (Birch et al., 2017).

During the last decades, the introduction of innovative relief representation techniques has significantly enhanced the effectiveness of geomorphological mapping. The Red Relief Image Map (RRIM) is a 3D visualization method for topography by employing the chroma of red to represent slopes and the brightness to convey ridge—valley values derived from digital elevation models (DEMs) (Kaneta, 2008). RRIM effectively surmounts the limitations of conventional visualization techniques, such as scalability issues, dependence on light direction, the need for stereoscopes and filters, and the ability to encapsulate intricate topographical details within a single image map (Chiba & Hasi, 2016). This innovative approach is underpinned by patented technology owned by Asia Air Survey, Co., Ltd., with patents registered not only in Japan but also in the United States, China, and Taiwan. Its applications span a wide array of topographical features, from volcanoes, landslides, and faults to archeological investigations, serving as a navigational aid for geomorphologists. RRIM initially emerged as a solution to visualize the outcomes of airborne LiDAR measurements, which yield vast datasets. Unlike traditional contour maps that struggle to faithfully represent data outside specific contour intervals, RRIM takes a different path, employing red chroma and brightness to intricately illustrate the ground surface (Daxer, 2020). What sets RRIM apart is its ability to capture minute details of the topography using digital elevation data and its reliance on a single map for 3D visualization (Kaneta, 2008). In the RRIM scheme, a redder hue signifies steeper slopes, brightness indicates ridges, and darkness conveys valleys, enabling it to represent topography at various scales, including highly detailed microtopographical features (Terasawa & Takano, 2020) (Fig. 14.5).

Another recent advance is the introduction of Web GIS platforms in innovating representation of the geomorphological information. Web GIS platforms have the potential to significantly enhance traditional geomorphological maps by offering an interactive and dynamic way of accessing, analyzing, and sharing spatial data (Quesada-Román & Peralta-Reyes, 2023). Unlike static printed maps, web GIS allows users to access maps and associated data online, making it easier to keep information up-to-date (Paron & Claessens, 2011). These platforms provide a user-friendly interface that allows individuals to overlay various layers of information, such as topography, landforms, sediment distribution, and historical data. Users can interact with the maps, zoom in and out, measure

Red Relief Image Map protocol

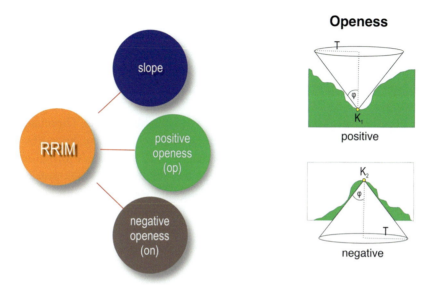

Figure 14.5 Conceptualization and methodological protocol of red relief image map. *(Modified by Chiba and Hasi (2016).)*

distances, and perform spatial analyses, all in real-time (Chandler et al., 2018). Furthermore, web GIS platforms enable collaboration and data sharing among researchers, policymakers, and the public, fostering a more comprehensive understanding of geomorphological features and changes over time. This integration allows for the creation of detailed, dynamic, and interactive digital maps that provide a comprehensive view of the delta's geomorphology (Griffiths et al., 2011). They can be updated in real time, enabling timely responses to changing conditions.

Furthermore, the incorporation of machine learning and artificial intelligence in geomorphological mapping allowed for the automated identification and classification of landforms, sediment types, and changes in the delta's morphology (Bishop et al., 2012). By analyzing vast datasets, AI algorithms can detect subtle variations and trends that might be missed by the human eye. This not only enhances the accuracy of geomorphological mapping but also allows for the extraction of valuable insights for predicting and managing the impact of natural and anthropogenic factors on delta ecosystems (Gustavsson et al., 2006).

In conclusion, geomorphological mapping has evolved significantly in its methodologies and capabilities, enhancing the precision, timeliness, and utility of delta monitoring. The integration of GIS, high-resolution remote sensing, drones, artificial intelligence, and advanced modeling techniques has ushered in a new era of data collection and analysis. These advancements not only enable us to understand the intricate dynamics of river deltas but also to proactively manage and conserve these landforms, ensuring their long-term sustainability and the well-being of the communities that rely on them. As we continue to refine and expand our geomorphological mapping techniques, we are better equipped to tackle the challenges posed by a changing world and its impact on delta environments.

Case study: Geomorphology of the Pinios River delta, Central Greece (Karymbalis et al., 2016).

The Pinios River delta is situated on the western shore of the Thermaikos Gulf in Thessaly, Central Greece. Covering an area of 69 km^2, it's characterized by its gently curved shoreline and low tidal range of less than 20 cm. The delta's formation is attributed to a combination of factors, including the local climate with high precipitation between November and February, erodible geological formations in the catchment area, and steep slopes upstream. These conditions lead to substantial sediment transport from the river into the coastal zone.

The receiving basin's characteristics, such as the mild tidal range and the presence of waves and longshore currents, facilitate sediment accumulation and delta formation over time. Human activities have primarily expanded residential and agricultural areas, with an 80% increase in recent decades. While major human interventions like artificial channel alignment and dam construction have not significantly impacted the delta, these changes reflect the evolving land use within the deltaic plain, highlighting the importance of considering human influences on delta dynamics.

The geomorphological mapping of the Pinios River delta involved a multiphase process. Initially, essential data sources were collected, including topographic diagrams, aerial photographs, geological maps, and more. These data were organized into a geographic information system (GIS) spatial geodatabase. Thematic layers, such as coastline, contour lines, isobaths, elevation points, and geological formations, were created through digitization.

A DEM of the delta was generated using contour data and drainage networks. Various land surface parameters, such as slope, gradient, aspect, curvature, and roughness, were calculated and used for the morphological sketch. These derivatives aided in mapping fluvial, coastal, and aeolian landforms, including abandoned channels, point bar ridges, beach ridges, and crevasse splays.

Older images, like aerial photos from 1945, proved more effective in mapping specific landforms. Lithological and structural features around the delta were derived from geological maps, supplemented by field observations.

The interpretation of morphological sketches was stored as vector data, classifying landforms into three main categories: fluvial, coastal, and aeolian. Some features were digitized in Google Earth and converted to shapefiles. Manual interpretation and contextual information played vital roles in identifying and outlining individual landforms.

Field control was executed to validate findings and correct any discrepancies. A hand-held differential GPS system ensured precise location data for these landforms. In the final mapping phase, the GIS database was integrated with GPS data, facilitating validation and classification accuracy. The end result was the production of the final geomorphological map of the delta, ensuring the correct inclusion and location of all existing landforms within the area.

The geomorphological survey led to the identification and mapping of three main sets of landforms and deposits: (i) fluvial, (ii) coastal, and (iii) aeolian (Fig. 14.6).

2.2.3.1 Fluvial landforms

Numerous abandoned distributaries and meandering channels were recognized across the delta plain. These meander bends showed meander scroll terrain patterns, created by the emplacement of point bar deposits. Changes in the river course and avulsions have led to

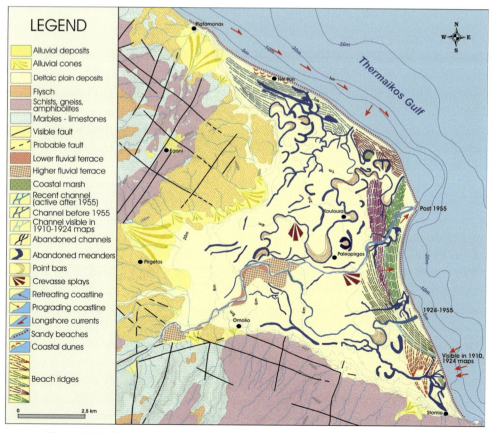

Figure 14.6 Geomorphological map of Pinios river delta (Karymbalis et al., 2016).

the formation of new distributary channels. Two crevasse splays were identified, indicating past levee breaches, which resulted in sediment deposition.

2.2.3.2 Coastal landforms

The lower Pinios River delta's coastal section features linear, sandy beach ridges called cheniers, signifying recent progradation of the delta. The particle size distribution of surficial sediments along the shoreline is relatively uniform. The analysis of shorelines from 1945, 1995, and 2013 revealed that approximately 50% of the delta coastline is currently eroding. Changes in coastline positions are influenced by sediment redistribution from former river mouths, resulting in straightening of the shoreline. The retreat of the coast around abandoned mouths and the progradation of the coast around the active mouth are significant features of this delta's coastal dynamics.

2.2.3.3 Aeolian landforms

A narrow strip of sand dunes, not exceeding 1.5 m in height, is located behind the beach at Nei Pori, indicating aeolian processes shaping the northern part of the delta.

2.2.3.4 Land-use change

Comparative analysis of historical aerial photos and recent satellite images revealed substantial land-use changes in the past 68 years. The coastal areas, previously natural grasslands and agricultural lands, have transitioned into tourist resorts, and residential areas have expanded significantly. Cottage-housing development has occurred along the coastline, especially in the south delta. Meanwhile, extensive reclamation work has converted interdistributary areas into arable land in the upper delta plain, leading to the disappearance of various landforms.

These land-use changes, including urbanization, tourism development, and agricultural expansion, have significantly transformed the Pinios River delta's landscape, impacting the preservation and visibility of various landforms. Additionally, the delta faces challenges from future sea-level rise, which may accelerate coastline retreat and lead to the loss of delta plain areas.

The geomorphological survey of the Pinios River delta, encapsulated in the 1:15,000 geomorphological map, uncovered valuable insights. It detailed the river's changing course, with a significant northward shift in 1955. The lower delta's key feature is a series of sandy beach ridges (cheniers), showcasing gradual delta expansion. Analyzing shoreline changes since 1945, it's evident that 50% of the modern delta coastline is currently experiencing erosion, with progradation around the present river mouth and reworking near the most recently abandoned mouth. Coastal transformations mainly result from wave action and sediment redistribution.

Considering the projected sea-level rise linked to global temperature increases, the delta's susceptibility to erosion is a major concern. Approximately 6.5 square kilometers of land, including valuable agriculture and tourism areas, lie below 0.5 m above sea level. The economic impact of increased erosion in this region could be significant.

2.2.4 Interferometric synthetic aperture radar

Interferometric synthetic aperture radar, commonly abbreviated as InSAR, is a remote sensing technique used to measure ground deformation over large areas with high precision (Zhang et al., 2019). It relies on the use of radar waves emitted by a satellite or aircraft to create detailed images of the Earth's surface and monitor changes in topography and motion (Li & Damen, 2010). InSAR is particularly useful for monitoring subsidence, uplift, and lateral movement of deltaic areas, making it valuable for various applications, including geology, environmental monitoring, and infrastructure assessment (Zhang et al., 2015).

InSAR uses a radar instrument on a satellite or aircraft to transmit electromagnetic waves (microwaves) toward the Earth's surface. These waves bounce off the surface and return to the instrument (Wang et al., 2017). To create precise measurements, InSAR often involves multiple acquisitions of radar data over the same area at different times. This can be days, months, or even years apart. InSAR relies on a technique called interferometry. During the multiple data acquisitions, the radar waves interact with the Earth's surface, and variations in the distance traveled by the radar waves result in changes in phase (Saleh & Becker, 2018). By comparing the phase differences between the radar images from different acquisition times, InSAR can detect even very small displacements of the Earth's surface, often on the order of millimeters (Minderhoud et al., 2020). InSAR processing software calculates the differences in phase between the radar images to create what's known as an interferogram, which represents the deformation that has occurred between the two acquisition times (de Wit et al., 2021). This deformation information is often represented as color-coded fringes, with each fringe corresponding to a specific amount of deformation.

Applications of InSAR are diverse and include monitoring subsidence caused by groundwater extraction, tracking tectonic plate movements, assessing landslides, monitoring the stability of infrastructure like bridges and dams, and studying volcanic activity, among many others. It can also detect subsidence due to sediment compaction, lateral erosion, and shifts in river channels making it an excellent tool for river delta monitoring.

Case study: Time Series Synthetic Aperture Radar Interferometry for Ground Deformation Monitoring over a Small Scale Tectonically Active Deltaic Environment (Mornos, Central Greece) (Parcharidis et al., 2011).

This case study explores the use of Time Series Synthetic Aperture Radar Interferometry (TS-SAR) to monitor ground deformation in the Mornos delta, located in central Greece, over the period of 1992—2009. The study's primary objectives include detecting spatial-temporal deformations, distinguishing between natural compaction and human-induced deformations, and investigating vertical subsidence related to buried fault zones and underwater mass movements. The Mornos delta, fed by the Mornos River, is a key water source for the Athens area and is situated in the tectonically active NW Gulf of Corinth.

The Mornos delta is a Gilbert-type fan-shaped alluvial land covering an area of 28 km^2 (Piper et al., 1990). It is primarily influenced by fluvial sediment supply and wave activity, with development dating back to the Late Holocene. The delta's formation was facilitated by favorable geological, lithological, and climate conditions in the drainage and receiving basin. The Mornos River flow, regulated by a dam since 1980, supplies Athens with potable water.

Subsurface Stratigraphy.

The delta's subsurface stratigraphy is characterized by the presence of compaction-free Mesozoic substratum overlaid by Late Holocene sediment layers. These layers consist of fluvial deltaic sediments with varying thicknesses, ranging from 60 to 150 m. Sedimentary sequences within the delta exhibit heterogeneity, featuring a combination of coarse and fine-grained materials. The delta hosts three distinct aquifers: the quaternary, the flysch, and the karstic limestone aquifers (Louis et al., 1993).

The study employs Single Look Complex (SLC) scenes from ERS-1,2, and ENVISAT satellites, using VV polarization and C-band frequencies. A DEM from the Shuttle Radar Topography Mission with a spatial resolution of approximately 90 m is also utilized. Precise orbit vectors enhance the accuracy of the satellite's orbit, and the Interferometric Point Target Analysis (IPTA) algorithm is used for subsidence estimation.

To investigate sediment compaction's influence on subsidence, borehole data from 17 locations within the delta plain are considered. These data were collected during a late 1970s geophysical study conducted by the Ministry of Agriculture in response to dam construction. The study creates a correlation between fine sediment percentages in the borehole logs and mean annual subsidence rates.

The analysis of Persistent Scatterers Interferometry (PSI) reveals that the majority of subsidence targets are concentrated within the Mornos delta. Subsidence is attributed mainly to natural sediment compaction, with areas featuring coarser sediments showing lower rates of compaction. Additionally, mountainous areas surrounding the delta exhibit stability or slight uplift.

Differential vertical movements result from aseismic slip along buried fault zones. The presence of an NE–SW trending normal fault beneath the Skala torrent, along with offshore fault scarps, influences vertical deformations. For the period 1992–2000, maximum subsidence rates reach approximately 27.2 mm/year, while uplift rates range from +2 mm/year to +4 mm/year (Fig. 14.7). The average estimated rate uncertainty is around 60.2 mm/year.

Statistical analysis of deformation rates shows that subsidence rates are generally higher in the delta plain compared to settlements at the delta's edge, which have a more stable

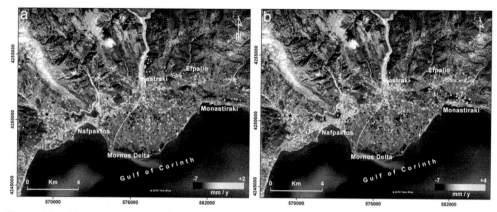

Figure 14.7 Linear component of ground deformation over Mornos delta broader area for the period 1992—2000 (a) and 2002—09 (b). Point targets are plotted on a high-resolution satellite image. Star symbol represents the reference point (Parcharidis et al., 2011).

subsidence pattern. The highest subsidence is observed at the old river mouth location (Bouka Karahassani), attributed to permanent marshes and fine sediments. Anthropogenic modifications, such as the construction of a dam and artificial channel confinement, play a role in shoreline retreat and accelerated subsidence.

The study demonstrates the efficacy of PSI in monitoring ground deformation in the Mornos delta. Subsidence primarily results from natural sediment compaction, aseismic slip along buried fault zones, and submarine mass movements on steep delta slopes. The cessation of sediment supply due to human activities also contributes to subsidence and shoreline erosion. Future efforts should focus on quantifying the impact of individual deformation factors through modeling and simulations to facilitate mitigation and response actions.

2.3 Integration of multisource data

2.3.1 Fusion of remote sensing and in situ data

River deltas often require multidimensional monitoring approaches. The fusion of remote sensing and in situ data is at the forefront of this effort, combining the advantages of satellite technology and ground-based measurements to provide a holistic understanding of delta dynamics (Qu et al., 2012).

Even though remote sensing, through high-resolution satellite imagery, LiDAR, and radar-based measurements provide a broad overview of an extended area, in situ data are indispensable to grasp the finer details (Hillen et al., 2014). In this context, ground-based measurements, such as water quality samples, sediment cores, and weather station data, offer a granular perspective. They validate remote sensing observations and provide real-time information for critical parameters.

The fusion of these data sources is a complex process that involves aligning and integrating datasets of varying resolutions, formats, and scales. However, this integration allows researchers and decision-makers to create a more detailed and accurate representation of the delta's physical and environmental characteristics (Teillet et al., 2002). It aids in monitoring land-use changes, tracking water quality, and understanding the evolution of geomorphological features, all of which are vital for delta management.

The fusion of remote sensing and in situ data is not only beneficial for environmental monitoring but also for addressing challenges such as flood prediction, natural resource management, and disaster risk reduction. It is a powerful tool that provides a comprehensive and multifaceted perspective on river delta ecosystems, supporting sustainable management strategies.

2.3.2 Data assimilation techniques

River delta monitoring often relies on numerical models to simulate and predict environmental changes. Data assimilation techniques have an important role in enhancing the accuracy and reliability of these models by integrating diverse data sources (Dorigo et al., 2007).

These techniques involve the incorporation of observational data, including remote sensing measurements and in situ observations, into numerical models. By fusing these datasets, data assimilation bridges the gap between real-world measurements and model simulations (Bertino et al., 2003). This process ensures that the models' predictions are consistent with observed data, resulting in more accurate assessments of delta dynamics.

Data assimilation is particularly valuable in predicting changes in sediment transport, water levels, and ecological patterns (Nichols, 2003). It facilitates the assimilation of satellite-derived information on water quality and land use, as well as field measurements of sediment properties and river flow rates. Through this integration, models can project the impact of various scenarios, such as land-use changes or sea-level rise, on delta ecosystems and their vulnerabilities (Ruiz et al., 2013). This approach enhances the ability to forecast and manage river deltas effectively and allows decision-makers to make informed choices based on the most up-to-date and validated data.

3. Key takeaways

This chapter has provided a comprehensive exploration of river delta monitoring, emphasizing their significance as dynamic landforms. Key takeaways include the intrinsic importance of river deltas in Earth and environmental monitoring, the complexities of delta dynamics, the wide variety of parameters measured during monitoring, and the important role of methodological approaches.

Continued research in river delta monitoring and mapping is essential for advancing our understanding of these environments. Ongoing studies promise enhanced data collection techniques, advanced modeling and simulation tools, improved data integration, and a deeper understanding of the impact of climate change, sea-level rise, and human activities.

The methodologies and insights developed in this chapter extend beyond river deltas, with applications in various environmental contexts. They have implications for addressing global challenges and fostering interdisciplinary collaboration among scientists, researchers, and policymakers to inform decisions on environmental conservation. In closing, river delta monitoring offers not only a lens into these unique ecosystems but also contributes to the larger mission of preserving the Earth's environmental integrity.

References

Abou Samra, R. M., & Ali, R. R. (2021). Applying DSAS tool to detect coastal changes along Nile Delta, Egypt. *Egyptian Journal of Remote Sensory Space Science, 24*, 463—470. https://doi.org/10.1016/j.ejrs.2020.11.002

Anthoff, D., Nicholls, R. J., & Tol, R. S. J. (2010). The economic impact of substantial sea-level rise. *Mitigation and Adaptation Strategies for Global Change, 15*, 321—335. https://doi.org/10.1007/s11027-010-9220-7

Apandi, E. B. T., Salleh, S. A., Rahim, H. A., & Adnan, N. A. (2022). Mapping of coastline changes in mangrove forest using digital shoreline analyst system (DSAS). *IOP Conference Series: Earth and Environmental Science, 1067*, 012036. https://doi.org/10.1088/1755-1315/1067/1/012036

Arkema, K. K., Guannel, G., Verutes, G., Wood, S. A., Guerry, A., Ruckelshaus, M., Kareiva, P., Lacayo, M., & Silver, J. M. (2013). Coastal habitats shield people and property from sea-level rise and storms. *Nature Climate Change, 3*, 913—918. https://doi.org/10.1038/nclimate1944

Baig, M. R. I., Ahmad, I. A., null, S., Tayyab, M., & Rahman, A. (2020). Analysis of shoreline changes in vishakhapatnam coastal tract of Andhra Pradesh, India: An application of digital shoreline analysis system (DSAS). *Annals of GIS, 26*, 361—376. https://doi.org/10.1080/19475683.2020.1815839

Bakus, G. J., Nishiyama, G., Hajdu, E., Mehta, H., Mohammad, M., Pinheiro, U., dos, S., Sohn, S. A., Pham, T. K., Yasin, Z. B., Shau-Hwai, T., Karam, A., & Hanan, E. (2007). A comparison of some population density sampling techniques for biodiversity, conservation, and environmental impact studies. *Biodiversity and Conservation, 16*, 2445—2455. https://doi.org/10.1007/s10531-006-9141-7

Balasubramanian, M. (2019). Economic value of regulating ecosystem services: A comprehensive at the global level review. *Environmental Monitoring and Assessment, 191*, 616. https://doi.org/10.1007/s10661-019-7758-8

Barbier, E. B., Hacker, S. D., Kennedy, C., Koch, E. W., Stier, A. C., & Silliman, B. R. (2011). The value of estuarine and coastal ecosystem services. *Ecological Monographs, 81*, 169—193. https://doi.org/10.1890/10-1510

Berardino, P., Fornaro, G., Lanari, R., & Sansosti, E. (2002). A new algorithm for surface deformation monitoring based on small baseline differential SAR interferograms. *IEEE Transactions on Geoscience and Remote Sensing, 40*, 2375—2383. https://doi.org/10.1109/TGRS.2002.803792

Berg, M., Stengel, C., Trang, P. T. K., Hung Viet, P., Sampson, M. L., Leng, M., Samreth, S., & Fredericks, D. (2007). Magnitude of arsenic pollution in the mekong and red river deltas — Cambodia and Vietnam. *Science of the Total Environment, 372*, 413—425. https://doi.org/10.1016/j.scitotenv.2006.09.010

Bertino, L., Evensen, G., & Wackernagel, H. (2003). Sequential data assimilation techniques in oceanography. *International Statistical Review, 71*, 223—241. https://doi.org/10.1111/j.1751-5823.2003.tb00194.x

Besset, M., Anthony, E. J., & Bouchette, F. (2019). Multi-decadal variations in delta shorelines and their relationship to river sediment supply: An assessment and review. *Earth-Science Reviews, 193*, 199—219. https://doi.org/10.1016/j.earscirev.2019.04.018

Bianchi, T. S., & Allison, M. A. (2009). Large-river delta-front estuaries as natural "recorders" of global environmental change. *Proceedings of the National Academy of Sciences, 106*, 8085—8092. https://doi.org/10.1073/pnas.0812878106

Birch, S. P. D., Hayes, A. G., Dietrich, W. E., Howard, A. D., Bristow, C. S., Malaska, M. J., Moore, J. M., Mastrogiuseppe, M., Hofgartner, J. D., Williams, D. A., White, O. L., Soderblom, J. M., Barnes, J. W., Turtle, E. P., Lunine, J. I., Wood, C. A., Neish, C. D., Kirk, R. L., Stofan, E. R., Lorenz, R. D., & Lopes, R. M. C. (2017). Geomorphologic mapping of titan's polar terrains: Constraining surface processes and landscape evolution. *Icarus, 282*, 214—236. https://doi.org/10.1016/j.icarus.2016.08.003

Bishop, M. P., James, L. A., Shroder, J. F., & Walsh, S. J. (2012). Geospatial technologies and digital geomorphological mapping: Concepts, issues and research. *Geomorphology, Geospatial Technologies and Geomorphological Mapping Proceedings of the 41st Annual Binghamton Geomorphology Symposium, 137*, 5—26. https://doi.org/10.1016/j.geomorph.2011.06.027

Boak, E. H., & Turner, I. L. (2005). Shoreline definition and detection: A review. *Journal of Coastal Research*, 688—703. https://doi.org/10.2112/03-0071.1, 2005.

Bocco, G., Mendoza, M., & Velázquez, A. (2001). Remote sensing and GIS-based regional geomorphological mapping—a tool for land use planning in developing countries. *Geomorphology, 39*, 211—219. https://doi.org/10.1016/S0169-555X(01)00027-7

Bradshaw, E., Woodworth, P. L., Hibbert, A., Bradley, L. J., Pugh, D. T., Fane, C., & Bingley, R. M. (2016). A century of sea level measurements at Newlyn, Southwest England. *Marine Geodesy, 39*, 115—140. https://doi.org/10.1080/01490419.2015.1121175

Breitburg, D. L., Hondorp, D. W., Davias, L. A., & Diaz, R. J. (2009). Hypoxia, nitrogen, and fisheries: Integrating effects across local and global landscapes. *Annual Review of Marine Science, 1*, 329—349. https://doi.org/10.1146/annurev.marine.010908.163754

Camarretta, N., Harrison, P., A., Lucieer, A., M. Potts, B., Davidson, N., & Hunt, M. (2020). From drones to phenotype: Using UAV-LiDAR to detect species and provenance variation in tree productivity and structure. *Remote Sensing, 12*, 3184. https://doi.org/10.3390/rs12193184

Chandler, B. M. P., Lovell, H., Boston, C. M., Lukas, S., Barr, I. D., Benediktsson, Í.Ö., Benn, D. I., Clark, C. D., Darvill, C. M., Evans, D. J. A., Ewertowski, M. W., Loibl, D., Margold, M., Otto, J.-C., Roberts, D. H., Stokes, C. R., Storrar, R. D., & Stroeven, A. P. (2018). Glacial geomorphological mapping: A review of approaches and frameworks for best practice. *Earth-Science Reviews, 185*, 806—846. https://doi.org/10.1016/j.earscirev.2018.07.015

Chen, G., Zhao, K., Lu, Y., Zheng, Y., Xue, M., Tan, Z.-M., Xu, X., Huang, H., Chen, H., Xu, F., Yang, J., Zhang, S., & Fan, X. (2022). Variability of microphysical characteristics in the "21·7" Henan extremely heavy rainfall event. *Science China Earth Sciences, 65*, 1861—1878. https://doi.org/10.1007/s11430-022-9972-9

Chen, J., Yi, S., Qin, Y., & Wang, X. (2016). Improving estimates of fractional vegetation cover based on UAV in alpine grassland on the Qinghai—Tibetan Plateau. *International Journal of Remote Sensing, 37*, 1922—1936. https://doi.org/10.1080/01431161.2016.1165884

Chiba, T., & Hasi, B. (2016). Ground surface visualization using red relief image map for a variety of map scales. *The International Archives of the Photogrammetry, Remote Sensing and Spatial Information Sciences, XLI-B2*, 393—397. https://doi.org/10.5194/isprs-archives-XLI-B2-393-2016

Daxer, C. (2020). *Topographic openness maps and red relief image maps in QGIS*. https://doi.org/10.13140/RG.2.2.18958.31047

Day, J. W., Agboola, J., Chen, Z., D'Elia, C., Forbes, D. L., Giosan, L., Kemp, P., Kuenzer, C., Lane, R. R., Ramachandran, R., Syvitski, J., & Yañez-Arancibia, A. (2016). Approaches to defining deltaic sustainability in the 21st century. *Estuaries Coastal Shelf Science, Sustainability of Future Coasts and Estuaries, 183*, 275—291. https://doi.org/10.1016/j.ecss.2016.06.018

de Wit, K., Lexmond, B. R., Stouthamer, E., Neussner, O., Dörr, N., Schenk, A., & Minderhoud, P. S. J. (2021). Identifying causes of urban differential subsidence in the Vietnamese mekong delta by combining InSAR and field observations. *Remote Sensing, 13*, 189. https://doi.org/10.3390/rs13020189

Dellepiane, S., De Laurentiis, R., & Giordano, F. (2004). Coastline extraction from SAR images and a method for the evaluation of the coastline precision. *Pattern Recognition Letters, Pattern Recognition for Remote Sensing (PRRS 2002), 25*, 1461—1470. https://doi.org/10.1016/j.patrec.2004.05.022

Demek, J., & Embleton, C. (1978). *Guide to medium-scale. Geomorphological mapping* [WWW Document]. URL https://www.schweizerbart.de/publications/detail/isbn/9783510650859/Guide_to_Medium_Scale_Geomorphological_Mapping.

Dorigo, W. A., Zurita-Milla, R., de Wit, A. J. W., Brazile, J., Singh, R., & Schaepman, M. E. (2007). A review on reflective remote sensing and data assimilation techniques for enhanced agroecosystem modeling. *International Journal of Applied Earth Observatuion. Geoinformation, Advances in Airborne Electromagnetics and Remote Sensing of Agro-Ecosystems, 9*, 165−193. https://doi.org/10.1016/j.jag.2006.05.003

Dramis, F., Guida, D., & Cestari, A. (2011). Chapter three - nature and aims of geomorphological mapping. In M. J. Smith, P. Paron, & J. S. Griffiths (Eds.), *Developments in earth surface processes, geomorphological mapping* (pp. 39−73). Elsevier. https://doi.org/10.1016/B978-0-444-53446-0.00003-3

Edmonds, D. A., Chadwick, A. J., Lamb, M. P., Lorenzo-Trueba, J., Murray, A. B., Nardin, W., Salter, G., & Shaw, J. B. (2022). Morphodynamic modeling of river-dominated deltas: A review and future perspectives. In *Treatise on geomorphology* (pp. 110−140). Elsevier. https://doi.org/10.1016/B978-0-12-818234-5.00076-6

Edmonds, D. A., & Slingerland, R. L. (2007). Mechanics of river mouth bar formation: Implications for the morphodynamics of delta distributary networks. *Journal of Geophysics Research in Earth Surface, 112*. https://doi.org/10.1029/2006JF000574

Elbing, B. R., Mäkiharju, S., Wiggins, A., Perlin, M., Dowling, D. R., & Ceccio, S. L. (2013). On the scaling of air layer drag reduction. *Journal of Fluid Mechanics, 717*, 484−513. https://doi.org/10.1017/jfm.2012.588

Elser, J. J., Bracken, M. E. S., Cleland, E. E., Gruner, D. S., Harpole, W. S., Hillebrand, H., Ngai, J. T., Seabloom, E. W., Shurin, J. B., & Smith, J. E. (2007). Global analysis of nitrogen and phosphorus limitation of primary producers in freshwater, marine and terrestrial ecosystems. *Ecology Letters, 10*, 1135−1142. https://doi.org/10.1111/j.1461-0248.2007.01113.x

Fagherazzi, S., Mariotti, G., Wiberg, P. L., & Mcglathery, K. J. (2013). Marsh collapse does not require sea level rise. *Oceanography, 26*, 70−77.

Folk, R. L., & Ward, W. C. (1957). Brazos River bar (Texas); a study in the significance of grain size parameters. *Journal of Sedimentary Research, 27*, 3−26. https://doi.org/10.1306/74D70646-2B21-11D7-8648000102C1865D

Gao, B. (1996). NDWI—a normalized difference water index for remote sensing of vegetation liquid water from space. *Remote Sensing of Environment, 58*, 257−266. https://doi.org/10.1016/S0034-4257(96)00067-3

Griffiths, J. S., Smith, M. J., & Paron, P. (2011). Chapter one - introduction to applied geomorphological mapping. In M. J. Smith, P. Paron, & J. S. Griffiths (Eds.), *Developments in earth surface processes, geomorphological mapping* (pp. 3−11). Elsevier. https://doi.org/10.1016/B978-0-444-53446-0.00001-X

Gustavsson, M., Kolstrup, E., & Seijmonsbergen, A. C. (2006). A new symbol-and-GIS based detailed geomorphological mapping system: Renewal of a scientific discipline for understanding landscape development. *Geomorphology, 77*, 90−111. https://doi.org/10.1016/j.geomorph.2006.01.026

Hillen, F., Höfle, B., Ehlers, M., & Reinartz, P. (2014). Information fusion infrastructure for remote-sensing and in-situ sensor data to model people dynamics. *International Journal of Image Data Fusion, 5*, 54−69. https://doi.org/10.1080/19479832.2013.870934

Hinkel, J., Lincke, D., Vafeidis, A. T., Perrette, M., Nicholls, R. J., Tol, R. S. J., Marzeion, B., Fettweis, X., Ionescu, C., & Levermann, A. (2014). Coastal flood damage and adaptation costs under 21st century sea-level rise. *Proceedings of the National Academy of Sciences, 111*, 3292−3297. https://doi.org/10.1073/pnas.1222469111

Just, D., & Bamler, R. (1994). Phase statistics of interferograms with applications to synthetic aperture radar. *Applied Optics, 33*, 4361−4368. https://doi.org/10.1364/AO.33.004361

Kaneta, S. (2008). *Red relief image map: New visualization method for three dimensional data.*

Karalis, S., Karymbalis, E., & Tsanakas, K. (2022). Estimating total sediment transport in a small, mountainous Mediterranean river: The case of Vouraikos River, NW Peloponnese, Greece. *Zeitschrift für Geomorphologie, 63*, 279−294. https://doi.org/10.1127/zfg/2021/0719

Karymbalis, E., Gaki-Papanastassiou, K., Tsanakas, K., & Ferentinou, M. (2016). Geomorphology of the Pinios River delta, Central Greece. *Journal of Maps, 12*, 12—21. https://doi.org/10.1080/17445647. 2016.1153356

Karymbalis, E., Gallousi, C., Cundy, A., Tsanakas, K., Gaki-Papanastassiou, K., Tsodoulos, I., Batzakis, D.-V., Papanastassiou, D., Liapis, I., & Maroukian, H. (2022). Long-term spatial and temporal shoreline changes of the Evinos River delta, Gulf of Patras, western Greece. *Zeitschrift für Geomorphologie, 63*, 141—155. https://doi.org/10.1127/zfg/2021/0684

Karymbalis, E., Valkanou, K., Tsodoulos, I., Iliopoulos, G., Tsanakas, K., Batzakis, V., Tsironis, G., Gallousi, C., Stamoulis, K., & Ioannides, K. (2018). Geomorphic evolution of the Lilas river fan delta (central Evia island, Greece). *Geosciences, 8*, 361. https://doi.org/10.3390/geosciences8100361

Kemp, W. M., Sampou, P., Caffrey, J., Mayer, M., Henriksen, K., & Boynton, W. R. (1990). Ammonium recycling versus denitrification in Chesapeake Bay sediments. *Limnology and Oceanography, 35*, 1545—1563. https://doi.org/10.4319/lo.1990.35.7.1545

Klees, R., & Massonnet, D. (1998). Deformation measurements using SAR interferometry: Potential and limitations. *Geologie en Mijnbouw, 77*, 161—176. https://doi.org/10.1023/A:1003594502801

Kumar Das, S., Sajan, B., Ojha, C., & Soren, S. (2021). Shoreline change behavior study of Jambudwip island of Indian Sundarban using DSAS model. Egypt. *Journal of Remote Sensing and Space Science, 24*, 961—970. https://doi.org/10.1016/j.ejrs.2021.09.004

Lee, E. M. (2001). Geomorphological mapping. *Geological Society London Engineering Geology Special Publications, 18*, 53—56. https://doi.org/10.1144/GSL.ENG.2001.018.01.08

Lee, H., & Liu, J. G. (2001). Analysis of topographic decorrelation in SAR interferometry using ratio coherence imagery. *IEEE Transactions on Geoscience and Remote Sensing, 39*, 223—232. https://doi.org/10.1109/36.905230

Lee, J., & Jurkevich, I. (1990). Coastline detection and tracing in SAr images. *IEEE Transactions on Geoscience and Remote Sensing, 28*, 662—668. https://doi.org/10.1109/TGRS.1990.572976

Li, X., & Damen, M. C. J. (2010). Coastline change detection with satellite remote sensing for environmental management of the Pearl River Estuary, China. *Journal of Marine Systems, 82*, S54—S61. https://doi.org/10.1016/j.jmarsys.2010.02.005

Lorenzen, C. J. (1967). Determination of chlorophyll and pheo-pigments: Spectrophotometric equations. *Limnology and Oceanography, 12*, 343—346. https://doi.org/10.4319/lo.1967.12.2.0343

Louis, J., F., Karantonis, G, A., & Voulgaris, N, S. (1993). The contribution of resistivity to the determination of aquifer parameters in Mornos Valley (Greece). In *55th EAEG meeting. Presented at the 55th EAEG meeting, European association of geoscientists and engineers, Stavanger, Norway.* https://doi.org/10.3997/2214-4609.201411684

Magliulo, P., & Valente, A. (2020). GIS-based geomorphological map of the calore river floodplain near benevento (southern Italy) overflooded by the 15th October 2015 event. *Water, 12*, 148. https://doi.org/10.3390/w12010148

Manfré, L. A., Hirata, E., Silva, J. B., Shinohara, E. J., Giannotti, M. A., Larocca, A. P. C., & Quintanilha, J. A. (2012). An analysis of geospatial technologies for risk and natural disaster management. *ISPRS International Journal of Geo-Information, 1*, 166—185. https://doi.org/10.3390/ijgi1020166

Maroukian, H., & Karymbalis, E. (2004). Geomorphic evolution of the fan delta of the Evinos river in western Greece and human impacts in the last 150 years. *Zeitschrift für Geomorphologie*, 201—217. https://doi.org/10.1127/zfg/48/2004/201

Metternicht, G., Hurni, L., & Gogu, R. (2005). Remote sensing of landslides: An analysis of the potential contribution to geo-spatial systems for hazard assessment in mountainous environments. *Remote Sensing of Environment, 98*, 284—303. https://doi.org/10.1016/j.rse.2005.08.004

Middelburg, J. J., Soetaert, K., & Herman, P. M. J. (1997). Empirical relationships for use in global diagenetic models. *Deep-Sea Research Part A Oceanographic Research Papers, 44*, 327—344. https://doi.org/10.1016/S0967-0637(96)00101-X

Milliman, J. D., & Meade, R. H. (1983). World-wide delivery of river sediment to the oceans. *The Journal of Geology, 91*, 1—21. https://doi.org/10.1086/628741

Milliman, J. D., & Syvitski, J. P. M. (1992). Geomorphic/tectonic control of sediment discharge to the ocean: The importance of small mountainous rivers. *The Journal of Geology, 100*, 525—544. https://doi.org/10.1086/629606

Minderhoud, P. S. J., Hlavacova, I., Kolomaznik, J., & Neussner, O. (2020). Towards unraveling total subsidence of a mega-delta — the potential of new PS InSAR data for the Mekong delta, in: Proceedings of IAHS. In *Presented at the TISOLS: The Tenth International Symposium on land subsidence — living with subsidence - Tenth International Symposium on land subsidence, Delft, The Netherlands, 17—21 may 2021, Copernicus GmbH* (pp. 327—332). https://doi.org/10.5194/piahs-382-327-2020

Moilanen, A., Franco, A. M. A., Early, R. I., Fox, R., Wintle, B., & Thomas, C. D. (2005). Prioritizing multiple-use landscapes for conservation: Methods for large multi-species planning problems. *Proceedings of the Royal Society B Biological Sciences, 272*, 1885—1891. https://doi.org/10.1098/rspb.2005.3164

Mora, C., Wei, C.-L., Rollo, A., Amaro, T., Baco, A. R., Billett, D., Bopp, L., Chen, Q., Collier, M., Danovaro, R., Gooday, A. J., Grupe, B. M., Halloran, P. R., Ingels, J., Jones, D. O. B., Levin, L. A., Nakano, H., Norling, K., Ramirez-Llodra, E., … Yasuhara, M. (2013). Biotic and human vulnerability to projected changes in ocean biogeochemistry over the 21st century. *PLoS Biology, 11*, e1001682. https://doi.org/10.1371/journal.pbio.1001682

Moussaid, J., Fora, A. A., Zourarah, B., Maanan, Mehdi, & Maanan, Mohamed (2015). Using automatic computation to analyze the rate of shoreline change on the Kenitra coast, Morocco. *Ocean Engineering, 102*, 71—77. https://doi.org/10.1016/j.oceaneng.2015.04.044

Nicholls, R. J., Hoozemans, F. M. J., & Marchand, M. (1999). Increasing flood risk and wetland losses due to global sea-level rise: Regional and global analyses. *Global Environmental Change, 9*, S69—S87. https://doi.org/10.1016/S0959-3780(99)00019-9

Nicholls, R. J., Hutton, C. W., Lázár, A. N., Allan, A., Adger, W. N., Adams, H., Wolf, J., Rahman, M., & Salehin, M. (2016). Integrated assessment of social and environmental sustainability dynamics in the Ganges-Brahmaputra-Meghna delta, Bangladesh. *Sustainability of Future Coasts and Estuaries, 183*, 370—381. https://doi.org/10.1016/j.ecss.2016.08.017

Nichols, N. K. (2003). Data assimilation: Aims and basic concepts. In R. Swinbank, V. Shutyaev, & W. A. Lahoz (Eds.), *Data assimilation for the earth system, NATO science series* (pp. 9—20). Netherlands, Dordrecht: Springer. https://doi.org/10.1007/978-94-010-0029-1_2

Oke, T. R. (1995). The heat Island of the urban boundary layer: Characteristics, causes and effects. In J. E. Cermak, A. G. Davenport, E. J. Plate, & D. X. Viegas (Eds.), *Wind climate in cities, NATO ASI series* (pp. 81—107). Netherlands, Dordrecht: Springer. https://doi.org/10.1007/978-94-017-3686-2_5

Palace, M. W., Sullivan, F. B., Ducey, M. J., Treuhaft, R. N., Herrick, C., Shimbo, J. Z., & Mota-E-Silva, J. (2015). Estimating forest structure in a tropical forest using field measurements, a synthetic model and discrete return lidar data. *Remote Sensing of Environment, 161*, 1—11. https://doi.org/10.1016/j.rse.2015.01.020

Parcharidis, I., Kourkouli, P., Karymbalis, E., Foumelis, M., & Karathanassi, V. (2011). Time series synthetic aperture radar interferometry for ground deformation monitoring over a small scale tectonically active deltaic environment (Mornos, Central Greece). *Journal of Coastal Research, 29*, 325—338. https://doi.org/10.2112/JCOASTRES-D-11-00106.1

Paron, P., & Claessens, L. (2011). Chapter Four - makers and users of geomorphological maps. In M. J. Smith, P. Paron, & J. S. Griffiths (Eds.), *Developments in earth surface processes, geomorphological mapping* (pp. 75—106). Elsevier. https://doi.org/10.1016/B978-0-444-53446-0.00004-5

Pettorelli, N., Laurance, W. F., O'Brien, T. G., Wegmann, M., Nagendra, H., & Turner, W. (2014). Satellite remote sensing for applied ecologists: Opportunities and challenges. *Journal of Applied Ecology, 51*, 839—848. https://doi.org/10.1111/1365-2664.12261

Pierdicca, N., Pulvirenti, L., & Chini, M. (2013). *Dealing with flood mapping using SAR data in the presence of wind or heavy precipitation, in: SAR image analysis, modeling, and techniques XIII. Presented at the SAR image analysis, modeling, and techniques XIII.* SPIE. https://doi.org/10.1117/12.2030105

Piper, D. J. W., Kontopoulos, N., Anagnostou, C., Chronis, G., & Panagos, A. G. (1990). Modern fan deltas in the western Gulf of Corinth, Greece. *Geo-Marine Letters, 10*, 5—12. https://doi.org/10.1007/BF02431016

Polyakov, V., Kartoziia, A., Nizamutdinov, T., Wang, W., & Abakumov, E. (2022). Soil-geomorphological mapping of Samoylov Island based on UAV imaging. *Frontiers in Environmental Science, 10.*

Poulos, S., Collins, M. B., & Ke, X. (1993). Fluvial/wave interaction controls on delta formation for ephemeral rivers discharging into microtidal waters. *Geo-Marine Letters, 13,* 24–31. https://doi.org/10.1007/BF01204389

Qu, Y., Zhang, Y., & Wang, J. (2012). A dynamic Bayesian network data fusion algorithm for estimating leaf area index using time-series data from in situ measurement to remote sensing observations. *International Journal of Remote Sensing, 33,* 1106–1125. https://doi.org/10.1080/01431161.2010.550642

Quesada-Román, A., & Peralta-Reyes, M. (2023). Geomorphological mapping global trends and applications. *Geographies, 3,* 610–621. https://doi.org/10.3390/geographies3030032

Ruiz, J. J., Pulido, M., & Miyoshi, T. (2013). Estimating model parameters with ensemble-based data assimilation: A review. *Journal of the Meteorological Society of Japan, 91,* 79–99. https://doi.org/10.2151/jmsj.2013-201

Sánchez-García, E., Balaguer-Beser, Á., Almonacid-Caballer, J., & Pardo-Pascual, J. E. (2019). A new adaptive image interpolation method to define the shoreline at sub-pixel level. *Remote Sensing, 11,* 1880. https://doi.org/10.3390/rs11161880

Sala, O. E., Stuart Chapin, F., Iii, Armesto, J. J., Berlow, E., Bloomfield, J., Dirzo, R., Huber-Sanwald, E., Huenneke, L. F., Jackson, R. B., Kinzig, A., Leemans, R., Lodge, D. M., Mooney, H. A., Oesterheld, M., Poff, N. L., Sykes, M. T., Walker, B. H., Walker, M., & Wall, D. H. (2000). Global biodiversity scenarios for the year 2100. *Science, 287,* 1770–1774. https://doi.org/10.1126/science.287.5459.1770

Saleh, M., & Becker, M. (2018). New estimation of Nile Delta subsidence rates from InSAR and GPS analysis. *Environmental Earth Sciences, 78,* 6. https://doi.org/10.1007/s12665-018-8001-6

Seijmonsbergen, A. C. (2013). *The modern geomorphological map.* Amsterdam: Elsevier. https://doi.org/10.1016/B978-0-12-374739-6.00371-7

Shanker, P., & Zebker, H. (2007). Persistent scatterer selection using maximum likelihood estimation. *Geophysical Research Letters, 34.* https://doi.org/10.1029/2007GL030806

Smith, S. V., & Hollibaugh, J. T. (1993). Coastal metabolism and the oceanic organic carbon balance. *Review of Geophysics, 31,* 75–89. https://doi.org/10.1029/92RG02584

Smith, Mike J., Paron, Paolo, & Griffiths, James S. (2011). *Geomorphological mapping, methods and applications.* Elsevier.

Syvitski, J. P. M., Kettner, A. J., Overeem, I., Hutton, E. W. H., Hannon, M. T., Brakenridge, G. R., Day, J., Vörösmarty, C., Saito, Y., Giosan, L., & Nicholls, R. J. (2009). Sinking deltas due to human activities. *Nature Geoscience, 2,* 681–686. https://doi.org/10.1038/ngeo629

Syvitski, J. P. M., Vörösmarty, C. J., Kettner, A. J., & Green, P. (2005). Impact of humans on the flux of terrestrial sediment to the global coastal ocean. *Science, 308,* 376–380. https://doi.org/10.1126/science.1109454

Tang, Y., Chen, F., Yang, W., Ding, Y., Wan, H., Sun, Z., & Jing, L. (2022). Elaborate monitoring of land-cover changes in cultural landscapes at heritage sites using very high-resolution remote-sensing images. *Sustainability, 14,* 1319. https://doi.org/10.3390/su14031319

Teillet, P. M., Gauthier, R. P., Chichagov, A., & Fedosejevs, G. (2002). *Towards integrated earth sensing: The role of in situ sensing.* https://doi.org/10.4095/219959

Terasawa, H., & Takano, Y. (2020). *Studying the utilization of UAV laser survey and a Red Relief Image Map (RRIM) in the construction of a large accelerator.*

Tessler, Z. D., Vörösmarty, C. J., Grossberg, M., Gladkova, I., Aizenman, H., Syvitski, J. P. M., & Foufoula-Georgiou, E. (2015). Profiling risk and sustainability in coastal deltas of the world. *Science, 349,* 638–643. https://doi.org/10.1126/science.aab3574

Tian, B., Wu, W., Yang, Z., & Zhou, Y. (2016). Drivers, trends, and potential impacts of long-term coastal reclamation in China from 1985 to 2010. *Estuarine, Coastal and Shelf Science, 170,* 83–90. https://doi.org/10.1016/j.ecss.2016.01.006

Tran, H., Tran, T., & Kervyn, M. (2015). Dynamics of land cover/land use changes in the Mekong delta, 1973–2011: A remote sensing analysis of the tran van Thoi District, Ca Mau Province, Vietnam. *Remote Sensing, 7,* 2899–2925. https://doi.org/10.3390/rs70302899

Turner, R. E., & Rabalais, N. N. (2003). Linking landscape and water quality in the Mississippi river basin for 200 years. *BioScience, 53*, 563–572. https://doi.org/10.1641/0006-3568(2003)053[0563:LLAWQI]2.0.CO;2

Walling, D. E. (1999). Linking land use, erosion and sediment yields in river basins. In J. Garnier, & J.-M. Mouchel (Eds.), *Man and river systems: The functioning of river systems at the basin scale, developments in Hydrobiology* (pp. 223–240). Netherlands, Dordrecht: Springer. https://doi.org/10.1007/978-94-017-2163-9_24

Wang, H., Feng, G., Xu, B., Yu, Y., Li, Z., Du, Y., & Zhu, J. (2017). Deriving spatio-temporal development of ground subsidence due to subway construction and operation in delta regions with PS-InSAR data: A case study in Guangzhou, China. *Remote Sensing, 9*, 1004. https://doi.org/10.3390/rs9101004

Wulder, M. A., Hilker, T., White, J. C., Coops, N. C., Masek, J. G., Pflugmacher, D., & Crevier, Y. (2015). Virtual constellations for global terrestrial monitoring. *Remote Sensing of Environment, 170*, 62–76. https://doi.org/10.1016/j.rse.2015.09.001

Zebker, H. A., & Villasenor, J. (1992). Decorrelation in interferometric radar echoes. *IEEE Transactions on Geoscience and Remote Sensing, 30*, 950–959. https://doi.org/10.1109/36.175330

Zhang, B., Wang, R., Deng, Y., Ma, P., Lin, H., & Wang, J. (2019). Mapping the Yellow River Delta land subsidence with multitemporal SAR interferometry by exploiting both persistent and distributed scatterers. *ISPRS Journal of Photogrammetry and Remote Sensing, 148*, 157–173. https://doi.org/10.1016/j.isprsjprs.2018.12.008

Zhang, J.-Z., Huang, H., & Bi, H. (2015). Land subsidence in the modern Yellow River Delta based on InSAR time series analysis. *Natural Hazards, 75*, 2385–2397. https://doi.org/10.1007/s11069-014-1434-7

Zhang, Y. (2011). Coastal environmental monitoring using remotely sensed data and GIS techniques in the Modern Yellow River delta, China. *Environmental Monitoring and Assessment, 179*, 15–29. https://doi.org/10.1007/s10661-010-1716-9

Zhu, Z., & Woodcock, C. E. (2014). Continuous change detection and classification of land cover using all available Landsat data. *Remote Sensing of Environment, 144*, 152–171. https://doi.org/10.1016/j.rse.2014.01.011

Further reading

Day, J. W., Britsch, L. D., Hawes, S. R., Shaffer, G. P., Reed, D. J., & Cahoon, D. (2000). Pattern and process of land loss in the Mississippi delta: A spatial and temporal analysis of wetland habitat change. *Estuaries, 23*, 425–438. https://doi.org/10.2307/1353136

Nittrouer, C. A., & DeMaster, D. J. (1996). The Amazon shelf setting: Tropical, energetic, and influenced by a large river. *Continental Shelf Research, 16*, 553–573. https://doi.org/10.1016/0278-4343(95)00069-0

Ryu, J.-H., Kim, C.-H., Lee, Y.-K., Won, J.-S., Chun, S.-S., & Lee, S. (2008). Detecting the intertidal morphologic change using satellite data. *Estuarine, Coastal and Shelf Science, 78*, 623–632. https://doi.org/10.1016/j.ecss.2008.01.020

Temmerman, S., Meire, P., Bouma, T. J., Herman, P. M. J., Ysebaert, T., & De Vriend, H. J. (2013). Ecosystem-based coastal defence in the face of global change. *Nature, 504*, 79–83. https://doi.org/10.1038/nature12859

Turner, W., Rondinini, C., Pettorelli, N., Mora, B., Leidner, A. K., Szantoi, Z., Buchanan, G., Dech, S., Dwyer, J., Herold, M., Koh, L. P., Leimgruber, P., Taubenboeck, H., Wegmann, M., Wikelski, M., & Woodcock, C. (2015). Free and open-access satellite data are key to biodiversity conservation. *Biological Conservation, 182*, 173–176. https://doi.org/10.1016/j.biocon.2014.11.048

mSphere. (2021). *Effects of recreational boating on microbial and meiofauna diversity in coastal shallow ecosystems of the Baltic Sea* [WWW Document], n.d. URL https://journals.asm.org/doi/full/10.1128/msphere.00127-21.

Science. (2015). *Profiling risk and sustainability in coastal deltas of the world* [WWW Document], n.d. URL https://www.science.org/doi/full/10.1126/science.aab3574?casa_token=9u2gnA7EfpUAAAAA%3ANt6ch55NKbsrvtcNghrj_OiSi7QVRoi-WE-fcguMrZWEQo8RwypP_xrnTLU06mOBILaITy4vBheRYw.

SECTION 4

Urban environment

CHAPTER 15

Dynamic modeling of urban hydrology in a geographic information system-setting with TFM-DYN

Andreas Persson[1,2], Petter Pilesjö[1,2] and Abdulghani Hasan[3]

[1]Lund University GIS Centre, Lund, Sweden; [2]Department of Physical Geography and Ecosystem Science, Lund University, Lund, Sweden; [3]Department of Landscape Architecture, Planning and Management, Swedish University of Agricultural Sciences, Alnarp, Sweden

1. Introduction

Urban hydrology has been an issue ever since the first dwellings needed to get water into and out of an area with a concentrated population. The development of in and out water of an urbanized area changes over time, and this change has been rapid over the last 200 years with population growth and changes in activities. In 1889, Emil Kuichling described the sewage system needs of a populated area in relation to the rainfall (Kuichling, 1889). This was proposed as "The rational method" and is still in use today. The changes to the hydrological system in a developed area have been described synoptically from several authors over the years. The input, pathways, storages, and recipients of water are changed in a multitude of ways with both more and less water in different parts of the system. This process of urbanization is for example described as an "hydrological headache" by Lindh (1972), with accurate schematic figures, valid still today. The modifications of the hydrological processes in urban areas are described by Walesh (1989, pp. 297–313), concluding that the processes of interception and infiltration are altered in the way that surface runoff is increased from 20% of the precipitation in an natural forested catchment to over 90% in an urban catchment.

Shaw et al. (2011) refer to the rational method and describe the urban catchment modifications and how these lead up to a necessary system for leading water out of the area. However, climate change high intensity storm events will result in precipitation amounts that the sewage pipe systems are not dimensioned for, leading to sewage capacity collapse and output from inlets resulting in surface flooding from storm water systems. This may be severe in itself, but with combined wastewater and storm water sewage systems, the damage and health hazard may be hard to handle.

Sustainable urban drainage systems (SUDSs) are approaches to take care of the water and relieve the storm water system in a sustainable way, often by using natural processes in, for example, infiltrations ponds, rain gardens, and swales, but also technical solutions such as detention ponds, pervious surfaces, and green roofs. There are practices and

Geographical Information Science
ISBN 978-0-443-13605-4
https://doi.org/10.1016/B978-0-443-13605-4.00004-7

systems related to SUDS such as LID (low impact development), BMP (best management practice), GI (green infrastructure), and more (Fletcher et al., 2015).

The implementation of such measures is a collaboration between several practitioners, including, among others, planners, engineers, horticulturists, and physical geographers. Where to apply these structures in the urban landscape requires a thorough investigation on the urban water system. The origin of extensive water in the urban landscape may be several. Flooding is often dived in the categories pluvial, fluvial, coastal, and urban. These also represent the types of flooding in urban areas and in some sense also the origin of the flood. Floods are built up from an input of water to the system that directly or indirectly produce too much water for the urban landscape and anthropogenic system to handle. Examples are cloud bursts (pluvial), rivers that breach their banks (fluvial), and storm surges (coastal). Urban flooding is often associated with the failure of the urban drainage system. However, this failure has an origin in itself and is interlinked with one or several of the other categories.

Today flood risk management practices do include multiple actors and stakeholders. The aim of minimizing damage and control the water input in urban areas requires planning with competence from disciplines ranging across society. The data needed and the analyses performed also need to be shared across organisational boundaries (Sörensen et al., 2016). Mapping in geographic information system (GIS) of local infrastructure's societal interdependencies in relation to risk management was performed by Johnson and McLean (2008) and by Wolthusen (2005), who included infrastructure protection from surface water movements. GIS is a versatile tool for both data management, analysis, visualisations, and communication.

Urban flood management does include a large number of measures that can be implemented at different sites along flow paths to relive the storm water pipe system, many of them often referred to as blue-green solutions. Depending on the upstream urban landscape and the origin of water solutions can be vegetated areas to increase infiltration, solid water detention structures in lowered urban squares, small detention tanks upstream in the catchment, rain gardens, green roofs, and green walls (C40 Knowledge Hub.Org). These structures aim to divert water into detention or change its path through the urban landscape. In order to do so, a thorough mapping and analysis process must precede the implementation. GIS analysis is especially suitable for such analyses since the most important data, elevation data for gravitational movement of the water, is in a vast majority of cases stored as raster data (see below).

1.1 Flow algorithms

Since GIS' beginning, flow routing and flood mapping have been carried out in raster-based data (Beven, 1996; Zhou et al., 2008). How these data, especially the digital elevation model (DEM) data, do approximate the real surface and how the movements/flows are transferred to a model will affect the outcome of the terrain modeling

(Holmgren, 1994; Tang et al., 2013). The most traditional approach is the single flow direction (SFD), which only move water from one cell in a grid to one possible neighbor, usually the one with the lowest elevation value. This means that one out of eight directions is used for mass movement (flow) (Tarboton, 1997). In the multiple flow direction (MFD) algorithms, the mass movements can be divergent, and the flow can potentially be split up to all eight neighboring cells (Freeman, 1991; Gallant & Hutchinson, 2011; Pilesjö & Hasan, 2014; Quinn et al., 1991). When slopes produce converging flow, as happens in concave slopes, the SFD algorithms produce relevant results, but when there is divergent flow, as on convex slopes, they often produce parallel flow with unrealistic patterns (Pilesjö et al., 1998). Since the data structure in grid format is part of the problem creating problems also for MFD algorithms, mainly because movements inside the focal cell are not well covered in a grid format, another type of algorithm is needed. With a combination of a grid and a TIN (triangular irregular network), Pilesjö and Hasan (2014) proposed a new MFD-algorithm called TFM (triangular form-based multiple flow algorithm). TFM considers the movements inside the focal cell and then distributes water to neighboring cells using triangles in a created TIN which is based on the grid center point elevation values. The focal cell is, hence, divided into eight triangles or facets. Each facet has a defined aspect and a constant slope gradient which makes it easy to distribute water packages both within the focal cell and to the neighboring cells.

1.2 Static flood mapping

Much of the flood analysis is made with a flow accumulation model that show how many upstream cells that flow to each cell in the grid. This flow accumulation result is a static map which show both the process and the end of a flow event at the same time. This is because flow that is counted as accumulated in a cell does not necessarily stay in that cell but may flow further downstream in the landscape. So, the flow accumulation numbers give information on the number of upstream cells that theoretically can contribute with flow to a specific cell. The lower a cell lies in the landscape, the higher number of contributing cells. As high numbers generally are assumed to be inundated or flooded areas, this is problematic. In reality, a cell with a high number may have lower ground below and is then not flooded at all.

A prerequisite when estimating flow accumulation is a flow direction calculation which show the paths the flow takes. In the common static flood mapping, calculations are made once, and paths are then as static as the flow accumulation, based on the ground levels throughout the analysis. This means that even though the conditions often change during a flow event, for example, pluvial (precipitation) and/or fluvial (river flooding) input to the landscape being analyzed, the results will not cover the changes. Furthermore, trying to estimate inundation depths at sites where flow accumulation is high is hard. In the static flood mapping, there are no input amounts, that is, no precipitation volumes or fluvial input volumes, and it is only the cell areas along flow paths that are summed up. Secondly, the ground processes influencing the volume and velocity, namely infiltration and friction, are not included in a flow accumulation analysis. A dynamic model is preferred to provide more realistic results by accounting for the factors mentioned above.

1.3 Dynamic flood mapping

A dynamic model will react to previous stages continuously. This is also what happens in the landscape when a storm event passes. The amounts of water and the ground condition change over time, which will vary the outcome of the process of flow, for example, the pathways and velocities.

TFM-DYN should therefore be seen as the next step in evolution from static flow accumulation estimations into dynamic hydrological modeling. It is using variable precipitation as input and incorporates functions for friction and infiltration to produce results of water depths, flow velocity, inundated areas as well as flow pathways and drainage area for each point in the landscape. All water volumes in the simulation, moved and infiltrated over time, are used as input to the storm water network and saved per timestep, with the possibility to be visualized in a GIS. All input and output data are GIS layers compatible with standard GIS software for handling, modification, and visualization.

2. Methods

2.1 The TFM algorithm

The triangular form-based multiple flow (TFM) algorithm was developed in order to enhance accuracy of modeled flow routing patterns in gridded GIS modeling (Pilesjö & Hasan, 2014). It implements the possibility of flow from a focal/center cell to all the eight neighboring cells in a 3×3 window. Below follows a summary of the method, based on Pilesjö and Hasan (2014).

The grid cells within the DEM, with each cell treated as a "central cell" during the estimation of flow from that specific cell, are subsequently subdivided into triangular facets characterized by consistent slopes and aspects (as depicted in Fig. 15.1). Leveraging these facets, it is possible to trace the surface flow path originating from each facet in a unique manner, subsequently redistributing water to other facets included in the central cell, and/or to one or more of the eight cells surrounding the central cell.

Upon the establishment of a facet, the slope and aspect values remain uniform across the surface. The coordinates of the three vertices of the triangular facet, as illustrated in Fig. 15.1, are denoted as C1(x1, y1, z1), C2(x2, y2, z2), and M(x3, y3, z3). The facet takes shape as a planar structure according to the following equation:

$$z = f(x, y) = ax + by + c \qquad (15.1)$$

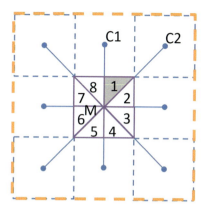

Figure 15.1 In a 3 by 3 cell window, the center cell is divided into eight triangular facets (1—8). Each facet is formed from three points; one is the center cell (M), and the two others are two adjacent cells (e.g., C1 and C2). *(From Pilesjö and Hasan (2014).)*

where *a*, *b*, and *c* can be derived as follows:

$$\left.\begin{aligned} a &= \frac{(y_1 - y_3)(z_1 - z_2) - (y_1 - y_2)(z_1 - z_3)}{(x_1 - x_2)(y_1 - y_3) - (x_1 - x_3)(y_1 - y_2)} \\[2mm] b &= \frac{(x_1 - x_2)(z_1 - z_3) - (x_1 - x_3)(z_1 - z_2)}{(x_1 - x_2)(y_1 - y_3) - (x_1 - x_3)(y_1 - y_2)} \\[2mm] c &= z_1 - ax_1 - by_1 \end{aligned}\right\} \tag{15.2}$$

If we let $q = f_y = \frac{\partial f}{\partial y} = b p = f_x = \frac{\partial f}{\partial x} = a$ p and q denote the gradients in the W—E and N—S directions, respectively, then according to Eq. (15.1), we get:

$$p = f_x = \frac{\partial f}{\partial x} = a, \quad \text{and} \quad q = f_y = \frac{\partial f}{\partial y} = b \tag{15.3}$$

The slope (β) and aspect (α) of the facet can then be derived as:

$$\left.\begin{aligned} \beta &= \arctan\sqrt{p^2 + q^2} = \arctan\sqrt{a^2 + b^2} \\[2mm] \alpha &= 180° - \arctan\frac{q}{p} + 90° \frac{p}{|p|} = 180° - \arctan\frac{b}{a} + 90° \frac{a}{|a|} \end{aligned}\right\} \tag{15.4}$$

The aspect angle is calculated clockwise from north.

In the vicinity of the central point (M, depicted in Fig. 15.1) of the focal cell, eight planar triangular facets are systematically constructed. By utilizing these eight triangular facets, the current grid cell (central cell) is subdivided into eight triangular facets, as indicated by numbers 1 through 8 in Fig. 15.1. The slope and aspect (slope direction) of each of these triangular facets are established as previously described. The configuration of the present grid cell is collectively represented by the cumulative surface area of these eight triangular facets. Each facet possesses an area equivalent to one-eighth of the cell size and symbolizes the proportion of water attributed to that specific facet. The path of overland flow from each triangular facet is determined by routing it either to another facet or neighboring cell, or allowing it to remain within the same triangular facet, depending upon the prevailing slope direction.

In the subsequent sections, we use triangular facet number one, as illustrated in Fig. 15.1, as an example to elucidate the method for estimating flow routing within the eight facets. It should be noted that a similar methodology is employed when estimating flow from any of the other eight facets (facets 2 through 8) to their adjacent facets.

As depicted in Fig. 15.2, the aspect value governs three distinct possibilities for flow routing. Water can be channeled directly from a facet to an adjacent cell, constituting the *stay* alternative, where there is no further routing to other facets. Alternatively, all the water within a facet may be directed to a single neighboring facet, termed the *move* option. In some cases, water on a facet can be distributed to both a neighboring facet and a neighboring cell, or even partitioned among two neighboring facets, designated as the *split* pathway. The specifics of these three routing alternatives are expounded upon in the subsequent discussion.

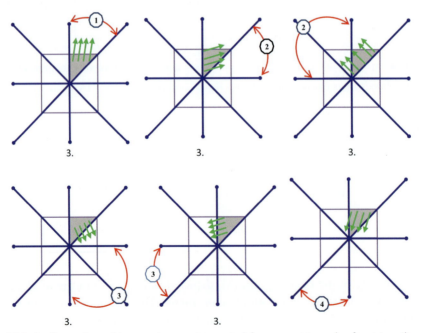

Figure 15.2 An illustration of how water can be routed from one triangular facet to other facets. Different aspect values lead to different actions (stay, move and split). *(From Pilesjö and Hasan (2014).)*

1. If the aspect value of facet 1 is between 0 and 45 degrees (see Case 1, Fig. 15.2), the amount of water (directly proportionate to the area of the facet) to be transported from this facet to a neighboring cell will *stay* as it is (1/8 of a cell).

2. If the aspect value is between 90 and 180 degrees, or between 225 and 270 degrees (see Case 3, Fig. 15.2), then all the water on the facet will be *moved* to one adjacent facet; that is, to facet number 2 and/or facet number 8. This means that the amount of water to be routed from facet 1 to a neighboring cell in these cases will be 0, and this water (1/8) will instead be added to facet number 2 or 8, respectively.

3. If the aspect value is between 45 and 90 degrees, or between 270 and 360 degrees (see Case 2, Fig. 15.2), then the water on facet 1 will be *split*, and partly routed into one neighboring facet according to a vector split (e.g., Pilesjö, 2008). The amount of water to be later routed to a neighboring cell will initially stay on the facet, while the other portion will be moved to a neighboring facet. If the aspect value is between 180 and 225 degrees (see Case 4, Fig. 15.2), then the water on facet 1 will be *split* between two adjacent facets; one portion will be added to the amount of water (proportional to the area) of facet number 2, while another portion will be added to facet number 8. The resulting amount of water of facet number 1 will in this case be 0.

Note, in the case of two facets sloping toward each other, that is, contradictive slopes, no water is routed between the two facets, but it will instead follow the break line between the two facets, either to a neighboring cell or toward the center of the cell, depending on surrounding elevations.

Since water can traverse between multiple facets (for instance, from facet 1 to facet 2, then to facet 3 and facet 4, and ultimately to neighboring cells), the process of flow routing between facets persists until all water has reached the designated outflow facet(s) within the cell. Iteratively, any remaining water is transferred to neighboring cells. This culminates in the internal redistribution of water within the cell, finally suggesting the destination (i.e., neighboring cell(s)) to which the water will be routed.

Subsequent to conducting flow routing for each cell, the next step involves the directed transfer of water to adjacent cell(s). Within one or more of the eight facets, water is in a state of *waiting*, poised for routing toward neighboring cell(s). If the cell delineates a concave topographical feature, it is conceivable that a solitary facet holds the entire volume of water, directly corresponding to the area of a single cell (designated as 1). Conversely, in the case of a convex surface (compare to a pyramid), it is possible for *waiting* water to be distributed among all eight facets.

Each facet has two neighboring cells (e.g., see cells C1 and C2 for facet number one in Fig. 15.1), and the accumulated water will now be distributed to these. Three different cases can then be identified:

1. If one neighboring cell is lower in elevation than the center cell (where the facet is located) and the other cell is higher or equal to the center cell, then the water accumulated in the facet will all be distributed to the lower cell.

2. If both neighboring cells are lower in elevation than the center cell, then the accumulated water in the facet will be proportionally distributed to both lower cells to slope ($x = 1$).

$$f_i = \frac{(\tan \beta_i)^x}{\sum\limits_{j=1}^{2} (\tan \beta_j)^x} \quad \text{for all } \beta > 0 \tag{15.5}$$

where i,j = flow directions (1 ... 2) to lower cells, f_i = flow proportion (0 ... 1) in direction i, $\tan \beta_i$ = slope gradient between the center cell and the cell in direction i, and x = variable exponent.

3. If both neighboring cells are higher or equal in elevation than the central cell and there are lower elevation cells surrounding the center cell, then the accumulated water in the facet will be proportionally distributed to these lower cells.

Flow distribution to adjacent cells is computed for all cells within the DEM, except for those situated at the border. Commencing from cells lacking incoming water, signifying the peaks, flow, measured in cell units, is projected to descend within the catchment. The outcomes are determined as flow accumulation values for each cell, excepting those positioned along the border of the DEM.

2.2 Dynamic flow estimation

Below we describe how the TFM algorithm can be applied in a dynamic flow model, denoted TFM-DYN. Based on Nilsson et al. (2021), we explain how flow depth and flow velocity can be estimated by combining TFM with basic physical flow estimations in time and space. Heterogeneous rainfall and ground characteristics are used, in order to describe the spatial/temporal variation in precipitation intensity, infiltration rate, and surface roughness. This enables the model to estimate water depths (m), volumes (m^3), and flow velocities (m/s). Matrices for precipitation (P), surface roughness (n), and infiltration (I) are included in the model framework.

An introduced time dimension captures the dynamic changes of the runoff process during and after a rainfall. For the discretization of time, a time step Δt is set, and iterations are done from $t = 0$ until the desired end time. All cells have a zero-water volume at $t = 0$ (if not, e.g., a pond or a river). The end time is set by the user, and can go on after the rainfall has stopped. The model includes five spatially distributed raster data layers, namely the DEM, precipitation, infiltration, inlets/outlets, and distributed roughness values (Manning's constant, n). Inlets and outlets are water paths that are not related to the elevation model, for example, sewage system and overflow drainage to ground. A synopsis of the modeling processes is illustrated in Fig. 15.3.

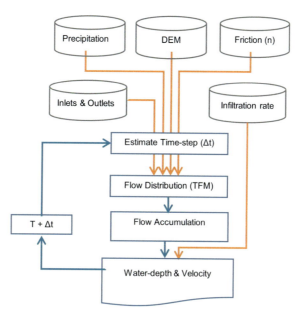

Figure 15.3 A schematic diagram to illustrate the modeling chain and workflow of the TFM-DYN model (Nilsson et al., 2021).

To model one time step, the procedure starts by letting it rain and infiltrate in each cell. This is carried out by adding precipitation to, and subtracting infiltration from, the current water volume in each cell.

The water volumes are estimated to be evenly distributed over each cell's surface. At each time step, the precipitation depth is added to the cells, and then the water is distributed to neighboring cells. The infiltration is deducted from the accumulated water at the end of each time step, depending on infiltration rate. Using the TFM algorithm, the flow directions are estimated from the elevation in each processing cell to its eight neighboring cells. It should be noted that the dynamic model incorporates water depths that are changing dynamically. Water can, for example, fill up sinks and flow out of them in new directions when distributing water (i.e., adding the water depth to the ground elevation). Therefore, new flow distributions are calculated at every time step based on the water surface elevation (z'), defined as the ground elevation (z) plus the water depth (d) (see Fig. 15.4). The water depth is always considered equal to a cell's water volume divided by the cell area.

The water is modeled to flow out of the processing cell, toward its neighbors according to the TFM flow distribution algorithm, estimating potential flow in all directions k (1−8) for a given t. The water is assumed to flow from the processing cell's center, over a surface with constant slope, to the next cell center. The amount of water routed from a

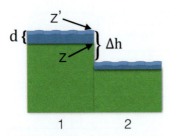

Figure 15.4 An illustration of a case where water is to be routed from one cell (1) to a downstream neighboring cell (2). For the processing cell (1), the ground elevation is z and z' is the water surface elevation (Nilsson et al., 2021).

cell, in one time step, will depend on the estimated water velocity, which is used to estimate how far the water flow will reach during one time step.

In order to estimate slope, and then velocity, the difference in water surface elevation between the processing cell and the downstream cell, denoted Δh, shown in Fig. 15.4, is calculated.

The slope is the water surface slope in direction k at time t, which is estimated from the difference in water surface elevations (Δh) in the k directions (1—8), divided by the distance between cell centers (L_k). The water velocity (v) for water flowing out of a cell is estimated as an average water velocity, calculated in each of the eight directions (k).

The volume of water available for routing in direction k, to a neighboring cell, is determined by how large proportion of the processing cell's water surface is above the water surface of the neighboring cell, as shown in Fig. 15.5. The water volume available for direction k is the processing cell's water volume multiplied by the water depth proportion (e.g., a or b in Fig. 15.5), and then multiplied by the proportion to be routed in this direction, as defined by the TFM algorithm.

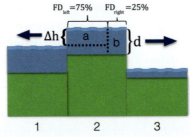

Figure 15.5 Available water in the processing cell to be routed *left* (a) and *right* (b). TFM has estimated maximum 75% of the water to be routed *left* and maximum 25% to be routed *right*. The available water depth for routing *left* is Δh, while the full depth is available to be routed *right* since the ground elevation in cell (2) is above the water surface elevation in cell (3) (Nilsson et al., 2021).

Also the flow velocity (v) influences the amount of water routed. A higher velocity will result in more water making it to the neighboring cell during the time step (Δt). We thus have to multiply by this factor, defined as the ratio between the distance the water flows in one time step ($v*\Delta t$) and the distance to the next cell center (L_k), when estimating the water volume to be distributed in a certain direction.

The time step regulates, in combination with slope, how far water flows from the processing cell at each iteration. It is thus of highest importance to set the time step (Δt) not to allow the model to estimate flow farther than one cell in one time step. An alternative way of avoiding the flow to reach further than one cell is to set a maximum velocity. The max velocity setting can be used to prevent the steepest features in a DEM from causing very small time steps, increasing the processing time considerably.

To summarize, at each time step t, the water flowing out of each cell is estimated in the eight possible routing directions (k). Then the new water volumes in each cell are estimated as the water volume from the last time step, plus precipitation, minus infiltration, minus the sum of water flowing out to neighboring, plus the sum of water flowing in from neighboring cells.

The model estimates water depths and velocities for all cells inside the border of the DEM, at every time step, whereas the border cells' water volumes are set to zero at every time step, not to affect the outflow from cells neighboring the border cells. The results attained from the model are the water depths and flow velocities in all cells at chosen time steps t, as well as maximum, minimum, and average values.

2.3 Experiment—Urban modeling
2.3.1 Test area
A test area in the middle of Lund was chosen due its diversity of permeable areas such as parks and green stretches together with a diversity of built up areas with mainly impermeable hard surfaces. The area covered include both commercial office areas with large parking lots and a large hospital area, both with mainly hardened surfaces. There is also the old university area in park surroundings, villa estates with large gardens, and a cemetery, all with permeable surface except for the buildings and some paved streets and paths. The center of the study area is situated at 55.71N, 13.20E, approximately where the Lund University GIS Centre is located. The area simulated is 2 by 2 km. There is a slight topographic gradient toward the south ranging from 81 to 44 m above the local vertical datum.

2.4 Data
2.4.1 Topographic data
The topographic data are based on LIDAR sampled points, collected from low altitude aircrafts with high resolution (± 5 cm) in elevation. The point density was on average 12 returns per square meter. The laser point cloud was filtered, and the vegetation was

removed, while hard structures as buildings, walls, curb stones were kept in the data. This means that also the building roofs were included. A 1 × 1 meter resolution DEM was generated using an inverse distance interpolation algorithm with the 12 nearest neighbors to each cell and a k = 2 as a weight. Each cell value includes returns mainly from that specific square meter. An advantage of creating the DEM from the original data is that the roof slope gradient is calculated on a square meter basis as well and is not generated from any generalized vector data.

2.5 Infiltration and friction

A vector ground cover dataset from the National Swedish Survey (Lantmäteriet) was used to generate two files, friction and infiltration. Infiltration was generalized to two types, namely permeable and impermeable areas. The permeable areas were set to be able to infiltrate 35 mm/h. These areas include lawns, tree rows, flower beds, gravel pathways, and graves. Impermeable areas were set to be able to infiltrate 4 mm/h. This because even in hardened surfaces, there are often cracks and gaps where water may infiltrate.

Friction was also generalized into the same categories. Permeable areas were given a high Manning constant, 0.05, due to the vegetation and gravel roughness. Impermeable surfaces, hardened surfaces such as paved areas, concrete, and roofs were given a Manning constant of 0.013.

2.6 Storm water inlets

A dataset with the positions of all storm water inlets, that is, gutter inlets to the municipality drainage system was retrieved from the municipality of Lund. A tabular value of the inlet type capacity was retrieved in reports from the Swedish Water Services association (Svenskt Vatten).

2.7 Precipitation

A designed precipitation event was created based on 100-year return period statistics from SMHI (The Swedish Meteorological and Hydrological Institute). The precipitation event was divided into seven 15 minute periods with intensities ranging from 52 mm/hour to 96 mm/hour. The simulation ended with two 15-minute periods without any precipitation.

3. Results

3.1 Water depth

Water depth raster files were saved every 300 second (every fifth minute) in the simulation time. The water depth maps generated from these differ over time where the first ones show only very shallow depths in the area. After 15—20 minutes, there were surface

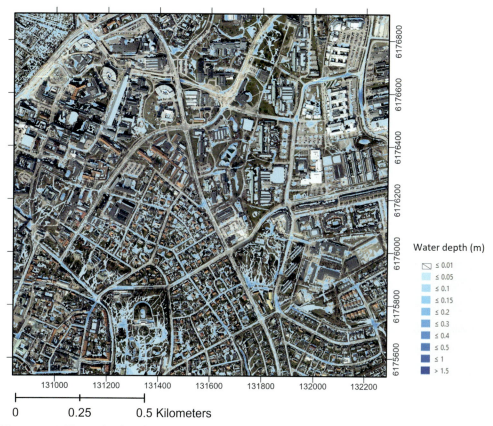

Water depth (m)

▱	≤ 0.01
	≤ 0.05
	≤ 0.1
	≤ 0.15
	≤ 0.2
	≤ 0.3
	≤ 0.4
	≤ 0.5
	≤ 1
	> 1.5

0   0.25   0.5 Kilometers

Figure 15.6 Water depths after 20 minutes simulated rain. *(Projection: SWEREF99 TM 13 30. Background aerial photograph: © Lantmäteriet.)*

water depths up to and above 50 cm in low laying sites and where flow paths come together (Fig. 15.6).

Toward the end of the rain event after 2 hours, the water depths in the landscape have increased to range up to and over 1 m in low points and along buildings where water is trapped on the upstream side (Fig. 15.7).

At the end of the simulation, a period of 30 minutes without precipitation was included. Fig. 15.8 depicts the last 5 minutes of the simulation, water depths have decreased due to less input from rain and a continuous infiltration and inflow in the storm water sewage system.

Each cell's maximum depth is represented in Fig. 15.9, called the global maximum depths. It should be noted that the time for each cell's maximum depth is different. These times area also recorded in a separate raster file.

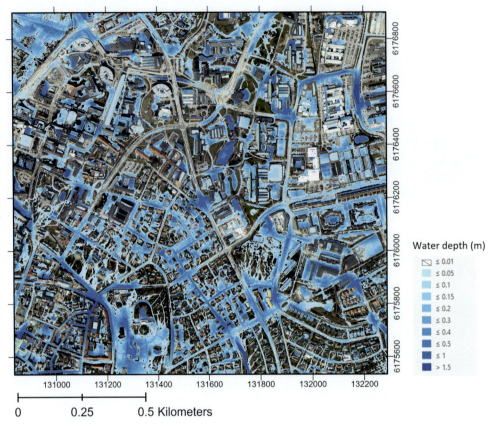

Figure 15.7 Water depths after 2 hours simulated rain. *(Projection: SWEREF99 TM 13 30. Background aerial photograph: © Lantmäteriet.)*

3.2 Water velocity per depth per time step and global max

Water velocity is a function of the friction and the hydraulic radius, calculated from the water depth differences from which a pressure head is estimated (Nilsson et al., 2021). The velocities therefore vary in the image, especially for the areas where water is accumulated. In steep areas, for example, roofs, the velocities are more or less constant since the water depths does not build up. The global maximum velocities are presented in Fig. 15.10.

3.3 Storm water system removal

For each of the storm water inlets, a position table with its capacity is set up and an extra removal in the cell corresponding to its position is handled by TFM-DYN. In the simulation, the removal was set to 3.62 L per second. For each of the inlets, a table

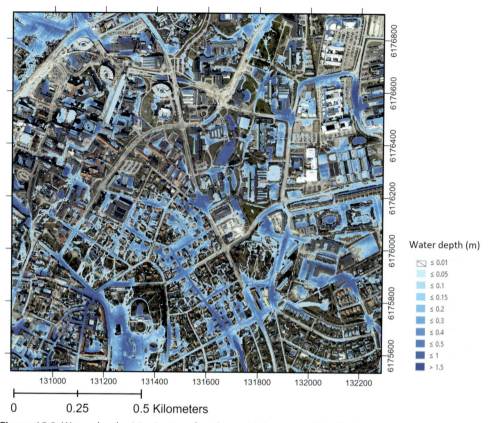

Figure 15.8 Water depths 25 minutes after the precipitation ended. *(Projection: SWEREF99 TM 13 30. Background aerial photograph: © Lantmäteriet.)*

with the accumulated water removal corresponding to the data saving period, here 300 s is saved.

3.4 Overall results from the simulation

Each iteration that the simulation tool makes is saved in a log file which present the calculated data for the whole landscape. In Fig. 15.11, the last iteration of the simulation with the above presented data is shown. Here the all the saved information is available.

4. Discussion

Topographical data: Topography plays a central role in hydrological models, where the quality of topographical data becomes a critical determinant of simulation accuracy.

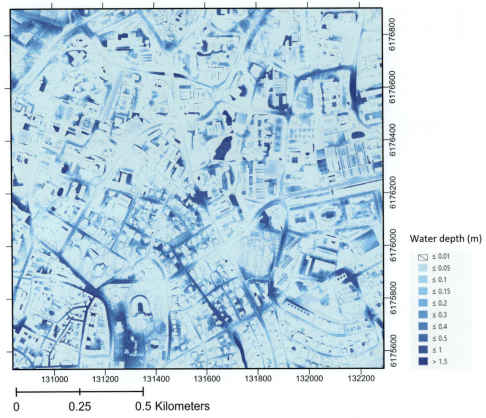

Figure 15.9 Global maximum water depths. *(Projection: SWEREF99 TM 13 30.)*

Typically, a DEM is employed for simulations, and DEM generation is frequently based on LiDAR data obtained from various platforms such as drones, airplanes, and helicopters. The accuracy of these data varies depending on the altitude and platform used, influencing point density and distribution (Hasan et al., 2012). The choice of interpolation technique for converting elevation points into a raster DEM further impacts the quality of the DEM (Persson et al. 2000, 2012). Considerations including the nature of the input data and the user's familiarity with the interpolation method, especially when employing geostatistical techniques such as kriging should also be taken into account. The DEM's resolution, determined during the interpolation process, significantly affects simulation outcomes. Coarse resolutions may result in the loss of objects that influence flow, often on the centimetre scale, and impact slope gradient calculations. Objects that may divert flow, often in the range of centimeters may disappear if a too coarse resolution is chosen. Slope gradient calculations differ a lot between different resolutions which also affect the result (Tang, 2013). In the case presented in this study, the LIDAR accuracy was very

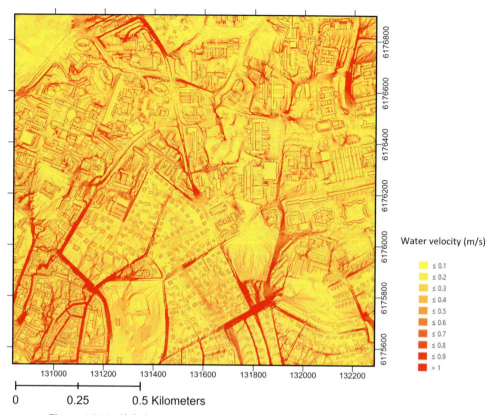

Figure 15.10 Global maximum water velocities. *(Projection: SWEREF99 TM 13 30.)*

high since a relatively low altitude of 1500 m was used for the data sampling. The inverse distance weighted average interpolation method was chosen because the number of laser returns, that is, sampling points, for each cell was so high (12) that the benefits of a geostatistical approach were deemed to be unnecessary.

4.1 Friction

In urban landscapes, a variety of surface materials come into play, each with its specific effects on water flow. To address this, roughness numbers and constants (e.g., Manning constants) are drawn from literature, tailored to engineering, natural sciences, and agriculture. Translating land use data into surface material types and selecting appropriate roughness values may be used to create a friction dataset. In this study, the urban landscape was categorized into two friction types: high friction for permeable areas with grass or gravel and low friction for impermeable features such as roads and rooftops. While such generalizations may have limitations, they are sufficient for this methodological presentation.

Figure 15.11 Log file excerpt of the last stage in the simulation.

```
###########
Time:                              06:12:20
State nr                           321081
Precipitation nr                   8
Simulation time passed[s]          8100.081066
Simulation time passed [h:m:s]     02:25:00
Time passed [s]                    137314.370872
Time passed [h:m:s]                14:08:34
Remaining simulation time [s]      137314.370872
Remaining simulation time [h:m:s]  00:00:00
Estimated remaining time [s]       -1.374256
Estimated remaining time [h:m:s]
Average water depth[m]             0.096791
Max water depth[m]                 17.464818
Total water volume[m3]             386391.958356
Average pressure head[m]           0.012479
Max pressure head[m]               0.582091
Avg effective vel [m/s]            0.038339
Avg mathematical vel [m/s]         0.038339
Max effective vel [m/s]            35.154066
Max mathematical vel [m/s]         35.154066
Nr flowing cells                   3985786
Flow ratio                         0.998442
Boundary water removed [m3]        149988.827821
Inlet system absorbed water [m3]   123534.343903
Outlet system added water [m3]     0.000000
###########
```

4.2 Infiltration

Determining infiltration in an urban setting poses challenges. Impermeable areas still allow for some infiltration through material or cracks. Permeable areas' topsoil may exhibit considerable variations without documented records since the urban subsurface has undergone alterations, deviating from the natural, preurbanization soil profile. This have led to a heterogeneous patchwork of different subsurface characteristics, also lacking records. The solution involves seeking preurbanization data as a baseline and conducting measurements in typical areas for informed generalizations. In the simulation discussed, two types of infiltration were implemented: low, but not zero, for impermeable areas and higher rates for permeable areas.

4.3 Precipitation

Precipitation stands as the primary input for surface flow and is introduced as the initial water type in the TFM-DYN tool. Precipitation data typically originate from national weather services, local or regional agencies measuring and recording precipitation. Design rain patterns can also be introduced to TFM-DYN depending on the application. There is a possibility to use either tabular input or raster file input of the precipitation data depending on if a homogenous or heterogenous rain pattern is sought for. In our case, the rain pattern was derived from local data to match a 2-hour rain event with a 100-year return period.

4.4 Gutter inlet capacity

The capacity of stormwater sewage inlets is documented, specifying the positions of inlets and their flow capacity in liters per second. The Swedish standard dictates that a storm-water sewage system should be able to handle a flow equivalent to a 10-year return period rain event.

4.5 Validation

Validating a scientific model for urban areas proves challenging due to the limited availability of discharge measurements in sewer inlets. Some locations have flow meters in inlets within restricted areas, offering potential testing opportunities for this recent model, which is an ongoing project. Additionally, proxy data for inundations can be used for a general assessment of the model's performance concerning depth and flow pathways. Such proxies may include insurance data or claims submitted to municipalities or water service companies responsible for stormwater management. Photos of inundated areas, coupled with recorded precipitation patterns, can serve as further validation. Infiltration in the ground presents difficulties in validation, given the diverse soil properties characteristic of urban environments.

4.6 Advantages of using GIS data

Directly utilizing GIS data for simulations presents several benefits. It reduces setup time for simulations when input data are readily available and simplifies the process of making alterations to the data for testing scenarios in the urban landscape using GIS software. A low-level programming language and expert programming have resulted in a simulation time which is notably expedited in comparison to other software.

GIS capabilities when it comes to visualisations give the user options on how to present the results. It may be as maps but can also be an animation of the dynamic processes using a sequence of the created raster files. The interval of saved raster images is set by the user and may be as low as every simulated second, only restricted by memory.

Data and results are today often presented in 3D, sometimes included in digital twins. It enhances spatial understanding by adding depth and perspective, enabling more accurate analyses. 3D representations improve communication, making complex geographic information more accessible to a wider audience. They support better decision-making in fields like urban planning, environmental management, and disaster risk, as 3D models provide a realistic view of landscapes and infrastructure. The water depth data can be presented combined with 3D data in a GIS for such visualisations (Fig. 15.12).

Figure 15.12 View with water depth data and 3D buildings superimposed. *(Projection: SWEREF99 TM 13 30. 3D building data ©Lund Municipality. Background aerial photograph: © Lantmäteriet.)*

5. Conclusion

At the Lund University GIS Centre, the development of the TFM has resulted in a model that surpasses others by delivering a more realistic representation of flow patterns. TFM handles multiple flow paths while considering the influence of slope gradient and the landscape subsection geometry, be it concave, planar, or convex (Pilesjö & Hasan, 2014).

Urban flooding and the risks with this have been noted in the literature for more than 5 decades and hydrological modeling is now playing a central role in risk mapping and decision support. The TFM-DYN model represents a significant advancement from static flow accumulation estimations into dynamic hydrological modeling (Nilsson et al., 2021). It incorporates variable precipitation in time and space as input together with high-resolution topography and includes functions for friction and infiltration to produce results such as water depths, flow velocity, inundated areas, flow pathways, and drainage areas for each point in the landscape. All water volumes in the simulation, whether moved, infiltrated, or input into the stormwater network, are recorded per time step and can be visualized in a GIS. All output data are saved in GIS layers, which can be visualized and managed in standard GIS software for further analyses.

The versatility of this model extends beyond urban hydrological issues. Its application spans over a wide range of hydrological analyses in diverse landscapes.

References

Beven, K. J. (1996). A discussion of distributed hydrological modelling. In M. B. Abbott, & J. C. Refsgaard (Eds.), *Distributed hydrological modelling* (pp. 255−278). Dordrecht: Kluwer Academic Publishers.

C40 Knowledge Hub.org. https://www.c40knowledgehub.org/s/article/How-to-reduce-flood-risk-in-your-city?language=en_US. (Accessed 22 September 2023).

Fletcher, T. D., Shuster, W., Hunt, W. F., Ashley, R., Butler, D., Arthur, S., Trowsdale, S., Barraud, S., Semadeni-Davies, A., Bertrand-Krajewski, J.-L., Mikkelsen, P. S., Rivard, G., Uhl, M., Dagenais, D., & Viklander, M. (2015). SUDS, LID, BMPs, WSUD and more − the evolution and application of terminology surrounding urban drainage. *Urban Water Journal, 12*, 525−542. https://doi.org/10.1080/1573062X.2014.916314

Freeman, T. G. (1991). Calculating catchment area with divergent flow based on a regular grid. *Computers and Geosciences, 17*, 413−422.

Gallant, J. C., & Hutchinson, M. F. (2011). A differential equation for specific catchment area. *Water Resources Research, 47*, Article 2009WR008540. https://doi.org/10.1029/2009WR008540

Hasan, A., Pilesjö, P., & Persson, A. (2012). On generating digital elevation models from LiDAR data − resolution versus accuracy and topographic wetness index indices in northern peatlands. *Geodesy and Cartography, 38*, 57−69. https://doi.org/10.3846/20296991.2012.702983

Holmgren, P. (1994). Multiple flow direction algorithms for runoff modelling in grid based elevation models: An empirical evaluation. *Hydrological Processes, 8*, 327−334.

Johnson, C. W., & McLean, K. (2008). Tools for local critical infrastructure protection: Computational support for identifying safety and security interdependencies between local critical infrastructures. In *Presented at the 3rd IET international conference on system safety 2008*.

Kuichling, E. (1889). The relation between the rainfall and the discharge of sewers in populous districts. *Transactions of the American Society of Civil Engineers, 20*, 1−56.

Lindh, G. (1972). Urbanization: A hydrological headache. *Ambio, 1*, 185−201. https://www.jstor.org/stable/4311983.

Nilsson, H., Pilesjö, P., Hasan, A., & Persson, A. (2021). Dynamic spatio-temporal flow modeling with raster DEMs. *Transactions in GIS, 26*, 1572−1588. https://doi.org/10.1111/tgis.12870

Persson, A., Hasan, A., Tang, J., & Pilesjö, P. (2012). Modelling flow routing in permafrost landscapes with TWI: An evaluation against site-specific wetness measurements: TWI in permafrost peatlands. *Transactions in GIS, 16*, 701−713. https://doi.org/10.1111/j.1467-9671.2012.01338.x

Persson, D. A., & Pilesjö, P. (2000). Digital elevation models in precision farming: Sensitivity tests of different sampling schemes and interpolation algorithms for the surface generation. In *Presented at the second international conference on geospatial information in agriculture and forestry* (pp. 214−221).

Pilesjö, P. (2008). An integrated Raster-TIN surface flow algorithm. In Q. Zhou, B. Lees, & G. Tang (Eds.), Lecture Notes in Geoinformation and Cartography. In *Advances in Digital Terrain Analysis* (pp. 237−255). Berlin: Springer.

Pilesjö, P., & Hasan, A. (2014). A triangular form-based multiple flow algorithm to estimate overland flow distribution and accumulation on a digital elevation model. *Transactions in GIS, 18*, 108−124. https://doi.org/10.1111/tgis.12015

Pilesjö, P., Zhou, Q., & Harrie, L. (1998). Estimating flow distribution over digital elevation models using a form-based algorithm. *Geographic Information Sciences, 4*, 44−51.

Quinn, P., Beven, K., Chevallier, P., & Planchon, O. (1991). The prediction of hillslope flow paths for distributed hydrological modelling using digital terrain models. *Hydrological Processes, 5*, 59−79.

Sörensen, J., Persson, A., Sternudd, C., Aspegren, H., Nilsson, J., Nordström, J., Jönsson, K., Mottaghi, M., Becker, P., Pilesjö, P., Larsson, R., Berndtsson, R., & Mobini, S. (2016). Re-thinking urban flood management—time for a regime shift. *Water, 8*, 332. https://doi.org/10.3390/w8080332

Shaw, E.,M., Beven, K.J., Chappel, N.,A., & Lamb, R. (2011). *Hydrology in practice* (4th ed.). Spon Press.

Tang, J., Pilesjö, P., & Persson, A. (2013). Estimating slope from raster data — a test of eight algorithms at different resolutions in flat and steep terrain. *Geodesy and Cartography, 39*, 41—52. https://doi.org/10.3846/20296991.2013.806702

Tarboton, D. G. (1997). A new method for the determination of flow directions and upslope areas in grid digital elevation models. *Water Resources Research, 33*, 309—319. https://doi.org/10.1029/96WR03137

Walesh, S. G. (1989). *Sedimentation basin design*. Urban Surface Water Management.

Wolthusen, S. D. (2005). GIS-Based command and control infrastructure for critical infrastructure protection. In *First IEEE international workshop on critical infrastructure protection (IWCIP'05)*. *Presented at the first IEEE international workshop on critical infrastructure protection (IWCIP'05)* (pp. 40—50). Darmstadt, Germany: IEEE. https://doi.org/10.1109/IWCIP.2005.12

Zhou, Q., Lees, B., & Tang, G. (Eds.). (2008). *Advances in digital terrain analysis, Lecture notes in geoinformation and cartography*. Berlin: Springer.

CHAPTER 16

Geospatial tools for cultural heritage education: innovations in teaching and learning through massive open online courses

Maria Pigaki[1] and Carmen Garcia[2]

[1]Department of Geography and Regional Planning, National Technical University of Athens, Athens, Greece; [2]Department of Geography and Land Planning, Universidad de Castilla-La Mancha, Albacete, Spain

1. Introduction

This paper presents a didactic proposal aimed at introducing geotechnologies for the research and study of cultural heritage (CH), through a MOOC (massive open online course) designed for this purpose. The first part emphasizes the changes and challenges facing education in cultural heritage, which justify the need to adapt current teaching-learning strategies to the characteristics of the contemporary social context. In the second part, we analyze the methodological foundations of this proposal intended to develop spatial thinking. Particular attention is devoted to the cross-cutting learning strategy used (case studies). The main results are then presented, with a synthesis of the design and content of the course, created as an open educational resource (OER), thanks to the MINERVA project (Mapping Cultural Heritage. Geosciences Value in Higher Education). Finally, the contributions to the acquisition of spatial skills applied to CH are evaluated.

2. The complexity of the teaching-learning process of cultural heritage

The appearance of technological applications and digital tools of increasingly accessible use has facilitated *spatial literacy*, a type of learning that is necessary to develop critical thinking applied to space, understood in all its depth, because, as Santos contends, it is constituted not only by things but also by relationships. Space must be considered an inseparable set comprising, on the one hand, a certain arrangement of geographical objects (both natural and social), and on the other, the life that fills and animates them, society in movement (Santos, 1996, p. 28). The material configuration of space, through a set of forms, is the expression of the society that manifests itself through them. Therefore, society cannot be separated from space and, in this sense, the idea that society becomes space can be understood.

Geographical Information Science
ISBN 978-0-443-13605-4
https://doi.org/10.1016/B978-0-443-13605-4.34001-2

This training implies a set of skills that are essential for the development of our daily activities, which, in addition, can be applied to highly diverse fields, including that of cultural heritage. In education, we are currently immersed in a process of applying information and communication technology (ICT) to highly varied areas of knowledge (even to some that seem more alien to this type of tool). This means going beyond the goal of learning technologies to focus on learning *with* technologies, which has given rise to the name of knowledge and learning technologies (KLTs).

In the case of cultural heritage, analyzing, managing, and disseminating its now fully recognized spatial dimension require digital skills related to spatial thinking; skills which must be included in the teaching-learning process.

2.1 The spatial dimension in the concept of cultural heritage

The definition of what we understand as heritage has broadened over time, encompassing more types of assets, and adding the spatial or environmental component (Feria, 2012). Since the mid–20th century, an evolution has occurred, changing the static vision of heritage (fundamentally considered as a set of material assets, of which the most outstanding are those of an architectural nature) as being located in space, which, in turn, is understood as a support or passive container of cultural heritage. Over the decades, other elements and values (such as intangible assets) have been added to its definition, while the importance of its spatial dimension has been recognized. Space is beginning to be considered not only as a support but also as a basic factor, both conditioning and expressing social processes.

Thus, we arrive at the current moment in which a double process can be observed: on the one hand, heritage is beginning to be seen as a constitutive part of a more complex spatial system and as something that is built in the dynamics of the present. On the other hand, the perceptions of individuals or groups are considered. Hence, the notion of landscape as the territory perceived by the population, resulting from the interaction of man with the environment over time, being a legacy with patrimonial values, makes full sense at this time (Fig. 16.1).

2.2 Changes and challenges in the educational context

Using spatial thinking to perform a wide range of everyday tasks (from finding the nearest school to parking a car) is not the same as the concept of spatial literacy. The latter is defined as the type of skills used to describe and analyze the spatial patterns of people, places, and environments on Earth. These are skills that must be developed by students that are interested in society and its reflection in space (i.e., those who deal with cultural heritage). However, there is a lack of focus on spatial thinking in studies of cultural heritage, and the social sciences and humanities in general, as has been recognized by various authors (Hespanha, 2009).

Figure 16.1 Pond built as part of the hydraulic system in the cultural landscape of Riópar (Spain) (case study). *(Source: Own preparation.)*

In the educational context, three changes can be highlighted, which, among others, are shaping the current stage of development. They are transformations common to those occurring in other areas and are the result of the evolution of society and technology.

First, everyone today has free access to large amounts of information. Data of all types (economic, social, demographic, etc.) are immediately available from the very moment they are produced. This information comes from different sources; some are formal (academic or institutional, for example), and others are of an informal or voluntary nature. Among the latter, the so-called neogeography has prospered.

Second, another characteristic is the availability of technological tools that are also freely accessible (in the form of specialized software). These are increasingly important in the teaching-learning process and are part of OERs, which are now notably more varied and complete. They have been defined as accumulated assets that enable the development of individual or social capacities for understanding and action. There is a great diversity of OER, the characteristics of which include their being freely used without interfering with the use of others, while providing nondiscriminatory access to the information and services they generate. Depending on the degree of openness, the possibility of their being modified by users can also be added (Tuomi, 2007, p. 35). These include geospatial tools, dedicated to the spatial treatment of data (such as GIS, among others), which are of particular interest for questions related to cultural heritage.

Third, the possibility for any person to rapidly create and disseminate content (which does not guarantee it is necessarily accurate or reliable) is another of the characteristics of our time. There are countless examples of blogs, wikis, and other procedures that serve this purpose.

Each of these changes emerge as challenges to which the education system must respond. The first great challenge is related to the need to improve the capacity to manage and optimize the flow of information that students receive so they are not overwhelmed by the large amount of data, opinions, theories, etc., available. The selection of relevant and reliable information is undoubtedly one of the challenges that teachers and students face. This initial task, which requires specific procedures and criteria, is necessary to leverage the potential that new technologies make available to the current educational system.

Another challenge is to promote the training of young people in technological tools and to improve their digital skills (be these knowledge, attitudes, or abilities). The EU has been engaged in the task of improving digital skills for many years, with a view to completing what is known as the digital transformation of the world. In this sense, it is one of the eight key competencies contemplated in lifelong learning. The Digital Education Action Plan, launched as a result of the Gothenburg Social Summit (in 2017), addresses the same objective. Here, we will focus specifically on the teaching-learning of geotechnologies, since they improve skills not only in spatial thinking but also in spatial literacy (or critical thinking about space as an expression of society).

The third challenge refers to the quality and depth of the content on which academic curricula are focused, forcing a reflection on the ultimate purpose of the teaching-learning process and of the teaching activity in question. In the case of the sciences that deal with cultural heritage, which include geographic space understood as a manifestation of society, it is necessary to focus on critical thinking on key notions like that of geographic space. The teaching-learning process is not restricted only to the acquisition of technological skills, even though these are absolutely necessary. Such skills should contribute to the ultimate goal of fostering the development of autonomous and critical thinking. In short, technological tools, in their different categories, are essential in current educational practice. This involves the use of both low-tech tools, that is, traditional ones (such as a simple paper or pencil) and high-tech applications (Excel, PowerPoint, Photoshop …), since all of them can be used to support spatial thinking.

2.3 The need to develop spatial thinking for cultural heritage

Although academic progress necessarily entails the use of a growing range of spatial competencies, spatial thinking is insufficiently recognized (from the initial levels of education) and given scant value, and therefore has a limited presence in school curricula (Sinton et al., 2013).

Some authors differentiate between spatial thinking (which can be considered as a set of abilities to visualize and interpret spatial concepts) and geospatial thinking, understood as a specialized form of spatial thinking that focuses on the patterns and processes that take place on the Earth's surface. Geospatial reasoning skills are high-level cognitive processes that provide the means to manipulate, analyze, interpret, and explain information, solve problems, or make decisions at geographic scales (Baker et al., 2014, p. 3), and also include visualization capabilities.

The use of geospatial technologies (and therefore GIS) involves both spatial thinking, in a broad sense, and geospatial thinking. Despite the lack of research supporting and verifying the relationship between the development of geospatial thought and the use of geotechnologies, a reciprocal relationship between the two can be recognized (Sinton et al., 2013). The acquisition of basic geospatial thinking skills, such as defining a problem, proposing solutions, or interpreting results, are prerequisites for solving spatial problems. The reverse is also true, since geotechnologies facilitate the set of activities that are necessary to identify spatial patterns, such as the visualization of spatial relationships or analysis and query of geospatial data. Maps and representations from GIS, for example, allow us to obtain information that would be difficult or impossible to acquire from direct experience and that is necessary to find spatial patterns. The idea of the world is today mediated by the technology available, since we learn to think about it through the images and content it makes available to us, which influences spatial thinking, in general and geospatial thinking, in particular.

3. Theoretical and methodological framework

The four core geotechnologies are GIS, remote sensing, global positioning systems (GPS), and digital globes. What is the value of this set of tools in the research, management, and dissemination of cultural heritage? This section presents the potential of applying these tools to cultural heritage. The methodological foundations of the MINERVA project, which focuses on promoting spatial thinking in this field of knowledge, is also analyzed.

3.1 Potential of geospatial tools for cultural heritage

In order to systematize the functionality of heritage technologies, it is possible to identify several types of tasks, which, in fact, are not independent of each other and give rise to three interrelated groups: (a) those dedicated to inventorying/documenting heritage (including actions such as registering, locating, identifying); (b) those that are oriented toward acquiring, processing, and analyzing patrimonial data; (c) those aimed at disseminating data on heritage, on its characteristics, and generating new experiences about it.

(a) New technologies and digitization play an essential role in the present and future of the conservation and dissemination of cultural heritage in general. In fact, the European Union has promoted the *declaration of cooperation on advancing the digitization of cultural heritage*, a document signed in 2019 by 25 member states, Norway, and the United Kingdom. The digitization of cultural objects has great potential to generate social and economic benefits for activities related to tourism, education, and the creative sectors (European Commission, 2019).

Therefore, for example, in the last 10 years, a large number of heritage inventories have been created that use GIS (such as the georeferenced catalog of the cultural heritage of Andalusia, Spain) (Junta de Andalucía, 2023), or the Mapping Ancient Athens project, a web mapping platform that locates and characterizes the archeological remains in this city identified and published from the 19th century onwards (Dipylon, 2023).

(b) Complementary to the previous ones are the tasks used to acquire and analyze data on heritage. There are numerous examples of works that show the possibilities offered by remote sensing systems, or GIS for studies on different heritage assets (archeological, artistic, landscape, etc.).

(c) On numerous occasions, cultural heritage is deteriorated or has even been destroyed. Geotechnologies help us to document and give new life to what is lost or deteriorated. Digital representations of cultural heritage can be made that may serve as digital surrogates for the real world. They may not only be used for leisure, play, or personal enjoyment but also permit viewing (under optimal conditions) and scientific study without the need for physical experience of the object or place. In this task, technologies such as photogrammetry, 3D modeling, GPS, and various types of software that allow for the generation of much more efficient and precise 2D and 3D information than a few years ago are extremely useful.

3.2 Significance statement

Spatial processes concern the locations of objects, their shapes, their relations, and the paths they take as they move in space and time. Recognition of spatial knowledge in the CH field enriches the traditional educational focus on developing literacy and skills to include a cognitive domain that is particularly relevant to achievement in both education and research.

How should geospatial tools be used in cultural heritage education in all its various aspects? Which didactic process is most effective in developing students' spatial reasoning? What actions must be applied? Which methods should be used?

The following sections of this chapter describe the process that refers to (a) the cognitive process of understanding space and (b) the relationship between spatial thinking and didactic methods that guides two strategies for exploiting these approaches in a MOOC environment. That is, Strategy 1 involves the development of spatial cognition using geoscience components that result in criteria and process. Strategy 2 involves spatializing the CH curriculum, via didactic material, using a case study approach suited to spatial thinking, including spatial language, maps, diagrams, graphs, analogical comparison, and argumentation.

For the first strategy, we analyze the cognitive process (1) across different teaching-learning approaches in e-learning environment, and (2) for students of different initial spatial skills backgrounds. Referring to the latter, we explain the didactic material of the MOOC as a cognitive process of spatial awakening in CH education, drawing on the strong link to the curriculum and the inherently spatial nature of our world to move toward to a new paradigm. More holistically and qualitatively, we determine the steps in which CH students engage via spatial activities within and across the respective goals of the development of the spatial thinking approach in CH education. This helps students to allocate study time reasonably and enhance self-regulated learning (SRL) and the self-perceived process.

SRL means that all students need to be able to use goal-setting, self-monitoring, effective self-instruction, and self-reinforcement. Self-perceived overall competence is defined as a personal characteristic that reflects an individual's global expectation or belief in their ability to accomplish tasks (Eccles & Gootman, 2002; Pajares & Schunk, 2005). The importance of self-perceived overall competence in positive youth development has been demonstrated and has been shown to serve to improve students' learning effectiveness and performance.

3.3 Spatial thinking process

Spatial thinking is an intellectual process supported by one or many "pictures" of space. Consequently, we cannot ignore the fact that developing spatial thinking can be facilitated by the use of various "pictures" of space. As a result, in order to approach and then describe space, there is a need for methods and tools that record and depict it, aiming to familiarize students with the three major components of spatial thinking, namely, cognition *about*, *in*, and *with* space.

Spatial thinking includes processes that support knowing, understanding, and planning. An expert spatial thinker visualizes relations, imagines transformations from one scale to another, mentally rotates an object to look at its other sides, creates a new viewing angle or perspective, and remembers images in places and spaces. Spatial thinking begins with the ability to use space as a framework. Furthermore, an object can be specified relative to the observer, to the environment, to its own intrinsic structure, or to other objects in the environment. Each instance requires the adoption of specific spatial frames of reference or context. In other words, the spatial context is critical because it is the space in which we find the data that ultimately determine its interpretation. In fact, we are thus actually mentioning three spatial contexts, physical space, intellectual space, and representational space, which require a student's cognitive transposition in two linked steps.

In the case of physical space, where space functions as an object of knowledge, there are two cognitive transpositions that create the background for spatial thought. The first of these is related to the passage from the identification of the objects and places *about* space into their physical context, referring to understanding. Intellectual space, as an instructive spatial tool, activates the analysis of spatial characteristics and their relationships, which, in turn, promotes student's participation in the process of understanding *in* space. The third stage is related to the passage from the intellectual space to the representational space, through images and symbols as an artifact of knowing, which activates the didactic process of applying knowledge, promoting the implementation of different scenarios and solutions *with* space.

3.4 Case study approach as a cross-cutting learning method

The previous description shows that the spatial thinking process requires different teaching-learning approaches through cognitive processes for students' spatial thought development. Additionally, it is a fact that the emergence of MOOC as a course delivery technology has changed the landscape of distance education. A case study as a learning experience of students in a CH curriculum is implemented, with an emphasis on approaches to students' spatial knowledge construction. In other words, it focuses on a constructivist approach that aims to produce new understanding or knowledge in CH, in a spiral manner. In fact, on the one hand, the case study is an essential method increasing students' knowledge construction when linked to other method, while, on the other, the case study approach acts as an organic pedagogical component of the assimilation of the research question, which is influenced by the attitudes, behavior, and learning approaches of the student members in the team. Thus, it develops student's capacities in order to enhance their SRL and self-perceived overall competence.

The case study method enables a researcher to closely examine the data within a specific context. In most cases, this method selects a small geographical area or a very limited number of criteria as the subjects of study. Case studies, in their true essence, explore and study a case as a lever method investigating contemporary real-life phenomenon through detailed contextual analysis of a limited number of events or conditions, and their relationships.

> "A case study is an empirical inquiry that investigates a contemporary phenomenon within its real-life context, especially when the boundaries between the object of study and context are not clearly evident. It copes with the technically distinctive situation in which there will be many more variables of interest than data points, and as one result relies on multiple sources of evidence, with data needing to coverage in a triangulating fashion, and as another result benefits from the prior development of theoretical propositions to guide data collection and analysis"
>
> **Dul and Hak (2008, p. 4).**

According to this overview, self-perceived overall competence means that all students need to be able to process according to goal-settings, self-monitoring, effective use of self-instructions, and self-reinforcement. For this purpose, case study methods are activated with a spiral form of learning in order to help students to acquire knowledge *about* space, to understand *in* space, and to apply *with* space. All the above obviously demonstrates that, to enhance spatial thinking, it is necessary to engage different strategies to achieve effective educational planning.

The case study method is implemented as an artifact to enhance the teaching-learning approach in both the CH and geotechnology fields. This method is linked to a theoretical framework (Tellis, 1997) aiming to achieve various learning methods according to the level of knowledge and the goal setting, with these being inquiry-based learning, problem-based learning, and research-based learning. Accordingly, the Minerva didactic proposal is designed around a set of case studies based on PBL methodology, which activates three levels of spatial thinking, that is, the case study as a lever of knowing, *about* space, an intellectual spatial procedure of understanding, *in* space, and a representational space that refers to a "logic chain" of applying *with* space.

Moreover, a set of didactic material was designed that contains the theory part via video lecture and a step-by-step training, to accomplish assignments such as guided practice, that is, guided exercises embedded with guided instructions. At the end of each part of a set of activities, a questionnaire is provided to be filled in by the students. At this stage, the case studies designed are focused as follows: an introduction that presents the context, the aim of the case, description of the skills involved and level of competences, the workflow, and finally the set of data needed.

3.4.1 The case study as a lever of knowing: about space

This type of case study seeks to instil curiosity about a specific question in a student. It aims to conceptualize the process of knowing in spatial terms expressed through two questions: "What is it?" and "Where is it?" That is, through queries *about* space, the aim is to help CH students to participate in two related functions with respect to managing/understanding geospatial data by observing and recalling. By doing so, they will be able to answer the fundamental questions of the previously mentioned knowledge process, as a result of having the necessary spatial skills. In fact, geoscience studies require the use of numerous spatial tools and techniques included under the term geographic skills. In this process, CH students are able to ask geographic questions, acquire spatial information, and organize it. The case study activities established draw on inquiry-based learning (IBL). This is an active student-centered form of instruction in which students are required to present a solution to a clearly defined authentic problem. This method emphasizes knowledge related to "how," which means how spatial knowledge is acquired. These cases study activities are designed for the purpose of "reading" a real spatial problem with a well-constructed questions through basic spatial skills, in order to enhance a student's basic spatial skills (Fig. 16.2).

3.4.2 The case study as an intellectual process: in space

Case study activities, as teaching-learning strategies, focus on relating data to real-life problems, using various sources of documentation. This process requires understanding

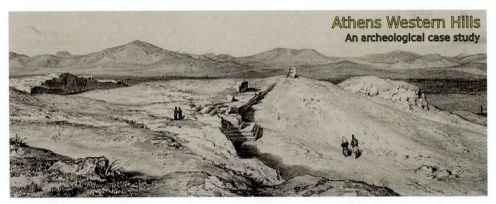

Figure 16.2 A case study about archeology in Athens. *(Source: Geotechnologies for cultural heritage MOOC.)*

space and its features and beyond; it is a process that links data to its spatial context. It is based on the potential of problem-based learning, that is, strengthening the teaching of real-life problems by analyzing a "structured," "guided," and "open" inquiry set of activities. It initiates the relation between teaching and research that leads to a reconceptualization of IBL, which, in turn, activates the process of understanding *in* space. This process is expressed by, at least, three questions: "Why is it there?", "How did it get there?", "What is its significance?" In terms of CH instruction, therefore, the focus is on students being able to attribute meaning to what they observe and mainly to explain it; in other words, to understand the subject matter of CH linked to its spatial context. Project-based learning gets students directly involved and actively working toward solutions (Fig. 16.3).

3.4.3 *The case study as a logical chain:* with *space*

This process is expressed by the commonly posed question: "How can spatial knowledge and understanding be used to enhance research in CH?" This process involves problem-solving processes, formulating hypotheses, and generating solutions to problems and, in general, applying geographic knowledge to real-life situations in a practical manner. This, of course, involves the learning of all the analytical concepts and principles that provide students with the ability to comprehend and find connections between diverse elements of geographic information and to use that information to explain patterns and processes in space. As a result, CH students comprehend and deal with complex contemporary issues by learning skills and understanding their subject matter to solve real problems, working *with* space. Additionally, in doing so, they interpret the results through their own hypotheses. Research-based learning promotes a student-centered approach that can strengthen the links between teaching and research. Thus, doing with understanding means activities that link research and project-based learning. In other words, they can

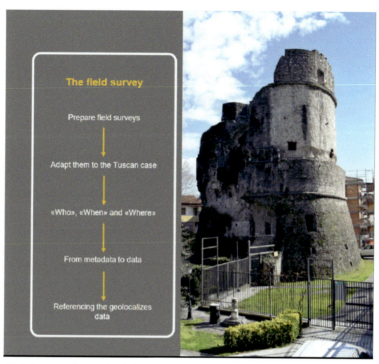

Figure 16.3 Training activity methodology in a case study. *(Source: Geotechnologies for cultural heritage MOOC.)*

learn to think critically and creatively. Students are provided with the opportunity to use the information they have collected to compare and contrast, interpret, apply, analyze, synthesize, open discussions, formulate a "chain" of logical procedures, summarize results, solve problems, make decisions, and verify key questions (Fig. 16.4).

4. Planning a MOOC on geotechnologies for cultural heritage

In order to use a MOOC and its content as a key topic, there are many references that describe the tools and techniques for empowering learners in a new pedagogical context. However, in the case of CH, there are challenges pertaining to teaching-learning approaches as well as others related to university curricula and life-long learning programs. In fact, the challenge lays in the pedagogical content, which requires a different process. On the one hand, the knowledge of geotechnologies must be tackled, and, on the other, spatial software, such as GIS, must be used as an interdisciplinary tool for teaching CH. This process is expressed by the following questions: "How can GIS benefit teachers of CH to enhance their students' learning? and How can spatial knowledge and understanding be used to solve problems in CH?"

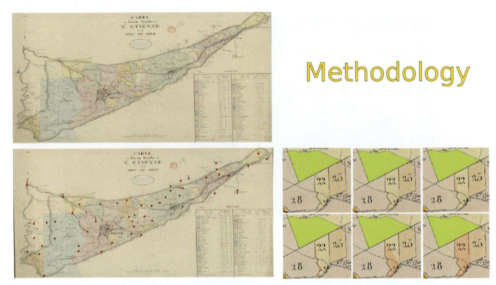

Figure 16.4 Case study: Loire coal mining basin. *(Source: Geotechnologies for cultural heritage MOOC.)*

This part is devoted to presenting the MINERVA project framework of teaching geotechnologies in CH as a tool to invigorate instruction and provide innovative content delivery in CH teaching-learning. The project uses these technologies as a means to increase student's spatial perception and reasoning in a MOOC environment. Nonetheless, using GIS in CH teaching does not actually constitute an innovation, but, if approached as a cognitive object, and as an object of spatial knowledge, it represents a viable and optimal methodology for CH study and research. That is, using GIS as an experiential tool can create a positive effect on the development of spatial thinking, reasoning, and interpreting in the field of cultural heritage. As suggested by Kolb, such experiential learning, which involves concrete experience, observation, and experimentation, is a feature of training provision (Kolb, 1984).

4.1 A cultural heritage MOOC framework

The possibilities of geographic information system environments are being activated, and different tools are offered to facilitate interaction in CH fields. In fact, the pedagogic aims of this whole-group hands-on experience are conceived to familiarize CH learners with software functionality by performing analytic tasks using sample data and to encourage them to reflect on its potential role for their own projects. For this purpose, the MINERVA project made a choice to use open-source and free-of-charge tools. This is justified because, first, it allows for better adoption by a larger community of teachers and students, and second, because Quantum GIS (the software selected) is a powerful desktop GIS for creating, editing, visualizing, analyzing, and publishing geospatial

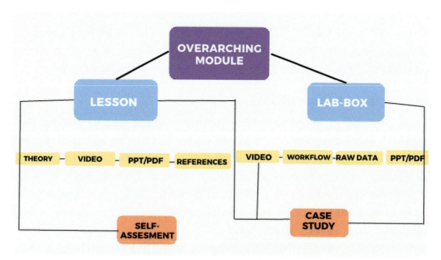

Figure 16.5 Teaching and learning workflow. *(Source: Own preparation.)*

information, providing a wide range of functionalities in both vector and raster modes. Fig. 16.5 provides an overview of the existing open didactic material available on each module of the Geotechnologies for Cultural Heritage course, in the web-based lessons. The teaching and learning material are provided in different languages (e.g., English, French, Greek, Italian, and others) and at different levels (such as Beginner 1, Beginner 1.2, and Advanced 2). References and resources used in the MINERVA MOOC and open data sources have been compiled by MINERVA project participants. Resources in different languages have been included and references to relevant "non-open" resources can also be found.

More specifically, the overarching modules allow a lecturer or a student to create conditional pathways through embedded material[1]: (1) The lessons' content is designed according to the main components of geotechnologies. Each lesson consists of videos (with subtitles in multiple languages), guided exercises (with downloadable data and documentation), bibliographical references, and, when useful, in-depth discussions (web-community). (2) Learning activities for each overarching module are designed with the purpose of listening and attempting to understand the theory part via video lectures that lead to assignments (such as guided practice, guided exercises, and guided instructions), and ending with questionnaire and a number of possible answers. (3) LAB BOX contains different case studies embedded in the content of lessons, as well as the overarching modules. (4) A self-assessment questionnaire is offered at the

---

[1]All the content of MOOC *Geotechnologies for Cultural Heritage* is available at: https://moodle.minerva. identitaculturale.eu/.

end of the module. Depending on a student's choice of answers, they either progress to the next module or are taken back to a previous one. When students correctly answer a set of question, they are redirected to the next module based on their answer. In other words, a bank of questions is available in order to evaluate students' process.

4.2 Teaching and learning pathways

The didactic material included in the overarching modules aims to guide a new curriculum approach in teaching and learning CH studies in order to add value to this field in the digital era. The content of the modules is designed with the aim of facilitating teaching and learning activities in two directions: the first, as a whole MOOC with graded didactic material for teachers and students, and the second, as a micro-MOOC, with modules used according to their own objective, that is, they can also be a supplement for stand-alone learning (Fig. 16.6).

More specifically, the CH MOOC design contains five overarching modules highlighting geosciences processes in the CH field. These are: (a) Introduction, the introductory module aims to introduce GIS as a powerful tool for monitoring, managing and

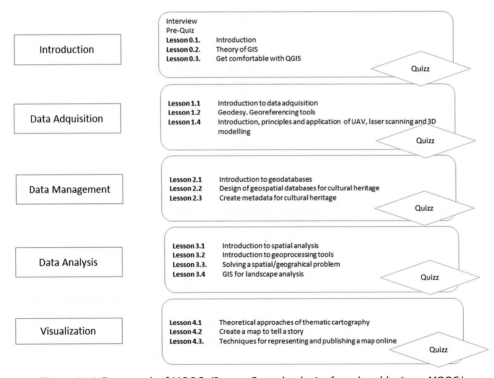

Figure 16.6 Framework of MOOC. *(Source: Geotechnologies for cultural heritage MOOC.)*

enhancing cultural and landscape heritage; (b) Data acquisition, which focuses on producing a geospatial database; (c) Data management, describing the ability to store, manage, and integrate geospatial data; (d) Data analysis, dedicated to explaining the key concepts of spatial analysis; and (e) Data visualization, which is devoted to the visualization of geographic data, that is, the use of visual representations to make their spatial context and spatial relationships visible. A pretraining questionnaire is provided in order to evaluate a student's background to guide them in an overarching module according to the respective levels, as mentioned above.

Each module includes several sections: (a) Modular description, outlining the content of the lessons of the module. (b) Video lecture, which is used to explain the theoretical part of a topic. (c) Teaser video, a trailer that provides information about the module content in order to create anticipation, spread awareness, and encourage enthusiasm about the topic. (d) Interview, which is an interactive, introductory, teacher-colleague activity to open the discussion about asking questions and defining problems in the CH field. (e) Interactive video, which supports guided instructions and guided exercises. (f) Guided practice, being an interactive exercise between the teacher and students. After the teacher introduces new learning, the student practice process it begins by engaging them in a similar task by simultaneously providing simple data. (g) Guided exercise, which is a process-oriented guided inquiry learning (POGIL) activity that develops, implements, complements, summarizes, and enables application of the material contained in the previously viewed lecture, which students will complete later. (k) Guided instructions, which are step-by-step instructions in software operations for case studies and guided exercises. (m) Quizzes: these practice quizzes allow students to increase critical thinking, and get into the habit of innovative learning. They integrate different types of mechanics into the learning process and will be discussed in the next section.

4.3 Assessment

The main goal of the MOOC (Bozkurt et al., 2017) is, therefore, to develop and integrate quality approaches, new pedagogies, and organizational mechanisms, with a strong focus on the learning processes (methodologies and assessments).

The Minerva MOOC was designed, organized, and evaluated in different steps during the project, in order to foster high-quality open education and learning in Europe and worldwide, in the quest for teaching and learning geotechnologies to facilitate a new generation in CH. This was done considering both the learning processes and technological factors that influence the efficient retention and completion of the MOOC by participants.

Concerning the learning process, self-assessment is an important aspect of educational pedagogy because it is important for students to monitor and know whether they are really acquiring knowledge. Furthermore, the assessment of the MOOC refers to different criteria, both internal and external. On the one hand, it addresses the respective didactic material of the teaching and learning process, and on the other hand, it refers to the MOOC effectiveness according to overview criteria such as motivation, usability, and interactivity.

4.3.1 Learning system assessment criteria

Self-assessment is an essential part of any type of training. The right kind of assessment strategies help teachers/students to obtain the appropriate feedback. The problem of question answering can be described according to a number of different dimensions, each making the problem more, or less, complex (Hirschman & Gaizauskas, 2001; Harabugiu et al., 2003; Moldovan et al., 2003). The level of spatial approach is distinguished by their overall purpose, or level of "understanding" (Hirschman et al., 1999).

To fulfill this objective, one of the strategies for self-assessment, in this instance, is to deploy quizzes for summative assessment per module. More specifically, the Minerva classification is based on Bloom's taxonomy, known as the "Taxonomy of Learning Objectives." These cognitive domains are aligned with the spatial process activated by the GIS processes, while their subcategories require cognitive "actions" to achieve the learning objectives in both specific fields, which in turn underlines sensorimotor strategies. With respect to this goal, the MOOC evaluation system is applied to the level of learning objectives, that is, knowing, reasoning, applying, and integrating skills, and goal setting.

The Moodle environment offers the possibility of creating more question banks, and, thus, all of them or a part of them can be used at any time they are needed. Different types of question answering approaches initiate different knowledge acquisition, which, in spatial terms, is expressed by "who," "what," "when," "where," "why," and "what if" related both to the Cultural Heritage and Geotechnologies fields. The required answers are broadly linked to the respective lessons, to references, or to links to the web. Finally, all the answers to the questions are saved for future analysis. Thus, an analysis can be made regarding the evolution of the responses of participants over time.

During the process of developing the MINERVA MOOC, an evaluation was carried out (Lu et al., 2021). This focused, on the one hand, on student-oriented research, using learners' perceived benefits to determine the learning design through didactic methods and material (Jung et al., 2019; Lowenthal et al., 2018), while, on the other hand, it provided an assessment of the MOOC environment, in order to overcome the challenges, as well as to take advantage of future opportunities in MOOC development (Donitsa-Schmidt & Topaz, 2018; Gan, 2018).

5. Instead of an epilogue

The MINERVA project, MappINg Cultural HERitage. Geosciences VAlue in Higher Education, focuses on two interlinked teaching-learning strategies, namely, internal and external. The first triggers methods for the purpose of enhancing spatial thinking by providing motivation thought a corpus of real-life problems. Therefore, case studies and interviews, based on pedagogical methods (inquiry-based learning, problem-based learning, and project-based learning), are used to enhance students' achievement and perspectives, according to their respective goal settings. The second refers to the usability of the MOOC, which is activated by the easy access to complete course content and by making full use of platform resources.

In conclusion, this project promotes and develops innovative methods and tools for teaching, learning, and assessing geotechnologies in higher education contexts. In particular, it supports teachers, educators, and learners to use geotechnologies in creative, collaborative, and efficient ways in the field of cultural heritage. MINERVA addresses the updating and development of digital learning materials, open pedagogies, and tools in the field of geotechnologies for CH, focusing on OERs through a free MOOC in its web learning platform. In general, MINERVA supports the effective use of digital technologies (especially GIS) through different methods. Introducing geographic information systems into the didactical process also develops learners' spatial thinking. Hence, it is important for a MOOC environment to take into consideration the factors that influence student learning in their particular setting and to provide them with all the necessary tools to nurture deep learning approaches in PBL methods. The didactic material implemented, and the tools support our teaching and learning strategy, integrate the influences of cognitive factors through the proposed pathways, and play a major role in enhancing students' overall competences and skills in the subject matter.

Acknowledgments

We would like to thank all the partners involved in the MINERVA project (2020-1-IT02-KA203-079559), funded by Erasmus+, within which this paper has been carried out.

University of Florence	Margherita Azzari
	Pauline Deguy
	Paola Zamperlin
	Giorgio Barbato
	Lucilla Spini
	Carmelo Pappalardo
	Vincenzo Bologna

Continued

Dipylon	Maria Pigaki
	Maria Karagiannopoulou
	Evi Sempou
	Eleni Mougiakou
	Yannis Paraskevopoulos
Research Centre of the Slovenian Academy of Sciences and Arts	Jani Kozina
	Rok Ciglič
	Erik Logar
	Peter Repolusk
University of Castilla—La Mancha	Carmen Garcia Martinez
	Francisco Javier Jover Martí
	Juan Antonio García González
	Francisco Cebrián Abellán
	Irene Sánchez Ondoño
	Pablo Valero Tocohulat
University of Jean Monnet-Etienne	Pierre-Olivier Mazagol
	Olivier Leroy
	Michel Depeyre
University of Niš	Vesna Lopičić
	Jasmina Đorđević
	Milan Đorđević
	Nikola Stojanović
	Vladimir Aleksić

References

Baker, T. R., Battersby, S., Bednarz, S. W., Bodzin, A. M., Kolvoord, B., Moore, S., Sinton, D., & Uttal, D. (2014). A research agenda for geospatial technologies and learning. *Journal of Geography*. https://doi.org/10.1080/00221341.2014.950684

Bozkurt, A., Akgün-Özbek, E., & Zawacki-Richter, O. (2017). Trends and patterns in massive open online courses: Review and content analysis of research on MOOCs (2008-2015). *International Review of Research in Open and Distance Learning, 18*(5), 118—147. https://doi.org/10.19173/irrodl.v18i5.3080

Dipylon Society. (2023). *Mapping ancient Athens.* https://map.mappingancientathens.org/.

Donitsa-Schmidt, S., & Topaz, B. (2018). Massive open online courses as a knowledge base for teachers. *Journal of Education for Teaching, 44*(5), 608—620. https://doi.org/10.1080/02607476.2018.1516350

Dul, & Hak. (2008). *Case study methodology in business research.* Elsevier Ltd.

Eccles, J., & Gootman, J. A. (Eds.). (2002). *Community programs to promote youth development.* Washington, DC: National Academy Press.

European Commission. (2019). *Shaping Europe's digital future.* https://digital-strategy.ec.europa.eu/en/news/eu-member-states-sign-cooperate-digitising-cultural-heritage.

Feria, J. M. (Ed.). (2012). *Territorial heritage and development.* Leiden: CRC Press.

Gan, T. (2018). Construction of security system of flipped classroom based on MOOC in teaching quality control. *Educational Sciences: Theory and Practice, 18*(6), 2707—2717. https://doi.org/10.12738/estp.2018.6.170

Harabugiu, S. M., Maiorano, S. J., & Pasca, M. A. (2003). Open-domain textual question answering techniques. *Natural Language Engineering, 9*(3), 231—267.

Hespanha, S. R., Goodchild, F., & Janelle, D. G. (2009). Spatial thinking and technologies in the undergraduate social science classroom. *Journal of Geography in Higher Education, 33*(Suppl. 1), S17—S27. https://doi.org/10.1080/03098260903033998

Hirschman, L., & Gaizauskas, R. (2001). Natural language question answering: The view from here. *Natural Language Engineering, 7*(4), 275—300. https://doi.org/10.1017/S1351324901002807

Hirschman, L., Light, M., Breck, E., & Burger, J. D. (1999). Deep read: A reading comprehension system, ACL. In *Proceedings of the 37th annual meeting of the association for computational linguistics* (pp. 325—332). https://aclanthology.org/P99-1042.

Jung, E., Kim, D., Yoon, M., Park, S., & Oakley, B. (2019). The influence of instructional design on learner control, sense of achievement, and perceived effectiveness in a supersize MOOC course. *Computers and Education, 128*, 377—388.

Junta de, Andalucía (2023). *Catálogo general del patrimonio histórico andaluz.* https://www.juntadeandalucia.es/organismos/turismoculturaydeporte/areas/cultura/bienes-culturales/catalogo-pha.html#toc-localizaci-n-de-bienes-protegidos.

Kolb, D. A. (1984). *Experiential learning: Experience as the source of learning and development.* Englewood Cliffs, NJ: Prentice Hall.

Lowenthal, P., Snelson, C., & Perkins, R. (2018). Teaching massive, open, online, courses (MOOCs): Tales from the front line. *International Review of Research in Open and Distance Learning, 19*(3). https://doi.org/10.19173/irrodl.v19i3.3505

Lu, M., Cui, T., Huang, Z., Zhao, H., Li, T., & Wang, K. (2021). A systematic review of questionnaire-based quantitative research on MOOCs. *International Review of Research in Open and Distance Learning, 22*(2), 285—313. https://doi.org/10.19173/irrodl.v22i2.5208

Moldovan, D., Clark, C., Harabagiu, S., & Maiorano, S. (2003). Cogex: A logic prover for question answering. In *Proceedings of HLT-NAACL, 2003, main papers, pp.87-93. Edmonton.*

Pajares, F., & Schunk, D. H. (2005). Self-efficacy and self-concept beliefs: Jointly contributing to the quality of human life. In H. W. Marsh, R. G. Craven, & D. M. McInerney (Eds.), *International advances in self-research Vol. II* (pp. 95—122). Greenwich: Age Publishing.

Santos, M. (1996). *Metamorfosis del espacio habitado, Barcelona, Oikos-Tau.*

Sinton, D., et al. (2013). *The people's guide to spatial thinking.*

Tellis, W. (1997). Introduction to case study. *Qualitative Report, 3*(2).

Tuomi, I. Open educational resources. What they are and why do they matter. Report prepared for the OECD 2007. http://www.meaningprocessing.com/personalPages/tuomi/moreinfo.html.

Further reading

Anderson, L. W., & Krathwohl, D. R. (2001). *A taxonomy for learning, teaching, and assessing: A revision of Bloom's taxonomy of educational objectives* (New York: Longman Editor).

BCcampus OpenEd. (2022). What is open education? Retrieved from https://open.bccampus.ca/what-is-open-education/.

Bitelli, G., Rinaudo, F., González-Aguilera, D., & Grussenmeyer, P. (2020). *[e-Book] data acquisition and processing in cultural heritage.* MDPI. https://www.mdpi.com/books/pdfdownload/book/2098.

Carbonell, J., Harman, D., Hovy, E., Maiorano, S., Prange, J., & Sparck-Jones, K. (2000). *Vision statement to guide research in Question and Answering (Q&A) and text summarization, Technical report, NIST.* Retrieved from: http://www-nlpir.nist.gov/projects/duc/papers/Final-Vision-Paper-v1a.pdf.

CHAPTER 17

Urbanization trends from global to the local scale

Gourgiotis Anestis and Demetris Stathakis
Department of Planning and Regional Development, University of Thessaly, Volos, Greece

1. Intoduction

Urbanization is an ancient phenomenon, linked to the development of human societies. The first cities date back to the Neolithic period, and appeared in the fertile regions of the Middle East, such as Mesopotamia and Egypt. In this way, people tended to group together near waterways and fertile land.

Over the centuries, urbanization continued, particularly in Europe and Asia, and cities became the center of cultural, intellectual, and economic life. In the 20th century, urbanization accelerated in developing countries as a result of industrialization and economic growth. At the same time, cities became more centralized as a result of globalization, which enabled exchanges between different regions.

Today, urbanization is a worldwide phenomenon that continues apace, with no region escaping the trend. At the same time, this trend is producing challenges such as poverty, pollution, and climate change, which require effective urban planning and sustainable city management.

This article focuses on the phenomenon of urbanization as a whole, analyzing the phenomenon itself, the reasons behind it, and its consequences. It also focuses on the phenomenon of urbanization in Europe and Greece in particular. The role of geographic information science (GIS) and in specific the new possibility in studying urbanization via earth observation is also discussed. Remote sensing is a powerful source of information to study how cities expand in a synoptic and systematic manner, frequently in the absence of official data that sometimes is impossible to produce, given the rapid pace of the phenomenon as well as its magnitude is some places. The increasing wealth of information is coupled by advancing image processing methods. These methods are progressively able to process very large volumes of remote sensing data (big data) and assist monitoring and decision making. A characteristic example is the advancement of classification methods that were initially based on statistical premises but soon shifted to advanced machine learning techniques. Neural networks hold a prominent role among these techniques. Neural networks are based on the discovery that input data can be effectively mapped to output classes using an in-between structure that is formed of nodes and connections organized in distinct "hidden" layers (Stathakis, 2009). It has

Geographical Information Science
ISBN 978-0-443-13605-4
https://doi.org/10.1016/B978-0-443-13605-4.00010-2

been found that neural networks mimic the way that humans learn by examples and produce accurate results even when statistical conditions can not be fulfilled (e.g., small populations of samples). The advancement of neural networks combined with other very efficient and complementary methods such as genetic algorithms and fuzzy sets paved the way for the development of today's artificial intelligence (AI) techniques (Stathakis & Vasilakos, 2006).

2. Methods and materials

This paper concerns a critical analysis on the issue of urbanization and explores a number of research questions concerning the sources of urban population growth, the way in which urbanization takes place, the ways in which urbanization spreads, the degree of urbanization of different areas, and the urbanization trends in Europe and in Greece. The data to answer the above research questions came from international literature and research.

3. Results

3.1 The phenomenon of urbanization—Territorial inequalities

Urbanized spaces are becoming both the habitat of the largest proportion of human beings and the social and political sphere of global scale within which we will have to learn, more and more, to grasp the evolutions and to regulate the various problems arising. Globalization, that is to say the institution of the world as a social space on a planetary scale, is deployed by and for urbanization. The process and trends of urbanization traditionally are hard to research and understand due to the lack of data as well as the different data specifications of across countries. The evolution of remote sensing has drastically changed that, progressively providing a wide range of earth observation data. Commercial earth observation data are available since the 1970s at medium spatial resolution that became high during the 1980s and very high (meter level) by 2000. As a result, global urbanization studies relying on earth observation data emerged (Schneider et al., 2009; Seto et al., 2012). At the same time, major advancements in geoinformatics offered new algorithms for the processing of the vast amounts of earth observation data. Notably, the classification of remote sensing data improved by advanced machine learning algorithms (Stathakis, 2009; Brown et al., 2008; Stathakis & Kanellopoulos, 2008; Stathakis & Perakis, 2007; Stathakis & Vasilakos, 2006). Recently, the segment of radar remote sensing has also substantially evolved, both in terms of spatial resolution improvements (comparable to optical remote sensing) as well as in processing methods and algorithms (such as interferometry). The advances in radar remote sensing open the way for monitoring and analyzing urbanization in the third dimension which by means of passive remote sensing is tedious (Letsios et al. 2023a, 2023b).

The horizon of societies is urban. Urbanization is indeed a global phenomenon, which concerns all scales. The concurrent way of life and employment generated new relationships with space but also with the degree of mobility upon space, replacing to an up to a degree the classic urban versus rural space antithesis. Françoise Choay (1994) named this phenomenon that shapes the fuzzy urban forms toward the end of 20th as well as in the beginning of the 21st century, as the "domination of the urban and the death of the city." Progressively, a substantial part of European rural space exhibit characteristics that up to recently where found only in urban areas.

This phenomenon was amplified via the trend to establish residential areas out of large cities that become popular and dominated during the recent decades, based on the north-European counterurban lifestyle (Leontidou & Couch, 2007) that is internationally refereed to also as "Californian" or "rural-urban" lifestyle (Kaufmann, 2000).

This is the primary source of the production of a novel space, that of urbanized rural areas, with the residential pattern of countercity (Mathieu, 1996). Other researchers such as Bocz et al. (2008) describe the same pattern as "urban lifestyle in a rural environment," assuming that this is the pattern that characterizes the present phase of a cycle that it closing (Anagnostou, 2010).

Urban is marked by the unlimited. Gone are the days of finite spaces—the peaceful days when the countryside and the city were clearly distinguished, clearly separated by stable boundaries. Unlimited is not a physical marker but a symptom of the rise of the principle of generalized connection between all urban realities. Unlimited allows mobilization, that is, the systematization of mobility as a basic principle of the urban (as well as a social, political, and even cultural value). But this mobility would not reach this importance if it were not accompanied by the digitization of societies. The traditional opposition between town and country is becoming a social mythology, because contemporary urbanization is composed of highly complex dynamics. The urban is now digital, but also material.

Another characteristic is the hyperspatialization, that is, the increasingly universal capacity (we connect at any time, in any place, for any actor) that characterizes the urban. The urban is marked by both accumulation and overabundance, but also and above all by the systematization of connection (all spaces are linked to all spaces, just as one goes from one Internet site to another via a hyperlink).

Cosmopolitanization is another striking phenomenon. Local populations now come from everywhere. Such openness, which makes every urban ensemble a cosmopolitan world, is not likely to diminish.

Another phenomenon of urbanization is segregation, which refers to problems of spatial distribution of social groups and individuals. There is no urban situation in the world where the segregation factor does not appear.

Finally, a last element is that the contemporary urban organization accumulates an increasingly impressive power. It is a question of technological power, economic power, financial power, political power, and cultural power, all of which combine to make the urban environment dominant and an unsurpassable reference. However, it is striking to note that the vulnerability of urban systems grows in fair proportion to this power. The global urban system thus accumulates both the greatest power and the greatest fragility; one does not go without the other, each feeds the other (Lussault, 2010).

On the other hand, the significant population growth caused by urbanization has major impacts on people, the environment, and development (Li et al., 2021), at local, regional, and global scales (Grimm et al., 2008; Nagendra et al., 2018). These impacts include air quality, water pollution (Hoekstra et al., 2018; Lin & Zhu, 2018), land use (Song et al., 2018), biodiversity loss (Newbold et al., 2015), and ecosystem degradation (Seto et al., 2012; DeFrie et al., 2010; Haddad et al., 2015), while carbon emissions from cities and associated land use changes are a major contributor to climate change (Wigginton et al., 2016). These urban growth dynamics are not homogeneous, and patterns of urban expansion vary considerably among developing countries (Gollin et al., 2016).

Never before such a large proportion of humanity lived in cities and in this ever expanding urban landscape in which metropolis play a determinant role. Despite the fact that they contain less than a forth of the global human workforce, the 300 largest cities produce almost half of the global Gross Domestic Product (GDP) and contribute for 67% of its growth (Kihlgren Grandi, 2021).

This important economic dynamic that cities poses makes them particularly attractive, mainly for migration flows both in national and international level (Bouchet et al., 2018).

However, beyond the economic dynamics, the demographic growth of large urban centers has gained such dimensions that even the term metropolis itself (coming from the ancient Greek words "mother city") is considered to be limited. Consequently, with increasing frequency, the term megalopolis, a term applied to the 33 urban conglomerations that each contains more that 10 million inhabitants (Bilsky et al., 2016), and, in the cases of urban areas that consist of multiple metropolis, such as the corridors between Boston and Washington or between San Francisco and San Diego "metropolitan regions" (Castells, 2010).

This phenomenon, having initially affected Europe and north America, then spread in Asia and Africa. According to the United Nations (UN), it is expected that India, China, and Nigeria will represent 35% of global increase of urban population in the year 2050 (ONU, 2019).

Currently, the main urban areas globally are located in Asia—Tokyo (37 million inhabitants), Delhi (29 million inhabitants), and Shanghai (26 million), as well as the metropolitan region of "Jing-Jin-Ji," that includes Pekin, Tianjin, and the Hubei region (110 million inhabitants in total). As far as Africa is concerned, Kinshasa with 15 million inhabitants has become the largest francophone metropolis, whereas Lagos could, if current trends persist, be placed first globally, in terms of population, with 88 million inhabitants (Hoornweg et Kevin Pope, 2014).

According to OECD data, global population has been changing rapidly. It increased from 4 billion in 1975 to 7.3 billion in 2015 and is projected to reach 9.1 billion in 2050. Between 1975 and 2015, the total population in cities more than doubled from 1.5 billion (37% of the world's population) to 3.5 billion (48% of the world's population). This increase is projected to continue with a further increase to 5 billion by 2050. The data show that urbanization is slowing down. In particular, up to 2015, the city population share increased by almost 3 percentage points in a decade, while up to 2050, it would be less than 2 percentage points (OECD Urban Studies/European Commission, 2020).

Focusing also on the rural area, it can be seen that the rural population share has been shrinking, from 30% in 1975 to 24% in 2015. The population in towns and semidense areas lost 1 percentage point per decade between 1975 and 2015 is projected to continue to do so up to 2050.

The reducing population shares in rural areas and towns and semidense areas obscures that the total population in these areas is not shrinking. Population in towns and semidense areas increased from 1.3 billion to 2.1 billion between 1975 and 2015 and is projected to reach 2.3 billion by 2050. Rural areas also experienced population growth between 1975 and 2015 from 1.2 billion to 1.7 billion, which is projected to increase to 1.9 billion by 2050. While over the next decades, the population outside cities will be increasing, it will do so at a slowing rate. Population growth in cities is also slowing down but less noticeably so.

According to the United Nations' population data spanning from 1950 to 2100, a notable trend emerges: the urban population has experienced a markedly swifter growth compared to its rural counterpart. This observation, derived from UN statistics in 2018, underscores the significant global shift towards urbanization over the past century and into the foreseeable future (UN, 2018). The UN data show urban population quadrupling between 1975 and 2050, while whereas OECD report (OECD Urban Studies/European Commission, 2020), estimates that the city population would triple over that period and the population in towns and semidense areas would increase by only 75%. Finally, the UN data show an increase in the rural population of only by 20% over the 1975–2015 period and rural population declining from 2025 onwards. In contrast, OECD report (OECD Urban Studies/European Commission, 2020) shows an increase of 50% and no population decline for the 2015–50 period (OECD Urban Studies/European Commission, 2020).

Urbanization, and especially the degree of urbanization, varies according to income group. Characteristically, low-income cities have the lowest percentage of population in cities and the highest in rural areas. The difference in urbanization between middle and high-income countries is small, but despite this small difference, it reflects the change in the type of urbanization. With the development of countries, moving to the city from the surrounding area is a common phenomenon. This has the effect of increasing the labor market of a city without increasing the population of the city itself. The inclusion of this commuting zone reveals that high income countries have a significantly higher proportion of the population in their cities and commuting zones than middle-income countries.

Globally, the vast majority of the population lives on a small fraction of land and as a result, the distribution of land across all degrees of urbanization is highly skewed. Almost all land is classified as rural. In 1975, rural areas covered 99.2% of the world's land. Over the next 40 years, this figure fell slightly to 98.5%. Estimates suggest that this percentage will fall to 98.3% by 2050.

In contrast, cities occupied only 0.2% of the total area in 1975. The 2 billion increase in urban population meant that cities needed more space to live and work. As a result, the share of land covered by cities increased to 0.5% in 2015. Projections show that this share will grow at a slower rate, reaching 0.7% in 2050.

Similar to the evolution of cities, the total population of cities and semiurban areas increased significantly between 1975 and 2015. As these people also needed space, the area of cities and semidense areas doubled from 0.5% to 1%. Estimates suggest that this area will increase only slightly by 2050 (OECD Urban Studies/European Commission, 2020).

According to the OECD report, there are three sources of urban population growth: (a) towns growing into cities; (b) city expansion; and (c) city densification (Table 17.1). Towns can grow into cities (as defined by the degree of urbanization) by reaching a population of at least 50,000 inhabitants. City expansion occurs through the building of new dense neighborhoods at the edge of the city or the densification of existing suburbs. City densification means that the population grows within the initial boundary of the city (OECD Urban Studies/European Commission, 2020).

Urban densification accounts for 50%–60% of global urban population growth. Increasing density within the original city limits—by definition—does not require additional land. However, additional investment is still needed to provide more housing, more jobs, and more services. Combined, the expansion of cities and the densification of capitals has led, globally, to a slight increase in the average population density of cities.

Table 17.1 Source of city population and area growth, percentage 1875–2015.

	Towns growing into cities (%)	City expansion (%)	City densification (%)	Total (%)
Population changes				
1975–90	23.9	26.4	49.7	100
1990–2000	18.3	29.3	52.4	100
2000–15	15.5	24.8	59.7	100
Area change				
1975–90	30.5	69.5	0	100
1990–2000	22.8	77.2	0	100
		77.4		
2000–15	22.6		0	100

From EC and OECD calculations based on Florczyk, A., et al. (2019). GHSL Data Package. http//dx.doi.org/10.2760/06297, in OECD Urban Studies (2020).

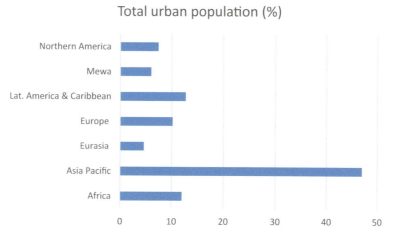

Figure 17.1 Total urban population. *(Reproduced from Gold IV (2016).)*

Nevertheless, large discrepancies remain in the population density of cities between income groups and across world regions. Cities in low-income countries are four times denser than those in high-income countries. The population density of cities in North America is less than 2000 inhabitants, while in Sub-Saharan Africa and South Asia, it is around 8000 inhabitants per km^2.

It is particularly important to highlight the fact that the global population is distributed across a wide range of settlement sizes. This urban–rural continuum ranges from a village with a few hundred inhabitants to mega-cities of more than 10 million inhabitants. Within this continuum, different countries use different thresholds to distinguish urban from rural. These thresholds range from 200 in Denmark to 100,000 in China.

According to OECD, the degree of urbanization essentially agrees on the classification of cities above 300,000 inhabitants. According to the UN World Urbanization Prospects, 1772 cities had at least 300,000 inhabitants in 2015. Of these cities, 1662 or 95% matched a city as defined by the degree of urbanization (OECD Urban Studies/European Commission, 2020).

As illustrated in Fig. 17.1, the Asia–Pacific region dominates the global urban system with 47% of the world's urban population. The second largest region by number of urban residents is Latin America and the Caribbean, with 13% of the world's urban population. Africa follows Latin America with 12% of the world's urban population. This is followed by Northern America with 7.4%, Middle East and West Asia (MEWA) with 6%, and finally Eurasia with 4.6% of the world's urban population.

3.2 Urbanization dynamics in the EU

In European Union's member states, countries of the European Free Trade Association, and the United Kingdom, half of the population lives within 15% of the total surface and 80% of the population is living within half of the total surface.

The differences due to population density are clear not only among densely populated urban areas and sparsely populated rural areas but also between countries as well. The spatial patterns vary based on each regions' particular characteristics and are classified as isolated municipalities, small- and medium-sized cities, as well as larger cities and urban regions to aggregated urban and metropolitan regions and large-scale and multiregional to transborder urban areas.

As illustrated in Fig. 17.2, Western Europe has the highest proportion of urban population with 37.7%, followed by Southern Europe with 27.1%, Northern Europe with 20.4%, and Eastern Europe with 14.8%.

Of particular importance are the larger, distinct densely populated cities and urban areas that assimilate urban functions as "regiopolises," that is, regional population centers between rural areas and metropolitan regions.

Larger cities and the metropolitan regions are connected with their suburbs via lines that follow the topography of transpiration axes that take the form of densely populated paths.

Overall, urbanization will increase in the form of uncontrolled urban sprawl. The main reasons for urbanization are the migration of people from rural areas to cities and the use of more living space per inhabitant. The recent economic crisis has led to the relaxation of urban planning regulations. The change of land use from agricultural to other land uses is to be expected in all European countries (ESPON, 2014).

According to ESPON data on land use changes across the ESPON territory over the period 2000–18, it is found that slightly less than 2.87 million hectares of land changed from one main category to another (see Table 17.2) or about 0.6% of the surface area of ESPON space. Almost half (1.26 million ha or 44%) concerned a conversion to urban land. As a result, artificial land cover increased from 19.2 million to 22.6 million hectares,

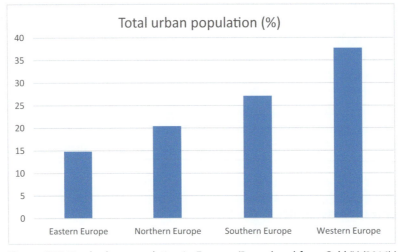

Figure 17.2 Total urban population in Europe. *(Reproduced from Gold IV (2016).)*

Table 17.2 Sum of land-use change (ha) in ESPON countries between 2000 and 2018.

……..… ….To From	Urban	Artificial (nonurban)	Agricultural	Terrestrial nature	Wetlands and water bodies	Total[a]
Urban		3040	33,000	12,120	9030	57,2190
Artificial (nonurban)	8760		52,400	50,030	19,8120	131,010
Agricultural	990,540	162,330		456,680	97,320	1,706,860
Terrestrial nature	246,970	97,800	325,140		125,900	795,820
Wetlands and water bodies	16,300	4620	45,130	111,500		177,500
Grand total[a]	1,262,570	267,790	455,680	630,320	252,070	2,868,440

[a]Total may not add up exactly due to a rounding to the nearest 10 ha.
Reproduced from ESPON (2020a, 2020b).

the vast majority of which (18.5 million in 2000 and 21.8 million in 2018) concerned uses such as homes, businesses, and infrastructure; the rest regarded mineral extraction and dump sites (ESPON, 2020a, 2020b).

According to ESPON, about 1.26 million hectares of land were converted to urban use in the 2000–18 period. If we subtract the land already registered as construction sites (450,000 ha) and add the land converted from construction sites to urban (350,000 ha), we get a total urbanization figure of 1,166,000 ha in the ESPON space. Of this, 35% became urban fabric (predominantly residential), 37% industrial (including business parks, shopping centers, and offices), 17% infrastructure (including airports), and 11% urban green. Urban development mainly occurred on agricultural land (78%), although in Scandinavian countries (except Denmark), Croatia, Greece, Iceland, and Portugal came more at the expense of terrestrial nature. Some NUTS 3 in Austria and the UK (Scotland) also saw a majority of new urban land coming from natural areas.

In particular, and with regard to deurbanized areas, ESPON data show that of the total of the 176,000 ha of deurbanizing land in the total ESPON space, the majority (69%) concerned transitions away from "artificial land" like mines and dump sites. This land was converted in equal proportions to agriculture and terrestrial nature, and a smaller part to water-related nature. Over half of these conversions took place in four countries: Germany (21%), Spain (15%), the UK (10%), and Poland (9%). In total, 8.6 times more land was converted to urban/artificial use than vice versa. Only in Romania (−0.8%) and Bulgaria (−0.1%) did the share of urban land decrease between 2000 and 2018 as a whole (ESPON, 2020a, 2020b).

It is also noted that the content and pace of urban land conversion also varied in Europe.

A typical example of limited land use change was seen in several countries such as Spain and Ireland during the recent economic crisis. In contrast, urbanization in Poland has taken on a significant dimension (almost tripled) since joining the EU. Between 2000 and 2018, nearly 20% of all Europe's urbanization occurred in Spain, followed by France with 15%. In the last period from 2012 to 2018, the UK took the lead; over one-fifth of all changes were registered here, followed by France with again 15% and Poland with approximately 13% (ESPON, 2020a, 2020b).

According to the ESPON, the rate of urbanization varies geographically across the EU. Analysis at NUTS 3 level in Poland, France, Spain, and Turkey reveals high rates of conversion. A significant rate of suburbanization around cities such as Prague, Budapest, and several Polish cities is also evident (Map 17.1).

3.3 The peculiarities of the phenomenon of urbanization in Greece

In Greece, the per capita land consumption is lower than the OECD average. Among the 2000 and 2012, there were notable increases in the total share of developed land, as well as in the in per capita land consumption. In the same period, a strong pattern of

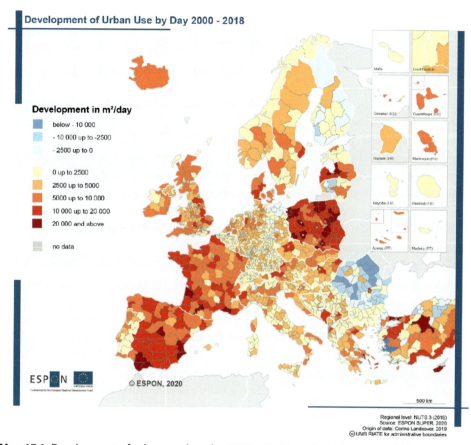

Map 17.1 Development of urban use bay day 2000—18. *(Reproduced from ESPON (2020a, 2020b).)*

suburbanization emerged. Population in the commuting zones of urban areas experienced significant growth, while decreased in the urban cores. This was partly offset by an increase in the share of developed land in the commuting zones. Outside the major urban areas, Greece is characterized by a relatively low proportion of developed land, as well as forest land (OECD, 2017).

Urbanization trends in Greece are dominated by uncontrolled expansion (Tsilimigkas et al., 2016; Triantakonstantis & Stathakis, 2015; Stathakis and Tsilimigkas, 2015; Stathakis & Tsilimigkas, 2013). The urban landscape exhibits typical sprawl characteristics. Cities expand to the immediate outlying areas in the form of low density built-up development. A strong sprawling factor is the road network that reduces travel distances and attracts ribbon development for both housing and retail uses. Sometimes this expansion forms extensive corridors that unifies smaller settlements in its path and practically generates a linear city. In Greece, the modes of travel other than the automobile is not quite as developed and used. Therefore it is the automobile that predominantly guides urban expansion. An additional very strong factor of ribbon development in Greece is the coastline. Sea-side plain areas attract substantial interest for the development of second-houses and of touristic facilities, primarily small-scale investments. Second-house areas sometimes, when travel distances to core urban areas are not prohibited, are later de facto transformed to regular residential areas where people commute to work.

Sprawl also takes the from of leap frog development where urban gaps are left unused because development moves to further out areas for several reasons. Land speculation is one of the reasons. Owners hold land in order to achieve higher selling prices at later stages of development. The pattern of open spaces however is random and clearly sub-optimal in planning terms. It usually is very difficult to ex post form a pattern of urban space with acceptable levels of access to open and common spaces.

Often the phenomenon of urban sprawl is also reinforced by the anarchic localization of industry or by the phenomenon of informal industrial concentrations. That is areas, very close to urban areas, with increased concentration of business activities, which have been established without spatial planning, without provision of the necessary infra-structure, resulting in the creation of urban planning and environmental problems.

The configuration of the existing spatial model for the development of industrial ac-tivities in Greece is considerably influenced by the peculiar—and very particular to Greece—established practice of urban development and construction outside the approved urban plans (Economou, 2000; Dimitropoulou, 2020).

Industrial development in Greece has typically evolved using an ad hoc location of manufacturing facilities, with no overall planning and arrangements based on land-use plans (General Secretariat for Industry, 2018; Greek Government Gazette 151/AAP/ 2009). Delays in introducing an integrated territorial development have significantly impacted the sustainable development of productive activities and, to some extent, this is still ongoing.

A comprehensive analysis of existing industrial areas at a national level concluded that Greece does not have a central mechanism for recording and monitoring key develop-ment and management attributes of organized receptors (Gourgiotis et al., 2021).

In Greece, an additional spatial pattern of sprawl is very common, that is scattered development. Random development of sites frequently totally disconnected with any preexisting already urbanized area. The sole driving force of this development is private initiative. Land is developed in a spontaneous fashion, either on preexisting land of owners but frequently on land that is specifically acquired to be developed. The coastline again attracts scattered development not only on close proximity to the sea but also in further out areas overlooking the sea, that is, with nice views to the sea. High-quality natural environment such as forested areas is also a major factor for spontaneous site development.

A side effect of uncontrolled horizontal expansion of cities is the suboptimal use of land. One aspect of this problem has to do with extensive areas within the borders of urban areas that remain undeveloped. The reasons for that include unresolved ownership ambiguities, speculation for better land prices in the future, planning inefficiencies that leave out space, etc. In any case, the fact is that uncontrolled expansion results in reducing the pressure to uniformly develop urbanize areas generating an almost endless pool of available land for development in the outlying areas. At the same time, areas within the city may be developed at suboptimal densities. Compact development at sufficient densities has been associated with agglomeration economies and economies of scale which generate the conditions for faster amortization of investments for public utilities, protecting the environment via shorter commuting times and thus less greenhouse emissions, easier to develop mass transportation networks and to provide valid alternatives to automobile use, better chances for land use mix, etc.

Overall horizontal urban expansion should be considered only after vertical expansion has reached a satisfactory level. Otherwise, horizontal expansion results in waste of land which is a valuable resource. Uncontrolled expansion is also associated with land take. In other words, the expansion is happening on land that previously had some other land use that is now lost. Typically urban expansion is realized at the expanse of agricultural land. This is because urban and agricultural lands have some shared qualities. For example both need to avoid large slopes on the ground. Also, generally, both can not overlap with floodplains.

Urban sprawl causes irreversible land take. It is not only that valuable land is lost but also that rural and natural areas are progressively fragmented by isolate development. Fragmentation by itself greatly reduces the ecological value and generates a standstill as owners adopt a waiting stance in the anticipation of the further urbanization of their areas. Agricultural revenue in most cases does not have the capacity to compete with the revenue generated by urban land uses.

A typical example, according to Tsilimigkas and Derdemezi (2019), who studied the ex-urban built up area expansion on Santorini island. The methodology followed was the digitization of built-up areas, because it is considered the most appropriate method for investigating the topic. Although it is particularly time-consuming, it can be much more accurate than standard remote sensing methods.

To identify built-up areas, the buildings were digitized from the orthophoto maps that were provided by the National Cadastre and Mapping Agency (NCMA, 2018). The orthophotos were georeferenced in GGRS84, and the shooting became from 2007 until 2009. The nonurban areas have spatial resolution 50 cm and geometrical accuracy on the ground RMSExy $< = 1.41$ m (NCMA, 2018). During the process of georeference, there were difficulties in distinguishing between built-up and soil sealed but nonbuilt-up areas. Generally, the cells that are presented as a white roof in the orthophoto maps were considered to be built-up areas. Nevertheless, the digitalization was based on the shape too, since many roofs are not white but gray or brown. The difficulty in recognizing each building is greater in the amphitheatrical construction of Santorini (Tsilimigkas & Derdemezi, 2019).

The above analysis has shown the excessive built-up area expansion that takes place on Santorini island. Built-up areas outside the settlements boundaries are 0.09 km^2, which means approximately 2.88% higher than the built-up areas within settlements.

There are many traditional settlements on Santorini that justify the big percentage of buildings that are within their boundaries. Caution should be given to construction outside the limits of traditional settlements but within a proximity area to them, due to the fact that this kind of built-up area dispersion could put a great deal of pressure on traditional settlements (Map 17.2b). Some nominated settlements either are parts of archeological sites or contain archeological sites. The case of buildings that are outside settlements boundaries but inside archeological sites (Map 17.2b) poses a risk that archeological sites can be degraded.

The Natura 2000 area has a small percentage of construction which is not only because of the special conditions for interventions that exist as a protected area but also because the Natura 2000 area covers the mountainous part of the island (Map 17.2c). The percentage of buildings that are placed on prominent areas is not negligible, that is, 0.5 km^2, which means approximately 16.67 km^2 (Map 17.2c). Buildings in prominent areas that do not follow the prevailing architecture and the local scale can be main cause of negative effects on landscape qualities (Tsilimigkas & Derdemezi, 2019).

The significant impacts of residential expansion in the island area are therefore identified, as sensitive sociospatial systems characterized by fragile natural and cultural environments are affected.

Remote sensing plays an important role in monitoring sprawl in a spatially explicit manner since national statistics often lack this dimension. The growth of cities can be monitored and evaluated using optical earth observation data. The most extensive archive of such imagery is that of the LANDSAT program that provides data since 1972. Urban areas can be extracted using segmentation or classification techniques in order to create a time-series for analysis.

More recently, optical imagery taken in the night has also been proven valuable to monitor urbanization. Night-time imagery has the benefit, compared to day-time images, that urbanized areas in most parts of the globe become immediately evident by the presence of light (Stathakis et al., 2018). Preprocessed composite data, with the various sources of noise removed, have greatly amplified the possibilities. Time-series can now quite easily be formed using annual, monthly or even 24 hours composites. It is now well established that the total amount of light present in an area is strongly correlated with population and the economic development (Stathakis et al., 2015; Tselios et al., 2019; Tselios & Stathakis, 2018).

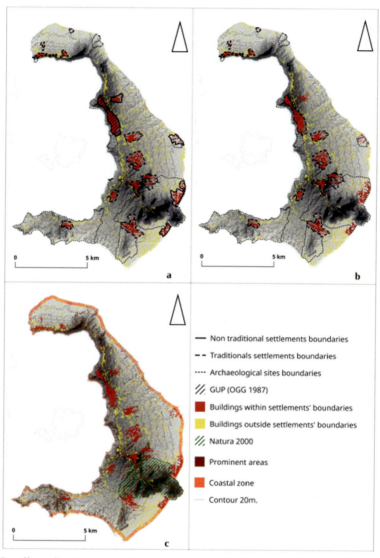

Map 17.2 The effect of built-up area of Santorini in areas of cultural or environmental interest. *(Source: Tsilimigkas and Derdemezi (2019).)*

The two main sources of night-lights are the DSMP/OLS sensor and recently the SUOMI/VIIRS sensor. The DSMP/OLS sensor provided data for the period of 1992—2013. These are annual composites that have been processed to remove ephemeral lights. Also, an intercalibration procedure is typically applied to the data, as a preprocessing step, in order to derive a comparable and meaningful time-series (Stathakis, 2016). However, the data are severely limited by the very low radiometric resolution (6 bits) but provide a quite solid basis to monitor urbanization with a sufficient spatial (1 km) and a very suitable temporal resolution (1 year). The successor mission was that of SUOMI/ VIIRS. This sensor provides data since 2012 (minimal overlap with OLS) at a slightly improved spatial resolution (0.75 km) but with a great improvement in radiometric and temporal resolutions. In specific, the radiometric resolution of VIIRS is now sufficient to capture almost all the range of values in urban areas (12 bits), thus avoiding the saturation problem that made urban cores appear as uniform in OLS. A further great advancement has been the substantial improvement of its temporal resolution. Initially monthly composites were made available that permitted studying the seasonality of urban areas (Stathakis & Baltas, 2018). Then 24 hours data ("daily") became available which open the way for a totally new spectrum of applications, including closely monitoring day-time population.

Therefore, night-time imagery has an unprecedented capacity in terms of temporal resolution to monitor and study the effects of urbanization almost globally. Because of this temporal capacity, new applications have emerged that facilitate the monitoring of population movements in short intervals, either for special events or for vocations, etc. It is expected that with the currently under development and testing night-light sensors that offer high spatial resolution, in addition to the very high temporal resolution already in place, will pave the way for new research topics aiming at explaining the daily function of cities (Stathakis et al., 2021; Stathakis & Baltas, 2018).

4. Conclusion remarks

The future cannot be predicted, but it can be prepared for. If there is one thing that is certain, beyond the subtleties of statistics, it is that tomorrow's world will be more urban. Many of the challenges of the 21st century lie in cities. The powerful global urbanization dynamic is giving rise to as many concerns in terms of sustainable development as it is to hopes for emancipation and improved living conditions.

The situation is obviously varied. Urbanization in a number of developed countries has reached very high levels, which are not likely to be exceeded everywhere. Conversely, urbanization in many developing countries is set to continue.

The implications of this phenomenon are profound for society, the environment, the economy, and everyday life. It is a complex phenomenon with significant advantages and disadvantages. Managing urbanization is a complex challenge that requires careful planning, sustainable management of resources and consideration of the needs of urban populations. Managing this process in a sustainable and equitable way is essential to ensure a prosperous urban future for future generations.

GIS and in particular remote sensing have made significant contributions to the way we perceive, manage, and evaluate the world around us.

Remote sensing, as a method of collecting data from remote sources, has enabled the analysis of soil, water supply, climate change, and many other fields. The outlook for the future is encouraging. It is expected that the role of GIS, and in particular remote sensing, will continue to evolve and will provide the opportunity to create databases that will allow us to predict future trends in phenomena such as urban development. The integration of AI and machine learning will accelerate the process of analyzing and drawing conclusions from geographic data.

References

Anagnostou, S. (2010). *Contemporary forms of urbanization and the issue of its statistical monitoring. The proposal of a relevant methodological model for Greece.* University of the Aegean.

Bilsky, E., Ciambra, A., & Guérin et Ludovic Terren, M. (2016). *Gold IV: Co-Creating the urban future.* Barcelone: Cités et gouvernements locaux unis (CGLU).

Brown, M., Lary, D., Vrieling, A., Stathakis, D., & Mussa, H. (2008). Neural networks as a tool for constructing continuous NDVI time series from AVHRR and MODIS. *International Journal of Remote Sensing, 29*(24), 7141−7158.

Bocz, G. A., Nilsson, C., & Pinzke, S. (2008). Periurbanity − a new classification model in rural futures: Dreams, dilemmas, dangers. University of Plymouth.

Bouchet, M., Liu, S., & et Nader Kabbani, J. P. (2018). *Global metro monitor.* Washington, DC: Metropolitan Policy Program, Brookings Institution.

Castells, M. (2010). Globalisation, networking, urbanisation: Reflections on the spatial dynamics of the information age. *Urban Studies, Thousand Oaks (Californie), 47*(13).

Choay, F. (1994). Le règne de l'urbain et la mort de la ville. In J. Dethier, & A. Guiheux (Eds.), *(dir.), La ville, art et architecture en Europe, 1870−1993* (pp. 26−35). Paris: Editions du Centre Pompidou.

Couch, C., Leontidou, L., & Petschel-Held, G. (Eds.). (2007). *Urban sprawl in Europe: Landscapes, land-use change and policy.* Oxford: Blackwell.

DeFries, R. S., Rudel, T., Uriarte, M., & Hansen, M. (2010). Deforestation driven by urban population growth and agricultural trade in the twenty-first century. *Nature Geoscience, 3*, 178.

Dimitropoulou, C. (2020). *On the location of industrial activities.* Available online: https://nomosphysis.org.gr/12329/peri-xorothetisis-biomixanikon-drastiriotiton-noembrios-2011/. (Accessed 20 December 2020).

Economou, D. (2000). Spatial planning system: Greek reality and international experience. *Review of Social Research., 1*, 3−57. https://doi.org/10.12681/grsr.993

ESPON. (2020a). *Applied research, main report, SUPER—Sustainable urbanization and land-use practices in European regions.*

ESPON. (2020b). *Atlas for the territorial Agenda 2030.*

ESPON. (2014). *Territorial scenarios and visions for Europe, "Making Europe open and polycentric" vision and roadmap for the European territory towards 2050.*

General Secretariat for Industry. (2018). *Business plan for the development of business parks in the Greek territory, volume: Summary-conclusions.* Available online: http://www.ggb.gr/sites/default/files/basic-page-files/ΚΕΙΜΕΝΟ_ΣΥΝΟΨΗΣ-ΣΥΜΠΕΡΑΣΜΑΤΑ.pdf#overlay-context=el/node/1274.

Gold IV. (2016). *Fourth global report on decentralization and local democracy. Co-creating the urban future the agenda of Metropolises, cities and territories.* UCLG.

Gollin, D., Jedwab, R., & Vollrath, D. (2016). Urbanization with and without industrialization. *Journal of Economic Growth, 21*, 35−70.

Gourgiotis, A., Kyvelloy, S., & Lainas, I. (2021). Industrial location in Greece: Fostering green transition and synergies between industrial and spatial planning policies. *Land, 10*, 271. https://doi.org/10.3390/land10030271

Greek Government Gazette 151/AAP/2009, Special Spatial Planning Framework for Industry (SSPFI). Available online: http://ypen.gov.gr/wp-content/uploads/2020/11/FEK151_AAP_2009_Viomixania.pdf. Accessed 15 November 2020.

Grimm, N. B., Faeth, S. H., Golubiewski, N. E., Redman, C. L., Wu, J., Bai, X., & Briggs, J. M. (2008). Global change and the ecology of cities. *Science, 319*, 756–760.

Haddad, N. M., Brudvig, L. A., Clobert, J., Davies, K. F., Gonzalez, A., Holt, R. D., Lovejoy, T. E., Sexton, J. O., Austin, M. P., Collins, C. D., et al. (2015). Habitat fragmentation and its lasting impact on Earth's ecosystems. *Science Advances, 1*, e1500052.

Hoekstra, A. Y., Buurman, J., & van Ginkel, K. C. H. (2018). Urban water security: A review. *Environmental Research Letters, 13*, 053002.

Hoornweg et Kevin Pope, D. (2014). *Socioeconomic pathways and regional distribution of the world's 101 largest cities*. Global Cities Institute Working Papers, Université de Toronto.

Kaufmann, V. (2000). *Mobilité quotidienne et dynamiques urbaines*. Lausanne: Presses Polytechniques et Universitaires Romandes.

Kihlgren Grandi, L. (2021). Le siècle métropolitain. In *Le monde diplomatique, les Villes avenir de l'humanité*.

Letsios, V., Faraslis, I., & Stathakis, D. (2023a). Monitoring building activity by persistent scatterer interferometry. *Remote Sensing, 15*(4), 950.

Letsios, V., Faraslis, I., & Stathakis, D. (2023b). Multi-temporal PSI analysis and burn severity combination to determine ground-burned hazard zones. *Remote Sensing, 15*(18), 4598.

Li, G., Fang, C., & Qi, W. (2021). Different effects of human settlements changes on landscape fragmentation in China: Evidence from grid cell. *Ecological Indicators, 129*, 107927.

Lin, B., & Zhu, J. (2018). Changes in urban air quality during urbanization in China. *Journal of Cleaner Production, 188*, 312–321.

Lussault, M. (2010). L'urbanisation Horizon du monde. In *DATAR Revue d'études et de prospective N° 1—2e semestre 2010, Territoires 20140, Aménager le changement*.

Mathieu, N. (1996). Rural et urbain. Unité et diversité dans les évolutions des modes d'habiter. In M. Jollivet, & N. Eizner (Eds.), *(dir.) L'Europe et ses campagnes,Paris* (pp. 187–205).

Nagendra, H., Bai, X., Brondizio, E. S., & Lwasa, S. (2018). The urban south and the predicament of global sustainability. *Nature Sustainability, 1*, 341–349.

NCMA (National Cadastre and Mapping Agency). (2018). *Viewing orthophotos*. Greek cadastre.

Newbold, T., Hudson, L. N., Hill, S. L. L., Contu, S., Lysenko, I., Senior, R. A., Börger, L., Bennett, D. J., Choimes, A., Collen, B., et al. (2015). Global effects of land use on local terrestrial biodiversity. *Nature, 520*, 45–50.

OECD. (2017). *Land-use planning systems in the OECD: Country fact sheets*.

OECD Urban Studies/European Commission. (2020). Cities in the world: A new perspective on urbanisation. In *OECD Urban Studies*. Paris: OECD Publishing. https://doi.org/10.1787/d0efcbda-en.

Organisation des Nations unies (ONU). (2019). *World urbanization prospects: The 2018 revision*. New York: Département des affaires économiques et sociales.

Schneider, A., Friedl, M., & Potere, D. (2009). A new map of global urban extent from MODIS satellite data. *Environmental Research Letters, 4*(4), 044003.

Seto, K. C., Güneralp, B., & Hutyra, L. R. (2012). Global forecasts of urban expansion to 2030 and direct impacts on biodiversity and carbon pools. *Proceedings of the National Academy of Sciences, 109*, 16083–16088.

Song, X.-P., Hansen, M. C., Stehman, S. V., Potapov, P. V., Tyukavina, A., Vermote, E. F., & Townshend, J. R. (2018). Global land change from 1982 to 2016. *Nature, 560*, 639–643.

Stathakis, D., Liakos, L., & Baltas, P. (2021). COVID-19 pandemic assessment by night-lights. *IEEE International Geoscience and Remote Sensing Symposium (IGARSS)*, 6801–6804.

Stathakis, D., & Vasilakos, A. (2006). Comparison of computational intelligence based classification techniques for remotely sensed optical image classification. *IEEE Transactions on Geoscience and Remote Sensing, 44*(8), 2305–2318.

Stathakis, D., & Tsilimigkas, G. (2013). Applying urban compactness metrics on pan-European datasets, International Archives of the Photogrammetry. *Remote Sensing and Spatial Information Sciences—ISPRS Archives, 40*(4W1), 127–132.

Stathakis, D., & Tsilimigkas, G. (2015). Measuring the compactness of European medium-sized cities by spatial metrics based on fused data sets. *International Journal of Image and Data Fusion, 6*(1), 42−64.

Stathakis, D. (2009). How many hidden layers and nodes? *International Journal of Remote Sensing, 30*(8), 2133−2147.

Stathakis, D. (2016). Intercalibration of DMSP/OLS by parallel regressions. *IEEE Geoscience and Remote Sensing Letters, 13*(10), 1420−1424.

Stathakis, D., & Kanellopoulos, I. (2008). Global elevation ancillary data for land-use classification using granular neural networks. *Photogrammetric Engineering and Remote Sensing, 74*(1), 55−63.

Stathakis, D., & Perakis, K. (2007). Feature evolution for classification of remotely sensed data. *IEEE Geoscience and Remote Sensing Letters, 4*(3), 354−358.

Stathakis, D., Liakos, L., Chalkias, C., & Pafi, M. (2018). A photopollution index based on weighted cumulative visibility to night lights. In *Proceedings of SPIE 10793, remote sensing technologies and applications in urban environments III, 1079304, 9 October 2018.* https://doi.org/10.1117/12.2500212

Stathakis, D., & Baltas, P. (2018). Seasonal population estimates based on night-time lights. *Computers, Environment and Urban Systems, 68*, 133−141.

Stathakis, D., Tselios, V., & Faraslis, I. (2015). Urbanization in European regions based on night lights. *Remote Sensing Applications: Society and Environment, 2*, 26−34.

Triantakonstantis, D., & Stathakis, D. (2015). Examining urban sprawl in Europe using spatial metrics. *Geocarto International, 30*(10), 1092−1112.

Tselios, V., & Stathakis, D. (2018). *Exploring regional and urban clusters and patterns in Europe using satellite observed lighting.* Environment and Planning B: Urban Analytics and City Science.

Tselios, V., Stathakis, D., & Faraslis, I. (2019). Concentration of populations and economic activities, growth and convergence in Europe using satellite-observed lighting. *Geocarto International.*

Tsilimigkas, G., & Derdemezi, E. T. (2019). *Unregulated built-up area expansion on Santorini island, Greece European planning studies.* ISSN: 0965-4313 (Print) 1469-5944 (Online) Journal homepage: https://www.tandfonline.com/loi/ceps20.

Tsilimigkas, G., Stathakis, D., & Pafi, M. (2016). Evaluating the land use patterns of medium-sized Hellenic cities. *Urban Research and Practice, 9*(2), 181−203.

UN. (2018). *World urbanization prospects.* United Nations.

Wigginton, N. S., Fahrenkamp-Uppenbrink, J., Wible, B., & Malakoff, D. (2016). *Cities are the future.* Washington, DC, USA: American Association for the Advancement of Science.

Further reading

https://population.un.org/wup/Publications/Files/WUP2018-Report.pdf.

CHAPTER 18

Mapping and monitoring night light pollution

Christos Chalkias and Chrysovalantis Tsiakos

Department of Geography, School of Environment, Geography and Applied Economics, Harokopio University, Athens, Greece

1. Introduction

Nocturnal light pollution, often simply referred to as light pollution or nighttime light pollution, is a pervasive and concerning environmental problem that has gained increasing attention in recent years. It is excessive, misdirected, or intrusive artificial light emitted by human activities, particularly outdoor lighting, that disturbs the natural darkness of the night sky, which in turn negatively impacts the environment, human health, and ecosystems (Cinzano et al., 2001).

Night light pollution is a complex and multifaceted problem that is described by a number of alternative names or related terms that are often used interchangeably, each emphasizing certain aspects of the problem. These terms collectively encompass the negative effects of artificial lighting on the nocturnal environment and human well-being and highlight the various dimensions of this phenomenon. Researchers in the field of light pollution use a variety of the terms mentioned below to describe and study different aspects of the problem:

Skyglow is a term that characterizes the brightening of the night sky over populated areas due to the scattering and reflection of artificial light by particles in the air and atmospheric molecules, thus diminishing the visibility of stars and celestial objects (Cinzano et al., 2001; Falchi et al., 2011; Spoelstra et al., 2017).

In the urban context, **urban sky glow** specifically refers to this phenomenon, predominantly driven by the extensive use of artificial lighting in urban areas and especially on the consequences of urbanization (Falchi et al., 2016; Jechow et al., 2020).

Light trespass is the unwanted or excessive spillover of artificial light from one property or area to another, resulting in light pollution and disturbance that often affects the neighborhood (Balsky & Terrich, 2020; Tong et al., 2023). Aspects of architectural design, fixture placement, and the choice of lighting technology to minimize light spillover when planning new facilities and infrastructure are also considered (Kim & Kim, 2021).

Glare is the discomfort and vision impairment caused by excessive brightness, which often results in temporary visual impairment referred as glare. Glare is related to the perception, measurement, and control of glare in outdoor environments, including road lighting and other outdoor applications (Atılgan et al., 2012).

Geographical Information Science
ISBN 978-0-443-13605-4
https://doi.org/10.1016/B978-0-443-13605-4.00006-0

The term **clutter** describes the presence of excessive and confusing groupings of bright, competing lights, particularly in urban settings, causing visual confusion and discomfort (Hölker et al., 2016).

Visual pollution serves as a broader concept encompassing various forms of environmental pollution, including light pollution, which degrades the visual quality of an area due to unwanted or excessive visual elements, including artificial lights (Ahmed et al., 2019; Chmielewski et al., 2018; Wakil et al., 2019).

Artificial sky brightness signifies the increase in night sky brightness attributed to artificial lighting, often measured as a deviation from the natural darkness of the night sky (Cox et al., 2022; Green et al., 2022; Linares et al., 2018).

Nocturnal luminance quantifies the brightness of the night sky at specific locations or across regions during nighttime hours, considering the combined effect of natural and artificial light sources, frequently utilized in light pollution studies (Hu et al., 2016).

Night light pollution has significant and wide-ranging impacts on various areas such as human health, ecology, and astronomy. More specifically, excessive artificial light at night can disrupt sleep and circadian rhythms, suppress the production of melatonin, and contribute to mental health problems, various types of infectious diseases, cancer, and obesity, as well as lead to complications during pregnancy (Bożejko et al., 2023). In addition, artificial light at night can have serious ecological impacts, affecting wildlife behavior and creating imbalances. Nocturnal animals rely on darkness to perform essential activities such as foraging, navigation, and reproduction. Light pollution can disrupt these behaviors, leading to altered migration patterns, impaired mating rituals, and increased vulnerability to predators (Cox & Gaston, 2023). This in turn can affect biodiversity and ecosystem health. Light pollution obscures the night sky and makes it difficult for astronomers to observe celestial objects. This reduced visibility limits our understanding of the universe and hinders scientific progress (Gallaway, 2010). Astronomical observatories must invest in costly equipment and technology to mitigate the effects of light pollution, diverting resources from research and education.

In conclusion, "night light pollution" has far-reaching and detrimental impacts on human health, ecology, and astronomy. It disrupts our natural sleep patterns, harms wildlife and ecosystems, and obstructs our ability to explore and understand the cosmos. Addressing light pollution through responsible lighting practices, such as shielding lights and implementing lighting ordinances, is essential to mitigate these adverse effects and preserve our connection to the natural nighttime environment.

2. Data and methods for mapping night light pollution

2.1 Data collection

One of the best-known sources for providing accurate artificial nighttime light data is the Defense Meteorological Satellite Program (DMSP), a joint program of the United States

Air Force and the National Oceanic and Atmospheric Administration (NOAA) that operates a constellation of satellites to monitor Earth's weather, climate, and environment. DMSP satellites have been in orbit since the 1960s and have provided valuable data for a wide range of scientific applications. The Operational Linescan System (OLS) is a sensor on DMSP satellites that is used to image cloud distribution and temperature. OLS has two spectral channels: a visible and near–infrared channel (VNIR) and a thermal infrared channel. The VNIR channel is sensitive to wavelengths of 0.4—1.1 μm, while the thermal infrared channel is sensitive to wavelengths of 10.0—13.4 μm.[1]

In 1992, the National Geophysical Data Center (NGDC) initiated the development of a digital archive for OLS data, leading to the generation of global cloud-free night light products from 1994 onwards. OLS data capture occurs through two spatial analysis modes: fine data, characterized by a spatial resolution of 0.56 km, and "smoothed" data, with a spatial resolution of 2.7 km. The latter results from on-board calibration of full resolution data, involving the application of a median filter to each 5×5 cell neighborhood, aimed at reducing satellite memory requirements (Doll, 2008a; Elvidge et al., 2001). This process is implemented to enhance efficiency, with full resolution data offering richer spatial information, including the detection of small lights that might be mistaken for noise in normalized data (Elvidge et al., 2001).

The OLS visible data capture emissions in the visible and near-infrared spectrum, including reflections from the sun or moon off clouds and various features. It also detects ground-based sources like fires, city lights, natural gas flaring, and other upper atmospheric phenomena such as the northern lights (Elvidge et al., 2023).

The Suomi National Polar-orbiting Partnership (NPP) weather satellite, and specifically its Visible Infrared Imaging Radiometer Suite (VIIRS) component constitutes another valuable data source for monitoring nighttime light emissions.[2] VIIRS, a scanning radiometer, captures visible and infrared images along with radiometric measurements of land, atmosphere, cryosphere, and oceans. With a swath width of 3060 km at a nominal altitude of 829 km, VIIRS achieves full Earth coverage within a day. Operating with 22 spatially registered spectral bands covering wavelengths from 0.412 to 12.01 μm, including 16 moderate resolution bands (M-bands) at 750 m spatial resolution, five imaging resolution bands (I-bands) at 375 m at nadir, and one panchromatic day-night bands (DNBs) with a 750 m spatial resolution across the scan (National Oceanic and Atmospheric Administration-NOAA, 2013). VIIRS DNB can collect global low-light imaging data and thus, be a suitable tool to monitor cyclical changes caused by human activities such as holiday lighting and seasonal migrations (Elvidge et al., 2017) (Fig. 18.1).

---

[1] https://ngdc.noaa.gov/
[2] https://ngdc.noaa.gov/eog/viirs/download_ut_mos.html

Figure 18.1 Earth at night, Africa and Europe. *(This image is based on Black Marble 2016 annual composite map generated using VIIRS Data; Source: From NASA.)*

Luojia 1-01 is a new satellite that has been designed to capture high-resolution night-time light imagery. Luojia 1-01 provides nighttime light imagery with a spatial resolution of 130 m, low positioning accuracy (ranging from 490 to 930 m), and a revisit time of 15 days (Ou et al., 2019). Compared to VIIRS, Luojia 1-01 can achieve finer spatial details of artificial nighttime light due to on-board calibration, while its narrow spectral range, the lack of accurate ground-based measurements for radiometric calibration, and the effects of clouds and moonlight in the quality of the image are still limitations for its widespread applicability (Jiang et al., 2018).

Recent work has exploited high-resolution imagery provided by Yangwang-1 ("Look Up 1") satellite, which comes with a small optical space telescope whose visible band sensor can also collect night light imagery. Yangwang-1 exhibits comparable to other artificial remote sensing data, such as VIIRS, with some aspects even surpassing them. Notably, Yangwang-1 boasts a superior spatial resolution of 38 m at the nadir while also demonstrating high radiometric consistency with VIIRS (Zhu et al., 2022).

Apart from satellite-based approaches, different platforms such as high-altitude balloons have been utilized for inexpensive monitoring of nighttime lights emissions at

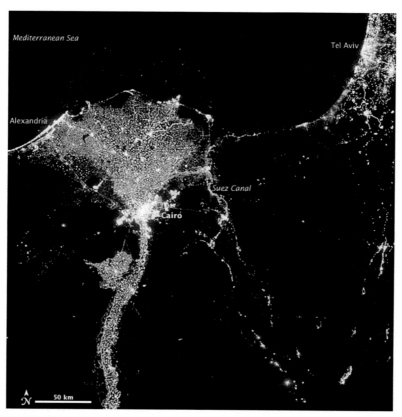

Figure 18.2 City lights illuminate the Nile. *(Source: From NASA, image acquired October 13, 2012.)*

higher spatiotemporal resolution (Walczak et al., 2023). The use of in situ-based networks such as photometers can also improve the provision of more accurate results regarding light pollution monitoring, given also the fact that VIIRS data provide under estimations and actual values could only be elicited by potentially knowing the types of outdoor lighting (Wallner et al., 2023). Mander et al. (2023) categorize ground-based sensors for mapping light pollution to (i) panchromatic tools that include sky quality meters (i.e., low-cost point-and-shoot devices), smartphone apps, luminance and illuminance meters, and wearable devices and (ii) imaging sensors that include digital cameras, imaging apps, spectrometers, and hyperspectral cameras.

Last but not least, collecting data from local authorities, municipalities, and community surveys can provide valuable information about the types of lighting fixtures in use, their intensity, and potential sources of light pollution (Beaudet et al., 2022) (Fig. 18.2).

2.2 Analysis and models

Before night time light imagery can be analyzed, it is important to preprocess the data to correct for noise and artifacts. This includes:

- Radiometric calibration: This process corrects for the different sensitivities of the different detectors on the satellite sensor (Doll, 2008b).
- Cloud masking: Clouds can obscure artificial light signals, so it is important to identify and mask out clouds in the imagery (Elvidge et al., 1999).
- Stray light removal: Stray light is scattered light from other sources (such as the sun or the moon) that can contaminate night time light signals. Stray light removal algorithms can be used to reduce the impact of stray light (Lee & Cao, 2016).
- Geometric correction: This process corrects for geometric distortions in the imagery caused by the satellite's orbit and sensor geometry (Zhao et al., 2015).

Once the night time lights imagery has been preprocessed, it can be analyzed to extract useful information. Some common analysis methods include:

- Index-based analysis: Different indexes have been computed according to literature, so as to quantify the level of light pollution in a particular area (Fang et al., 2023; Saraiji & Oommen, 2012; Zhou et al., 2023). The use of indexes allows to establish relationships between different factors and the extent of light pollution, while also allowing for the comparison of light pollution levels across different regions.
- Time series analysis: Time series analysis can be used to track changes in night time lights over time. This can be useful for monitoring urbanization, economic growth, and disaster response (Zhao et al., 2020).
- Spatial analysis: Spatial analysis can be used to identify spatial patterns in night time lights imagery. This can be useful for identifying urban centers, transportation networks, and other human features (Chang et al., 2022).
- Statistical analysis: Statistical analysis can be used to correlate nighttime lights data with other socioeconomic data, such as population density, gross domestic product, and crime rates (Zheng et al., 2023). This can be useful for developing predictive models and understanding the relationship between nighttime light and human activities (Kyriakos & Vavalis, 2023).
- Model-based analysis: Model-based analysis involves the use of models, that is, radiative transfer models (Wallner & Kocifaj, 2019), light pollution models (Linares et al., 2018), etc., to simulate and analyze the propagation of light in a given environment. This approach provides a more in-depth understanding of light distribution and its impact on the surrounding areas, contributing valuable insights to the study of nighttime lights (Chalkias et al., 2006).

3. Applications and case studies

Night light imagery, which is usually captured using remote sensing technologies such as satellites and sensors, has a wide range of applications in various fields. Here are some important applications of night light imagery:

3.1 Light pollution assessment

Night light images are used to assess and map the extent of light pollution in urban and rural areas. They help identify regions with excessive artificial lighting and contribute to efforts to mitigate the negative effects of light pollution on human health, ecology, and astronomy.

Cinzano et al. (2001) using radiometrically calibrated night images and modeling the propagation of light into the atmosphere through the various scattering patterns created the world's first Atlas with the brightness of the night sky due to artificial light. Therefore, it has been observed that many areas which should appear dark because of the absence of terrestrial light sources have in fact been affected by light pollution from neighboring luminous areas. Walczak et al. (2023) used imaging systems equipped on an aerial platform for the collection of night light pollution data while also proceeding with the association of the collected measurements with different land use/cover types. Wu et al. used hyperspectral data to assess light pollution in conjunction with species spectral sensitivity as well as the resulting eco-impacts on human circadian rhythm and photopic vision and plant photosynthesis (Wu et al., 2023). Linares et al. (2023) applied a numerical model in order to simulate the human-induced sky radiance and create sky brightness maps, while the methodology outputs were tested with other related measurements and data products (Linares et al., 2023).

3.2 Urban planning and development

Urban planners use nighttime light imagery to monitor urban growth and assess the impact of development on lighting patterns. Nightlight radiance can be used to calculate indices (e.g., Comprehensive Night Light Index) that enable efficient assessment of urban development trends and urbanization processes at different scales (Wu & Zhang, 2023) and the calculation of simulation models for urban expansion (Chen et al., 2023).

Hsu et al. (2013) proved that nighttime lights can provide a good estimation of in-use steel stock of buildings and civil engineering infrastructure at prefecture and national levels. Also, Hattori et al. (2013) monitored the continuous change of the in-use steel stock in buildings and civil engineering infrastructure, while nighttime light imagery can be used as a good proxy for extracting urban built-up areas (Shi et al., 2014) as well as assessing urban sprawl (Xu et al., 2023).

Given the fact that night time light intensity can be associated with economic activity, it has been used along with other data sources for the production of poverty maps and subsequently for enabling efficient policy making and SDGs monitoring (Putri et al., 2023).

Furthermore, monitoring artificial light helps to optimize street lighting, reduce energy consumption, and improve urban design. Gyenizse et al. have monitored night light pollution using in situ and satellite-based observations to evaluate street lighting policies and the impact of available lighting (Gyenizse et al., 2022). The implementation of efficient lighting design principles is an essential parameter for ensuring sustainability through design in urban environments and should consider various social, safety, cultural, environmental, health, economic, and technological aspects (Zielinska-Dabkowska & Bobkowska, 2022). Tong et al. (2023) use a simulation model to evaluate the vertical distribution of outdoor lighting on buildings in a 3D environment. Sordello et al. (2022) propose a new form of green infrastructure that takes into account the quality of darkness to limit the impact of light pollution on biodiversity loss and landscape fragmentation.

3.3 Energy efficiency

Night light imagery can be used to identify areas with insufficient or excessive outdoor lighting. This information helps local authorities and businesses make informed decisions about upgrading to energy-efficient lighting systems, reducing energy costs and greenhouse gas emissions.

Since nightlight imagery can help monitor changes in energy consumption patterns, being indicative of economic activity and urbanization, different models have been toward predicting light pollution and subsequently assessing impact in protected areas (Riza et al., 2023).

Elvidge et al. (2010, pp. 102–123) introduced a method to assess the proportion of the population with access to electricity using night lights as an indicator. This involved integrating the spatial distribution of night lighting with population data to estimate electrification rates. Additionally, Letu et al. (2010) suggested that accurate identification of permanent lighting areas from night light data could enhance the precision of parameters used in estimating electricity consumption.

3.4 Public safety and security

Law enforcement and emergency services use night light imagery to monitor public safety and security. Night light imagery proves instrumental in identifying areas with inadequate lighting, a crucial factor in ensuring public safety. Studies, such as those by Xu et al. (2018) and Peng et al. (2022), investigate the effect and mechanism of nighttime light brightness on the urban crime rate and highlight the correlation between well-lit environments and reduced crime rates. The integration of advanced remote sensing technologies aids law enforcement in pinpointing locations where improved lighting infrastructure is essential, contributing to the creation of safer urban environments (Xu et al., 2018).

Night light imagery also serves as a valuable resource for detecting patterns of criminal activity. The work of Zhou et al. (2019) showcases the application of spatial analysis on nighttime lights to identify crime patterns, while other studies have used nighttime light imagery for novel spatiotemporal modeling and for predicting street crime at local scale (Yang et al., 2020).

Additionally, nighttime light imagery has been used as a proxy for the detection of conflict events or the identification of patterns of destruction (Sticher et al., 2023), while the potential of the imagery to be used for evaluating armed conflicts at global scale has been also investigated (Li et al., 2013).

3.5 Transportation planning

Night light imagery assists in transportation planning by identifying traffic patterns and congestion in urban areas. By capturing the illuminated arteries of urban landscapes, night light data provide a visual representation of the intensity and spatial distribution of traffic flow, aiding planners in assessing the dynamics of movement throughout the night (Verma & Dutta, 2017).

Nighttime light imagery has been used for providing reliable datasets for long-term monitoring of highway traffic of different grades. Results demonstrate that nighttime light data have a response to the vehicle lighting and traffic volume (Chang et al., 2020).

Studies have also attempted to estimate freight transport activity, showcasing that in comparison with conventional freight traffic estimation approaches, the freight traffic data calculated from artificial light imagery contain more spatial information (Lopez-Ruiz et al., 2019). These models enable transportation planners to anticipate congestion, implement proactive measures, and optimize road infrastructure to accommodate varying traffic loads.

3.6 Environmental research

Researchers use night light imagery to study the effects of artificial light on the behavior of wildlife and ecosystems. It helps to understand how light pollution disrupts nocturnal species, influences migration patterns, and contributes to ecological imbalances.

A strong correlation between night light pollution and its threats to seabird grounding locations has been established in literature, including the identification of seasonal patterns (Friswold et al., 2023; Heswall et al., 2022). Other studies analyzed the effects of light pollution on avian species diet (Morelli et al., 2023).

Davies et al. (2020) assessed the impact of light pollution to marine habitats through the combined application of radiative transfer modeling and mapping accounting for in situ measured optical seawater properties. Light pollution can also negatively affect the biological cycle of coral reefs, intensifying spawning events (Davies et al., 2023).

Depending on the spectral distribution of nighttime lights, different impacts can occur in tree species and specifically on the photosynthetic rate and oxidative stress of plants (Wei et al., 2023).

However, apart from the negative impacts, some recent studies also reported benefits for the embryonic development of benthic marine species by the blue wavelength of artificial night lights and showed higher hatching and metamorphosis rates in the species studied (Zhang, Gao, et al., 2023).

3.7 Astronomy and space exploration

Astronomers rely on night light imagery to identify optimal locations for observatories and research facilities. Large astronomical observatories are already significantly endangered by artificial light and some of them are very polluted (Falchi et al., 2022). Nighttime light pollution helps to select areas with minimal light pollution to improve the visibility of celestial objects. Studies have also attempted to predict the surface brightness of the night sky and thus assess the potential impact of the loss of visibility of astronomical objects due to increasing artificial light pollution (Wesołowski, 2023).

3.8 Monitoring natural disasters

Night light imagery can be used to assess the extent of power outages during natural disasters such as hurricanes, earthquakes, and wildfires. This information is valuable for emergency response and recovery efforts.

Studies by Cui et al. (2023) show the effectiveness of nighttime light imagery in mapping power outages caused by natural disasters. By analyzing changes in nighttime luminosity, the severity and spatial distribution of areas affected by power disruptions can be determined. This approach not only facilitates the rapid assessment of the disaster's impact on critical infrastructure but also helps prioritize areas for emergency response.

Zhang, Gao, et al. (2023), Zhang, Yin, et al. (2023), Qiang et al. (2020) assessed the potential of Black Marble nighttime light product for the identification and evaluation of damage resulting from hurricanes, tornadoes, and earthquakes. Nighttime lights exhibit notable effectiveness in recognizing hurricane damage in well-illuminated areas and demonstrate potential in identifying damage along tornado paths. However, the study acknowledges a limited correlation between the NTL change ratio and the extent of damage, highlighting the method's constraint in quantifying the severity of damage.

Similarly, the applicability of artificial light in measuring community resilience to natural disasters has also been evaluated by modeling the recovery patterns of postdisaster economic activities (Qiang et al., 2020). Images of night light are effective in tracking human dynamics during natural disasters and fill a critical gap in empirical data and methods for assessing resilience. Quantifying changes in nighttime radiation intensity provides decision makers and disaster responders with a reliable tool for developing resilient and sustainable communities while enabling efficient resource allocation and prioritization of critical infrastructure recovery.

3.9 Health studies

Public health researchers use night light imagery to investigate the impact of artificial light at night on human health. Exposure to artificial light is associated with a high risk of mental disorders, including anxiety and depression (Jin et al., 2023), while nighttime light emissions have also been positively associated with mortality and, in particular, nervous system disorders (Lu et al., 2023).

The effects of artificial light, especially blue light emitted by electronic devices and some streetlights, on the suppression of melatonin levels have also been analyzed by calculating specific indices and identifying critical zones in terms of their expected potential effects (Aubé et al., 2020). Excessive artificial light at night can disrupt the circadian rhythms that determine the temporal characteristics of human physiology, leading to sleep disturbances (Meléndez–Fernández et al., 2023).

Areas with high levels of light pollution seem to elevate the risk of other diseases such as acute lymphoblastic leukemia (Zhong et al., 2023), metabolic diseases and obesity (Wang et al., 2023; Badpa et al., 2023; Kim et al., 2023; Xu et al., 2024), allergic diseases (Hsu et al., 2013), and different types of cancer (Bujnakova Mlynarcikova et al., 2023). Artificial light at night has been also positively associated with systolic blood pressure (Shen et al., 2023), while in other cases linked with higher fetal abdominal circumference during pregnancy (Zhang, Yin, et al., 2023).

In other cases, the intensity of night light is used to model the impact of other processes such as urbanization (Khan et al., 2023) on public health or even the association of the latter between areas with different economic growth (Dang et al., 2023). Xiao et al. (2023) examined artificial light at night with a social vulnerability index, which is widely used to measure local-level imbalances in health behaviors and disease outcomes. The study concluded that various social and environmental factors can influence the distribution of light at night and that further research should investigate the health inequalities resulting from light pollution exposure and assess urban planning, housing, and land use policies to mitigate the associated health hazards for vulnerable populations.

3.10 Cultural and social studies

Night light imagery can reveal cultural and social trends related to human activities. They offer insights into nocturnal cultural events, religious customs, and urbanization patterns.

The nighttime light images were used to assess the impact of the COVID-19 pandemic on socioeconomic activity exchange. The reduction in the brightness of the night lights reflected the socioeconomic impact, with the closure of mosques and curfews having a significant impact at the beginning of the COVID-19 pandemic. The results underscored the sensitivity of nighttime lighting analysis to social and cultural contexts and emphasized the broader impact of prevention measures on society (Alahmadi et al., 2021).

Shi et al. (2023) focused on commercial ports, canal junctions, and interactive nighttime lights and found that the total number of night lights in commercial ports decreased due to the pandemic policy. Alahmadi et al. (2023) also investigated the utility of night lights data to study the impact of the COVID-19 pandemic on religious and socioeconomic activities. Nocturnal light images were found to effectively capture the recovery of brightness in the postpandemic period, highlighting the role of cultural context in interpreting the impact of COVID-19.

Guo et al. (2022) addressed the consequences of continuous urbanization and industrialization, leading to rural depopulation. The study introduced a novel approach using nighttime light images to identify potential population hollowing areas.

Livny (2021) proposed an innovative method for estimating religiosity during Ramadan in Turkey using satellite imagery. Nocturnal lights influenced by the behavior of fasting Muslims provided an unobtrusive measure of religiosity in small geographical units. The study highlighted the potential of using nocturnal lights to understand the interplay between religiosity and sociopolitical and economic processes, and provided a cost-effective alternative to traditional survey-based measures.

3.11 Tourism and dark sky tourism

Nighttime light emissions are highly correlated with tourist activity, and different combinations of models and datasets have been tested to investigate this relationship (Krikigianni et al., 2019). In addition, nighttime light remote sensing has been used for the identification of tourist sites with suitable infrastructure to support the development of the respective activity (Devkota et al., 2019).

Night light pollution has a negative impact on the visibility of the night sky, affecting not only living beings and the ecosystem but also different types of touristic activity.

Visiting places with preserved sky darkness or astronomy-related historical sites and observing astronomical events or celestial bodies, that is, astronomy tourism, is another activity affected by light pollution (Bjelajac et al., 2021). Thus, regions with minimal light pollution reflected in nighttime light images are often promoted as destinations for stargazing and astrotourism. Such areas also offer opportunities for environmentally friendly tourism and revenue generation (Priyatikanto et al., 2019).

3.12 Climate change research

Night light imagery can be useful for assessing air pollution monitoring as well as the impact of climate change mitigation efforts. Zheng et al. used data from nighttime light imagery together with emissions and building data and implemented linear regression models to monitor carbon dioxide emissions from buildings to support policy making for carbon neutrality (Zheng et al., 2022). Similarly, other studies (Mahmud et al., 2023) have assessed the contribution of socioeconomic activities to air pollution by using nighttime lights as a proxy for air pollution and found a positive association between air pollutants such as NO_2 and CO and nighttime light emissions. The nocturnal data, along with NO_2, have been also used to determine other related parameters such as changes in aerosol optical depth (Jiadan et al., 2023).

4. Concluding remarks—Future research directions

Future research should address areas of concern related to light pollution, including urban-rural differences, local variability, temporal dynamics, and unexplored geographic regions. Much of the research on light pollution to date has focused predominantly on urban areas, while rural and suburban regions have been relatively underresearched. To gain a comprehensive understanding of the topic, it is essential to investigate the differences in the extent and sources of light pollution between urban and rural areas. Light pollution exhibits nonuniform patterns and can vary significantly in small geographical areas. To gain insight into the effects on different communities and ecosystems, research should prioritize the development of methods to detect and analyze these local differences. While numerous studies have provided static snapshots of light pollution, there remains a significant gap in understanding its temporal dynamics. These include daily, seasonal, and long-term variations that still need to be thoroughly explored. A comprehensive assessment of light pollution requires a detailed study of its temporal changes. Currently, much of the research on light pollution focuses on Western countries, while there are few data and studies from other parts of the world. Extending research to previously understudied regions, such as the developing world, can provide a more global perspective on the issue. This broader geographical coverage will contribute to a more comprehensive understanding of the extent and impact of light pollution on a global scale.

References

Ahmed, N., Islam, M. N., Tuba, A. S., Mahdy, M. R. C., & Sujauddin, M. (2019). Solving visual pollution with deep learning: A new nexus in environmental management. *Journal of Environmental Management, 248*, Article 109253. Cited 27 times. Accessed via Scopus.

Alahmadi, M., Mansour, S., Dasgupta, N., Abulibdeh, A., Atkinson, P. M., & Martin, D. J. (2021). Using daily nighttime lights to monitor spatiotemporal patterns of human lifestyle under COVID-19: The case of Saudi Arabia. *Remote Sensing, 13*, 4633. https://doi.org/10.3390/rs13224633

Alahmadi, M., Mansour, S., Dasgupta, N., & Martin, D. J. (2023). Using nighttime lights data to assess the resumption of religious and socioeconomic activities post-COVID-19. *Remote Sensing, 15*, 1064. https://doi.org/10.3390/rs15041064

Atılgan, E., Trampert, K., & Neumann, C. (2012). Evaluation of discomfort glare from LED lighting systems. *Journal of Lighting Engineering, 14*, 17—27.

Aubé, M., Marseille, C., Farkouh, A., Dufour, A., Simoneau, A., Zamorano, J., Roby, J., & Tapia, C. (2020). Mapping the melatonin suppression, star light and induced photosynthesis indices with the LANcube. *Remote Sensing, 12*, 3954. https://doi.org/10.3390/rs12233954

Badpa, M., Schneider, A., Ziegler, A.-G., Winkler, C., Haupt, F., Wolf, K., & Peters, A. (2023). Outdoor light at night and children's body mass: A cross-sectional analysis in the Fr1da study. *Environmental Research, 232*, 116325. https://doi.org/10.1016/j.envres.2023.116325

Balsky, M., & Terrich, T. (2020). Light trespass in street LED lighting systems. In *Proceedings—2020 21st international scientific conference on electric power engineering, EPE 2020* (p. 9269248).

Beaudet, C., Tardieu, L., & David, M. (2022). Are citizens willing to accept changes in public lighting for biodiversity conservation? *Ecological Economics, 200*, 107527. https://doi.org/10.1016/j.ecolecon.2022.107527

Bjelajac, D., Đerčan, B., & Kovačić, S. (2021). Dark skies and dark screens as a precondition for astronomy tourism and general well-being. *Information Technology and Tourism, 23*, 19—43. https://doi.org/10.1007/s40558-020-00189-9

Bożejko, M., Tarski, I., & Małodobra-Mazur, M. (2023). Outdoor artificial light at night and human health: A review of epidemiological studies. *Environmental Research, 218*, 115049. https://doi.org/10.1016/j.envres.2022.115049

Bujnakova Mlynarcikova, A., & Scsukova, S. (2023). The role of the environment in hormone-related cancers. In R. Pivonello, & E. Diamanti-Kandarakis (Eds.), *Environmental endocrinology and endocrine disruptors. Endocrinology.* Cham: Springer. https://doi.org/10.1007/978-3-030-39044-0_17

Chalkias, C., Petrakis, M., Psiloglou, B., & Lianou, M. (2006). Modelling of light pollution in suburban areas using remotely sensed imagery and GIS. *Journal of Environmental Management, 79*(1), 57—63. https://doi.org/10.1016/j.jenvman.2005.05.015

Chang, P., Pang, X., He, X., Zhu, Y., & Zhou, C. (2022). Exploring the spatial relationship between nighttime light and tourism economy: Evidence from 31 provinces in China. *Sustainability, 14*, 7350. https://doi.org/10.3390/su14127350

Chang, Y., Wang, S., Zhou, Y., Wang, L., & Wang, F. (2020). A novel method of evaluating highway traffic prosperity based on nighttime light remote sensing. *Remote Sensing, 12*, 102. https://doi.org/10.3390/rs12010102

Chen, X., Wang, Z., Yang, H., Ford, A. C., & Dawson, R. J. (2023). Enhanced urban growth modeling: Incorporating regional development heterogeneity and noise reduction in a cellular automata model—A case study of Zhengzhou, China. *Sustainable Cities and Society, 99*, 104959. https://doi.org/10.1016/j.scs.2023.104959

Chmielewski, S., Samulowska, M., Lupa, M., Lee, D., & Zagajewski, B. (2018). Citizen science and Web-GIS for outdoor advertisement visual pollution assessment. *Computers, Environment and Urban Systems, 67*, 97—109. Cited 33 times. Accessed via Scopus.

Cinzano, P., Falchi, F., & Elvidge, C. D. (2001). The first world atlas of the artificial night sky brightness. *Monthly Notices of the Royal Astronomical Society, 328*, 689—707.

Cox, D. T. C., & Gaston, K. J. (2023). Global erosion of terrestrial environmental space by artificial light at night. *Science of the Total Environment, 904*, 166701. https://doi.org/10.1016/j.scitotenv.2023.166701

Cox, D. T. C., Sánchez de Miguel, A., Bennie, J., Dzurjak, S. A., & Gaston, K. J. (2022). Majority of artificially lit Earth surface associated with the non-urban population. *Science of the Total Environment, 841*, Article 156782. Cited 7 times. Accessed via Scopus.

Cui, H., Qiu, S., Wang, Y., Zhang, Y., Liu, Z., Karila, K., Jia, J., & Chen, Y. (2023). Disaster-caused power outage detection at night using VIIRS DNB images. *Remote Sensing, 15*, 640. https://doi.org/10.3390/rs15030640

Dang, J., Shi, D., Li, X., Ma, N., Liu, Y., Zhong, P., Yan, X., Zhang, J., Lau, P. W. C., Dong, Y., et al. (2023). Artificial light-at-night exposure and overweight and obesity across GDP levels among Chinese children and adolescents. *Nutrients, 15*, 939. https://doi.org/10.3390/nu15040939

Davies, T. W., Levy, O., Tidau, S., et al. (2023). Global disruption of coral broadcast spawning associated with artificial light at night. *Nature Communications, 14*, 2511. https://doi.org/10.1038/s41467-023-38070-y

Davies, T. W., McKee, D., Fishwick, J., et al. (2020). Biologically important artificial light at night on the seafloor. *Scientific Reports, 10*, 12545. https://doi.org/10.1038/s41598-020-69461-6

Devkota, B., Miyazaki, H., Witayangkurn, A., & Kim, S. M. (2019). Using volunteered geographic information and nighttime light remote sensing data to identify tourism areas of interest. *Sustainability, 11*, 4718. https://doi.org/10.3390/su11174718

Doll, C. N. H. (2008a). *CIESIN thematic guide to night-time light remote sensing and its applications.* Palisades, NY, USA: Center for International Earth Science Information Network (CIESIN), Columbia University.

Doll, C. N. H. (2008b). *CIESIN global nighttime lights time series. Socioeconomic data and applications center (SEDAC).* Columbia University.

Elvidge, C. D., Baugh, K., Zhizhin, M., Hsu, F. C., & Ghosh, T. (2017). VIIRS night-time lights. *International Journal of Remote Sensing, 38*(21), 5860−5879. https://doi.org/10.1080/01431161.2017.1342050

Elvidge, C., Baugh, K., Sutton, P., Bhaduri, B., Tuttle, B., Tilottama, G., Ziskin, D., & Erwin, E. (2010). *Who's in the dark: Satellite based estimates of electrification rates, earth observation group* (pp. 102−123). NOAA National Geophysical Data Center.

Elvidge, C. D., Baugh, K. E., Dietz, J. B., Bland, T., Sutton, P. C., & Kroehl, H. W. (1999). Radiance calibration of DMSP-OLS low-light imaging data of human settlements. *Remote Sensing of Environment, 68*(1), 77−88.

Elvidge, C. D., Imhoff, M. L., Baugh, K. E., Hobson, V. R., Nelson, I., Safran, J., Dietz, J. B., & Tuttle, B. T. (2001). Night-time lights of the world: 1994−1995. *ISPRS Journal of Photogrammetry and Remote Sensing, 56*(2), 81−99.

Elvidge, C. D., Zhizhin, M., Sparks, T., Ghosh, T., Pon, S., Bazilian, M., Sutton, P. C., & Miller, S. D. (2023). Global satellite monitoring of exothermic industrial activity via infrared emissions. *Remote Sensing, 15*, 4760. https://doi.org/10.3390/rs15194760

Falchi, F., Ramos, F., Bara, S., Sanhueza, P., Arancibia, J., Marcelo, Damke, G., & Cinzano, P. (2022). Light pollution indicators for all the major astronomical observatories. *Monthly Notices of the Royal Astronomical Society, 519*, 26−33. https://doi.org/10.1093/mnras/stac2929

Falchi, F., Cinzano, P., Duriscoe, D., Kyba, C. C. M., Elvidge, C. D., Baugh, K., ... Furgoni, R. (2016). The new world atlas of artificial night sky brightness. *Science Advances, 2*(6), e1600377.

Falchi, F., Cinzano, P., Elvidge, C. D., Keith, D. M., & Haim, A. (2011). Limiting the impact of light pollution on human health, environment and stellar visibility. *Journal of Environmental Management, 92*, 2714−2722.

Fang, L., Wu, Z., Tao, Y., & Gao, J. (2023). Light pollution index system model based on Markov random field. *Mathematics, 11*, 3030. https://doi.org/10.3390/math11133030

Friswold, B., Idle, J. L., Learned, J., Penniman, J., Bolosan, T., Cotin, J., Young, L., & Price, M. R. (2023). From colony to fallout: Artificial lights pose risk to seabird fledglings far from their natal colonies. *Published in Conservation Science and Practice.* https://doi.org/10.1111/csp2.13000

Gallaway, T. (2010). On light pollution, passive pleasures, and the instrumental value of beauty. *Journal of Economic Issues, 44*(1), 71−88. JSTOR . (Accessed 12 November 2023).

Green, R. F., Luginbuhl, C. B., Wainscoat, R. J., & Duriscoe, D. (2022). The growing threat of light pollution to ground-based observatories. *Astronomy and Astrophysics Review, 30*(1), Article 1. Cited 24 times. Accessed via Scopus.

Guo, B., Bian, Y., Pei, L., Zhu, X., Zhang, D., Zhang, W., Guo, X., & Chen, Q. (2022). Identifying population hollowing out regions and their dynamic characteristics across central China. *Sustainability, 14*, 9815. https://doi.org/10.3390/su14169815

Gyenizse, P., Soltész, E., Lóczy, D., Kovács, J., Nagyváradi, L., Elekes, T., Nagy, G., Bodzac, S., Németh, G., & Halmai, Á. (2022). Light pollution mapping in Pécs city with the help of SQM-L and VIIRS DNB: The effect of public luminaire replacements on the sky background of the urban sky. *Geographica Pannonica, 26*(4), 334−344.

Hölker, F., et al. (Eds.). (2016). *Mapping the night: An Atlas of the artificial sky brightness.*

Hattori, R., Horie, S., Hsu, F.-C., Elvidge, C. D., & Matsuno, Y. (2013). Estimation of in-use steel stock for civil engineering and building using night-time light images. *Resources, Conservation and Recycling.* https://doi.org/10.1016/j.resconrec.2013.11.007

Heswall, A., Miller, L., McNaughton, E. J., Brunton-Martin, A. L., Cain, K. E., Friesen, M. R., & Gaskett, A. C. (2022). Artificial light at night correlates with seabird groundings: Mapping city lights near a seabird breeding hotspot. *PeerJ, 10*, e14237. https://doi.org/10.7717/peerj.14237

Hsu, F.-C., Elvidge, C. D., & Matsuno, Y. (2013). Exploring and estimating in-use steel stocks in civil engineering and buildings from night-time lights. *International Journal of Remote Sensing, 34*, 490–504.

Hu, R.-F., Hegadoren, K. M., Wang, X.-Y., & Jiang, X.-Y. (2016). An investigation of light and sound levels on intensive care units in China. *Australian Critical Care, 29*(2), 62–67. Cited 21 times. Accessed via [Source].

Jechow, A., & Hölker, F. (2020). Evidence that reduced air and road traffic decreased artificial night-time skyglow during COVID-19 lockdown in Berlin, Germany. *Remote Sensing, 12*, 3412. https://doi.org/10.3390/rs12203412

Jiadan, D., Liqiao, T., Fang, C., Xiaobin, C., Xiaoling, C., Qiangqiang, X., & Xinghui, X. (2023). Spatiotemporal variations of aerosol optical depth over Ukraine under the Russia-Ukraine war. *Atmospheric Environment, 314*, 120114. https://doi.org/10.1016/j.atmosenv.2023.120114

Jiang, W., He, G., Long, T., Guo, H., Yin, R., Leng, W., Liu, H., & Wang, G. (2018). Potentiality of using Luojia 1-01 nighttime light imagery to investigate artificial light pollution. *Sensors, 18*, 2900. https://doi.org/10.3390/s18092900

Jin, J., Han, W., Yang, T., Xu, Z., Zhang, J., Cao, R., Wang, Y., Wang, J., Hu, X., Gu, T., He, F., Huang, J., & Li, G. (2023). Artificial light at night, MRI-based measures of brain iron deposition and incidence of multiple mental disorders. *Science of the Total Environment, 902*, 166004. https://doi.org/10.1016/j.scitotenv.2023.166004

Khan, J. R., Islam, M. M., Faisal, A. S. M., et al. (2023). Quantification of urbanization using night-time light intensity in relation to women's overnutrition in Bangladesh. *Journal of Urban Health, 100*, 562–571. https://doi.org/10.1007/s11524-023-00728-9

Kim, K.-H., & Kim, G. (2021). Using simulation-based modeling to evaluate light trespass in the design stage of sports facilities. *Sustainability, 13*(9), 4725.

Kim, M., Vu, T. H., Maas, M. B., Rosemary, I. B., Michael, S. W., Till, R., Daviglus, M. L., Reid, K. J., & Zee, P. C. (March 2023). Light at night in older age is associated with obesity, diabetes, and hypertension. *Sleep, 46*(3). https://doi.org/10.1093/sleep/zsac130. zsac130.

Krikigianni, E., Tsiakos, C., & Chalkias, C. (2019). Estimating the relationship between touristic activities and night light emissions. *European Journal of Remote Sensing, 52*(Suppl. 1), 233–246. https://doi.org/10.1080/22797254.2019.1582305

Kyriakos, C., & Vavalis, M. (2023). Business intelligence through machine learning from satellite remote sensing data. *Future Internet, 15*, 355. https://doi.org/10.3390/fi15110355

Lee, S., & Cao, C. (2016). Soumi NPP VIIRS day/night band stray light characterization and correction using calibration view data. *Remote Sensing, 8*, 138. https://doi.org/10.3390/rs8020138

Letu, H., Hara, M., Tana, G., & Nishio, F. (2010). A saturated light correction method for DMSP/OLS night-time satellite imagery. *IEEE Transactions on Geoscience and Remote Sensing, 50*, 389–396.

Li, X., Chen, F., & Chen, X. (Oct. 2013). Satellite-observed nighttime light variation as evidence for global armed conflicts. *Ieee Journal of Selected Topics in Applied Earth Observations and Remote Sensing, 6*(5), 2302–2315. https://doi.org/10.1109/JSTARS.2013.2241021

Linares, H., Masana, E., Ribas, S. J., García-Gil, M., Aubé, M., de Sánchez, M. A., & Simoneau, A. (2023). Assessing light pollution in vast areas: Zenith sky brightness maps of Catalonia. *Journal of Quantitative Spectroscopy and Radiative Transfer, 309*, 108678. https://doi.org/10.1016/j.jqsrt.2023.108678

Linares, H., Masana, E., Ribas, S. J., Garcia - Gil, M., Figueras, F., & Aubé, M. (2018). Modelling the night sky brightness and light pollution sources of Montsec protected area. *Journal of Quantitative Spectroscopy and Radiative Transfer, 217*, 178–188. Cited 8 times. Accessed via Scopus.

Livny, A. (2021). Can religiosity be sensed with satellite data? An assessment of luminosity during Ramadan in Turkey. *Public Opinion Quarterly, 85*(Issue S1), 371–398. https://doi.org/10.1093/poq/nfab013

Lopez-Ruiz, Hector, G., Nezamuddin, Nora, Hassan, Reema Al, & Muhsen, Abdelrahman (2019). *Estimating freight transport activity using nighttime lights satellite data in China, India and Saudi Arabia. KS-2019-MP07. Riyadh*. KAPSARC. https://doi.org/10.30573/KS–2019-MP07

Lu, Y., Yin, P., Wang, J., et al. (2023). Light at night and cause-specific mortality risk in Mainland China: A nationwide observational study. *BMC Medicine, 21*, 95. https://doi.org/10.1186/s12916-023-02822-w

Mahmud, H., Mitra, B., Uddin, M. S., Hridoy, A. E., Aina, Y. A., Abubakar, I. R., Rahman, S. M., Tan, M. L., & Rahman, M. M. (2023). Temporal assessment of air quality in major cities in Nigeria using satellite data. *Atmospheric Environment X, 20*, 100227. https://doi.org/10.1016/j.aeaoa.2023.100227

Mander, S., Alam, F., Lovreglio, R., & Ooi, M. (2023). How to measure light pollution—A systematic review of methods and applications. *Sustainable Cities and Society, 92*, 104465. https://doi.org/10.1016/j.scs.2023.104465

Meléndez-Fernández, O. H., Liu, J. A., & Nelson, R. J. (2023). Circadian rhythms disrupted by light at night and mistimed food intake alter hormonal rhythms and metabolism. *International Journal of Molecular Sciences, 24*, 3392. https://doi.org/10.3390/ijms24043392

Morelli, F., Tryjanowski, P., Ibáñez-Álamo, J. D., et al. (2023). Effects of light and noise pollution on avian communities of European cities are correlated with the species' diet. *Scientific Reports, 13*, 4361. https://doi.org/10.1038/s41598-023-31337-w

National Oceanic and Atmospheric Administration-NOAA. (2013). *Technical report NESDIS 142, visible infrared imaging radiometer suite (VIIRS) sensor data record (SDR) user's guide version 1.3.* Available at: https://ncc.nesdis.noaa.gov/documents/documentation/viirs-users-guide-tech-report-142a-v1.3.pdf.

Ou, J., Liu, X., Liu, P., & Liu, X. (2019). Evaluation of Luojia 1-01 nighttime light imagery for impervious surface detection: A comparison with NPP-VIIRS nighttime light data. *International Journal of Applied Earth Observation and Geoinformation, 81*, 1—12. https://doi.org/10.1016/j.jag.2019.04.017

Peng, C., Sun, W., & Zhang, X. (2022). Crime under the light? Examining the effects of nighttime lighting on crime in China. *Land, 11*, 2305. https://doi.org/10.3390/land11122305

Putri, S. R., Wijayanto, A. W., & Pramana, S. (2023). Multi-source satellite imagery and point of interest data for poverty mapping in East Java, Indonesia: Machine learning and deep learning approaches. *Remote Sensing Applications: Society and Environment, 29*, 100889. https://doi.org/10.1016/j.rsase.2022.100889

Qiang, Y., Huang, Q., & Xu, J. (2020). Observing community resilience from space: Using nighttime lights to model economic disturbance and recovery pattern in natural disaster. *Sustainable Cities and Society, 57*, 102115. https://doi.org/10.1016/j.scs.2020.102115

Riza, L. S., Putra, Z. A. Y., Firdaus, M. F. Y., Trihutama, F. Z., Izzuddin, A., Utama, J. A., Samah, K. A. F. A., Herdiwijaya, D., Rinto Anugraha, N. Q. Z., & Mumpuni, E. S. (2023). A spatiotemporal prediction model for light pollution in conservation areas using remote sensing datasets. *Decision Analytics Journal, 9*, 100334. https://doi.org/10.1016/j.dajour.2023.100334

Priyatikanto, R., Admiranto, A. G., Putri, G. P., Elyyani, E., Maryam, S., & Suryana, N. (2019). Map of sky brightness over greater Bandung and the prospect of astro-tourism. *The Indonesian Journal of Geography, 51*(2), 190—198.

Saraiji, R., & Oommen, M. S. (2012). Light pollution index (LPI): An integrated approach to study light pollution with street lighting and façade lighting. *Leukos, 9*, 127—145. https://doi.org/10.1582/leukos.2012.09.02.004

Shen, M., Li, Y., Li, S., Chen, X., Zou, B., & Lu, Y. (2023). Association of exposure to artificial light at night during adolescence with blood pressure in early adulthood. *Chronobiology International, 40*(10), 1419—1426. https://doi.org/10.1080/07420528.2023.2266485

Shi, K., Huang, C., Yu, B., Yin, B., Huang, Y., & Wu, J. (2014). Evaluation of NPP-VIIRS night time light composite data for extracting built-up urban areas. *Remote Sensing Letters, 5*, 358—366.

Shi, K., Cui, Y., Wu, S., & Liu, S. (2023). The impact of COVID-19 pandemic on socioeconomic activity exchanges in the Himalayan region: A satellite nighttime light perspective. In *IEEE geoscience and remote sensing letters* (Vol 20, pp. 1—5). https://doi.org/10.1109/LGRS.2023.3291438. Art. no. 2503105.

Sordello, R., Busson, S., Cornuau, J. H., Deverchère, P., Faure, B., Guetté, A., Hölker, F., Kerbiriou, C., Lengagne, T., Le Viol, I., Longcore, T., Moeschler, P., Ranzoni, J., Ray, N., Reyjol, Y., Roulet, Y., Schroer, S., Secondi, J., … Vauclair, S. (2022). A plea for a worldwide development of dark infrastructure for biodiversity—Practical examples and ways to go forward. *Landscape and Urban Planning, 219*, 104332. https://doi.org/10.1016/j.landurbplan.2021.104332

Spoelstra, H., van Grunsven, R. H. A., Ramakers, J. J., & Ferguson, K. B. (2017). Skyglow extends into the world's key biodiversity areas. *Animal Conservation, 23*(2), 153—159.

Sticher, V., Wegner, J. D., & Pfeifle, B. (May 29, 2023). Toward the remote monitoring of armed conflicts. *PNAS Nexus, 2*(6), Article pgad181. https://doi.org/10.1093/pnasnexus/pgad181. PMID: 37378391; PMCID: PMC10291284.

Tong, J. C. K., Wun, A. H. L., Chan, T. T. H., Lau, E. S. L., Lau, E. C. F., Chu, H. H. K., & Lau, A. P. S. (2023). Simulation of vertical dispersion and pollution impact of artificial light at night in urban environment. *Science of the Total Environment, 902*, 166101. https://doi.org/10.1016/j.scitotenv.2023.166101

Verma, N., & Dutta, M. (2017). Contrast enhancement of night time imagery for traffic activity understanding. In *2017 international conference on computing methodologies and communication* (pp. 809–813). Erode, India: ICCMC). https://doi.org/10.1109/ICCMC.2017.8282578

Wakil, K., Naeem, M. A., Anjum, G. A., Waheed, A., Thaheem, M. J., ul Hussnain, M. Q., & Nawaz, R. (2019). A hybrid tool for visual pollution assessment in urban environments. *Sustainability, 11*(8), Article 2211. Cited 15 times. Accessed via Scopus.

Walczak, Ken, Wisbrock, Lauren, Tarr, Cynthia, Gyuk, Geza, Amezcua, Jose, Cheng, Cynthia, Cris, Joshua, Jimenez, Claudia, Mehta, Megan, Mujahid, Aisha, Pritchard, Liberty, Suquino, Kly, & Turkic, Laris (2023). Quantifying nighttime light emission by land use from the stratosphere. *Journal of Quantitative Spectroscopy and Radiative Transfer, 310*, 108739. https://doi.org/10.1016/j.jqsrt.2023.108739

Wallner, S., & Kocifaj, M. (2019). Impacts of surface albedo variations on the night sky brightness—A numerical and experimental analysis. *Journal of Quantitative Spectroscopy and Radiative Transfer, 239*, 106648. https://doi.org/10.1016/j.jqsrt.2019.106648

Wallner, S., Puschnig, J., & Stidl, S. (2023). The reliability of satellite-based light trends for dark sky areas in Austria. *Journal of Quantitative Spectroscopy and Radiative Transfer, 311*, 108774. https://doi.org/10.1016/j.jqsrt.2023.108774

Wang, X., Ma, X., Yu, Z., et al. (2023). Exposure to outdoor artificial light at night increases risk and burden of metabolic disease in Ningxia, China. *Environmental Science and Pollution Research, 30*, 87517–87526. https://doi.org/10.1007/s11356-023-28684-6

Wei, Y., Li, Z., Zhang, J., et al. (2023). Influence of night-time light pollution on the photosynthesis and physiological characteristics of the urban plants Euonymus japonicus and Rosa hybrida. *Ecological Process, 12*, 38. https://doi.org/10.1186/s13717-023-00449-6

Wesołowski, M. (2023). The increase in the surface brightness of the night sky and its importance in visual astronomical observations. *Scientific Reports, 13*, 17091. https://doi.org/10.1038/s41598-023-44423-w

Wu, E., Xu, W., Yao, Q., Yuan, Q., Chen, S., Shen, Y., Wang, C., & Zhang, Y. (2023). Spectral-level assessment of light pollution from urban façade lighting. *Sustainable Cities and Society, 98*, 104827. https://doi.org/10.1016/j.scs.2023.104827

Wu, X., & Zhang, Y. (2023). Coupling analysis of ecological environment evaluation and urbanization using projection pursuit model in Xi'an, China. *Ecological Indicators, 156*, 111078. https://doi.org/10.1016/j.ecolind.2023.111078

Xiao, Q., Lyu, Y., Zhou, M., Lu, J., Zhang, K., Wang, J., & Bauer, C. (2023). Artificial light at night and social vulnerability: An environmental justice analysis in the U.S. 2012–2019. *Environment International, 178*, 108096. https://doi.org/10.1016/j.envint.2023.108096

Xu, H., Hu, S., & Li, X. (2023). Urban distribution and evolution of the Yangtze river economic belt from the perspectives of urban area and night-time light. *Land, 12*, 321. https://doi.org/10.3390/land12020321

Xu, Y., Fu, C., Kennedy, E., Jiang, S., & Owusu-Agyemang, S. (2018). The impact of street lights on spatial-temporal patterns of crime in Detroit, Michigan. *Cities, 79*, 45–52. https://doi.org/10.1016/j.cities.2018.02.021

Yang, B., Liu, L., Lan, M., Wang, Z., Zhou, H., & Yu, H. (2020). A spatio-temporal method for crime prediction using historical crime data and transitional zones identified from nightlight imagery. *International Journal of Geographical Information Science [Online ahead of print]*. https://doi.org/10.1080/13658816.2020.1737701

Xu, Y.-J., Xie, Z.-Y., Gong, Y.-C., Wang, L.-B., Xie, Y.-Y., Lin, L.-Z., Zeng, X.-W., Yang, B.-Y., Zhang, W., Liu, R.-Q., Hu, L.-W., Chen, G., & Dong, G.-H. (2024). The association between outdoor light at night exposure and adult obesity in Northeastern China. *International Journal of Environmental Health Research*. https://doi.org/10.1080/09603123.2023.2165046

Zhang, M., Gao, X., Luo, Q., Lin, S., Lyu, M., Luo, X., Ke, C., & You, W. (2023). Ecological benefits of artificial light at night (ALAN): Accelerating the development and metamorphosis of marine shellfish larvae. *Science of the Total Environment, 903*, 166683. https://doi.org/10.1016/j.scitotenv.2023.166683

Zhang, L., Yin, W., Yu, W., Wang, P., Wang, H., Zhang, X., & Zhu, P. (2023). Environmental exposure to outdoor artificial light at night during pregnancy and fetal size: A prospective cohort study. *Science of the Total Environment, 883*, 163521. https://doi.org/10.1016/j.scitotenv.2023.163521

Zhao, N., Liu, Y., Hsu, F.-C., Samson, E. L., Letu, H., Liang, D., & Cao, G. (2020). Time series analysis of VIIRS-DNB nighttime lights imagery for change detection in urban areas: A case study of devastation in Puerto Rico from hurricanes Irma and Maria. *Applied Geography, 120*, 102222. https://doi.org/10.1016/j.apgeog.2020.102222

Zhao, N., Zhou, Y., & Samson, E. (2015). Correcting incompatible DN values and geometric errors in nighttime lights time-series images. *IEEE Transactions on Geoscience and Remote Sensing, 53*, 2039–2049. https://doi.org/10.1109/TGRS.2014.2352598

Zheng, Q., Seto, K. C., Zhou, Y., You, S., & Weng, Q. (2023). Nighttime light remote sensing for urban applications: Progress, challenges, and prospects. *ISPRS Journal of Photogrammetry and Remote Sensing, 202*, 125–141. https://doi.org/10.1016/j.isprsjprs.2023.05.028

Zheng, Y., Ou, J., Chen, G., Wu, X., & Liu, X. (2022). Mapping building-based spatiotemporal distributions of carbon dioxide emission: A case study in England. *International Journal of Environmental Research and Public Health, 19*, 5986. https://doi.org/10.3390/ijerph19105986

Zhong, C., Wang, R., Morimoto, L. M., et al. (2023). Outdoor artificial light at night, air pollution, and risk of childhood acute lymphoblastic leukemia in the California linkage study of early-onset cancers. *Scientific Reports, 13*, 583. https://doi.org/10.1038/s41598-022-23682-z

Zhou, H., Liu, L., Lan, M., Yang, B., & Wang, Z. (2019). Assessing the impact of nightlight gradients on street robbery and burglary in Cincinnati of Ohio state, USA. *Remote Sensing, 11*, 1958. https://doi.org/10.3390/rs11171958

Zhou, Z.-H., Cao, H.-L., Feng, T.-Y., & Zhu, J.-M. (2023). Optimization algorithms for light pollution management based on TOPSIS-non-linear regularization model. *Frontiers in Energy Research, 11*, 1242010. https://doi.org/10.3389/fenrg.2023.1242010

Zhu, X., Tan, X., Liao, M., Liu, T., Su, M., Zhao, S., Xu, Y. N., & Liu, X. (2022). Assessment of a new fine-resolution nighttime light imagery from the Yangwang-1 ("Look up 1") satellite. *IEEE Geoscience and Remote Sensing Letters, 19*, 1–5. https://doi.org/10.1109/LGRS.2021.3139774

Zielinska-Dabkowska, K. M., & Bobkowska, K. (2022). Rethinking sustainable cities at night: Paradigm shifts in urban design and city lighting. *Sustainability, 14*, 6062. https://doi.org/10.3390/su14106062

CHAPTER 19

Crowdsourcing applications for monitoring the urban environment

Mariana Vallejo Velázquez[1], Antigoni Faka[2] and Ourania Kounadi[1]
[1]University of Vienna, Department of Geography and Regional Research, Vienna, Austria; [2]Department of Geography, School of Environment, Geography and Applied Economics, Harokopio University, Athens, Greece

1. Introduction

The constant changes in social and spatial structures have transformed cities into complex landscapes. In addition to the sustained expansion of major cities, there has been a notable surge in urban population over the past three decades, particularly in regions with historically low urban population ratios. This trend is anticipated to endure in the foreseeable future, reflecting the ongoing shift toward urbanization (Hall & Barrett, 2018). As urban areas evolve, the development process must continuously adapt and respond to various factors, such as population dynamics, environmental changes, and emerging technologies. As part of this ongoing evolution process, there has been a notable change in the role of inhabitants from passive observers to active participants. One significant change is the increasing inclusion of citizens' perspectives, knowledge, and participation in decision-making and shaping the urban policies. This has been possible due to the decentralization of public policy that acknowledge the importance of involving a broader range of actors beyond traditional governmental authorities. By incorporating the insights and experiences of local communities, urban governance can better align with the needs of its residents.

Beyond political aspects and decision-making, citizens have also become actively engaged in scientific endeavors. This concept is not new, there have been historical examples of citizen involvement such as volunteers reporting weather conditions in the 1890s, conducting bird surveys in the early 1900s, or participating in the study of surface water currents in the 1960s (Njue et al., 2019). However, with advancements in current technologies, citizen participation has been made possible on an unprecedented and massive scale. Multiple studies have shown the population's ability to contribute to scientific research in various ways, as the local population possesses a deep understanding of their immediate environment and can offer valuable insights from an expert standpoint. Therefore, citizens have been involved in applied research efforts through their active or passive participation such as volunteering in data gathering, sharing local information, volunteer mapping and reporting events; therefore, this practice is known as citizen science.

Geographical Information Science
ISBN 978-0-443-13605-4
https://doi.org/10.1016/B978-0-443-13605-4.00015-1

From a geographic perspective, different yet overlapping concepts have emerged to describe the involvement of individuals in these scenarios, such as public participation geographic information systems (PPGIS), participatory sensing, and volunteered geographic information (VGI). These concepts can vary in terms of the data collected, methods utilized, and types of participants involved. Nevertheless, they all share a common origin, namely crowdsourcing. By harnessing the collective knowledge in contributions with scientists, authorities, and stakeholders, these approaches tap into the potential of collaborative efforts to generate geographic information that can be applied to benefit the society.

The concept of "crowdsourcing" encompasses various definitions, depending on the specific disciplines, domains, and practices in which it is utilized. However, its essence is shared across researchers, authorities, and industries as a means of the vast potential of the collective intelligence and contributions of large groups of individuals. It is a termed coined by Jeff Howe, who mainly developed it from the business perspective and was presented in his article titled "The Rise of Crowdsourcing" published in *Wired Magazine* in June 2006. Howe defined it as:

> *Crowdsourcing represents the act of a company or institution taking a function once performed by employees and outsourcing it to an undefined (and generally large) network of people in the form of an open call. This can take the form of peer-production (when the job is performed collaboratively) but is also often undertaken by sole individuals. The crucial prerequisite is the use of the open call format and the large network of potential laborers*
>
> **Howe (2006)**

The term "crowdsourcing" has found widespread application across various disciplines, each with its own unique characteristics. In the field of geospatial sciences, in the present context, a key aspect of this concept revolves around the gathering of information and knowledge. Given the nature of this field, the inclusion of spatial reference in the collected data becomes particularly crucial. Consequently, three discernible approaches to crowdsourced geodata collection can be identified: *"event logging," "sensor data collection,"* and *"media data harvesting."*

The event logging approach involves individuals using online platforms or mobile applications to report the occurrence of an event, whether it is in real time or afterward. Participants provide location information and, occasionally, additional details about the incident (Hamrouni et al., 2020; Susanto et al., 2017). Regarding the second, sensor data collection, it involves the gathering of data on a specific topic through the use of sensors. This form of crowdsourcing is known as *"crowdsensing"* and relies on low-cost sensors and smartphones as the primary tools. Smartphones have evolved into multifunctional devices capable of collecting data, thanks to the integration of various built-in sensors, including GPS, camera, microphone, accelerometer, and gyroscope (Bale et al., 2019; Becker et al., 2013; Hu et al., 2017; Kanjo, 2010). These sensors enable the capturing and recording of information, which, when combined with a diverse range of developed applications, allows for data analysis and transmission to a server for further processing. The third approach, media data harvesting, relies on various methods of extracting data from digital platforms, such as social networks; this includes extracting timestamp and geolocation information from posts and comments (Shen & Karimi, 2016; Wang et al., 2018; Xu et al., 2020). Taking into account that the data can be originated from diverse sources, it becomes crucial to emphasize the significance of ensuring the reliability and quality of such data.

This classification of approaches encompasses the broad range of crowdsourcing data collection methods utilized in spatial sciences. Geographic studies involve the integration of qualitative and quantitative data, depending on the nature of the subject being studied. However, due to the intricate interplay between social and physical aspects, crowdsourcing emerges as a valuable and innovative resource in this field. It enables the collection of diverse data from a wide range of participants, allowing for a comprehensive understanding of spatial phenomena. By harnessing the power of collective intelligence, crowdsourcing offers real-time, high-resolution, and dynamic data on various topics. This wealth of information contributes to the advancement of spatial studies, facilitating in-depth analysis and the exploration of spatial patterns, trends, and relationships. Moreover, it promotes public engagement, encouraging active participation from individuals who have firsthand knowledge and experiences related to the studied environments, thus enriching the overall understanding of geographic phenomena.

Specifically, in urban environments, the objective of utilizing crowdsourcing data is to tackle the lack of local-level data by gathering information across a wide spatial coverage and involving individuals who have direct engagement with the environment being studied. This approach can contribute to a deeper comprehension of the real context in which people conduct their daily activities, leading to a broader understanding (Brown, 2015). This enhances our knowledge of the urban environment, enabling to make informed decisions and implement targeted interventions that address the unique challenges and dynamics of specific urban areas (Fig. 19.1).

Here are presented some examples of research applications in which crowdsourcing has been used within the physical and social sphere in urban environments. While they are not exclusive, they show general characteristics of what the use of crowdsourcing implies.

2. Monitoring physical aspects in urban environments

Traditionally, observations and measurements of geophysical variables have been dependent on specialized fixed sensors or high-precision instruments operated by field experts. However, this approach has certain limitations in terms of spatial coverage as the sensors are strategically placed in specific locations, which means they provide values at a regional scale that may not always meet the requirements of certain applications that demand localized data, such as the development of urban-scale models.

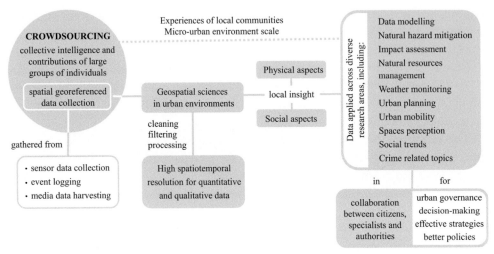

Figure 19.1 Crowdsourcing for physical and social applications in the urban environment.

The utilization of crowdsensing data gathering has been found to have a significant positive impact on data modeling. This can be primarily attributed to the high spatial and temporal resolution data it provides, surpassing the relatively limited measurements obtained from fixed sensors. Notably, when both sources are combined, the performance of a model is improved (Schneider, 2017; Yang & Ng, 2017). The collected data from the crowd serve as the input for model development that encompasses setup, calibration, and validation. On the other hand, the data can also be used as input for model utilization that refers to simulation, forecasting, and updating. Increased availability of information has the potential to yield a more comprehensive understanding and definition of natural processes, leading to advancements in practices related to natural hazard mitigation, impact assessment, infrastructure planning, and natural resources management (Zheng et al., 2018).

Crowdsourcing data collection of geophysical variables typically utilizes sensing methods that often involve low-cost sensors, smartphones, or basic test kits (Faka et al., 2021). In such cases, participants are usually required to undergo preliminary training to ensure adherence to specific protocols that maintain data quality standards. In other research scenarios, data collection may involve photographs, videos, or detailed written descriptions.

2.1 Weather monitoring

The widespread availability of affordable sensors in the market has empowered citizens to actively participate in monitoring various environmental variables. An exemplary application of this trend is the use of citizen weather stations (CWSs), which collect and report weather data. These data have the potential to significantly improve the accuracy of localized weather forecasts, especially at the neighborhood level (Nipen et al., 2020). Studies utilizing this data source have exhibited temporal and spatial variations. Some investigations have focused on shorter time frames, spanning a few months, while others have gathered weather conditions over several years. Furthermore, these initiatives can range in scale, from local community projects to regional undertakings, and even extend to global efforts led by federal agencies.

An exemplary global-scale project is the Weather Observations Website (WOW), launched in 2011 by the Met Office, the UK's national meteorological service. WOW serves as a platform that enables individuals to contribute real-time weather data obtained from CWS, as well as through direct observations reported by citizens, supported by visual evidence such as photographs or videos. By 2020, WOW collected a total of 1.5 billion reports with an average of 25 million per month (Met Office, 2022). Similarly, in 2012, the National Oceanic and Atmospheric Administration (NOAA), in collaboration with the University of Oklahoma and the Cooperative Institute for Mesoscale Meteorological Studies, launched the crowdsourcing weather reports app called mPING[1] (Meteorological Phenomena Identification Near the Ground). The data collected are used to complement the weather radar data to achieve more accurate and reliable weather forecasts (NOAA, 2016).

Utilizing crowdsourcing to gather meteorological data from the urban areas, particularly through the deployment of CWS, is a valuable approach for intraurban variability in climate research. However, a notable challenge associated with CWS is the inherent risk of compromised data quality, stemming from limitations in the low-cost sensors themselves, irregular calibration, and the variations in user implementations. Therefore, it becomes necessary to perform data cleaning, filtering, processing, and quality control on crowdsensed data, similar to any other dataset, in order to reduce noise. Studies have shown that implementing these processes enhances data quality and allows it to be utilized as inputs for further analysis (Chen et al., 2021; Nipen et al., 2020).

2.2 Hydrological monitoring

In the context of hydrological monitoring, there remains a necessity for ground-based observations that go beyond remote sensing techniques and automated or fixed monitoring stations. Over the past decade, there has been a notable rise in the number of studies incorporating citizen science into hydrological research. This increase can be attributed to the growing interest among individuals in the field of sustainable water resource management. While this trend has been more prominent in North America and Europe, there have also been, to a somewhat lesser extent, similar developments in certain countries in Asia and Africa (Njue et al., 2019). Collaborative efforts primarily

---

[1]Meteorological Phenomena Identification Near the Ground (mPING) https://mping.nssl.noaa.gov/

focus on collecting water level and water quality data, which are pivotal for the effective management of surface water systems.

In some cases, measuring water quality involves collecting water samples, while in others, it entails monitoring parameters directly at the site. It is worth noting that, the former usually demands additional efforts to ensure its success. For instance, in certain cases, data collection involves taking samples for subsequent laboratory analysis, necessitating the preservation of samples under specific environmental conditions (Breuer, 2015). Alternatively, water quality variables are measured in the field using simple test kits by taking samples.

In the realm of water level monitoring, a recent innovative approach is the implementation of open water levels (OWLs), a mobile application developed by Melanie Elias (2021). OWL utilizes a photogrammetric method within the app to detect water level measurements from photographs captured with a smartphone's camera. The captured images are georeferenced and overlaid onto a 3D surface model, allowing for the development of a spatiotemporal densification of hydrometric networks which in combination with weather data could prevent from flood risk.

2.3 Air quality

Air quality is a significant research area that involves the collection of sensing data through low-cost monitoring sensors (Gupta et al., 2018; Huang, 2019). Similar to the aforementioned application areas, one of the key objectives of these studies is to overcome the challenge of limited spatial coverage posed by static monitoring stations. Given the environmental and health effects of air pollution (Samulowska et al., 2021), there is a growing interest in data monitoring within this field. By actively participating in air quality data collection, individuals can raise awareness regarding the impact of air pollution on health, influencing people's exposure during outdoor activities and making them aware of the potential consequences for their well-being.

The primary objective is to measure the concentration levels of air pollutants, including those resulting from anthropogenic activities and natural processes such as pollen, dust storms, forest fires, and volcanic emissions. By assessing the interplay of both human-induced and natural factors, it is possible to determine the air quality of the region through local data collection.

2.4 Noise pollution

Noise pollution, along with air pollution, are two prevailing challenges in urban environments that exert a substantial influence on public health and overall well-being. The construction and operation of stationary monitoring units come at a high cost, making mobile sensing more prevalent in applications that demand dense spatial data and extensive area coverage, such as monitoring noise pollution. These applications leverage the widespread availability of mobile phones equipped with microphones and GPS receivers to enable users in capturing real-time noise data in different urban environments (Becker et al., 2013; Leao et al., 2014; Maisonneuve et al., 2010). With the proliferation of smartphones, individuals can download dedicated applications that turn their devices into noise monitoring tools. The collected data, along with geolocation information, can be shared with centralized platforms or research initiatives, creating extensive and dynamic noise maps (Das et al., 2019; Gelb & Apparicio, 2019). In addition to smartphones, external sensors have also been developed to enhance the accuracy and versatility of noise monitoring (Shakya et al., 2019). These sensors can be attached to mobile devices or integrated into wearable gadgets, providing more precise measurements and expanding the range of noise sources that can be recorded.

Crowdsourcing sensing data for noise pollution has proven to be an effective means of capturing a wide range of noise sources and their variations across different locations and times. This collective effort not only provides comprehensive noise data but also encourages citizen engagement and awareness about noise pollution issues. Aggregated and analyzed crowdsourced data can aid urban planners, researchers, and policymakers in making informed decisions related to noise mitigation strategies, urban design, and transportation planning. The wealth of data generated through crowdsourcing enables a deeper understanding of noise patterns, identification of noise hotspots, and evaluation of the effectiveness of noise reduction initiatives.

Here, several examples of crowdsourcing applications for monitoring various physical variables in urban areas were presented. These examples underscore the value of citizen science, particularly in terms of data collection. Despite concerns about data quality, ongoing progresses in data validation techniques, such as comparing crowdsensing data with official reference data (Chen et al., 2021), and methods to reduce potential bias (Samulowska et al., 2021), might contribute to the development of methodologies for addressing inaccuracies in crowdsourced data.

The integration of crowdsourced data with information from fixed sensors, along with appropriate filtering and processing methods, holds the key to significantly enhancing the overall reliability and accuracy of environmental information. By bridging spatial and temporal gaps in official monitoring networks, crowdsourcing data provides a more comprehensive view of environmental conditions. This extended coverage allows for the identification of localized variations, contributing to improved models, mapping, and predictions of environmental events at both small and large scales.

3. Monitoring social aspects in urban environments

When it comes to social monitoring, crowdsourcing data offers flexibility in terms of collection methods and the type of information gathered. The data primarily consist of qualitative insights rather than quantifiable metrics. The methods employed can vary, including social media analysis, participatory mapping, or citizen reporting through online platforms or mobile applications. By harnessing crowdsourcing data from local communities, valuable knowledge can be extracted, social trends can be uncovered, and emerging issues can be identified to inform policy implementation. Furthermore, although the gathered data are predominantly descriptive and nonmetric, in some cases, it can also complement quantitative datasets, enhancing the overall analysis.

3.1 Urban planning

Urban planning is undergoing a rapid transformation with the widespread application of crowdsourcing. Collaborative planning is progressively supplanting the conventional "top-down" approach observed in numerous countries, where authorities were solely responsible for decisions and directives. The utilization of crowdsourced data and knowledge presents a multitude of possibilities for application in urban planning, effectively shifting the decision-making process toward a data-driven development approach. According to a study by Liao et al. (2019), an analysis of crowdsourcing and urban planning research over the past decade revealed notable trends. Initially, the primary focus was on gathering data through diverse channels like social media, web platforms, and mobile phones. Subsequently, these extensive data were analyzed using machine learning algorithms. In the mid-2010s, researchers began placing greater emphasis on integrating citizen participation into the planning process, highlighting the management aspects involved. Additionally, the research observed a broadening scope of application over time, transitioning from localized studies to encompass more expansive urban areas within cities.

Within urban planning, social media is a valuable data source that gather information that is relevant to the study of social dynamics and interactions. Its analytics are centered around extracting insights that help identify patterns, facilitating a deeper understanding of group behavior, opinion and needs (Zachlod et al., 2022). This is especially crucial in the field of urban planning, where the focus lies in gaining knowledge about people's needs and thoughts regarding their environment. By engaging with local residents, who possess a better understanding of the prevailing situation, problems, and issues that require resolution, social media aids in addressing these challenges.

Social media can prove beneficial in four out of the five stages of the collaborative process: assessment, education, resolution, and implementation, leaving the organization stage as an exception (Lin, 2022). Within these stages, users can assist stakeholders by offering unique perspectives in problem identification and delimitation. They can facilitate the sharing of interests and ideas among community members. Additionally, it enables effective monitoring of solution implementation. However, it is important to keep in mind that the openness of social media platforms exposes a diverse and heterogeneous range of data, including written posts, comments, images, and videos. As a result, it becomes crucial to carefully manage the process of data filtering, selection, and analysis.

In this sense, it is important to differentiate between opinion and knowledge. From a critical perspective, Yankelovich proposed "that the quality of public opinion be considered good when the public accepts responsibility for the consequences of its views and poor when the public, for whatever reason, is unprepared to do so" (Yankelovich, 1991, p. 24). He defined mass opinion as a lower-quality form of public opinion characterized by inconsistency, volatility, and nonresponsibility. On the other hand, public judgment refers to the higher-quality opinion marked by stability, consistency, and a sense of responsibility. Although adopting a critical approach to establish a measure of the quality of public opinion, may not be valid in all terms, could be an initial approach to overcome the complex challenge of quality assessment.

3.2 Mobility

Another application of crowdsourcing within urban areas lies in the field of mobility. One of the mayor efforts is to provide updates on traffic conditions by capturing location and speed data using GPS and accelerometers through ubiquitous measurements done unawareness by the owner of a smartphone. This approach aligns with the concept of opportunistic sensing, as described by Campbell et al. (2008), where there is no direct involvement or interaction between the sensing system and the individual, in contrast to participatory sensing. When it comes to monitoring social aspects within mobility, the emphasis tends to lean toward participatory sensing. By integrating this method into traffic management, data collection surpasses the mere reporting of vehicle congestion. It encompasses capturing insights on traffic dynamics, event reporting, short-lived traffic incidents, as well as identifying instances of aggressive and risky driving behaviors (Predic & Stojanovic, 2015).

Current research has specifically targeted pedestrians to improve mobility in urban environments. This research emphasizes crucial factors such as the creation of pedestrian networks to enable efficient navigation (Zhou et al., 2021), promoting accessibility for inclusive design (Mobasheri et al., 2017), and optimizing infrastructure and urban features (Saha, 2019). One example is the OpenSidewalks (n.d.) platform launched by the Taskar Center for Accessible Technology at the University of Washington, this initiative serves as a collaborative pedestrian-centric feature mapping application, such as sidewalks. OpenSidewalks aims to address the inclusivity and accessibility of urban spaces.

Other studies are dedicated to exploring public transportation with the goal of fostering a more democratic and inclusive mobility planning process. These studies primarily concentrate on the initial phases of planning, including problem identification and conducting in-depth analyses of the current state of affairs. They emphasize the significance of public participation in order to ensure the success of transportation decision-making (Giuffrida et al., 2019).

Through collaborative efforts, researchers attempt to gain a comprehensive understanding of passenger flows, optimize route planning, and enhance overall efficiency and accessibility in public transportation systems to enhance urban design for a more user-centric experience.

3.3 Perception

Crowdsourcing has also been employed to gather information about aspects of people's perception, landscape preferences and behavior, addressing a wide range of topics aimed at obtaining data on their vision and understanding of their surroundings (Bubalo et al., 2019). These practices focus on collecting subjective data derived from individual cognitive processes. In line with other crowdsourcing applications, participants can engage in active or passive collaboration. In certain applications, users are requested to provide information through surveys using web platforms or mobile applications, and in other cases, data are extracted directly from social media using semantic analysis and natural language processing techniques, which involve analyzing tags or comments to extract information regarding emotions, perception, and opinion.

One application involves assessing people's perception of safety and fear of crime. One on-site method used for collecting data about the spatiotemporal context is ecological momentary assessments (Chataway et al., 2017). This approach allows for the evaluation of how different locations impact individuals' perception of crime. In the same regard, there are different methods for data gathering, like through collaborating mapping through digital sketch maps to identify the crime perception gap (Vallejo et al., 2020), that is, the disparity between perception and reality that could retract from the quality of life, change social behavior and spatial dynamics.

3.4 Vulnerability and risk assessment

Crowdsourcing has also emerged as a tool in natural hazard events within urban environments, not only for gathering data about physical variables but also for assessing the social impact of such events. This approach aims for more effective strategies and interventions to be developed to enhance the resilience and preparedness of urban areas.

The 2010 earthquake in Haiti marked a significant turning point in this regard. It transformed Ushahidi, originally a blog focused on reporting postelection violence in Kenya in 2008, into a crisis mapping platform. Through the utilization of free text messages, Haitian individuals were able to communicate their urgent needs, provide guidance on aid distribution, and share crucial information about the prevailing conditions in specific areas. These messages included location data, facilitating the coordination of humanitarian responses. Volunteers meticulously georeferenced the incoming reports, which were subsequently compiled on the platform for further on-ground action. This intricate initiative involved the collaboration of diverse governmental agencies, volunteers, authorities, and developers. Ushahidi has garnered recognition as a potent tool for real-time reporting, mapping, and visualization of crises or events (Norheim-Hagtun & Meier, 2010). Over a decade since its inception, Ushahidi has been deployed in more than 160 countries, with contributions from citizens worldwide, resulting in over 50 million reports (Ushahidi, 2018).

In the field of crowdsourcing applied to natural hazards, two primary types of data capture can be identified. The first type involves the collection of real-time information, both on-site and remotely, during hazardous events. This real-time data capture intends to provide immediate situational awareness and enable timely response and decision-making. The process of gathering on-event data front challenges due to the inherent risks and chaotic nature of the scenarios and will depend on the type and scale of the event. Technical difficulties often arise, particularly in situations where internet connectivity is inadequate or limited. These challenges can hinder the seamless collection and transmission of data from the affected areas.

The second type focuses on gathering information to enhance the planning and implementation of measures aimed at mitigating. This data collection effort seeks to gather insights and knowledge that can inform risk assessment, hazard mapping, and the development of effective strategies and policies. Within urban areas, the emphasis is placed on extracting and analyzing urban features to establish correlations with vulnerability to natural processes (Salvati et al., 2021) and conduct risk assessment (Maragno et al., 2020).

The application of crowdsourcing for social aspects of urban environments has shown its capacity to augment traditional approaches with collaborative methodologies. The integration of citizen participation allows to address complex urban challenges with more effective strategies and interventions in urban planning and management. Crowdsourcing has become a powerful asset in enhancing urban development, enabling public engagement, and fostering data-driven decision-making.

4. Limitation and challenges

Crowdsourcing has served as a flexible framework across various disciplines, adapting to the unique needs and workflows of each field, spanning from business to scientific endeavors. In the realm of geosciences, the availability of georeferenced data has opened up a load of opportunities for active and passive crowdsourcing within the urban research domain. Crowdsourcing is continuously evolving and the pace of technological developments introduces a dynamic set of challenges and limitations.

One of the primary challenges in active crowdsourcing is ensuring the **representativeness of contributions**, particularly in open-call online scenarios where there is limited control over the selection of participants (Brambilla & Pedrielli, 2020; Nelson et al., 2021; Niu & Silva, 2020). In many instances, these efforts result in nonprobability samples, which can lead to an overrepresentation or underrepresentation of specific age groups due to the digital divide. Consequently, the collected data may not adequately reflect the diversity of the actual population. When crowdsourcing is being used as a web survey, the sampling method must be chosen based on the objectives and target population to reduce errors in the results. Additionally, there are statistical correction methods that can be employed to mitigate the impact of sampling bias (Bethlehem, 2010), but still there will always be the exclusion of the population that has no access to a device with Internet connection or to the Internet itself.

As crowdsourcing research continues to grow, it is important to note the **geographical bias** of the current studies that focus on online platforms. Most of these platforms are based in North America and Europe, while only a fraction of them (10%) can be regarded as international (Fornaroli & Gatica-Perez, 2023). Hence, global participation and generalizability of findings is currently restricted. At the same time, other parts of the world can greatly benefit from crowdsourcing due to their expanding population and urbanization trends, such as, for example, the Global South (GS). Diop et al. (2022) examined crowdsourcing studies in the GS and suggest avoiding digital colonialism by collaborating with foreign experts, including a diverse population, and using accessible platforms. These solutions would prevent challenges of successful adoption of crowdsourcing in the GS and potentially minimize the existing geographical bias.

Moreover, **low participation rates** in web questionnaires present a significant challenge. Unlike face-to-face crowdsourcing methods, self-administered surveys are more susceptible to nonresponse. Although, research has shown that the provision of incentives can effectively increase participation rates, what sets geographical research apart is its inherent social impact. As a result, people tend to demonstrate a heightened willingness to contribute to activities that directly benefit the society or their communities (Jacobs et al., 2019). This can be an important factor in overcoming the challenges of low participation rates.

Data validation and quality must be also considered, as errors may emerge during the data collection process of acquiring and recording responses. In the case of crowdsensing, measurement errors may result from factors such as instrument calibration, the selection of the sampling site, or the sampling process itself. An advantage here lies in the fact that these measurements typically come with established parameters or valid ranges, for instance, weather monitoring with personal stations, making it relatively straightforward to identify erroneous samples. In other cases, such as when extracting geosocial media data, careful data cleaning, preprocessing, and examination is required and can be implemented via a Geo-topic-sentiment analytical framework (Wang et al., 2021). Generally speaking, some degree of uncertainty always remains in the sampling or data extractions procedure.

One of the advantages of crowdsourcing methods that involve close collaboration, such as workshops or face-to-face interactions, is that participants are guided through the data collection process. This guidance not only enhances data quality but also fosters a deeper understanding of the task's instructions. Participants can seek immediate clarification, leading to a more robust and accurate dataset. In contrast, online questionnaires and web platforms introduce a different **set of uncertainties**. Here, the level of uncertainty extends to how well participants comprehended the instructions for the task, including the utilization of data gathering techniques, especially in the context of spatial data. For instance, when individuals are asked to identify locations on a map, there is an inherent degree of uncertainty regarding the spatial accuracy of each response. This uncertainty can be attributed to factors such as the participant's geographical knowledge, the clarity of map symbols, and their interpretation of the task's spatial requirements. Furthermore, online crowdsourcing may introduce additional layers of uncertainty when considering the diversity of participants. Different levels of digital literacy, language proficiency, and cultural backgrounds can all contribute to variations in how participants engage with and interpret the tasks. Recognizing and managing these uncertainties becomes essential in ensuring the reliability of the collected data.

As the foundation of crowdsourcing is a collaborative work between the organizer and the participants, the core element that drives the success of crowdsourcing initiatives is active and **meaningful engagement**. Participants find engagement through the sense of purpose, intrinsic motivation, and the actively participating in research or projects they care about serve as driving forces. Additionally, the engagement encourages a sense of ownership, as participants feel like cocreators and collaborators rather than mere data providers. However, a common practice to guarantee engagement has been via incentives (e.g., monetary) or gaming mechanisms (Bastardo et al., 2023). Nevertheless, we shall reflect on how efforts to ensure a high participation rate of contributors may or may not affect the level of engagement, impact on the quality of responses, and thus on the results. Both organizers and participants must actively foster an environment that promotes ongoing interaction, communication, and mutual benefits. Clear communication channels, feedback loops, and well-defined objectives are crucial in maintaining engagement.

5. Conclusions

The knowledge of the crowd in collaboration with specialists and authorities has opened up the possibility of working together to improve the conditions of urban areas. Cities are complex spaces with both social and physical local dynamics. The knowledge of individuals as experts has allowed for the exploration of details in these microurban environments. While this is not extensive worldwide due to the diversity and distinctiveness that exists across different countries and regions, as these ideas and concepts can vary significantly because of the unique set of circumstances and contextual factors that shape their social dynamics and urban geographies.

Although various studies from different disciplines have confirmed that crowdsourcing is a useful source of knowledge and data, as long as it is appropriately utilized through preprocessing procedures and following methodologies that allow for quality evaluation, skepticism about its value still persists. Crowdsourcing can be viewed as an innovative concept that necessitates addressing challenges related to data bias, quality, and completeness. Consequently, many studies have opted to treat crowdsourcing as a complementary dataset to a primary data source, if available, especially when working with crowdsensing data.

Furthermore, the issue of citizen participation remains unresolved, necessitating the exploration of ways to involve a larger number of individuals in an accountable manner, particularly when it comes to active participation, which fosters a sense of personal responsibility among participants. Additionally, the aspect of data privacy has yet to be adequately addressed.

The engagement of the crowd not only enables a broader spatial and temporal coverage in the information provided but also enhances the overall scope of available knowledge. By harnessing the collective wisdom and contributions of a diverse group of individuals, a more comprehensive understanding of various phenomena can be attained.

References

Bales, E., Nikzad, N., Quick, N., Ziftci, C., Patrick, K., & Griswold, W. G. (2019). Personal pollution monitoring: Mobile real-time air quality in daily life. *Personal and Ubiquitous Computing, 23*, 309–328.

Bastardo, R., Pavão, J., & Rocha, N. P. (2023). Crowdsourcing technologies to promote citizens' participation in smart cities, a scoping review. *Procedia Computer Science, 219*, 303–311.

Becker, M., Caminiti, S., Fiorella, D., Francis, L., Gravino, P., Haklay, M. M., Hotho, A., Loreto, V., Mueller, J., Ricchiuti, F., Servedio, V. D. P., Sîrbu, A., & Tria, F. (2013). Awareness and learning in participatory noise sensing. *PLoS One, 8*(12), e81638.

Bethlehem, J. (2010). Selection bias in web surveys. *International Statistical Review, 78*(2), 161–188. https://doi.org/10.1111/j.1751-5823.2010.00112.x

Brambilla, G., & Pedrielli, F. (2020). Smartphone-based participatory soundscape mapping for a more sustainable acoustic environment. *Sustainability, 12*(19), 7899.

Breuer, L., Hiery, N., Kraft, P., Bach, M., Aubert, A. H., & Frede, H. G. (2015). HydroCrowd: A citizen science snapshot to assess the spatial control of nitrogen solutes in surface waters. *Scientific Reports, 5*. https://doi.org/10.1038/srep16503

Brown, G. (2015). Engaging the wisdom of crowds and public judgement for land use planning using public participation geographic information systems. *Australian Planner, 52*(3), 199–209. https://doi.org/10.1080/07293682.2015.1034147

Bubalo, M., van Zanten, B. T., & Verburg, P. H. (2019). *Crowdsourcing geo-information on landscape perceptions and preferences: A review*. Landscape and urban planning. Elsevier B.V. https://doi.org/10.1016/j.landurbplan.2019.01.001

Campbell, A. T., Eisenman, S. B., Lane, N. D., Miluzzo, E., Peterson, R. A., Lu, H., … Ahn, G. S. (2008). The rise of people-centric sensing. *IEEE Internet Computing, 12*(4), 12–21. https://doi.org/10.1109/MIC.2008.90

Chataway, M. L., Hart, T. C., Coomber, R., & Bond, C. (2017). The geography of crime fear: A pilot study exploring event-based perceptions of risk using mobile technology. *Applied Geography, 86*, 300–307.

Chen, J., Saunders, K., & Whan, K. (2021). Quality control and bias adjustment of crowdsourced wind speed observations. *Quarterly Journal of the Royal Meteorological Society, 147*(740), 3647–3664. https://doi.org/10.1002/qj.4146

Das, P., Talukdar, S., Ziaul, S., Das, S., & Pal, S. (2019). Noise mapping and assessing vulnerability in meso level urban environment of Eastern India. *Sustainable Cities and Society, 46*, 101416.

Diop, E. B., Chenal, J., Tekouabou, S. C. K., & Azmi, R. (2022). Crowdsourcing public engagement for urban planning in the Global South: Methods, challenges and suggestions for future research. *Sustainability, 14*(18), 11461.

Elias, M. (2021). *On the use of smartphones as novel photogrammetric water gauging instruments. Developing tools for crowdsourcing water levels.* PhD thesis. Technische Universität Dresden.

Faka, A., Tserpes, K., & Chalkias, C. (2021). Environmental sensing. In G. P. Petropoulos, & P. K. Srivastava (Eds.), *GPS and GNSS Technology in geosciences* (pp. 199–220). Elsevier.

Fornaroli, A., & Gatica-Perez, D. (2023). Urban crowdsourcing platforms across the world: A systematic review. *Digital Government: Research and Practice, 4*(3), 1–19.

Gelb, J., & Apparicio, P. (2019). Noise exposure of cyclists in Ho Chi Minh city: A spatio-temporal analysis using non-linear models. *Applied Acoustics, 148*, 332–343.

Giuffrida, N., Le Pira, M., Inturri, G., & Ignaccolo, M. (2019). Mapping with stakeholders: An overview of public participatory GIS and VGI in transport decision-making. *ISPRS International Journal of Geo-Information, 8*(4). https://doi.org/10.3390/ijgi8040198

Gupta, S., Pebesma, E., Degbelo, A., & Costa, A. C. (2018). Optimising citizen-driven air quality monitoring networks for cities. *ISPRS International Journal of Geo-Information, 7*(12). https://doi.org/10.3390/ijgi7120468

Hall, T., & Barrett, H. (2018). *Urban geography.* Taylor and Francis. https://doi.org/10.4324/9781315652597

Hamrouni, A., Ghazzai, H., Frikha, M., & Massoud, Y. (2020). A spatial mobile crowdsourcing framework for event reporting. *IEEE Transactions on Computational Social Systems, 7*(2), 477–491.

Howe, J. (2006). *Crowdsourcing: A definition.* Retrieved from http://crowdsourcing.typepad.com/cs/2006/06/crowdsourcing_a.html.

Hu, K., Rahman, A., Bhrugubanda, H., & Sivaraman, V. (2017). HazeEst: Machine learning based metropolitan air pollution estimation from fixed and mobile sensors. *IEEE Sensors Journal, 17*(11), 3517–3525.

Huang, J., Duan, N., Ji, P., Ma, C., Hu, F., Ding, Y., … Sun, W. (2019). A crowdsource-based sensing system for monitoring fine-grained air quality in urban environments. *IEEE Internet of Things Journal, 6*(2), 3240–3247. https://doi.org/10.1109/JIOT.2018.2881240

Jacobs, L., Kabaseke, C., Bwambale, B., Katutu, R., Dewitte, O., Mertens, K., … Kervyn, M. (2019). The geo-observer network: A proof of concept on participatory sensing of disasters in a remote setting. *Science of the Total Environment, 670*, 245–261. https://doi.org/10.1016/j.scitotenv.2019.03.177

Kanjo, E. (2010). NoiseSPY: A real-time mobile phone platform for urban noise monitoring and mapping. *Mobile Networks and Applications, 15*(4), 562–574.

Leao, S., Ong, K.-L., & Krezel, A. (2014). 2Loud?: Community mapping of exposure to traffic noise with mobile phones. *Environmental Monitoring and Assessment, 186*(10), 6193–6206.

Liao, P., Wan, Y., Tang, P., Wu, C., Hu, Y., & Zhang, S. (2019). Applying crowdsourcing techniques in urban planning: A bibliometric analysis of research and practice prospects. *Cities, 94*, 33–43.

Lin, Y. (2022). Social media for collaborative planning: A typology of support functions and challenges. *Cities, 125*. https://doi.org/10.1016/j.cities.2022.103641

Maisonneuve, N., Stevens, M., & Ochab, B. (2010). Participatory noise pollution monitoring using mobile phones. *Information Polity, 15*(1–2), 51–71.

Maragno, D., Fontana, M. D., & Musco, F. (2020). Mapping heat stress vulnerability and risk assessment at the neighborhood scale to drive Urban adaptation planning. *Sustainability, 12*(3). https://doi.org/10.3390/su12031056

Met Office. (2022). *Crowdsourcing for weather and climate monitoring.* Retrieved from https://blog.metoffice.gov.uk/2022/07/07/crowdsourcing-for-weather-climate-monitoring/.

Mobasheri, A., Deister, J., & Dieterich, H. (2017). Wheelmap: The wheelchair accessibility crowdsourcing platform. *Open Geospatial Data, Software and Standards, 2*(1). https://doi.org/10.1186/s40965-017-0040-5

Nelson, T., Ferster, C., Laberee, K., Fuller, D., & Winters, M. (2021). Crowdsourced data for bicycling research and practice. *Transport Reviews, 41*(1), 97–114.

Nipen, T. N., Seierstad, I. A., Lussana, C., Kristiansen, J., & Hov, Ø. (2020). Adopting citizen observations in operational weather prediction. *Bulletin of the American Meteorological Society, 101*(1), E43–E57. https://doi.org/10.1175/BAMS-D-18-0237.1

Niu, H., & Silva, E. A. (2020). Crowdsourced data mining for urban activity: Review of data sources, applications, and methods. *Journal of Urban Planning and Development, 146*(2), 04020007.

Njue, N., Stenfert Kroese, J., Gräf, J., Jacobs, S. R., Weeser, B., Breuer, L., & Rufino, M. C. (2019). *Citizen science in hydrological monitoring and ecosystem services management: State of the art and future prospects. Science of the total environment.* Elsevier B.V. https://doi.org/10.1016/j.scitotenv.2019.07.337

NOAA. (2016). *mPING weather app goes global.* Retrieved from https://inside.nssl.noaa.gov/nsslnews/2016/01/mping-weather-app-goes-global/.

Norheim-Hagtun, I., & Meier, P. (2010). Crowdsourcing for crisis mapping in Haiti. *Innovations: Technology, Governance, Globalization, 5*(4), 81–89. https://doi.org/10.1162/inov_a_00046

OpenSidewalks. (n.d.). Retrieved from https://www.opensidewalks.com/.

Predic, B., & Stojanovic, D. (2015). Enhancing driver situational awareness through crowd intelligence. *Expert Systems with Applications, 42*(11), 4892–4909. https://doi.org/10.1016/j.eswa.2015.02.013

Saha, M., Saugstad, M., Maddali, H. T., Zeng, A., Holland, R., Bower, S., … Froehlich, J. (2019). Project sidewalk: A web-based crowdsourcing tool for collecting sidewalk accessibility data at scale. In *Conference on human factors in computing systems - proceedings.* Association for Computing Machinery. https://doi.org/10.1145/3290605.3300292

Salvati, P., Ardizzone, F., Cardinali, M., Fiorucci, F., Fugnoli, F., Guzzetti, F., … Vujica, I. (2021). Acquiring vulnerability indicators to geo-hydrological hazards: An example of mobile phone-based data collection. *International Journal of Disaster Risk Reduction, 55.* https://doi.org/10.1016/j.ijdrr.2021.102087

Samulowska, M., Chmielewski, S., Raczko, E., Lupa, M., Myszkowska, D., & Zagajewski, B. (2021). Crowdsourcing without data bias: Building a quality assurance system for air pollution symptom mapping. *ISPRS International Journal of Geo-Information, 10*(2). https://doi.org/10.3390/ijgi10020046

Schneider, P., Castell, N., Vogt, M., Dauge, F. R., Lahoz, W. A., & Bartonova, A. (2017). Mapping urban air quality in near real-time using observations from low-cost sensors and model information. *Environment International, 106*, 234–247. https://doi.org/10.1016/j.envint.2017.05.005

Shakya, K. M., Kremer, P., Henderson, K., McMahon, M., Peltier, R. E., Bromberg, S., & Stewart, J. (2019). Mobile monitoring of air and noise pollution in Philadelphia neighborhoods during summer 2017. *Environmental Pollution, 255*, 113195.

Shen, Y., & Karimi, K. (2016). Urban function connectivity: Characterisation of functional urban streets with social media check-in data. *Cities, 55*, 9–21.

Susanto, T. D., Diani, M. M., & Hafidz, I. (2017). User acceptance of e-government citizen report system (a case study of City113 app). *Procedia Computer Science, 124*, 560–568.

Ushahidi. (2018). *10 years of global impact. An impact report celebrating 10 years of Ushahidi.* Retrieved from https://www.ushahidi.com/in-action/10-years-of-global-impact/.

Vallejo, M., Kounadi, O., & Podor, A. (2020). Analysis and mapping of crime perception: A quantitative approach of sketch maps. *AGILE: GIScience Series, 1*, 1–18. https://doi.org/10.5194/agile-giss-1-20-2020

Wang, R.-Q., Mao, H., Wang, Y., Rae, C., & Shaw, W. (2018). Hyper-resolution monitoring of urban flooding with social media and crowdsourcing data. *Computers & Geosciences, 111*, 139–147.

Wang, Y., Gao, S., Li, N., & Yu, S. (2021). Crowdsourcing the perceived urban built environment via social media: The case of underutilized land. *Advanced Engineering Informatics, 50*, 101371.

Xu, Z., Liu, Y., Yen, N. Y., Mei, L., Luo, X., Wei, X., & Hu, C. (2020). Crowdsourcing based description of urban emergency events using social media big data. *IEEE Transactions on Cloud Computing, 8*(2), 387–397.

Yang, P., & Ng, T. L. (2017). Gauging through the crowd: A crowd-sourcing approach to urban rainfall measurement and storm water modeling implications. *Water Resources Research, 53*(11), 9462–9478. https://doi.org/10.1002/2017WR020682

Yankelovich, D. (1991). *Coming to public judgment: Making democracy work in a complex world*. Syracuse University Press.

Zachlod, C., Samuel, O., Ochsner, A., & Werthmüller, S. (2022). Analytics of social media data — state of characteristics and application. *Journal of Business Research, 144*, 1064—1076. https://doi.org/10.1016/j.jbusres.2022.02.016

Zheng, F., Tao, R., Maier, H. R., See, L., Savic, D., Zhang, T., … Popescu, I. (2018). *Crowdsourcing methods for data collection in geophysics: State of the art, issues, and future directions. Reviews of geophysics*. Blackwell Publishing Ltd. https://doi.org/10.1029/2018RG000616

Zhou, B., Zheng, T., Huang, J., Zhang, Y., Tu, W., Li, Q., & Deng, M. (2021). A pedestrian network construction system based on crowdsourced walking trajectories. *IEEE Internet of Things Journal, 8*(9), 7203—7213. https://doi.org/10.1109/JIOT.2020.3038445

SECTION 5

Atmosphere/air environment

CHAPTER 20

Revealing air quality dynamics: Leveraging remote sensing and GIS for active monitoring and risk evaluation

Hamaad Raza Ahmad[1], Khalid Mehmood[2], Saifullah[1], Sadia Bibi[1], Muhammad Hassan Bashir[1] and Ayesha Siddique[1]

[1]Institute of Soil and Environmental Sciences, University of Agriculture, Faisalabad, Pakistan; [2]Institute of Environmental Health and Ecological Security, School of the Environment and Safety Engineering, Jiangsu University, Zhenjiang, Jiangsu, China

1. Introduction

Globally, air pollution is a leading public health concern and a major environmental issue and attaining growing interest of the scientific community (Dominski et al., 2021; Landrigan, 2017). The World Health Organization (WHO) estimates that fine particulate matter ($PM_{2.5}$) contributes to 4.2 million premature deaths in the world each year (WHO, 2018). It is apparent that citizens and policymakers are becoming increasingly aware of the importance of maintaining high air quality standards, but the current levels of air pollution are still above the limits set by the European Union and the WHO (EEA, 2019). The poor air quality especially among large urban agglomerations is of particular concern for developing countries and regions (Baklanov et al., 2016; Schneider et al., 2017). It is therefore crucial to formulate a comprehensive plan to address poor air quality. The development of such strategies requires a better understanding of pollution phenomena based on empirical evidence, specifically the observation of air quality on the ground levels (Demuzere et al., 2019; Kumar et al., 2015). This type of data and information can be used to increase public awareness. It can also help to individuals already aware of the problem to plan their daily activities in a manner that minimizes their exposure to different levels of particulate matter (PM) pollution.

Air quality observations focus primarily on monitoring airborne pollutants as these could have significant impacts on human health. Monitoring stations are either nonfunctioning, or largely absent to collect air pollutant data in many developing-countries, especially in Asia and Africa (Garcia et al., 2016). There are generally few ambient monitors even in high-income regions (Apte et al., 2017). It is frequently necessary to use fixed monitoring stations to accomplish this objective. Despite their exceptional precision, these monitoring stations cannot be deployed in abundance, primarily due to their high acquisition and maintenance costs. Therefore, these fixed monitoring stations have a limited spatial distribution of the data points concerning air quality (Santana

Geographical Information Science
ISBN 978-0-443-13605-4
https://doi.org/10.1016/B978-0-443-13605-4.00021-7

et al., 2021; Mehmood et al., 2021). Furthermore, such stations are equipped with expensive and bulky certified equipment, requiring trained personnel to maintain measurement accuracy. Unfortunately, environmental experts have experienced difficulty in determining the relationship between human activities, topography, and spatiotemporal distribution. The imprecise assessment of individual exposure to air pollution can produce biases especially in health risk assessments, which compromises epidemiological and toxicological studies' reliability (Peng & Bell, 2010).

The mapping of air quality is an essential tool for understanding and managing air pollution assessments. Through mapping, one can gain a better understanding of the distribution of air pollutants, identify areas with poor air quality, and assess the health and environmental impacts of air pollution through the collection and display of air quality data on a map. This chapter explores the critical elements of air quality by examining the sources of air pollutants and their effect on human health. Further, the commonly employed air quality procedures have been explored and introduced remote sensing methods used for air quality mapping, including features, advantages, and limitations of the various kinds of remote sensors. Moreover, geographic information systems (GIS) and their pivotal role in air quality mapping, emphasizing their significance in developing effective management strategies to mitigate air quality concerns have been discussed thoroughly.

2. Earth observation and GIS in science: Advantages and limitations

Earth observation (EO) and GIS are increasingly important for scientific and environmental research (Jiang et al., 2016). Integrating EO and GIS technology is advantageous for providing comprehensive, up-to-date geospatial information about the Earth's surface. EO satellites and sensors collect high-resolution images and data that allow scientists to monitor and analyze land cover changes, urban sprawl, deforestation, and climate patterns on a regional, national, and global scale (Woldai, 2020; Zhao et al., 2022). In contrast, GIS enables researchers to manage spatial data, analyze it, and visualize it to make informed decisions regarding natural resources, disaster planning, urban planning, and conservation (Psomiadis et al., 2020).

The air pollutants across multiple geographic locations are essential for spatial air quality mapping. This typically involves using computer models, sensor networks, and satellite data. Sources, including ground-based monitoring stations, remote sensing platforms, and mobile sensors, are used to collect information on pollutants such as $PM_{2.5}$, PM_{10}, NO_2, O_3, SO_2, and volatile organic compounds (VOCs). For instance, spatial mapping was conducted to analyze air quality across Europe. This type of geographically dispersed maps offers data for policymaking and comprehending the spatial characteristics of air quality trends (Denby et al., 2010). Huaman health, and ecosystems, are seriously endangered by air pollution, a critical environmental concern of our day. Using GIS, air quality data mapped to reveal important information about the geographical distribution of contaminants and their potential effects on human health through digitization. Data from Poland population and air pollution were analyzed in research by Gadka et al. using GIS. Their study found a link between PM concentrations and suicides, indicating that prolonged exposure to PM may raise the risk of mental health issues (Gładka et al., 2022). Yadav et al. (2022) employed GIS-based mapping in research to evaluate the geographic correlation between air pollution and emergency department visits for acute respiratory symptoms in Delhi, India. They discovered that greater $PM_{2.5}$ concentration levels were linked to increased ER visits for respiratory problems (Vos et al., 2015). In another GIS study, Verhaegh et al. (2019) examined the relationship between ambient air quality and microscopic colitis.

To analyze changes in the regional environment and assist regional environmental air quality management, the system merged GIS technology with B/S architecture (Yang et al., 2022). A study in Addis Abeba, Ethiopia, examined the relationship between urban air pollution, greenery, and public health (Bikis, 2023). The study employed GIS to map the availability of green space in the city and air quality levels ($PM_{2.5}$, PM_{10}, CO_2, AQI) at transportation hubs and manufacturing facilities. The findings revealed poor air quality levels around transport hubs and manufacturing facilities, with little green space in residential areas (Bikis, 2023).

Understanding the distribution and sources of air pollution through spatial air quality mapping using GIS has become more critical. A work by Ma & Tong (2022) provides an unprecedented attempt to map anthropogenic emissions of air pollutants throughout the contiguous United States (CONUS) at a 1 km spatial resolution. During 2017 National Emission Inventories from the United States Environmental Protection Agency (US EPA), the Neighborhood Emission Mapping Operation (NEMO) dataset was developed. The authors achieved fine-scale spatial allocation by distributing emission sources using spatial surrogates. NEMO offers fresh possibilities for investigations on environmental justice, exposure assessment, and fine-scale air quality modeling (Ma & Tong, 2022). Thepanondh and Toruksa (2011) researched ambient NO_2 concentrations in Rayong Province, Thailand, and assessed spatial variability and proximity analyses. They employed geostatistics and various interpolation methods to forecast yearly quantities of NO_2. Their findings showed the value of GIS-based prediction for assessing exposure mapping and identifying locations with high exposure levels and associated health problem (Mitmark & Jinsart, 2016). A GIS-based air quality map can then be created using the database to create accurate and detailed air quality maps. By spatially interpolating air quality measurements, continuous surface maps can be created that show pollution levels throughout entire regions. In areas without monitoring stations, GIS can also predict air quality by modeling pollutants' dispersion, taking into account factors such as wind patterns, terrain, and sources of emissions. Further, GIS-based air quality mapping can be used to identify pollution hotspots, assess exposure risks for vulnerable populations, and develop mitigation strategies that target those populations. As a result of the use of GIS for air quality mapping, we gain a deeper understanding of air pollution patterns.

Air pollution is the contamination of the indoor or outdoor environment by any chemical, physical, or biological agent that modifies the natural characteristics of the atmosphere (WHO, 2023). The sources of air pollution are multiple and context specific. The major anthropogenic outdoor pollution sources include residential energy for cooking and heating, vehicles, power generation, agriculture/waste incineration, and industry (Almetwally et al., 2020; Bai et al., 2019; Shi et al., 2019). Natural sources include wildfires, dust storms, and volcanic eruptions. Area sources include small businesses, gas stations, and residential wood combustion (Daellenbach et al., 2020). A majority of air toxics come from man-made sources, including motor vehicles, industrial facilities, and small "areas" of the environment. Air toxics are emitted by many types of stationary sources, including power plants, chemical plants, aerospace industries, and steel mills (US-EPA, 2012; WHO, 2023). A study conducted by Mehmood and colleagues on air quality of Iqbal Town, Faisalabad, Pakistan, concluded that the total suspended particulate (TSP) matter was more near the road and decreases with distance and the values were above the safe limit of 260 $\mu g/m^3$ proposed by US-EPA (2012) (Fig. 20.1).

Depending upon the industrial, transport, and agricultural development, different countries contribute differently in generating air pollution sources. This contribution may vary among countries or region. The transformation of industries and combustion in energy have caused 19% NO_2 and 61% SO_2 emissions, respectively. Similarly, about 4% of SO_2 and 49% NO_2 is contributed by road transport. Similar statistics about air pollution from road transport and industries were observed in the United Kingdom. However, in most Eastern European countries, road transportation contributed to about 10%—20% of NOx emissions due to lower traffic ratio (Anenberg et al., 2019). Since air pollution has no boundaries, and developing countries are paying less attention to manage air pollution, the levels of air pollutants in underdeveloped or developing countries are increasing with time. For example, GIS maps of aerosols density and opacity of Madian Town, Faisalabad, Pakistan, showed that the air quality of Faisalabad is poor mainly due to road traffic and industrial units (Fig. 20.2).

There are, however, some limitations associated with these technologies. EO data collection can be hindered by adverse weather conditions, cloud cover, and atmospheric conditions. The acquisition and processing of high-quality EO data can be expensive, posing a financial constraint for some applications. It is often difficult to manage GIS data and analyze it, as it requires specialized software and skilled personnel. Data collection and data processing errors can further affect the accuracy and precision of GIS analyses. However, EO and GIS technologies are still irreplaceable tools for understanding and addressing a wide range of Earth-related challenges, despite these limitations.

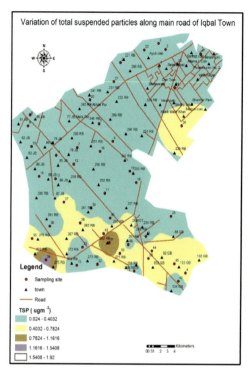
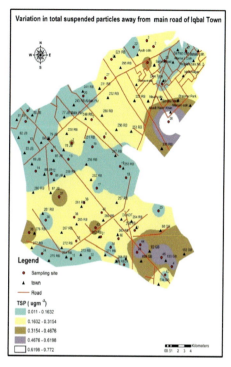

Figure 20.1 Spatial distribution of total suspended particulate matter in Madina town, Faisalabad, Pakistan. Maximum concentrations of total suspended particulate matter were observed around industrial clusters and busy junctions of the city (Mehmood, 2013).

3. Air quality parameters used for air quality mapping

Monitoring and mapping of air quality parameters not only helps to identify associated health risks that we can face but also their management to minimize the adverse effects. Following are the air quality parameters being used in air quality mapping to assess pollution level.

Particulate matter (PM): PM refers to suspended smaller particles present in air like dust, soot, and water droplets. According to US-EPA, these particles were categorized based on their size, as PM10 (10 μm or less) and $PM_{2.5}$ (2.5 μm or less). High volume air samples are used to collect the suspended particles present in the air. This instrument is used to determine TSP, PM_{10}, and $PM_{2.5}$ particulate concentrations. Portable low-volume air sampler is also used to collect PM pollution (Raynor et al., 2021).

Nitrogen dioxide (NO_2): The NO_2 is primarily emitted from vehicle exhaust and industrial processes. Urban traffic is the primary outdoor source, while indoor sources include cigarette smoke, kerosene oil, and coal burning (Daly & Zannetti, 2007). Several instruments are used to monitor NO_2 concentration in the air including gas sensors, diffusion tubes, chemiluminescence analyzers, photochemical converters, and satellite images.

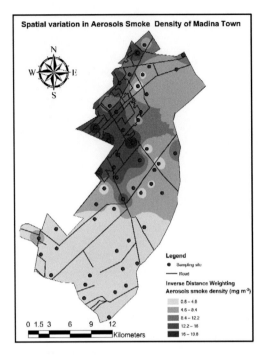
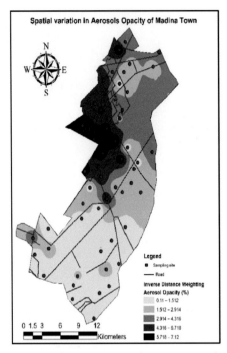

Figure 20.2 Spatial variation of aerosol smoke density and aerosols opacity in Madina town, Faisalabad, Pakistan. Maximum aerosol smoke density was found around busy commercial centers and traffic congested roads of the city (Khan, 2012).

Ozone (O_3): The O_3 has a dual nature; it protects earth from UV rays and becomes a harmful pollutant at the soil level, mainly originated from anthropogenic activities resulting from a series of reactions of nitrogen species with oxygen. Instruments used to monitor its concentration in the air are ozone analyzers and spectrometers. Ozone analyzers are specialized instruments designed to directly measure O_3 concentrations in ambient air. These instruments often employ electrochemical sensors or ultraviolet (UV) absorption techniques to detect and quantify different O_3 levels.

Carbon monoxide (CO): The CO is originating in the environment due to incomplete combustion of fossil fuels, and it becomes more toxic when reacting with other air pollutants, resulting in air quality degradation (Carter et al., 2017).

Sulfur dioxide (SO_2): The SO_2 emitted through power generation plants using sulfur-containing coals, vehicles running on diesel fuel, and volcanic eruptions degrade air quality (Santana et al., 2021). Commonly used instruments and methods for monitoring SO_2 are sulfur dioxide analyzer using UV absorption spectrum, gas chromatography, satellite remote sensing, and portable devices like WineScan SO_2 analyzer, SO_2 analyzer, and flue gas analyzer (Kumar et al., 2015).

4. Remote sensing and GIS techniques used for air quality mapping

Remote sensing is a powerful tool that makes use of electromagnetic radiations. With the help of these radiations, data on the status of air pollutants of various locations on the earth's planet can be easily recorded. In remote sensing, the status and geospatial variation in air pollutants can be captured by satellite images. Satellite images are recorded by satellite sensors that is the result of combined effects of the earth surface reflectance and atmosphere. The levels of air pollutants then can easily be correlated with other climatic parameters such as wind direction and speed, pressure, and temperature of air. Therefore, remote sensing has become a very popular technique with a range of extended benefits when compared with traditional air monitoring techniques. A great deal literature thus shows the increasing attention of scientific community in mapping air pollutants such as ozone and nitric oxides with the help of remote sensing. The digital maps not only help to indicate spatial and temporal variation but also the sources (power plants, industries, household, and vehicular) of air pollution. The information on sources of pollutants can help guide policymakers to control and minimize risk associated with air pollution.

Although the use of remote of sensing technique has seen an exponential increase, many challenges are there that can affect accuracy of data generated from the use of electromagnetic radiation. For example, tropospheric aerosols can significantly alter the earth radiation budget. However, quantification of the changes in radiation is difficult to monitor because atmospheric aerosol distributions vary considerably in type, size, space, and time (Koukouli, et al., 2022). Monitoring and modeling the subsequent risks, trends, and repercussions of air pollution are vitally critical steps in designing effective environmental and public health policies to decrease the harmful effects of air pollution. So, it is important to conduct precise exposure assessments at specific location or region. However, it is still challenging task to map the highly complex systems. There are few ground air quality monitoring stations in many parts of the world because of inadequate air quality monitoring infrastructure (Brauer et al. 2016). According to Forouzanfar et al. (2015), the "The Global Burden of Disease" (GBD) study, which used a regression calibration approach to integrate data from satellite remote sensing and chemical transport models. This allowed the researchers to develop a set of exposures metrics and capable of predicting a high-resolution structure (0.1 0.1), approximately 11×11 km at the equator. These exposure estimates were compared to population to estimate the impact of diseases. The GBD employed linear regression to calibrate a fused estimate of $PM_{2.5}$, which was derived by averaging satellite values and compound transfer patterns compared to areas taken on the ground. Measurements were regressed against this combined estimate and data from local monitoring systems for cells with a ground monitor (Brauer et al., 2016).

Similarly, Van Donkelaar et al. (2016) employed globally balanced analysis to improve measurement system of $PM_{2.5}$ allowed the calibration factors which varies significantly (between the observations and estimates) and used additional data on land utilization and the chemical composition of pollutants. Satellite-based aerosol optical depth (AOD) was used by Mohan et al. (2011) to model $PM_{2.5}$ in the Northeastern United States. In one study, Ali et al. (2021) used remote sensing-based evaluation to examine the environmental spatial heterogeneity of COVID-19 effects on air quality in Asia—Pacific top 20 urban areas before and during lockdown times. In comparison to prelockdown times, lockdown periods were associated with measurable decreases in AOD, SO_2, CO, and NO_2 concentrations (Abbas et al., 2021). Apte et al. (2017) conceived and deployed an accurate spatiotemporal monitoring and early warning platform for air pollution based on IPv6. In addition, several studies used machine learning approaches to model to map and forecast air quality (Garcia et al., 2016; Muhammad and Yan, 2015; Pak et al., 2005; Park et al., 2018). Overall, remote sensing techniques contributed in several way to map air quality features.

Remote sensing can also be used to assess how air quality regulations are being implemented and followed by emitters. The successfulness of air quality regulations can be easily evaluated with the help of remote sensing that provides data on air quality before and after regulations are put in place. Remote sensing can further be helpful in providing information to public on air quality status in a particular area. This information may be comprised of level, and variation in response to location, and time.

5. Conclusion

Research in the fields of environmental science, public health, and policymaking relies heavily on-air quality mapping. So, there are many limitations in usage to EO and GIS mapping like cloud cover, atmospheric interferences user expertise, and data fusion. We have highlighted its pivotal significance for understanding air pollution dynamics and its wide-ranging effects on human health and the environment through this comprehensive book chapter. In this concluding section, we summarize the key points and emphasize how important air quality mapping has become to society and science today. The air quality mapping process provides a comprehensive picture of pollutant concentrations across regions, transcending isolated measurements to provide a holistic understanding of air pollution dynamics. Taking on the complexities of modern air quality challenges requires this level of understanding. Using a combination of diverse data sources, advanced mathematical modeling techniques, and the central role of GIS, air quality mapping is accurate, precise, and actionable. As a result, human health and the environment are protected through evidence-based decision-making. Air quality mapping has a direct impact on public health, policy development, and environmental protection. These tools can be used to identify pollution hotspots and assess risk exposure to human health. Additionally, it contributes to better air quality regulation and management by guiding the formulation and evaluation of environmental policies. This approach transcends academic research and provides actionable strategies to combat air pollution and protect communities and ecosystems. Air quality mapping remains an invaluable tool in our quest for cleaner air and a healthier planet by upgrading data fusion and integration, satellite technology, and policy integration.

To address these challenges and enhance the effectiveness of air quality mapping through EO, a multifaceted approach is essential. Firstly, advancements in satellite technology should prioritize higher spatial and temporal resolutions, enabling a more granular and real-time assessment of air quality. Collaboration between space agencies, researchers, and policymakers is crucial to harmonize data collection methodologies and ensure the standardization of air quality monitoring practices globally. Integrating machine learning and artificial intelligence into data processing can improve the accuracy of EO-based air quality assessments by identifying patterns and trends that may be challenging for traditional methods. Furthermore, investing in a network of ground-based sensors and mobile applications for citizen science can complement satellite observations, providing localized and real-time data to validate and enhance the accuracy of EO-derived air quality maps. Public awareness campaigns can encourage active participation in such initiatives, fostering a sense of shared responsibility for environmental stewardship. In conclusion, the way forward in advancing EO for air quality mapping involves a combination of technological innovation, collaborative efforts, and community engagement. By addressing current challenges and adopting a holistic approach, we can leverage the full potential of EO to create more accurate, timely, and comprehensive air quality maps, ultimately contributing to effective environmental management and policy decision-making.

Funding

This work was financially supported by the Natural Science Foundation of China (No. 42350410448).

References

Abbas, J., Mubeen, R., Iorember, P. T., Raza, S., & Mamirkulova, G. (2021). Exploring the impact of COVID-19 on tourism: Transformational potential and implications for a sustainable recovery of the travel and leisure industry. *Current Research in Behavioral Sciences, 2*, 100033.

Ali, G., Abbas, S., Qamer, F. M., & Irteza, S. M. (2021). Environmental spatial heterogeneity of the impacts of COVID-19 on the top-20 metropolitan cities of Asia-Pacific. *Scientific Reports, 11*(1), 20339.

Almetwally, A. A., Bin-Jumah, M., & Allam, A. A. (2020). Ambient air pollution and its influence on human health and welfare: An overview. *Environmental Science and Pollution Research.* https://doi.org/10.1007/s11356-020-09042-2

Anenberg, S. C., Miller, J., Henze, D. K., Minjares, R., & Achakulwisut, P. (2019). The global burden of transportation tailpipe emissions on air pollution-related mortality in 2010 and 2015. *Environmental Research Letters, 14*(9), 094012.

Apte, J. S., Messier, K. P., Gani, S., Brauer, M., Kirchstetter, T. W., Lunden, M. M., … Hamburg, S. P. (2017). High-resolution air pollution mapping with google street view cars: Exploiting big data. *Environmental Science & Technology, 51*(12), 6999—7008.

Bai, L., Jiang, L., Yang, D-y, & Liu, Y-b (2019). Quantifying the spatial heterogeneity influences of natural and socioeconomic factors and their interactions on air pollution using the geographical detector method: A case study of the yangtze river economic belt, China. *Journal of Cleaner Production, 232*, 692—704.

Baklanov, A., Molina, L. T., & Megacities, G. M. (2016). Air quality and climate. *Atmospheric Environment, 126*, 235—249. https://doi.org/10.1016/j.atmosenv.2015.11.059

Bikis, A. (2023). Urban air pollution and greenness in relation to public health. *Journal of Environmental and Public Health, 2023*.

Brauer, M., Freedman, G., Frostad, J., Van Donkelaar, A., Martin, R. V., Dentener, F., ... Cohen, A. (2016). Ambient air pollution exposure estimation for the global burden of disease 2013. *Environmental Science & Technology, 50*(1), 79—88.

Carter, E., Norris, C., Dionisio, K. L., Balakrishnan, K., Checkley, W., Clark, M. L., Ghosh, S., et al. (2017). Assessing exposure to household air pollution: A systematic review and pooled analysis of carbon monoxide as a surrogate measure of particulate matter. *Environmental Health Perspectives, 125*(7), 076002.

Daellenbach, K. R., Uzu, G., Jiang, J., Cassagnes, L.-E., Leni, Z., Vlachou, A., Stefenelli, G., et al. (2020). Sources of particulate-matter air pollution and its oxidative potential in Europe. *Nature, 587*(7834), 414—419.

Daly, A., & Zannetti, P. (2007). Air pollution modeling—an overview. In *Ambient air pollution* (pp. 15—28).

Demuzere, M., Bechtel, B., Middel, A., & Mills, G. (2019). Mapping Europe into local climate zones. *PLoS One, 14*(4), e0214474. https://doi.org/10.1371/journal.pone.0214474

Denby, B., Sundvor, I., Cassiani, M., de Smet, P., de Leeuw, F., & Horálek, J. (2010). Spatial mapping of ozone and SO_2 trends in Europe. *Science of the Total Environment, 408*(20), 4795—4806.

Dominski, F. H., Branco, J. H. L., Buonanno, G., Stabile, L., da Silva, M. G., & Andrade, A. (2021). Effects of air pollution on health: A mapping review of systematic reviews and meta-analyses. *Environmental Research, 201*, 111487.

EEA. (2019). Air quality in Europe-2019 report. In *Paper read at 14th International conference on evaluation and assessment in software engineering (EASE)*. Luxembourg: Technical Report Publications Office of the European Union.

Forouzanfar, M. H., Alexander, L., Anderson, H. R., Bachman, V. F., Biryukov, S., Brauer, M., Burnett, R., Casey, D., Coates, M. M., & Cohen, A. (2015). Global, regional, and national comparative risk assessment of 79 behavioral, environmental and occupational, and metabolic risks or clusters of risks in 188 countries, 1990—2013: A systematic analysis for the global burden of disease study 2013. *Lancet, 386*, 2287—2323.

Gładka, A., Blachowski, J., Rymaszewska, J., & Zatoński, T. (2022). Investigating relationship between particulate matter air concentrations and suicides using geographic information system. *Psychology Health & Medicine, 27*(10), 2238—2245.

Garcia, J. M., Teodoro, F., Cerdeira, R., Coelho, R. M., Kumar, P., & Carvalho, M. G. (2016). Developing a methodology to predict PM10 concentrations in urban areas using generalized linear models. *Environmental Technology, 37*, 2316—2325.

Jiang, Y., Sun, M., & Yang, C. (2016). A generic framework for using multi-dimensional earth observation data in GIS. *Remote Sensing, 8*(5), 382.

Khan, M. A. (2012). *Spatial variation of smoke and heavy metals concentration and their effect on plants, soil and water of Madina Town, of Faisalabad and preparation of GIS maps*. Faisalabad, Pakistan: A thesis of MSc.(Hons) Agriculture, Institute of Soil and Environmental Science, University of Agriculture.

Koukouli, M. E., Pseftogkas, A., Karagkiozidis, D., Skoulidou, I., Drosoglou, T., Balis, D., ... Hatzianastassiou, N. (2022). Air quality in two northern Greek cities revealed by their tropospheric no2 levels. *Atmosphere, 13*(5), 840.

Kumar, P., Morawska, L., Martani, C., Biskos, G., Neophytou, M., DiSabatino, S., Bell, M., Norford, L., & Britter, R. (2015). The rise of low-cost sensing for managing air pollution in cities. *Environment International, 75*, 199—205.

Landrigan, P. J. (2017). Air pollution and health. *The Lancet Public Health, 2*, e4—e5.

Ma, S., & Tong, D. Q. (2022). Neighborhood emission mapping operation (NEMO): A 1-km anthropogenic emission dataset in the United States. *Scientific Data, 9*(1), 680.

Mehmood, K. (2013). *Spatial variability of aerosols and metal concentration in soils, water and plants*. Faisalabad, Pakistan: A thesis of MSc.(Hons) Agriculture, Institute of Soil and Environmental Science, University of Agriculture.

Mehmood, K., Bao, Y., Abbas, R., Saifullah, Petropoulos, G. P., Ahmad, H. R., Abrar, M. M., Mustafa, A., Abdalla, A., Lasaridi, K., & Fahad, S. (November 2021). Pollution characteristics and human health risk assessments of toxic metals and particle pollutants via soil and air using geoinformation in urbanized city of Pakistan. *Environmental Science and Pollution Research International, 28*(41), 58206–58220. https://doi.org/10.1007/s11356-021-14436-x

Mitmark, B., & Jinsart, W. (2016). Using GIS tools to estimate health risk from biomass burning in northern Thailand. *Athens Journal of Sciences, 3*(4), 285–296.

Mohan, K. M., Wolfe, C. D., Rudd, A. G., Heuschmann, P. U., Kolominsky-Rabas, P. L., & Grieve, A. P. (2011). Risk and cumulative risk of stroke recurrence: A systematic review and meta-analysis. *Stroke, 42*(5), 1489–1494.

Muhammad, I., & Yan, Z. (2015). Supervised machine learning approaches: A survey. *ICTACT Journal on Soft Computing, 5*(3).

Pak, E. P. A. (2005). Guideline for solid waste Management. *Pakistan Environmental Protection Agency, 7.*

Park, S. Y., Byun, E. J., Lee, J. D., Kim, S., & Kim, H. S. (2018). Air pollution, autophagy, and skin aging: Impact of particulate matter (PM10) on human dermal fibroblasts. *International Journal of Molecular Sciences, 19*(9), 2727.

Psomiadis, E., Charizopoulos, N., Efthimiou, N., Soulis, K. X., & Charalampopoulos, I. (2020). Earth observation and GIS-based analysis for landslide susceptibility and risk assessment. *ISPRS International Journal of Geo-Information, 9*(9), 552.

Raynor, P. C., Adesina, A., Aboubakr, H. A., Yang, M., Torremorell, M., & Goyal, S. M. (2021). Comparison of samplers collecting airborne influenza viruses: 1. Primarily impingers and cyclones. *PLoS One, 16*(1), e0244977.

Peng, R. D., & Bell, M. L. (2010). Spatial misalignment in time series studies of air pollution and. *Health Data Biostatistics, 11*(4), 720–740.

Santana, P., Almeida, A., Mariano, P., Correia, C., Martins, V., & Almeida, S. M. (2021). Air quality mapping and visualisation: An affordable solution based on a vehicle-mounted sensor network. *Journal of Cleaner Production, 315*, 128194. https://doi.org/10.1016/j.jclepro.2021.128194

Schneider, P., Castell, N., Vogt, M., Dauge, F. R., Lahoz, W. A., & Bartonova, A. (2017). Mapping urban air quality in near real-time using observations from low-cost sensors and model information. *Environment International, 106*, 234–247. https://doi.org/10.1016/j.envint.2017.05.005

Shi, Z., Vu, T., Kotthaus, S., Roy, M., Harrison, Grimmond, S., Yue, S., Zhu, T., et al. (2019). Introduction to the special issue "In-depth study of air pollution sources and processes within Beijing and its surrounding region (APHH-Beijing)". *Atmospheric Chemistry and Physics, 19*(11), 7519–7546.

Thepanondh, S., & Toruksa, W. (2011). Proximity analysis of air pollution exposure and its potential risk. *Journal of Environmental Monitoring, 13*(5), 1264–1270.

US-EPA (2012). *Revised air quality standards for particle pollution and updates to the Air Quality Index (AQI).* United States Environmental Protection Agency, Office of Air Quality Planning and Standards, EPA 454/ R99-010.

Van Donkelaar, A., Martin, R. V., Brauer, M., Hsu, N. C., Kahn, R. A., Levy, R. C., Lyapustin, A., Sayer, A. M., & Winker, D. M. (2016). Global estimates of fine particulate matter using a combined geophysical-statistical method with information from satellites, models, and monitors. *Environmental Science & Technology, 50*, 3762–3772.

Verhaegh, B. P., Bijnens, E. M., van den Heuvel, T. R., Goudkade, D., Zeegers, M. P., Nawrot, T. S., … Pierik, M. J. (2019). Ambient air quality as risk factor for microscopic colitis—A geographic information system (GIS) study. *Environmental Research, 178*, 108710.

Vos, T., Barber, R. M., Bell, B., Bertozzi-Villa, A., Biryukov, S., Bolliger, I., … Brugha, T. S. (2015). Global, regional, and national incidence, prevalence, and years lived with disability for 301 acute and chronic diseases and injuries in 188 countries, 1990–2013: A systematic analysis for the global burden of disease study 2013. *The Lancet, 386*(9995), 743–800.

Woldai, T. (2020). The status of earth observation (EO) & geo-information sciences in Africa—trends and challenges. *Geo-spatial Information Science, 23*(1), 107–123.

World Health Organization. (May 2, 2018). *9 out of 10 people worldwide breathe polluted air, but more countries are taking action available at* Accessed 16th Sep. 2023 https://www.who.int/news-room/detail/02-05-2018-9-out-of-10-people-worldwide-breathe-polluted-air-but-more-countries-are-taking-action.

World Health Organization (WHO). (September 26, 2023). *Air pollution.* https://www.who.int/health-topics/air-pollution.

Yadav, R., Nagori, A., Mukherjee, A., Singh, V., Lodha, R., Kabra, S. K., Yadav, G., Saini, J. K., Singhal, K. K., Jat, K. R., & Madan, K. (2022). Geographic information system-based mapping of air pollution & emergency room visits of patients for acute respiratory symptoms in Delhi, India (March 2018-February 2019). *Indian Journal of Medical Research, 156*(4—5), 648.

Yang, R., Hao, X., Zhao, L., Yin, L., Liu, L., Li, X., & Liu, Q. (2022). Design and implementation of a highly accurate spatiotemporal monitoring and early warning platform for air pollutants based on IPv6. *Scientific Reports, 12*(1), 4615.

Zhao, Q., Yu, L., Du, Z., Peng, D., Hao, P., Zhang, Y., & Gong, P. (2022). An overview of the applications of earth observation satellite data: Impacts and future trends. *Remote Sensing, 14*(8), 1863.

CHAPTER 21

Spatial distribution of noise levels in the Historic Centre of Athens in Greece using geoinformation technologies

Avgoustina I. Davri[1], George P. Petropoulos[1], Spyridon E. Detsikas[1], Kleomenis Kalogeropoulos[2] and Antigoni Faka[3]

[1]Department of Geography, Harokopio University of Athens, Athens, Greece; [2]Department of Surveying and Geoinformatics Engineering, University of West Attica, Athens, Greece; [3]Department of Geography, School of Environment, Geography and Applied Economics, Harokopio University, Athens, Greece

1. Introduction

In recent decades, a substantial portion of the world's population has gravitated toward urban areas, resulting in over 50% of the global population residing in bustling cities (Popa et al., 2022). The phenomenon of rapid urbanization carries profound implications for noise levels, thereby detrimentally affecting the well-being and overall quality of life for urban dwellers. The problem is exacerbated by the high population density, traffic, particularly so during rush hours, construction activities, and tourism (Oyedepo, 2012). As projections by the United Nations (UN) organization suggest that urbanization rates will continue to surge, with projections estimating more than 50% of the world's population will reside in urban areas, the issue of noise pollution emerges as one of the foremost challenges to the welfare of urban citizens.

Sound is an essential and integral part of people's everyday life contributing to the mood, psychology, and overall health of individuals. However, as Arnal et al. (2019) have showed, high frequency sounds are perceived by the brain as intolerable, disturbing, and even dangerous. According to the European Environmental Agency's (EEA) Report of 2019 (No 22/2019) investigating the impact of long-term exposure to noise, long-term exposure to noise can cause various health effects such as negative effects on the cardiovascular and metabolic systems, as well as cognitive dysfunction in children (Borchgrevink, 2003). Apart from this, current evidence shows that environmental noise contributes to 48,000 new cases of ischemic heart disease annually as well as 12,000 premature deaths. Therefore, noise can affect the quality of life and cause annoyance and sleep disturbance which is essential for resting the mind for optimal performance (Bustaffa, 2022). The National Institute for Occupational Safety and Health (NIOSH) in the United States points out that elevated noise levels can cause stress. This can, for example, cause changes in a pregnant woman's body and consequently affect the developing fetus. Still, it is scientifically documented that hearing loss is caused by intense noise. Note that hearing

Geographical Information Science
ISBN 978-0-443-13605-4
https://doi.org/10.1016/B978-0-443-13605-4.00012-6

impairment is still the most prevalent disability in Western societies (NIDCD, 2022; NIOSH, 2019). It is important to note that noise can be extremely detrimental to employees and the adjacent residential community too (Kaluarachchi et al., 2022).

Due to the serious implications of noise pollution, it is necessary to map noise to regulate or eliminate noise. Strategic noise maps are nowadays used as means for evaluating populations affected by environmental noise under the European Union's noise Directive 2002/49/EC, commonly known as the Environmental Noise Directive (END). The European Union has prioritized the reduction of the harmful effects of exposure to environmental noise by developing a common approach to avoid, prevent, and control noise nuisance. When possible, it advises using computer techniques rather than field observations (Al-Hajri et al., 2020; Faka et al., 2021; Kalogeropoulos et al., 2020; Karkaletsou, 2014; Ministry of Tourist, 2023, National Institute for Occupational Safety and Health (NIOSH), 2019, National Institute of Deafness and Other Communication Disorters (NIDCD), 2022; Petropoulos et al., 2015; Popa et al., 2022; Robertson, 1998; Srivastava et al., 2019; Tsatsaris et al., 2021).

Noise pollution monitoring employs two main methods: stationary monitoring units and portable procedures. Stationary monitoring units are installed at specific locations to collect noise data at regular intervals. On the other hand, portable procedures involve the use of dB meters along with standard Global Positioning Systems (GPS) or smartphones (Guillaume et al., 2016; Murphy & King, 2022). The use of mobile sensing has seen a rise in popularity for applications requiring dense spatial data and extensive area coverage, such as monitoring noise pollution (Faka et al., 2021). This trend is primarily driven by the significant costs associated with constructing and operating stationary monitoring units.

Geoinformatics and specifically Geographic Information System (GIS) consist an important tool in environmental sciences and their usefulness is increasingly adopted to study noise and its properties. The advantages are more than conventional methods such as field surveying (Petropoulos et al., 2015; Tsatsaris et al., 2021; Wieczorek & Delmerico, 2009). The advantages of this technology include the simplicity of searching, analyzing, and representing data, as well as the potential for automated cartography (ease of producing alternative cartographic options, simplicity of map production). Notable is also the integration of data from various sources, the simplicity of revisions and updates, and the convenience of information storage and retrieval (Kalogeropoulos et al., 2020). The purpose of GIS is to provide the required information by following a decision-making process. The information provided can either identify the problem or identify and develop the various alternatives to best approach the execution of a decision (Al-Hajri et al., 2020; Bryant, 2021; Tsai, 2008). Due to these advantages that arise as far as the collection and analysis of spatial data is concerned, GIS is regarded as one of the most effective tools for performing a spatial mapping of noise levels.

Noise pollution monitoring generates valuable data that can be utilized for noise pollution mapping, enabling the estimation of noise levels in areas lacking official recordings. Geostatistical interpolation methods have been widely adopted for this purpose, exhibiting promising results in various studies. Zuo et al. (2016) demonstrated the integration of GIS and crowdsourced data to generate interpolated noise maps, effectively providing spatial representations of noise pollution. In a similar context, Shim et al. (2016) utilized participatory noise data to create detailed maps for Seoul, distinguishing between temporal variations, specifically day-time and night-time, as well as weekdays and weekends. This approach, employing the inverse distance weighted (IDW) interpolation method, contributed to a comprehensive understanding of the noise pollution patterns in the city. Furthermore, Grubeša et al. (2018) applied the ordinary Kriging method in the case of Zagreb, which allowed for the accurate interpolation of noise values and yielded informative noise pollution maps. These geostatistical interpolation techniques have proven to be valuable tools for noise pollution mapping, offering insights into noise distribution over wide geographical areas, where official monitoring may not be feasible or sufficient.

In purview of the above, the scope of this study has been to obtain a spatial mapping of the noise levels in the center of Athens during the period May—August 2021 using GIS and geospatial interpolation techniques together with filed-based relevant measurements. In this context, the present study maps and analyzes noise in the form of thematic maps, and interprets the results. Subsequently, the resulted interpolated maps are correlated with regional characteristics of the study area, namely the building's height and population density. The study's last objective has been to develop a WebGIS application to allow the wider distribution of the study's main findings also accommodating some interoperability features with the end users of the platform.

2. Experimental setup

2.1 Study area

The most tourism-oriented part of the historic center of Athens was chosen as the study area (Fig. 21.1). The selected area is notable for its landmarks attracting numerous visitors daily (Hucal, 2022) while also being home to more than 600,000 permanent residents (EL.STAT., 2011). According to the Greek Ministry of Tourism, there were approximately 2,085,500 visitors to Greece in 2021, when this study was conducted, and 366,560 individuals visited Akropolis, between May and August of that year (Ministry of Tourists in Greece, 2023). Worldwide renowned landmarks such as the Ancient Agora, the Temple of Hephaestus, the Acropolis, and the Acropolis Museum are among the most touristic sites that are within the borders of this study area. In addition to the landmarks located within the study area, this area stands as an important economic and cultural center for the Greek capital city. Local shops and services are abundant with one of the largest and most vibrant shopping streets in Greece, "*Ermou*" street, being located there. Robertson (1998), Paraschou (2022) as well as one of the largest open-air market000s of Europe being located near "*Thiseion*" train station (Kitsantonis, 2022; Kwak, 2018). Last but not least, one of Greece's most recognized neighborhoods, "*Plaka,*" near the Acropolis, is also located within the borders of the study area, notable

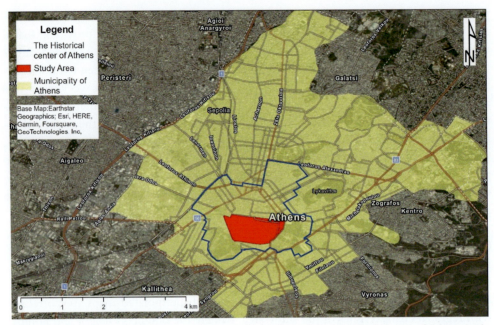

Figure 21.1 The study area location (*in red*) and the borders of the Historical Center of Athens in the Municipality of Athens (*in yellow*).

for its disorderly layout and architecture. Due to the economic and cultural importance of this very vibrant area, it is vital to monitor noise levels to protect the quality of life of both tourists and local residents.

2.2 Data collection

For the purposes of this study, noise levels were recorded using a digital Mengshen sound level meter dB. The instruments' measuring range is 30—130 dBA with ±1.5 dB accuracy and 0.1 dB resolution. The collected data were noted in the Fieldmove Clino app (Fig. 21.2) via a mobile which is relatively easy to use. More specifically, it is a digital compass with the ability to use the device's GPS position and orientation sensors for data download to an Android phone. It is designed to capture and store digital photos and text notes in a CSV file with georeferencing, just as the measurements in this paper were. It can also use the phone as a traditional handheld compass, sighting compass, and a digital compass inclinometer in the field to measure and capture the orientation of planar and linear features. This study employed stratified random sampling (random and systematic) to collect the noise level from various locations within the study area at regular intervals (Delmelle, 2009; Wong, 2005). The selected period for the collection of the noise data was between May and August of 2021, which is the period when the area receives the most visitors and tourists. A specific data collection route was carefully designed passing through the different land uses of the study area (Fig. 21.2).

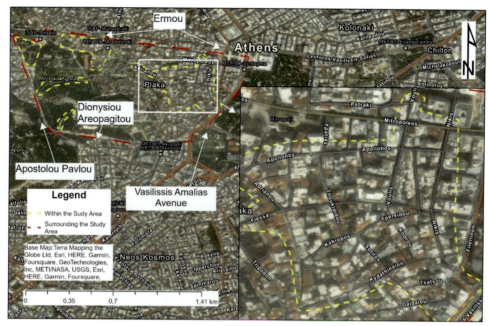

Figure 21.2 The area under study as well as the path of the samples that were collected during the field surveying.

The perimeter route includes areas between the streets *"Othonos," "Apollonos," "Andrianou," "Kekropos,"* and *"Kordou."* These areas contain sections of the Ancient Agora and the Open-Air Market of Monastiraki Square. The survey areas were chosen to cover places with diverse natural and urban characteristics. For example, *"Vasilissis Amalias"* Road was chosen to cover the road network, *"Monastiraki," "Thiseio,"* and *"Plaka"* because of the commercial and tourist shops and part of *"Syntagma"* to cover the residential network. The *"Ancient Agora"* was ultimately selected due to the wealth of vegetation that runs through it. The route is depicted in the figure below (Fig. 21.3).

During a period of four months, from May to August, when the highest number of tourists visit Athens, an amount of 16 visits were conducted (see Table 21.1 below). Each month, specifically, there were two weekday trips, one Saturday trip and one Sunday trip. In total, there are four Wednesdays, two Fridays, one Tuesday, one Monday, and four weekends. Specifically, there are four Wednesdays, two Fridays, one Tuesday, one Monday, and four weekends. The dates were chosen in order to observe the difference in the spatial capture of sound between two weekdays of a month, when there are fewer people walking around, and two weekend days, when there was a greater amount of people. In addition, the diversity of the acoustic data has increased. The sampling hours and days are shown in more detail below in Table 21.1, which is shown below.

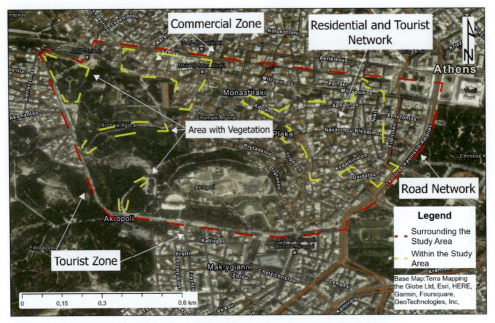

Figure 21.3 The data collection route followed in this study.

Table 21.1 Sampling days and hours May—August 2021.

Day	Hour [EEST time zone]	Day	Hour [EEST time zone]
May 12, 2021	12:30—15:25	July 19, 2021	10:40—13:45
May 14, 2021	12:20—15:20	July 21, 2021	10:50—14:00
May 15, 2021	10:50—13:40	July 24, 2021	10—45—13:55
May 16, 2021	13:00—16:10	July 25, 2021	12:00—15:00
June 8, 2021	11:00—14:00	August 18, 2021	10:30—13:10
June 9, 2021	10:50—13:15	August 20, 2021	11:00—13:45
June 12, 2021	12:00—15:00	August 21, 2021	11:30—14:00
June 13, 2021	13:00—15:40	August 22, 2021	11:00—14:00

In May, the Outdoor Market, which is part of the commercial zone in the North-West part of the research area, is not in the sampling route. From June onwards, it was decided to enter and map also another part located within it. In addition, in the area of *"Plaka,"* along *"Andrianou Street,"* it was determined to take measurements along a longer distance of the road due to the fact that it is a popular tourist route.

When the weather was favorable (free of wind and precipitation) and conducive to data collection, so that the results correspond as closely as possible to reality. It should be noted that during May and June in the vegetated areas, there were cicadas, which had a slightly increasing effect on the dB of the sound level meter.

3. Methodology

This section provides an overview of the methodology employed to satisfy the study objectives. Measurements with the noise level meter were taken at different times of the day, including both weekdays and weekends (as already discussed previously and summarized in Table 21.1 above). These measures aimed to capture a holistic understanding of the noise levels characterizing the study area. Once the noise data were collected, it underwent a meticulous processing phase. This involved organizing and structuring the collected measurements, ensuring data integrity and removing any outliers or inconsistencies. The data were then subjected to statistical analysis to derive meaningful insights. The analysis involved calculating descriptive statistics such as mean, standard deviation, and data range to understand the distribution and characteristics of the noise data.

To visualize and communicate the results effectively, a WebGIS app was created. This interactive platform allowed for the mapping of the interpolated noise values across the study area. By leveraging the capabilities of ArcGIS Online, the WebGIS app provided an optimal way to present the thematic noise maps. Users could interact with the maps, explore different locations, and gain a visual understanding of the spatial distribution of noise levels. This facilitated the interpretation of the results and enabled stakeholders to draw meaningful conclusions regarding noise patterns and potential mitigation strategies. The process followed to carry out this study is presented in Fig. 21.4.

3.1 Data preprocessing

For the preprocessing of the data, a number of steps were implemented. In the first stage of the present study preparation, a polygonal shapefile of the surveyed area was created. In the next stage of data processing, the Excel program was used to enter the coordinates and measurements in different columns. Then, the same program was used to calculate the mean (Mean) of the maximum (Max) and minimum (Min) values measured at each point while conducting the sampling. Then, weekdays and weekends of each month were categorized together, as well as whole months separately. For statistical processing, the mean, median, standard deviation, maximum, and minimum values of the above were calculated in the Jasp software (v.1.6), which are presented in histograms and graphs.

3.2 Spatial interpolation

Following the data preprocessing, the next step involved the spatial interpolation implementation in order to generate the noise maps from the point data. In implementing this, the first step was to import the final Excel files into Arcgis Pro 2.9 GIS software. Spatial interpolation estimates unknown variable values at any position using known observations. The data's collection and type necessitate this interpretation. The data do not fully cover the area under consideration; hence, interpolation is utilized when the aim is to

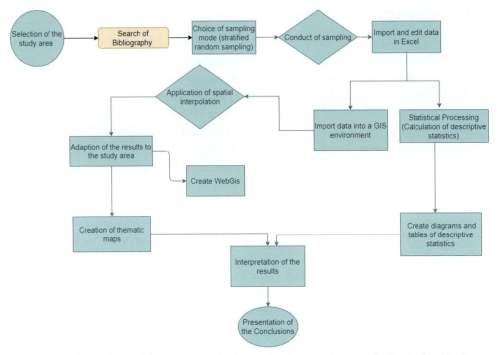

Figure 21.4 Flow chart of the overall methodology implemented to satisfy the study objectives.

convert point sets to continuous surfaces (Stivastava et al., 2019). Spatial interpolation methods vary. Global interpolation methods include classification using external information, trend surfaces on geometric coordinate, and regression models on surrogate attributes methods of spectral analysis. Local deterministic interpolation methods include, for example, Thiessen, linear and inverse distance weighting, and splines. These strategies just require deterministic or rudimentary statistical knowledge. On the other, approaches such as Kriging employ a spatial correlation–based geostatistical approach and require spatial autocorrelation statistics. When the feature matrix variation is irregular and the sample density is low, several methods are applied. Geostatistics estimates interpolation quality probabilistically. They are able to predict land blocks greater than the support. Geostatic algorithms can interpolate index functions using soft data, improving accuracy (Burrough, 1998).

Because of the complex statistical methods, the most valid results, and the type of data in this study, the Kriging method was chosen to process the samples (ÖzkanSertlek et al., 2019); the Kriging method uses the semicovariance to measure the spatially correlated component otherwise called spatial dependence (Can, 2014; Karkaletsou, 2014). The Kriging format chosen is ordinary Kriging. It focuses on the spatially correlated component and uses the fitted semivariogram directly for interpolation. The general equation for estimating a value at a point is:

$$Z_0 = \sum_{i=1}^{s} Z_x W_x \qquad (21.1)$$

where Z_0 is the estimated value, Z_x is the known value, W_x is the weight associated with point x, and s is the number of sample points used in the equation. The weights are derived from the model.

Two commonly used models are spherical and exponential models. A spherical model shows a progressive decrease in spatial dependence up to a certain distance, beyond which the spatial dependence decreases. An exponential model shows a less gradual pattern than a spherical model: spatial dependence decreases exponentially with increasing distance and disappears completely at infinite distance (Kang-Tsung, 2006; Esri, 2022a, 2022b, 2022c).

Neighborhood type was chosen standard because, with smooth type, it was observed that sometimes the root mean square grew larger and maximum neighbors were chosen 7 because the closer the points are to each other, the more common characteristics they have (Tobler, 1970).

For this study, two interpolation models were employed due to their relevance. The ordinary Kriging spatial interpolation method was selected to capture the overall sound distribution spatially. It was implemented separately for each month, as well as for weekends and weekdays. During the data analysis process using ArcGIS Pro, the Geostatistical Wizard was utilized, offering the ordinary Kriging method for spatial interpolation. This tool facilitated model optimization by minimizing the mean square error and focusing on key parameters specific to each model. The optimization process centered on the semibarogram model in Kriging, with particular emphasis on the range parameter. However, it is important to note that the optimization assumes isotropy and employs a default neighborhood search consisting of four domains. By calculating the mean square error across a range of values, the parameter with the lowest mean square error was chosen. Model parameterization was performed using the optimization option, aiming to minimize squared error. Ordinary Kriging is generally preferred as a method unless there is a need to eliminate the trend surface from the data, as it is desirable to maintain simplicity in the dataset. Estimating too many parameters can decrease the accuracy of the models. Thus, the goal was to keep the models as straightforward as possible by minimizing the number of parameters estimated. In general, the choice of utilizing the Ordinary Kriging method and optimizing its parameters through the Geostatistical Wizard in ArcGIS Pro aimed to achieve accurate spatial interpolation while maintaining simplicity in the dataset (Esri, 2022a, 2022b, 2022c).

3.3 Cross-validation of the interpolated maps

Cross-validation compares the fidelity of the generated surfaces' visual quality. Validation is done by removing a known point from the dataset and using the remaining points to predict it. This is done for each point to calculate the error of the model. Two common diagnostic statistics for estimating the error in the above process are RMSE (root mean square error) and standardized RMSE.

$$\text{RMSE} = \sqrt[2]{\frac{1}{n} \sum_{i=1}^{n} \left(z_{i,\text{act}} - z_{i,\text{est}} \right)^2}, \tag{21.2}$$

$$\text{Standarized RMSE} = \frac{\text{RMSE}}{S}, \tag{21.3}$$

where n is the number of points, $z_{i,act}$ is the known value of point i, $z_{i,est}$ is the estimated value of point i, and s is the standard error. A satisfactory interpolation method should yield a smaller RMSE. By extension, an optimal method should have the smallest RMSE, or the smallest average deviation between the estimated and known values at the sample points. Clearly, a better Kriging method should yield a smaller RMSE and a standardized RMSE closer to 1. If the standardized RMSE is 1, it means that the RMSE statistic equals standard error. Therefore, the estimated standard error is a reliable or valid measure of the uncertainty of the predicted values.

Particularly in regard to during the cross-validation procedure for interpolated maps, a number of evaluation techniques are used to assess the accuracy and dependability of the predictions. One such technique that is widely used for this purpose involves examining the fitted regression line through the dispersion of points on a graph. To further analyze the error associated with the predictions, an error plot is generated. The error plot is similar to the prediction plot, with the key distinction that it represents the discrepancy between the measured values and the predicted values. To obtain the error values, the measured values are subtracted from the corresponding predicted values. Additionally, these error values are divided by the estimated standard Kriging errors, resulting in the standard error plot. By examining the error plot, one can gain insights into the distribution and magnitude of the errors in the interpolated maps. Deviations from zero in the error plot indicate the presence of discrepancies between the predicted values and the actual measured values. The errors magnitude, as indicated by the vertical distances from the zero line in the error plot, provides an understanding of the level of deviation from the measured values.

By analyzing the fitted regression line, error plot, and standard error plot, the cross-validation procedure permits a thorough evaluation of the accuracy and dependability of the interpolated maps. This information helps to comprehend the extent of error and refine interpolation models to improve the overall quality of the maps (Esri, 2022a, 2022b, 2022c; Stone et al. 1974).

3.4 WebGIS development

An interactive WebGIS platform using ArcGIS Pro and ArcGIS Online was created as an innovative tool to present the produced maps of noise levels. ArcGIS Pro and ArcGIS Online are powerful tools for creating, analyzing, and sharing geographic information. With ArcGIS Online, users can easily publish and share maps, data, and applications over the web. One of the key functionalities of ArcGIS Pro is the ability to share maps to ArcGIS Online, making them accessible to a wider audience. This integration between ArcGIS Pro and ArcGIS Online enables users to create dynamic, interactive maps and share them with stakeholders, decision-makers, and the public.

When sharing a map to ArcGIS Online, ArcGIS Pro lets users bundle and add accompanying data including feature classes, tables, and base maps (Corbin, 2018). Once the map is published to ArcGIS Online, it may become accessible through various platforms and devices. Users can view and interact with the map using ArcGIS Online's web-based map viewer, which provides tools for zooming, panning, querying, and analyzing the map data. Additionally, ArcGIS Online supports embedding maps in websites, enabling seamless integration of maps into web applications, blogs, or other online platforms. This flexibility in sharing maps ensures that the information reaches the target audience, regardless of their technical expertise or the device they are using.

4. Results

4.1 Statistical exploration of the collective data

The collected noise data were subject to statistical analysis to derive meaningful insights and assess various characteristics of the noise levels. Several statistical measures were computed to summarize the data and provide a comprehensive understanding of its distribution and variability. The following is a summary of the statistical analysis that was carried out to the data for the measurements by weekdays and weekends for each month and for the whole month as a whole. The statistical analysis for the months of May and June is presented in Tables 21.2 and 21.3, accordingly.

The statistical analysis reveals consistent patterns across all four cases, as indicated by the relatively small values of the standard deviation. This implies that the data points tend to cluster closely around the mean, suggesting a normal distribution. Additionally, the small standard errors further support this finding, indicating a high level of precision in the measurements. Moreover, a noteworthy observation is that the means in all the tables are very similar to each other. This suggests a consistent average noise level across the different time periods or categories considered. The minimal variation in the means further reinforces the stability and reliability of the collected noise data. These findings collectively indicate that the noise data exhibit characteristics of a normally distributed dataset, with a tight clustering of values around the mean. The consistency in means across the tables suggests a consistent noise profile, providing confidence in the

Table 21.2 Statistical analysis for the months of May and June.

Descriptive statistics according to average value (dBA)	Wednesday– Friday of May	Saturday– Sunday of May	All four days of May	Tuesday– Wednesday of June	Saturday– Sunday of June	All four days of June
n	297	307	604	297	317	612
Median	59.4	60.2	60.0	62.3	60.3	61.3
Mean	58.5	59.8	59.1	60.5	60.4	60.5
Standard deviation	9.3	7.7	8.5	8.6	7.9	8.2
Minimum value	23.7	41.4	23.7	31.9	43.9	31.9
Maximum value	78.6	82.7	82.7	81.2	81.9	81.9

Table 21.3 Statistical analysis for the months of July to August.

Descriptive statistics according to average value (dBA)	Monday– Wednesday of July	Saturday– Sunday of July	All four days of July	Wednesday– Friday of August	Saturday– Sunday of August	All four days of August
Values	346	328	672	280	300	579
Median	64.1	64.3	64.1	63.6	63.3	63.4
Mean	63.4	64.2	63.8	62.9	63.3	63.1
Standard deviation	6.3	5.6	6	5.8	4.9	5.3
Minimum value	32.1	50.9	32.1	35.4	52	35.4
Maximum value	78.3	86.6	86.6	78.1	79.5	79.5

representativeness of the collected data and enabling reliable comparisons and analyses between different time periods or categories. Table 21.4 presents the descriptive statistics for all days of each month for the collected data.

The provided table presents a comparative analysis of the arithmetic mean, standard deviation, minimum, and maximum values for each month individually. By examining the table, distinct patterns and variations across the different months become apparent. It is observed that the average noise levels in July and August are higher compared to May and June. This indicates that during the summer months, noise levels tend to be generally elevated. Additionally, the data reveal that July and August exhibit a wider range of higher noise values, suggesting a greater occurrence of high-intensity noise events during this

Table 21.4 Descriptive Statistics for all days of each month.

Months	Mean	Standard deviation	Median	Minimum value	Maximum value
May	59.1	8.5	60	23.7	82.7
June	59.1	8.2	61.3	31.9	81.9
July	63.8	6	64.1	32.1	86.6
August	63.1	5.3	63.4	35.4	79.5

period. On the other hand, May and June display a wider range of lower noise values, indicating a relatively quieter environment during these months. Furthermore, May records the lowest minimum noise value, suggesting a lower threshold of noise occurrence during that month. Conversely, July showcases the highest maximum noise value, indicating instances of particularly high noise levels during that period. In terms of variability, the standard deviation is higher in May and June compared to July and August. This implies that the noise values in May and June display a broader spread or dispersion around the mean, while in July and August, the range of values between the median and the mean is narrower. This indicates a more compact distribution of noise values during the summer months, making the sample more representative of the overall noise conditions.

In general, the findings of the above table provide useful insights into the monthly variations in noise levels. The higher average values in July and August suggest a period of increased noise intensity, while the wider range of values in May and June indicates more diverse noise conditions. Understanding these month-to-month variations is crucial for assessing the impact of noise pollution, identifying potential noise hotspots, and formulating targeted mitigation strategies.

The averages of noise levels on weekdays and weekends reveal an ascending trend from May to August, with the peak occurring on a weekend in July. Although there are discernible differences, the variations between the average values are relatively minor. The data illustrate a gradual increase in the average noise levels as the summer months' progress. This indicates a potential correlation between the seasonality and noise intensity, with noise levels tending to rise over this period. Moreover, the highest average value is observed during a weekend in July, suggesting that weekends in the summer months may experience slightly higher noise levels compared to weekdays. Despite these fluctuations, the disparities between the average noise values on weekdays and weekends are relatively small. This implies that noise levels during weekdays and weekends exhibit similar patterns, with only marginal distinctions between the two. These slight variations may be attributed to factors such as variations in human activity, transportation patterns, or other environmental factors.

Table 21.5 shown below presents the descriptive statistics of the noise data acquired for the whole period from May to August. The descriptive statistics computed for the data collected from May to August reveal interesting observations. Firstly, it is apparent

Table 21.5 The main features of a dataset from May to August.

Descriptive statistics according to the average price (dBA)	Overall for all four months
Values	2466
Median	62.8
Mean	61.7
Standard deviation	7.4
Minimum value	23.7
Maximum value	86.6

that all the collected values exhibit a mean value of approximately 61.7 dB which is remarkably close to the median. This indicates a balanced distribution of data around the central tendency. Moreover, the relatively small standard deviation suggests that the data points are clustered closely around the mean and median. This indicates a narrow spread or dispersion of values, indicating a higher level of consistency in the measurements. The consistency between the mean and median, as well as the small standard deviation, imply that the majority of the collected measurements fall in close proximity to the central values. This suggests a concentration of noise levels around the mean and median, indicating a relatively stable noise pattern during the months of May to August. Considering these descriptive statistics helps in comprehending the distribution and variability of noise levels during this period. The close alignment of the mean and median, along with the small standard deviation, signifies a robust and consistent noise profile.

By considering these descriptive statistics, researchers and decision-makers can gain insights into the typical noise levels and their variability during the months of May to August. This knowledge aids in formulating appropriate noise management strategies and developing targeted interventions to minimize the impact of noise pollution.

The histogram displayed below (Fig. 21.5) represents the total measurements collected from May to August, offering insights into the distribution pattern. The histogram exhibits characteristics of a normal distribution, as evidenced by the close proximity between the mean and median values. This suggests that the frequency at which the data points appear is neither excessively high nor low, indicating a balanced distribution across the range of values. In addition to this, the histogram has the shape of a mesokurtic curve, which implies that it has a modest level of peakedness in comparison to a fully normal distribution. The similarity between the median and the mean reinforces this observation, indicating that the density of values tends to be concentrated toward the center of the distribution. In other words, there is a relatively high frequency of values around the mean, while the frequency decreases as we move away from the mean. These findings align with the concept of a typical bell-shaped curve, where the majority of the data points cluster around the mean, and the number of data points gradually decreases as we move toward the extremes of the distribution. The relatively even distribution of

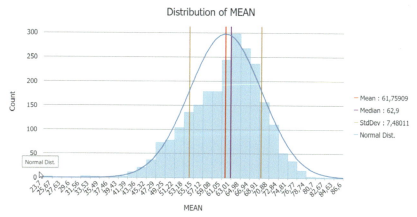

Figure 21.5 The histogram of measurements for May—June.

values suggests a relatively stable and balanced noise pattern throughout the May to August period (Altman & Bland, 1995). Such a distribution is shown in Fig. 21.5.

4.2 Cross-validation results

During the cross-validation process for the interpolated maps, various evaluation techniques are employed to assess the accuracy and reliability of the predictions. One such technique involves examining the fitted regression line through the dispersion of points on a graph. In this case, the regression line is depicted in blue, and its corresponding equation is provided just below the graph for reference. The regression line position in relation to the gray reference line provides valuable insights into the error magnitude. When the regression line is in close proximity to the gray reference line, it signifies that the error is minor. This observation indicates that the interpolated values closely align with the measured values, resulting in a high level of accuracy in the predictions. Fig. 21.6 presents such a paradigm of predicted values versus measured values for May—August measurements.

Table 21.6 displays various statistical measures for the interpolated maps, categorized by different time-periods within the context of evaluating the accuracy and reliability of the interpolated maps.

4.3 Noise maps from the spatial interpolation

Noise maps display spatial interpolation findings to help understand noise levels in an area. These maps show interpolated noise levels at various places, revealing noise distribution and intensity. Kriging can be used to extrapolate noise measurements from specific places to estimate noise levels over the research region. Noise maps are created using interpolated values. Noise maps show noise levels with color or contour shading. Lower noise levels are indicated with lighter colors, whereas higher noise levels are shown with

Figure 21.6 Plot of the cross-validation errors and standardized errors versus the measured values for May–August.

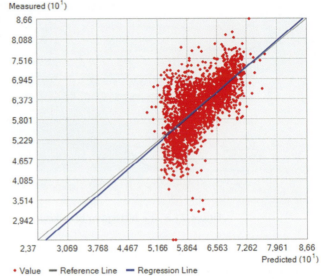

Regression function: 1,04452926693293 * x + -3,000381534883

Table 21.6 Cross-Validation results for all weekends, all weekdays, May–August and overall all the values.

	All weekends	All of the weekdays	May	June	July	August	Overall
Count	1206	1195	543	612	671	578	2466
Mean	0.019	0.077	0.016	0.018	0.022	0.026	0.068
Root-mean-square	5.47	6.08	4.64	4.93	4.91	4.08	5.69
Mean standardized	−0.0039	0.013	0.004	0.004	0.004	0.006	0.012
Root-mean-square standardized	1.1	1.06	1.15	1.08	1.05	1.02	1.04
Average standard error	4.94	5.71	4009	453	4.66	3981	5.46

darker hues. Contour lines can also mark map noise levels. These maps improve noise assessment and management. They identify noise hotspots and quieter locations. Noise maps can also identify regions that need noise mitigation or zoning. These maps also help stakeholders, legislators, and the public understand noise issues. They show noise distribution clearly, helping stakeholders make educated decisions and design effective noise reduction and management plans. Spatial interpolation-based thematic noise maps help comprehend and address noise pollution. They help understand noise patterns, make decisions, and reduce noise's influence on the environment and human well-being by visualizing noise levels (Meng 2013; Statuto et al., 2016).

The maps that follow (Fig. 21.10) illustrate the spatial distribution of the noise that was present in the study area and during the time period that the measurements were conducted (May 2021 to August 2021) (Fig. 21.7).

In particular, as it is observed in the figure above, in May, the analysis reveals that the southwest region of the study area has low sound measurements, whereas the northeast and a portion of the north have high sound intensity. The road network, "*Plaka,*" and

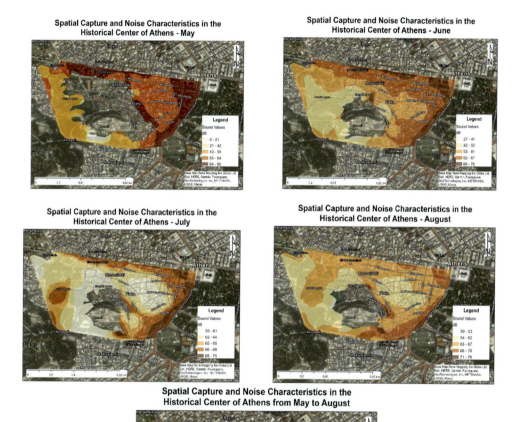

Figure 21.7 The noise maps for the months of May through August 2021.

"*Monastiraki*" have particularly high levels of sound intensity. Extremely low in the vicinity of "*Thiseion*," the "*Acropolis*," the "*Ancient Agora*," and a tiny portion of "*Andrianos Street.*"

In June, it is once again observed that the north-western region has low levels of intensity, while the south-eastern region has high levels of intensity. The lowest are located near the *Ancient Agora* and the *Acropolis*, while the highest are located near the "*Monastiraki*" metro station, "*Ermou Stree*t," "*Syntagma metro*" station, the road network, and "*Andrianou Street.*"

In July, it appears (Fig. 21.8) that centrally of the study area, that is, in "*Acropolis*," "*Plaka*" and "*Thiseio*," dB levels are quite low. On the contrary, in the *Ancient Market*, the *Acropolis metro*, the road network, the *Syntagma metro*, and "*Ermou Street*," the levels are quite high.

In general, in the month of August, the values observed are high. Especially near the "*Acropolis metro*" station, the road network, "*Ermou Street*" and the "*Ancient Agora.*" The volume decreases in part of the area of the thesaurus, near the "*Acropolis*" and in part of "*Plaka.*" The spatial mapping of the sound overall on all days sampled shows that very low sound levels were recorded in the SW and in a part of "*Plaka.*" The lowest values can be seen near the "*Ancient Agora*" and the "*Acropolis.*" On the other hand, high values are recorded SE, NE, and in particular, in the road network and Syntagma metro.

Regarding the main streets (Fig. 21.9) of the study area, it appears that the dB values are quite low on "*Apostolou Pavlou*" and "*Dionysiou Areopagitou*" streets. On the contrary, on "*Vasilissis Amalias Avenue*" and "*Ermou Street*," they are high with the highest at the beginning of the avenue.

Comparing the measurements taken on weekdays and weekends during the period May–June, it is clear that the area of "*Thiseion*" and "*Acropolis*" has lower intensity than the road network, "*Plaka*" and "*Ermou Street.*" Still, the intensities are higher on weekends as they reach 81 dB mainly on "*Ermou Street*" and the road network. Around the "*Acropolis*" and the "*Ancient Agora*," the lowest readings are recorded.

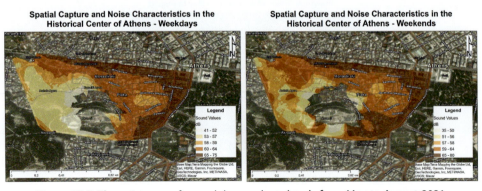

Figure 21.8 The noise maps for weekdays and weekends from May to August 2021.

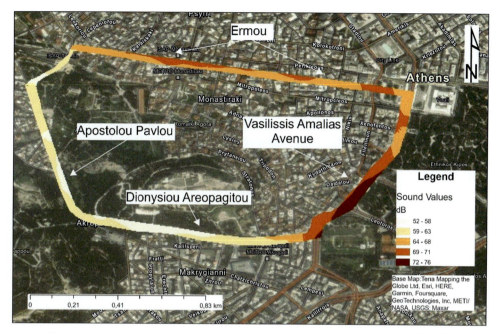

Figure 21.9 The noise map of the road network only for weekdays and weekends.

4.4 The study WebGIS

The next figure (Fig. 21.10) depicts the results of the aforementioned procedure described in Section 3.4. ArcGIS Online offers collaboration features that enhance the sharing and dissemination of maps. Users can invite others to view, edit, or contribute to the map, facilitating teamwork and data sharing within organizations. ArcGIS Online also supports the creation of interactive web applications, called Web AppBuilder, allowing users to build customized applications with specific functionality and design (Shellito, 2020).

The app that was developed in the framework of the present study is available at the weblink: https://arcg.is/1CnKCm.

5. Discussion

Initially, it is clear that the area with the lowest values in the measurements carried out is "*Thiseio,*" which is crossed by "*Apostolou Pavlou*" Street and where the "*Ancient Agora*" and the temple of "*Hephaestus*" are located. There is a lot of vegetation in this area. Afterward, from "*Apostolou Pavlou*" Street, at "*Dionysiou Areopagitou,*" the values of the readings increase approaching "*Vassilisis Amalias*" Avenue. This is where the highest measurements of the study area of about 75–80 dB are recorded. According to Directive 70/157/EEC on maximum permissible noise limits per type of vehicle, the limits range from 74 to 80 dB depending on the size and engine of each vehicle. Then, in "*Syntagma Square*" and "*Ermou*"

Figure 21.10 The development of the WebGIS page.

Street," which is crossed at its beginning by "*Stadiou*" Street, the values usually range from 65 to 75 dB and are considered quite high according to Presidential Decree 1180/1981, whereas in urban areas, the maximum permissible limit is set at 50 dB. As regards the residential and tourist network studied in the context of the present study, medium to low measurements were found in the section of *Syntagma* and "*Plaka.*" In particular, in the parts of the residential network close to the Avenue, the measurements were mostly medium level and with moving away from there the values decreased. In "*Plaka,*" which is mainly a destination for foreign tourists, the measurements were medium to high, again in accordance with Directive 1180/1991, which sets the upper limit at 45 dB in residential areas and 50 dB in urban areas. It is worth noting that the area investigated has three Metro stations and one Electric station. The "*Syntagma,*" "*Monastiraki,*" and "*Acropolis*" metro stations have in each case high intensity values, while the "*Thiseio*" station often does not. On the traffic side of things, crowds were prevalent after 10:30 a.m. roughly, especially on weekends on "*Ermou*" and "*Apostolou Pavlos*" streets.

It should be mentioned that in July and August, the cicadas in the green areas slightly increased the dB of the measurements, which can be seen from the statistical analysis and the thematic maps. Cicadas, however, each year affect the sound at certain times of the year where there are many trees. This is therefore a factor that should be studied. Expressing personal opinion, from conducting the sampling, the cicadas are considered less disturbing than the mob and vehicles.

On the statistical side, the measured data, those seem to be broadly in the range of 55–70 dB. According to the average and median analyzed for each week, they seem to increase in June and July and are lower in May. This may be due to the greater arrival of tourists in midsummer. Still, studying the thematic maps, the values are higher on weekends and reach 80 dB.

In addition, the GPS of the mobile phone had a slight deviation of 30—50m in some cases. This deviation was observed a few times for more than 50m, so the results were not greatly affected. The mobile made sampling easier in terms of time, weight, and data volume. In the data analysis process in ArcGIS Pro, the Geostatistical Wizard tool was used from which spatial interpolation with ordinary Kriging was applied. This tool offers the option to optimize the model. This is based on minimizing the mean square error and for each model it focuses on the main parameters. In Kriging, it is used to help fit the semibarogram model and focuses mainly on the range parameter. However, its optimization is done under the assumption that it is isotropic and the neighborhood search is default, with four domains. The mean square error for a range of values is calculated and the parameter with the smallest mean square error value is selected. The parameterization was performed using the model optimization option for these reasons. Because this mode selects the parameters that provide the finest results with the lowest RMS, In general, all parameterizations were made with a view to minimizing the squared error.

In general, ordinary Kriging is preferred as a method unless the trend surface must be removed from the data, because the data must be kept as simple as possible. The more parameters estimated, the less accurate the models become. The Kriging method provides the best interpolation results if the data are normally distributed. The first step in geostatistical analysis is to determine the normality of the data. Due to these factors, the Kriging method was utilized throughout the course of this research (Esri, 2022a, 2022b, 2022c).

The study's findings are depicted in Fig. 21.11, along with a comparison to the findings of the Greek Ministry of the Environment regarding the same region for which the Lden indicator was calculated in 2019.

The results in the two maps are similar. Near "*Apostolos Pavlos*" and "*Dionysiou Areopagitou*" streets, the sound intensities are low. In the north-east, near "*Vasilissis Amalias*" and "*Ermou*" streets, the intensities are quite high especially, on the road network. Overall, it is clear from both maps that in the areas dominated by nature and not by the residential and road network, dB is low.

Karkaletsou (2014), who also measured the noise at Kallithea, Greece, concluded that the dB levels on the road network are higher than those on the residential network. In addition, the results of the noise measurements in Isparta, Turkey, revealed that noise levels in remote areas of the city were substantially lower (Harman, 2016).

According to Greek Directive 1180/1981, the maximum permissible limits are 70 dB in legislated industrial areas, 65 dB in areas where the industrial element predominates, 55 dB in areas where the industrial and urban elements predominate equally, 50 dB in urban areas, and 45 dB in the residential network. It is observed that in the southwestern part of the research area ("*Thiseio*"—"*Apostolou Pavlou*" and "*Dionysiou Areopagitou*") low sound intensity prevails. In this area, there are mainly cafeterias, archeological sites, and rich vegetation to the right and left of the two roads mentioned above, which occupies a large area. In contrast, in areas with numerous shops and congested vehicles, sound

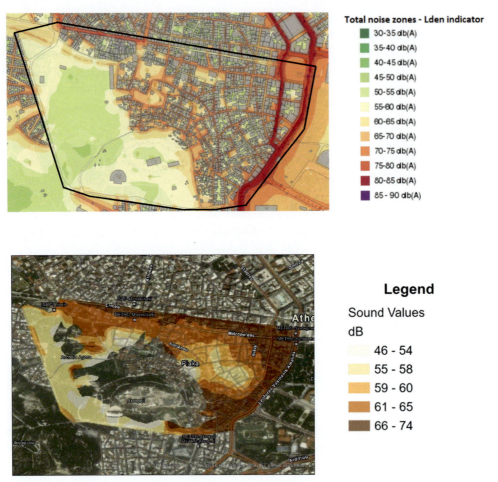

Figure 21.11 Comparison of the results of the present study with the noise map of the Ministry of environment and energy. *(From http://mapsportal.ypen.gr/maps/196/view.)*

intensities were high. In particular, in *Ermou* Street, which has too many shops for shopping combined with numerous cafes and restaurants, the intensities measured were over 65 dB. In addition, "*Vasilissis Amalias Avenue*" is overrun with cars and "*Plaka*" has a large number of tourist shops and restaurants that focus on tourists. The highest intensities were recorded on the avenue and near "*Plaka*," reaching up to 90 dB. Regarding the residential network, areas close to the road network had higher intensities than those further away. Therefore, the study made it possible to conclude that in areas where there is greenery and archeological sites, the sound values are low, while where there is strong human intervention, such as dense building and areas of human activity, the values recorded are higher.

6. Conclusions

This study's objective was to map the spatial distribution of noise characteristics for the historical center of Athens, Greece, using geoinformation technologies and field measurements from May to August 2021. A further objective has been to develop a WebGIS, as this interactive platform enables easy viewing of the thematic noise maps and facilitates the generation of conclusions through user engagement with the results.

The study revealed high sound levels of 65–75 dB in the road network and commercial areas. In contrast, in green spaces and archeological sites, sound levels were lower with a range of 45–65 dB. The Historic Center of the municipality of Athens is a busy area. It is a pole of attraction for tourists and residents of the Attica Basin and Greece in general. The reason is that in this area, there are many restaurants, cafes, commercial and tourist shops, hotels, and dozens of archeological sites and museums. Therefore, the investigation of the sound in this particular area, which is the aim of this study, is quite important. As such, the study provides useful information to policymakers in order to reduce the negative impact of noise pollution on residents' quality of life and health, whereas the WebGIS developed allowed for a much wider dissemination of the key study findings a larger audience.

The present study may be expanded in different directions. A possible one could be toward increasing the geographical scale and extending the study period, which would potentially yield more comprehensive and conclusive findings. Additionally, the noise maps developed could be compared against building height data in respective locations, which could provide insights into the influence of buildings on sound propagation. Furthermore, to effectively communicate the study findings, together with the WebGIS, a dashboard could be created which would allow to visually presenting the results and enhancing reader comprehension and accessibility. Some of those suggestions are subject of ongoing research by our group.

References

Al-Hajri, S. M., Petropoulos, G. P., & Markogianni, V. (2020). Seasonal variation of key environmental parameters in the Sea of Oman using EO data and GIS. *Environment, Development and Sustainability, 23*(4), 6021–6046. https://doi.org/10.1007/s10668-020-00860-5

Altman, D. G., & Bland, J. M. (1995). *Statistics notes: The normal distribution BMJ, 310*, 298. https://doi.org/10.1136/bmj.310.6975.298

Arnal, L. H., Kleinschmidt, A., Spinelli, L., Giraud, A., & Mégevand, P. (2019). The rough sound of salience enhances aversion through neural synchronisation. *Nature Communications, 10*(1), 1–12. https://doi.org/10.1038/s41467-019-11626-7

Borchgrevink, H. M. (2003). Does health promotion work in relation to noise? *Noise and Health, 5*(18), 25–30. Retrieved from: https://pubmed.ncbi.nlm.nih.gov/12631433/.

Bryant, J. (2021). Digital mapping and inclusion in humanitarian response. *The Humanitarian Policy Group (HPG)*. http://cdn-odi-production.s3.amazonaws.com/media/documents/Digital_IP_Mapping_case_study_web_EDSoP6n.pdf.

Burrough, P. A., & McDonnell, R. A. (1998). *Principle of geographic information systems*. New York: Oxford. https://www.researchgate.net/publication/37419765_Principle_of_Geographic_Information_Systems.

Bustaffa, E., Curzio, O., Donzelli, G., Gorini, F., Linzalone, N., Redini, M., Bianchi, F., & Minichilli, F. (2022). Risk associations between vehicular traffic noise exposure and cardiovascular diseases: A residential retrospective cohort study. *International Journal of Environmental Research and Public Health, 19*(16). https://doi.org/10.3390/ijerph191610034

Can, A., Dekoninckb, L., & Botteldooren, D. (2014). Measurement network for urban noise assessment: Comparison of mobile measurements and spatial interpolation approaches. *Applied Acoustics, 83,* 32–39. https://doi.org/10.1016/j.apacoust.2014.03.012

Corbin, T. (2018). *ArcGIS Pro 2.x Cookbook: Create, manage, and share geographic maps, data, and analytical models using ArcGIS Pro.* Packt Publishing.

Delmelle, E. (2009). 'Spatial sampling', the SAGE handbook of spatial analysis [online]. *183,* 206. διαθέσιμο από: View article (google.com) [τελευταία πρόσβαση 20 Φεβρουαρίου 2022].

Directive 2002/49/EC. Directive 2002/49/EC of the European Parliament and of the council of 25 June 2002 relating to the assessment and management of environmental noise. https://eur-lex.europa.eu/legal-content/EN/TXT/?uri=celex%3A32002L0049.

EL.STAT. (2011). *Population count.* https://www.statistics.gr/2011-census-pop-hous.

Esri. (2022a). *Cross validation (geostatistical analyst)—ARCGIS Pro | documentation.* https://pro.arcgis.com/en/pro-app/latest/tool-reference/geostatistical-analyst/cross-validation.htm.

Esri. (2022b). *Get started with geostatistical analyst in ArcGIS pro—ArcGIS Pro | documentation.* https://pro.arcgis.com/en/pro-app/latest/help/analysis/geostatistical-analyst/get-started-with-geostatistical-analyst-in-arcgis-pro.htm.

Esri. (2022c). *Kriging (spatial analyst)—ARCGIS Pro | documentation.* https://pro.arcgis.com/en/pro-app/latest/tool-reference/spatial-analyst/kriging.htm.

Faka, A., Tserpes, K., & Chalkias, C. (2021). Environmental sensing. In G. P. Petropoulos, & P. K. Srivastava (Eds.), *GPS and GNSS technology in geosciences* (pp. 199–220). Elsevier.

Grubeša, S., Petošić, A., Suhanek, M., & Ðurek, I. (2018). Mobile crowdsensing accuracy for noise mapping in smart cities. *Automatika, 59*(3–4), 286–293.

Guillaume, G., Can, A., Petit, G., Fortin, N., Palominos, S., Gauvreau, B., Bocher, E., & Picaut, J. (2016). Noise mapping based on participative measurements. *Noise Mapping, 3*(1). https://doi.org/10.1515/noise-2016-0011

Harman, B. I., Koseoglu, H., & Yigit, C. O. (2016). Performance evaluation of IDW, kriging and multiquadric interpolation methods in producing noise mapping: A case study at the city of Isparta, Turkey. *Applied Acoustics, 112,* 147–157. https://doi.org/10.1016/j.apacoust.2016.05.024

Hucal, S. (2022). *Iconic European cities: Athens.* dw.com. https://www.dw.com/en/iconic-european-cities-athens/g-59854139.

Kalogeropoulos, K., Stathopoulos, N., Psarogiannis, A., Pissias, E., Louka, P., Petropoulos, G. P., & Chalkias, C. (2020). An Integrated GIS-Hydro modeling methodology for surface runoff exploitation via Small-Scale reservoirs. *Water, 12*(11), 3182. https://doi.org/10.3390/w12113182

Kaluarachchi, M., Waidyasekara, K. G. A. S., & Rameezdeen, R. (2022). *Environment project and asset management, vol. Ahead-of-print, issue ahead-of-print* (pp. 277–292). https://doi.org/10.1108/BEPAM-04-2020-0071

Kang, Tsung C. (2006). *Introduction to geographic information systems* (3rd ed.). New York: McGraw-Hill.

Karkaletsou, A. E. (2014). Noise mapping using geographic information systems - case study of Kallithea. *Attica. Dissertation Study. [online].* https://estia.hua.gr/browse/16517.

Kitsantonis, N. (July 4, 2022). In *Athens, creativity in art, food and more rises.* The New York Times. https://www.nytimes.com/2022/06/30/travel/athens-greece.html?register=email&auth=register-email.

Kwak, C. (2018). *36 hours in Athens.* The New York Times. https://www.nytimes.com/2018/10/11/travel/what-to-do-in-athens.html.

Meng, Q., Liu, Z., & Borders, B. E. (2013). Assessment of regression kriging for spatial interpolation – comparisons of seven GIS interpolation methods. *Cartography and Geographic Information Science, 40*(1), 28–39. https://doi.org/10.1080/15230406.2013.762138

Ministry of Tourist. (2023). *Greek application of big data analytics in smart tourist.* https://public.tableau.com/views/Final_GR/sheet1?%3AshowVizHome=noFielmoveclino.

Murphy, E., & King, E. A. (2022). *Soundscapes and noise mapping. Environmental noise pollution* (2nd ed., pp. 257–277). https://doi.org/10.1016/B978-0-12-820100-8.00009-9

National Institute of Deafness and Other Communication Disorders (NIDCD). (2022). *Noise-induced hearing loss.* https://www.nidcd.nih.gov/health/noise-induced-hearing-loss.

National Institute for Occupational Safety and Health (NIOSH). (2019). *Noise – reproductive health.* https://www.cdc.gov/niosh/topics/repro/noise.html.

ÖzkanSertlek, H., Slabbekoorna, H., Catea, C., & Ainslie, M. A. (2019). Source specific sound mapping: Spatial, temporal and spectral distribution of sound in the Dutch North Sea. *Environmental Pollution, 247,* 1143–1157. https://doi.org/10.1016/j.envpol.2019.01.119

Oyedepo, S. (2012). *Noise pollution in urban areas: The neglected dimensions* (pp. 259–271). Nigeria: Environmental Research Journal, Covenant University Ota Ogun State. https://doi.org/10.3923/erj.2012.259.271

Paraschou, A., & Maistrou, H. (2022). Abandoned or degraded areas in historic cities: The importance of multifunctional reuse for development through the example of the historic commercial triangle (commercial triangle) of Athens. *Land, 11*(1), 114. https://doi.org/10.3390/land11010114

Petropoulos, G. P., Kalivas, D., Georgopoulou, I. A., & Srivastava, P. K. (2015). Urban vegetation cover extraction from hyperspectral imagery and geographic information system spatial analysis techniques: Case of Athens, Greece. *Journal of Applied Remote Sensing, 9*(1), 096088. https://doi.org/10.1117/1.jrs.9.096088

Popa, A. M., Onose, D. A., Sandric, I. C., Dosiadis, E. A., Petropoulos, G. P., Gavrilidis, A. A., & Faka, A. (2022). Using GEOBIA and vegetation indices to assess small urban green areas in two climatic regions. *Remote Sensing, 14*(19), 4888. https://doi.org/10.3390/rs14194888

Presidential Decree 1180/1981. Government Gazette A-293/6-10-1981 On the regulation of issues relating to the establishment and operation of industries, crafts of all kinds of mechanical installations and warehouses and the safeguarding of the environment in general. Retrieved from: https://www.e-nomothesia.gr/ka-biomixania-biotexnia/pd-1180-1981.html.

Robertson, N. (1998). The city center of archaic Athens. *Hesperia: The Journal of the American School of Classical Studies at Athens, 67*(3), 283–302. https://doi.org/10.2307/2668475

Shellito, B. A. (2020). *Discovering GIS and ArcGIS Pro* (3rd ed.). W. H. Freeman publishing.

Shim, E., Kim, D., Woo, H., & Cho, Y. (2016). Designing a sustainable noise mapping system based on citizen scientists smartphone sensor data. *PLoS One, 11*(9), e0161835.

Srivastava, P. K., Pandey, P. C., Petropoulos, G. P., Kourgialas, N. K., Pandley, S., & Singh, U. (2019). GIS and remote sensing aided information for soil moisture estimation: A comparative study of interpolation techniques resources. *MDPI, 8*(2), 70.

Statuto, D., Cillis, G., & Picuno, P. (2016). Analysis of the effects of agricultural land use change on rural environment and landscape through historical cartography and GIS tools. *Journal of Agricultural Engineering, 47*(1), 28–39. https://doi.org/10.4081/jae.2016.468

Stone, M. (1974). Cross-validatory choice and assessment of statistical predictions. *Journal of the Royal Statistical Society: Series B, 36*(2), 111–133. https://doi.org/10.1111/j.2517-6161.1974.tb00994.x

Tobler, W. (1970). A computer movie simulating urban growth in the Detroit region. *Economic Geography, 46*(Suppl. ment), 234–240. https://web.archive.org/web/20190308014451/http://pdfs.semanticscholar.org/eaa5/eefedd4fa34b7de7448c0c8e0822e9fdf956.pdf.

Tsai, C., Chen, C., Chiang, W., & Lin, M. (2008). Application of geographic information system to the allocation of disaster shelters via fuzzy models. *Engineering Computations, 25*(1), 86–100. https://doi.org/10.1108/02644400810841431

Tsatsaris, A., Kalogeropoulos, K., Stathopoulos, N., Louka, P., Tsanakas, K., Tsesmelis, D. E., Krassanakis, V., Petropoulos, G. P., Pappas, V., & Chalkias, C. (2021). Geoinformation technologies in support of environmental hazards monitoring under climate change: An extensive review. *ISPRS International Journal of Geo-Information, 10*(2), 94. https://doi.org/10.3390/ijgi10020094

Wieczorek, W. F., & Delmerico, A. M. (2009). Geographic information systems. *Computational Statistics,* *1*(2), 167. https://doi.org/10.1002/wics.21

Wong, D. W. S., & Lee, J. (2005). *Statistical analysis of geographic information with ArcView GIS and ArcGIS.* Wiley.

Zuo, J., Xia, H., Liu, S., & Qiao, Y. (2016). Mapping urban environmental noise using smartphones. *Sensors,* *16*(10), 1692.

CHAPTER 22

Exploring the effect of the first lockdown due to covid-19 to atmospheric NO$_2$ using Sentinel 5P satellite data, Google Earth Engine and Geographic Information Systems

Georgios Gkatzios[1], George P. Petropoulos[1], Spyridon E. Detsikas[1] and Prashant K. Srivastava[2]

[1]Department of Geography, Harokopio University of Athens, Athens, Greece; [2]Remote Sensing Laboratory, Institute of Environment and Sustainable Development, Banaras Hindu University, Varanasi, Uttar Pradesh, India

1. Introduction

1.1 Air pollution and its major consequences

The term air pollution is applied to any form of pollution caused by natural or man–made causes of ambient air (Hutton, 2011, pp. 4–10; Mehmood et al., 2021a). Generally, it is an imbalance of the atmospheric condition due to the introduction of foreign agents within the air, which interact with each other causing serious consequences (Admassu & Wubeshet, 2006). Specifically, based on the World Health Organization (WHO), air pollution is defined as a condition in which the ambient atmosphere exhibits a high concentration of materials that are harmful to both the natural and man–made environment (Bao et al., 2019; Shao et al., 2019; Yassi et al., 2001, pp. 1–51). Those specific materials that cause it are called air pollutants.

Air pollutants are responsible for a multitude of effects across the whole spectrum of the natural environment. They are responsible for various losses of vegetation species such as trees and plants as they destroy their foliage and tissues (Duvris, 2009). In addition, they contribute to the occurrence and deterioration of a range of harmful events such as photochemical smog, the urban heat island effect, and the greenhouse effect (Popa et al., 2022). The contribution of pollutants to climate change is responsible for the occurrence of violent and catastrophic rainfall leading to significant geographical losses and periods of prolonged drought.

As far as human health is concerned, it is observed that the effects of air pollutants are mainly related to the occurrence of respiratory problems, deterioration of the condition of people suffering from cardiorespiratory problems or allergies (Srivastava, 2021), and generally affect the human body at developmental, neurological, and reproductive levels

Geographical Information Science
ISBN 978-0-443-13605-4
https://doi.org/10.1016/B978-0-443-13605-4.00016-3

(Kapitsimadis, 2009). In the European Union, air pollution causes, on average, more than 1000 premature deaths per day (Wojciechowski et al., 2018). In Greece, mainly in urban areas such as Athens, Thessaloniki, Patras, Heraklion, etc., specific types of pollutants such as particulate matter (**PM_{10}**), nitrogen oxides (**NOx**), carbon monoxide (**CO**) and carbon dioxide (**CO_2**), ozone (**O_3**), sulfur dioxide (**SO_2**), and hydrocarbons have been recorded (Valanidis et al., 2015, pp. 1−21).

The recording of the number of harmful pollutants in an area gives a picture of its environmental status. This provides an important source of information on the consequences of a possible accumulation of pollutants, so that the state or government concerned can take the appropriate action. In addition, the lockdown due to Covid-19 provides a rare opportunity to assess the change in pollution (Su et al., 2021).

1.2 Nitrogen dioxide

Nitrogen oxides (**NO_X**) are the nitrogen monoxide (**NO**) and nitrogen dioxide (**NO_2**) found in the atmosphere. Nitrogen monoxide is an odorless and colorless gas unlike dioxide which in turn is a gas with a strong odor and brown, red, and yellow color. Nitrogen oxides are the most dangerous pollutants due to the devastating effects they cause to both the natural and man-made environment. In fact, NO_2 is one of the main pollutants affecting air quality degradation (Kurniawan et al., 2023). Also, the presence of NO_X in high concentrations contributes to the thinning and eventual destruction of the ozonosphere. Its role is particularly important since it protects all living organisms from the excessive effects of UV radiation. NO_X effects are more intense and common in urban areas due to the majority of anthropogenic activities in such areas (Kwak et al., 2016). An example of that would be a number of commercial and residential buildings located along a road to form a street canyon. A busy street canyon is frequently regarded as harmful to human health due to the combination of large traffic volume and its weak ventilation capability (Bagieński, 2015).

At high levels, it causes serious impacts on plants, several aquatic life forms, and humans. Now, in terms of human health, studies have shown that high concentrations of NOx cause problems in the blood, liver, spleen, and especially the lungs. More specifically, the reaction of oxides with ammonia, moisture, and other compounds creating nitric acid, which in turn attacks lung tissue causing severe respiratory effects. Also, the microparticles created, when they enter the blood stream, cause worsening of heart disease, make the respiratory system prone to microbial infections, and reduce life expectancy (Kapitsimadis, 2009). Finally, this group includes nitrous oxide (**N_2O**). N_2O is a recent concern due to the acidification caused by the greenhouse effect and its contribution to the destruction of upper atmospheric ozone.

Oxide formation is either natural or occurs through human processes (Kopparapu et al., 2021, pp. 1−15). Regarding the first case, atmospheric nitrogen does not react with oxygen under normal conditions. The reason for this is that the nitrogen molecule (N_2) consists of a strong triple bond that practically renders it inert. However, the existence of high temperatures usually due to the occurrence of electrical discharges such as thunder and lightning can causes this reaction between oxygen and nitrogen (Huang et al., 2017).

Anthropogenic sources of nitrous oxide production are divided into two categories— static and mobile sources (Wang et al., 2022):

1. **Mobile sources**: This type of source contributes a large proportion of the emission of nitrogen oxides and is divided into road mobile sources such as cars, buses, trucks (Beirle et al., 2003) and nonroad mobile sources such as trains, marine vessels, construction equipment (e.g., bulldozers, excavators).
2. **Static sources**: This type refers to nonmoving sources and is divided into two categories point sources and area sources.
   - The point source is essentially a single stationary source of oxide emission in an area such as a stack of smoke.
   - An area source is a set of multiple emission points located within an area. A typical example is a set of natural gas compression engines along a pipeline.

Other examples of static sources include chemical plants, refineries, cement, and power plants. These plants need special combustion systems such as boilers, burners, furnaces, and gas or diesel generators in order to perform the necessary processes. These systems are the sources of nitrous oxide production (Cofala & Syri, 1998, pp. 5—20).

1.3 Geoinformatics in the study of air pollution

Over the years, several activities have been carried out to study, analyze, and map environmental air pollution and its effects, through the use of various sciences. The most important of these sciences is geoinformatics, which encompasses a range of disciplines. Among these are remote sensing, Geographic Information Systems (GIS), and Global Positioning System (GPS) technology. These technologies all together are important tools in analyzing environmental pollution (Wang et al., 2022). Specifically, satellite remote sensing is a key pillar of Earth observation (EO) because it offers significant recording capabilities. These are the systematic coverage of the Earth's atmosphere and surface, the collection of data on global climate variables (e.g., pollutant concentration), and the ability to analyze the data at temporal, spectral, and spatial levels (Asimakopoulou et al., 2021).

There is a multitude of advantages that satellite remote sensing offers in the study of air pollution such as having a continuous data history providing a detailed picture of the change of pollution in an area (Brown et al., 2021; Mehmood et al., 2021b). It also provides the ability to record pollution data in order to understand its dynamics in the absence of ground data (Marais et al., 2014). Another example is the possibility of extensive spatial coverage at both the spatial and global level. Typically, ground-based pollution monitoring instruments (e.g., air stations) cover a relatively small recording area, extending a few meters from the instrument's point of deployment (Hadjimitsis et al., 2010; Mehmood et al., 2022).

In the last 25 years, the European Space Agency (ESA) has been launching satellites to monitor and analyze the composition of the Earth's atmosphere. Some examples are the "Global Ozone Monitoring Experiment" (GOME-1) on the ERS-2 satellite, GOME-2 on the MetOp satellite, the "Scanning Imaging Absorption spectrometer for Atmospheric Cartography" (SCIAMACHY) on the Envisat satellite, and the Sentinel-4 and Sentinel-5 missions (Rosmino, 2017). Subsequently, the use of GIS technology in analyzing the environmental pollution of an area is a highly effective method (Mehmood et al., 2022; Ngo et al., 2012; Shao et al., 2019). GIS can be used to design spatial patterns of pollution in order to capture it in a region. They also offer the possibility of correlating pollution with other characteristics of the area in question, such as population or land cover (Briggs, 2005; Li et al., 2022). Due to these advantages that arise as far as the collection and analysis of spatial data is concerned, GIS are regarded as one of the most effective tools for performing spatial mapping studies suitable for air pollution monitoring. For example, Pandey et al. (2013) used the geostatistical analysis capabilities offered by GIS to show the correlation between long-term exposure to air pollutants and the development of certain diseases such as asthma, bronchitis, and even cancer.

Nowadays, the plethora of the available EO datasets has also created inevitably new challenges in EO data processing. Consequently, cloud-based platforms such as Google Earth Engine (GEE) have emerged as a very valuable tool to handle the abundance of big EO datasets. These cloud-based platforms allow a cost and computational efficient way to process big EO data on the cloud environment, supporting at the same time the implementation of different degree of complexity image processing and analysis techniques useful in the study of air pollutant dynamics (Demertzi et al., 2023; Tselka et al., 2023; Yang et al., 2022).

1.4 Study objectives

Based on the above framework, the present study aims to perform a spatiotemporal mapping of specific pollutant NO_2 in the prefecture of Thessaloniki, Greece, through the use of EO and GIS technologies. Specifically, the objectives of the study are defined as follows: (1) To perform a spatial-temporal mapping and study of NO_2 concentrations and changes in the area of Thessaloniki before and during the lockdown due to Covid and (2) to investigate the correlation of the results with specific characteristics of the study area which are the population density and land use. The study period is divided into two main time subperiods, one year before and during the first lockdown due to Covid.

2. Study area

2.1 Description of the study area

The prefecture of Thessaloniki is located in the region of Central Macedonia and is the second largest prefecture in northern Greece in terms of area and population (Perra et al., 2017). It is bordered to the east by the Strymonikos Gulf or Orfanos Bay and to the west by the Thermaikos River. In the corresponding central-northern part of the prefecture is located the valley of Mygdonia, bordered by Lake Koroneia or Lake Lagada with an area of 57 km^2 and to the east of this area is Lake Volvi, which in turn is the second largest lake in Greece with an area of 72 km^2 (its waters reach the Strymonikos Gulf). The city of Thessaloniki is located in the western part of the corresponding regional unit in the cove of Thermaikos. It is built amphitheatrically on the slopes of Cedar Hill. To the east it is surrounded by the Sheikh Su forest, while the airport and Thermi are located to the south. To the southeast is the panorama area, while the industrial zone is located in Sindos. Finally, according to the 2021 census, the prefecture of Thessaloniki has 1,091,424 inhabitants, which testifies to the decrease of the population of the region by 1.7% in the last decade, since in 2011 it reached 1,110,551 inhabitants (a decrease of 19,127 inhabitants) (ELSTAT, 2021).

The area covering the whole of the prefecture has been selected as a study area. This is because the prefecture of Thessaloniki is the second largest in terms of population after Attica and, as already mentioned, NO_2 levels are inextricably linked to anthropogenic activities. Furthermore, within the prefecture two smaller areas were selected for further study and these are the municipality of Thessaloniki, which has the largest population compared to the rest of the municipalities and the industrial area in Sindos.

Spatial organization of the prefecture: The New Regulatory Plan of Thessaloniki (RSP) was approved in 2014. The RSP is a set of policy guidelines aimed at implementing measures for environmental protection, spatial organization, and economic prosperity of a region. Therefore, based on the 2014 RSP, the prefecture of Thessaloniki is divided into the following spatial zones shown in Fig. 22.1.

1. Metropolitan zone
2. Urban complex of Thessaloniki
3. Other area zone

2.2 Description of data

To assess the NO_2 concentration data during the examined time in the prefecture of Thessaloniki, Sentinel 5 Precursor data were used. The Sentinel 5P satellite is part of the European Earth Observation Program Copernicus and its mission is to study the quality and composition of the atmosphere worldwide. Sentinel-5 Precursor is a satellite launched on October 13, 2017 by the ESA to monitor air pollution. The onboard sensor is frequently referred to as TROPOMI (TROPOspheric Monitoring Instrument). Now, the TROPOMI instrument is a multispectral sensor that records reflectance of wavelengths

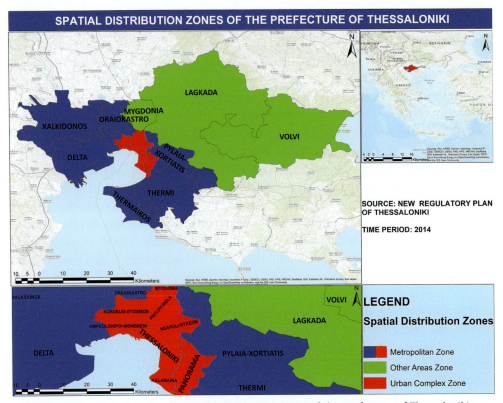

Figure 22.1 Map showing the spatial distribution zones of the prefecture of Thessaloniki.

important for measuring atmospheric concentrations of ozone, methane, formaldehyde, aerosol, carbon monoxide, nitrogen oxide, and sulfur dioxide, as well as cloud characteristics at a spatial resolution of 0.01 arc degrees. (Source: Sentinel-5 – Missions – Sentinel Online – Sentinel Online (copernicus.eu)) (Last access on July 20, 2023).

All of the S5P datasets, except CH_4, have two versions: near real-time (NRTI) and offline (OFFL). CH_4 is available as OFFL only. The NRTI assets cover a smaller area than the OFFL assets, but appear more quickly after acquisition. The OFFL assets contain data from a single orbit (which, due to half the earth being dark, contains data only for a single hemisphere).

Because of noise on the data, negative vertical column values are often observed in particular over clean regions or for low SO_2 emissions. It is recommended not to filter these values except for outliers, that is, for vertical columns lower than $-0.001 \ mol \ m^{-2}$.

The Sentinel 5P NO_2 concentration OFFL product was accessed using the GEE (Source: Sentinel-5P OFFL NO_2: Offline Nitrogen Dioxide | Earth Engine Data Catalog | Google for Developers) (Last access on December 15, 2022). From the set of products on the website, the product "Tropospheric Column NO_2 Number Density" is selected, which has NO_2 concentration data detected in the troposphere. This product is used due to the fact that NO_2 concentrations in the troposphere are significantly more affected by anthropogenic activity than the rest of the atmospheric layers.

In addition to the Sentinel-5P data, two other datasets were used as ancillary data. The first dataset included shapefiles of Greek prefectures, whose boundaries are defined on the basis of the national Kallikratis system Source: Digital Cartographic Data—ELSTAT (statistics.gr) (Last Access on December 13, 2022). Also, the shapefile of the industrial area of Sindos is used, which is constructed through the application "Google Earth Pro."

For the second objective of the research, the data of the land cover of Greece collected by the Sentinel 1 and Sentinel 2 satellites and available on the "Earth Engine Datasets Catalog" Website (Source: ESA WorldCover 10m v100 | Earth Engine Data Catalog | Google for Developers (Last access on December 12, 2022).

3. Methodology

3.1 Data preprocessing

In order to achieve the objectives of the research, specific steps are followed: Firstly, the product "Tropospheric Column NO_2 Number Density" in its original form it contains information on the concentration of NO_2 on a global scale with a period of record from June 28, 2018 to December 18, 2022 (during the time of selection). From this set of information, NO_2 concentrations should be extracted only for the study area in the desired time period, which is the month the first lockdown due to Covid-19 occurred for 2020 and the corresponding one for 2019.

The time period under study is the following:
- **04 April—04 May 2019/2020**: This month is the main study period for lockdown. Of course, the lockdown in Greece was implemented on March 23, 2020, but in this study, the first 10 days of the lockdown are excluded in order to show a substantial reduction in NO_2 in the results.

Finally, the product has as unit of measurement for NO_2 concentrations mol m^{-2}, which are converted to μmol m^{-2}, a procedure commonly performed in similar studies through the international literature in order to facilitate data comparison and analysis.

Based on the above framework, this product is subjected to a preprocessing, which is carried out through the online platform "Google Earth Engine" (GEE), a widely used cloud platform in EO data analysis for different disciplines (e.g., Demertzi et al., 2023; Tselka et al., 2023). In the present study, implementation of GEE of the preprocessing of the "Tropospheric Column NO_2 Number Density" in GEE is the export of two rasters containing the above desired information and transformations, for the selected study month for the years 2019/2020.

3.2 Data processing

Following the preprocessing of the datasets, the average monthly value of the NO_2 concentration for the study month for the years 2019 and 2020 was obtained for the prefecture of Thessaloniki. In addition, another raster file is created from the previous two, showing the change in NO_2 levels between the respective months for the two years. Specifically, the raster file with the 2020 values was subtracted in each case from the raster file with the 2019 values.

The final stage in the implementation of the first research objective is the statistical analysis of the data. First, the histograms of the monthly distribution of mean NO_2 concentrations for each of the months under study are produced. Then, the scatter plots between the respective months are generated in order to obtain the spatial autocorrelation of the values. At this point, in addition, the degree of correlation between the values is calculated using the nonparametric Spearman rho index. The equation of this correlation index is presented below:

$$r_s = 1 - \left(6 * \sum_{i=1} n * d^2\right) / n^3 - n \tag{22.1}$$

where: r_s: correlation coefficient

   d: difference between the ranked values of the samples
   (the concentrations of NO_2)
   n: number of sample observations

The set of index values is $[-1, 1]$ with $r_s = 1$ indicating absolute positive correlation, $r_s = -1$ indicating absolute negative correlation, and $r_s = 0$ indicating the absence of correlation.

In the next step, monthly distribution plots of the mean, maximum, and minimum NO_2 concentrations for each month are produced. The plots are generated via GEE and in each case give values for the whole study area and not for each pixel as in the generated raster images. In addition, the command in GEE "Interpolate Nulls" was used to remove null values. In the case of the minimum value, of course, very small NO_2 values (near zero) occurred, which GEE displayed as negative and then converted to zero to facilitate data analysis. Finally, a graph showing the average mean concentration for both years is also generated.

Figs. 22.2 and 22.3 below show the NO_2 distribution maps for the study month along with the corresponding histograms of values for the two years.

3.3 Spatial correlation of NO_2 distribution with site characteristics

The next stage of the methodological approach concerns the implementation of the study's second objective. Therefore, for its realization, the construction of maps showing land cover for the prefecture of Thessaloniki and the distribution of population density is required. For the construction of the first map, the product "ESA WorldCover 10m 2020" is utilized, which provides the land cover at a global level. It is then imported into GEE for preprocessing, through which the land cover for the prefecture of Thessaloniki is isolated. The next step is to export the file from GEE as a raster and import it into "ARCGIS," in order to perform the processing.

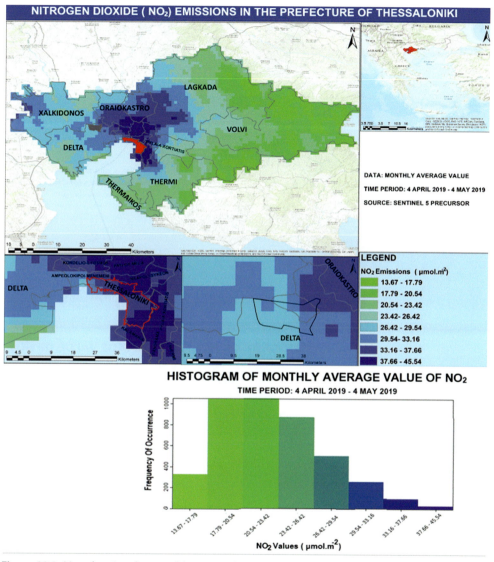

Figure 22.2 Map showing the monthly mean NO$_2$ concentration for the study month for 2019 along with the corresponding histogram of values.

In the following, only the shapefile of the prefecture of Thessaloniki will be used to construct the population density map. The population density for an area is calculated by the following mathematical formula.

$$\text{Population density} = \text{Total population}/\text{Total area (Km}^2) \qquad (22.2)$$

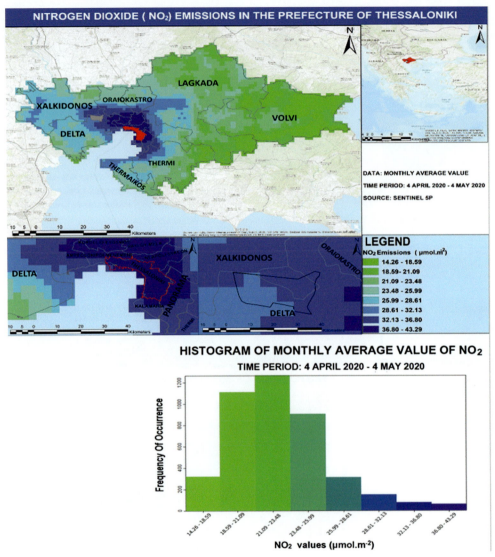

Figure 22.3 Map showing the monthly mean NO$_2$ concentration for the study month for 2020 along with the corresponding histogram of values.

Based on the above formula, the area in square kilometers of each municipal unit of the prefecture of Thessaloniki is calculated and added as a new layer to the attribute table of the prefecture's shapefile. Then, the number of inhabitants for each municipal unit is inserted, again as a new layer in the attribute table. Finally, Eq. (22.2) is generated through which the population density is calculated.

4. Results

4.1 Spatial distribution results of NO_2

This section presents the results of the spatial distribution of NO_2. Specifically, Fig. 22.5 below shows the difference map between the NO_2 values for the second month of the study alongside the corresponding scatter plot and mean value of the average NO_2 concentration.

From Fig. 22.4, the following can be observed: Positive and negative difference values appear in the whole of the studied region. The appearance of positive values indicates that 2019 showed higher NO_2 concentrations compared to 2020. The largest positive values range between 4.97 and 12.57 $\mu mol\ m^{-2}$ and are mainly found within the metropolitan zone in the municipality of Oraeokastro, the municipality of Chalkidonos and within the other zone mainly in the municipality of Lagada.

Then, the existence of areas with negative difference values indicates that those in 2020 during the lockdown had higher concentrations than in 2019. The largest absolute negative difference values range between 4.49 and 9.52 $\mu mol\ m^{-2}$ and are mainly located within the metropolitan area such as in the Urban Complex, the municipality

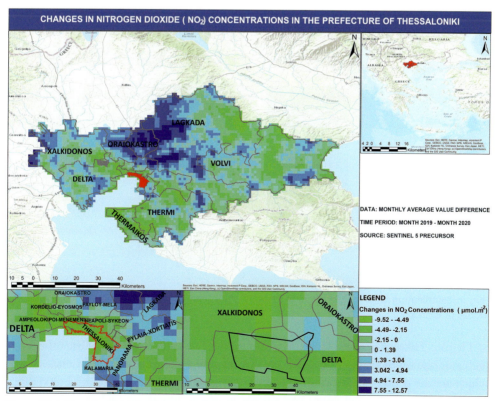

Figure 22.4 Map showing the change in the monthly mean NO_2 concentration of the study month for the years 2019–2020.

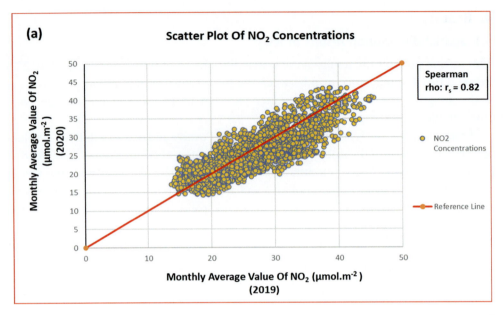

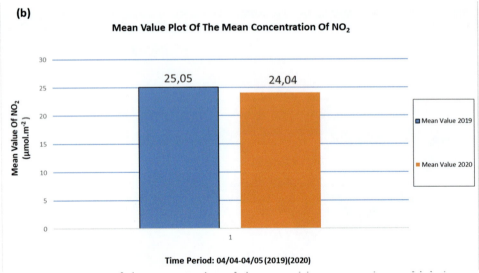

Figure 22.5 Presentation of the scatter plot of the monthly mean values of (a) the concentration of NO$_2$ in combination with (b) the average value of the mean concentration for the corresponding month of each year.

of Thermi, Thermaikos, and Delta. In addition, they also occur in areas of the other zone such as in the municipality of Lagada. Finally, regarding the two study areas under study, which are the center of Thessaloniki and the industrial area of Sindos, it is observed that in 2020, although the lockdown was already in place, they show higher NO$_2$ concentrations than in 2019.

Based on Fig. 22.5a, the distribution of values around the reference line does not give a clear picture of which year had the highest NO_2 concentrations. For this reason, the average value of the mean NO_2 concentration for the study month of each year was calculated, which is presented in Fig. 22.5b. Therefore, based on the latter graph, it can be seen that the mean concentration of 2019 reaches 25.05 $\mu mol\ m^{-2}$ and is relatively higher than the corresponding mean value of 2020, which reaches 24.04 $\mu mol\ m^{-2}$. The conclusion is that the study month showed higher NO_2 values for 2019 than 2020 having a very small difference.

The next objective is to test the correlation of the values. The first step is to calculate the probability value (P-value), which shows that P-value $= .000 < 0.001$, where 0.001 is the F level of significance defined for this study. The result shows the existence of correlation between the values of the two years. Next, Spearman's correlation index rho is calculated resulting in $r_s = 0.82$. This value indicates a strong positive monotonic correlation between the monthly values of the two years which means that when the 2019 values increase, the 2020 values tend to increase as well.

Below are the graphs of the average, maximum, and minimum values of NO_2 for the years 2019 and 2020 in the respective figures.

Fig. 22.6 shows that during the month under study for 2019 (a), rapid fluctuations in NO_2 values occur at certain periods. Specifically, this is the period between 06 April—15 April and 19 April −24 April. In fact, on 15 April and 20 April, there is a matching of all three values, while in the period 22 April −25 April, there is a matching between maximum and average values only.

In the case of the corresponding month for 2020 (b), large fluctuations are again observed in certain periods. The first period ranges between 07 April and 14 April where on 12 April the three values are matched. A similar fluctuation phenomenon occurred in the corresponding period of 06 April −15 April for 2019. Then, the second period of rapid fluctuation ranges between 01 May and 03 May, with the three values matching on 02 May. Also, in the period between 15 April and 30 April, the fluctuations are smooth. Finally, it appears that in both years, the concentration values do not show a large divergence between them.

4.2 Spatial correlation results of NO_2

This section presents the main results of the spatial correlation of the distribution of NO_2 concentrations with characteristics of the study area, namely population density and land use. Their distributions are presented in Figs. 22.7 and 22.8.

The data presented in Fig. 22.7 lead to the initial conclusion that the metropolitan area of the county has a higher population density overall than the rest of the area. Then, as regards the areas within the metropolitan, with the highest population density is found in the Thessaloniki Urban Complex. Based on the above framework and in combination with the data presented in the maps of distribution and variation of NO_2

(a)

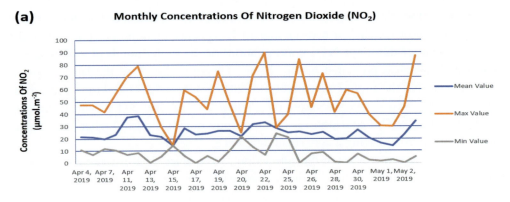

(b)

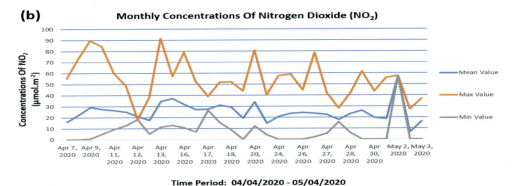

Figure 22.6 (a,b) Plots of mean, maximum, and minimum values of NO_2 concentration for the years 2019/2020.

concentrations for the month under study, which are presented in Figs. 22.2, 22.3 and 22.4, the following can be seen: In each case, the highest distribution values and absolute value differences of the mean NO_2 concentration are located within the metropolitan area of the county and mainly within the Urban Complex.

Next, the map of the land use distribution is shown below.

The map data in Fig. 22.8 show that the land use of the prefecture of Thessaloniki is distributed as follows: In the Other Area zone, nature plays the dominant role, with a small proportion of urban buildings and crops. In contrast, in the metropolitan zone, there is more urban land use with the Urban Complex of Thessaloniki having a dominant role in that part. There is also a high proportion of crops located in the metropolitan area. Therefore, based on the above context, it can be seen that the highest distribution values of the average NO_2 concentration are found within the metropolitan zone of the county and mainly within the Urban Complex. Similarly, the highest absolute value differences are also found within this zone.

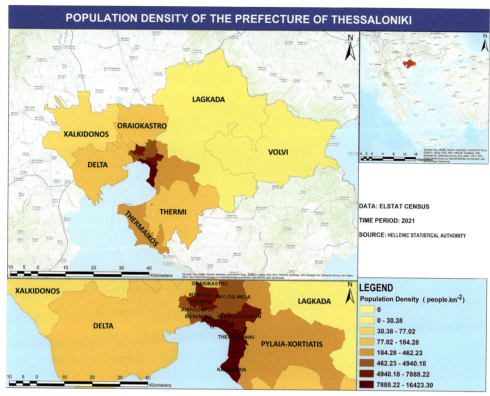

Figure 22.7 Map showing the population density distribution of the prefecture of Thessaloniki. *(Source: Hellenic Statistical Authority.)*

4.3 Discussion

In this section, the discussion of the results already presented in relation to the two research study objectives is carried out. For the first objective, it is found that the highest concentrations of NO_2 in each year are mainly located within the metropolitan area. This phenomenon is based on the fact that this zone has the highest population concentration. Specifically, high population levels correspond to a large number of anthropogenic activities such as increased use of transport, which in turn are a major source of nitrous oxide emissions (Beirle et al., 2003; Koukouli et al., 2022). For this reason, the area of Thessaloniki's Urban Complex shows the highest NO_2 concentrations within the zone. Subsequently, the existence of industrial activity in the Sindos area contributes to the increase in nitrous oxide emissions due to its various processes (e.g., combustion systems) (Kapitsimadis, 2009). Also, the high proportion of crops, located throughout the zone may be associated with the emission of significant amounts of NO_X through the use of agricultural activities (e.g., use of agricultural machinery).

Analysis results also showed that the lowest concentrations of NO_2 occur mainly within the rest area zone. This may be because this area has the lowest percentage of

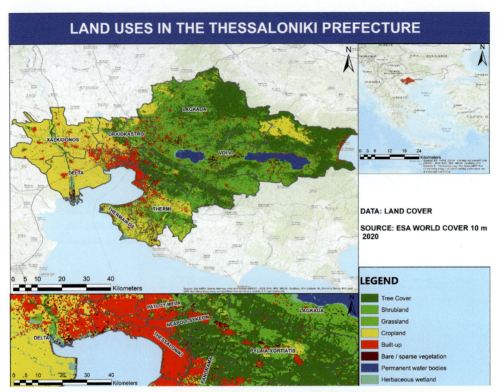

Figure 22.8 Map showing the land use distribution of the prefecture of Thessaloniki. *(Source: ESA WorldCover 10m 2020.)*

population and corresponding anthropogenic activities, and therefore the lowest sources of NO_X emissions. Nevertheless, there are cases where high values also occur in this area. This may be due to the winds blowing along the prefecture, which have carried a quantity of NO_2 from areas of higher concentrations (Bi et al., 2020; Cakir & Sita, 2020). Finally, it should be noted that in both years, the lower values have a higher frequency of occurrence than the high ones due to the fact that the latter are mainly restricted to the metropolitan area only, as mentioned above. The lower values, on the contrary, are distributed throughout the whole county.

For the two study areas, the center of Thessaloniki and the industrial area of Sindos, it is found that higher amounts of NO_2 occur in the city center than in the industrial area in both years. This finding is reasonable because NO_X emissions are more influenced by transport than by industrial activities.

Concerning the change in NO_2, the conclusion drawn is that NO_2 concentrations showed very little decrease between the two years, even though the lockdown was fully implemented in the month of 2020. Due to this fact it would be more logical that this month would show high decreases in NO_2 concentrations relative to the previous year.

However, a similar study conducted and recorded monthly average NO_2 concentrations using Sentinel 5P data reached a similar result (Koukouli et al., 2021). A possible reason for this situation may be the conduct of various agricultural activities, which are distributed over a wide area of the county. Even during the lockdown period such activities continue to take place resulting in the continuous emission of NO_X (Beirle et al., 2003). These emissions combined with others generated, even if in small amounts (e.g., due to some transportation use) are likely to have not allowed for the rapid reduction that would be expected.

Finally, at this point, it should be noted that the process of preparing for the aforementioned results has encountered certain obstacles. One of them is the impossibility of collecting data on the meteorological events prevailing in the respective time periods in the study area. Events such as high moisture levels in the atmosphere or the type of winds that prevailed in the area contribute to a more detailed approach to the distribution and variation of NO_2 concentration (Kwak et al., 2016). Similar results were presented in a similar study. Also, another limitation was the inability to find consistent ground-based measurement data for the 2019/2020 period to allow comparison of fluctuation trends between these data and the Sentinel 5P data for both years. It should also be noted at this point that the measurement units of the Sentinel 5P data are only mol m^{-2}, which does not allow for any effective comparison with the ground-based data for the region, which have a measurement unit of μg m^{-3}.

5. Final remarks

The present study aimed to map the distribution and variation of nitrogen dioxide concentration in the prefecture of Thessaloniki, Greece, in the period one year before and during the first lockdown using Sentinel-5 and other geospatial data. In this study, Sentinel 5P satellite logging data were used for the time period April 04, 2020—May 04, 2020 which is the main period of the lockdown due to covid in 2020 and the corresponding period one year before was utilized.

Atmospheric pollution is a major problem because of its impact on both the natural environment and human health. It is, therefore, important to study and analyze the factors and substances that contribute to the aggravation of this problem. These substances, or air pollutants, become harmful when their concentration reaches a certain level. The sources of pollutant emissions are inextricably linked to a variety of processes, some of which are natural and some of which are man-made, the latter being responsible for the majority of the production. Based on the results, it was found that there was a decrease in NO_2 concentrations between the respective month between the two years. In particular, a very small decrease in NO_2 values was observed, a result which, while seemingly an oxymoron, was also observed in other studies.

The results of the present study are very important in general because a detailed picture of the air quality of a region provides information on the factors that cause the fluctuations of pollutants and the consequences that may have been caused by their accumulation. In this way, any attempt to mitigate the effects and avoid future consequences can be made possible by taking appropriate measures.

For the above reasons, it is important to conduct more similar studies and to continue the present one. Of course, as far as this study is concerned, its conduct encountered some limitations that led to the absence of some information that could provide a more detailed picture of the distribution of NO_2 concentrations. Any future follow-up efforts should overcome these limitations.

Based on this framework, the creation and use of meteorological models is proposed for the preparation of similar environmental research. The use of such models will provide information on the respective conditions prevailing in the area of the prefecture of Thessaloniki during the study period (e.g., humidity levels, type and direction of winds). This information, after correlation with the NO_2 concentration data, will provide a detailed picture of the distribution of the pollutant over the whole range of the prefecture. Also, another parameter is to find continuous ground-based measurements for NO_2 over the whole study period, in order to compare them with the satellite data and analyze the discrepancies between the values, if any. Finally, it is of major importance to carry out similar studies with other metropolitan areas both in Greece and abroad as the main study area.

Last but not least, the methodological approach employed herein could be developed also in cloud computing platforms such as GEE, which can detect changes, map trends, and quantify differences in the spatial and temporal domain so it is considered an ideal and innovative tool for detecting and mapping changes in air pollutants.

References

Admassu, M., & Wubeshet, M. (2006). *Air pollution, lecture notes distributed in the topic environmental health science*. University of Gondar. on August 2006.

Asimakopoulou, P., Nastos, P., Vassilakis, E., Hatzaki, M., & Antonarakou, A. (2021). Earth observation as a facilitator of climate change education in schools: The teachers' perspectives. *Remote Sensing, 13*(8). https://doi.org/10.3390/rs13081587

Bagieński, Z. (2015). Traffic air quality index. *Science of the Total Environment, 505*(1), 606−614. https://doi.org/10.1016/j.scitotenv.2014.10.041

Bao, Y. L., Guan, Zhu, Q., Guan, Y., Lu, Q., Petropoulos, G. P., Che, H., Ali, G., Dong, Y., Tang, Z., Gu, Y., Tang, W., & Hou, Y. (2019). Assessing the impact of Chinese FY-3/MERSI AOD data assimilation on air quality forecasts: Sand dust events in Northeast China. *Atmospheric Environment, S1352−2310*(19), 30118−30119. https://doi.org/10.1016/j.atmosenv.2019.02.026

Beirle, S., Platt, U., Wenig, M., & Wagner, T. (2003). Weekly cycle of NO_2 by GOME measurements: A signature of anthropogenic sources. *Atmospheric Chemistry and Physics, 3*(6). https://doi.org/10.5194/acp-3-2225-2003

Bi, C., Chen, Y., Zhao, Z., Li, Q., Zhou, Q., Ye, Z., & Ge, X. (2020). Characteristics, sources and health risks of toxic species (PCDD/Fs, PAHs and heavy metals) in PM2.5 during fall and winter in an industrial area. *Chemosphere, 238*. https://doi.org/10.1016/j.chemosphere.2019.124620

Briggs, J. D. (2005). The role of GIS: Coping with space (and time) in air pollution exposure assessment. *Journal of Toxicology and Environmental Health, Part A, 68*(13-14), 1243−1271. https://doi.org/10.1080/15287390590936094

Brown, A., Hayward, T., Timmis, R., Wade, K., Pope, R., Trent, T., Boesch, H., & Guillo, R. B. (2021). *Satellite measurements of air quality and greenhouse gases: Application to regulatory activities.* Bristol, England: Environment Agency Horizon House, Deanery Road. Bristol BS1 5AH.

Cakir, S., & Sita, M. (2020). Evaluating the performance of ANN in predicting the concentrations of ambient air pollutants in Nicosia. *Atmospheric Pollution Research, 11*(12). https://doi.org/10.1016/j.apr.2020.06.011

Cofala, J., & Syri, S. (1998). *Nitrogen oxides emissions, abatement technologies and related costs for Europe in the RAINS model database* (pp. 5—20). International Institute for Applied System Analysis.

Demertzi, I. I., Detsikas, S. E., Tselka, I., Petropoulos, G. P., & Karymbalis, E. (2023). Deposition and erosion dynamics in Axios and Aliakmonas river deltas (Greece) with the use of Google earth engine and geospatial analysis tools. Chapter X, pp: Xx-xx. In N. Stathopoulos, A. Tsatsaris, & K. Kalogeropoulos (Eds.), *Geoinformatics for geosciences. Advanced geospatial analysis using RS, GIS and soft computing.* Publisher Elsevier, ISBN 9780323989831 (accepted).

Duvris, E. (2009). *Atmospheric pollution and its impact on climate change, thesis.* University of Piraeus, Department of Maritime Studies.

Hadjimitsis, D. G., Nisantzi, A., Themistocleous, K., Matsas, A., & Trigkas, V. (2010). Satellite remote sensing, GIS and sun–photometers for monitoring PM10 in Cyprus: Issues on public health. In *Proceedings of SPIE 7826, sensors, systems, and next-generation satellites XIV, 78262C, Toulouse, France, 13 October 2010.* https://doi.org/10.1117/12.865120

Hellenic Statistical Authority. (2022). *Residential population census results ELSTAT 2021, (online), Available at: 2021 population-housing census - ELSTAT (statistics.gr)* (Last Access on 15/12/2022).

Huang, H., Fu, D., & Qi, W. (2017). Effect of driving restrictions on air quality in Lanzhou, China: Analysis integrated with internet data source. *Journal of Cleaner Production, 142*(2), 1013—1020. https://doi.org/10.1016/j.jclepro.2016.09.082

Hutton, G. (2011). *Air pollution: Global damage costs of air pollution from 1900 to 2050, assessment paper, Copenhagen Consensus on human challenges* (pp. 4—10). https://doi.org/10.1017/CBO9780511493560.006

Kapitsimadis, F. (2009). *Exposure to air pollution and respiratory function. Study on postal delivery workers in the Athens area.* PhD thesis. Greece: National and Kapodistrian University of Athens.

Kopparapu, R., Arney, G., Haqq-Misra, J., Yaeger-Lustig, J., & Villanueva, G. (2021). *Nitrogen dioxide pollution as a signature of extraterrestrial technology, astrophysics: Earth and planetary astrophysics* (pp. 1—15). Cornell University. https://doi.org/10.3847/1538-4357/abd7f7

Koukouli, M.-E., Pseftogkas, A., Karagkiozidis, D., Skoulidou, I., Drosoglou, T., Balis, D., Bais, A., Melas, D., & Hatzianastassiou, N. (2022). Air quality in two northern Greek cities revealed by their tropospheric NO2 levels. *Atmosphere, 13*(5), 840. 2022. https://doi.org/10.3390/atmos13050840.

Koukouli, M.-E., Skoulidou, I., Karavias, A., Parcharidis, I., Balis, D., Manders, A., Segers, A., Eskes, H., & van Geffen, J. (2021). Sudden changes in nitrogen dioxide emissions over Greece due to lockdown after the outbreak of COVID-19. *Atmospheric Chemistry and Physics, 21*, 1759—1774. https://doi.org/10.5194/acp-21-1759-2021

Kurniawan, R., Alamsyah, R. A., Fudholi, A., Purwanto, A., Sumargo, B., Gio, U. P., Wongsonadi, K. S., & Susanto, E. A. (2023). Impacts of industrial production and air quality by remote sensing on nitrogen dioxide concentration and related effects: An econometric approach. *Environmental Pollution, 334*, 5—20. https://doi.org/10.1016/j.envpol.2023.122212

Kwak, H. K., Lee, H. S., Seo, M. J., Park, B. S., & Baik, J. J. (2016). Relationship between rooftop and on-road concentrations of traffic-related pollutants in a busy street canyon: Ambient wind effects. *Environmental Pollution, 208*, 185—197. https://doi.org/10.1016/j.envpol.2015.07.030

Li, M., Wu, Y., Bao, Y., Liu, B., & Petropoulos, G. P. (2022). Near-surface NO_2 concentration estimation by random forest modeling and Sentinel-5P and ancillary data. *Remote Sensing, 14*(15), 3612. https://doi.org/10.3390/rs14153612

Marais, E. A., Jacob, D. J., Wecht, K., Lerot, C., Zhang, L., Yu, K., Kurosu, T. P., Chance, K., & Sauvage, B. (2014). Anthropogenic emissions in Nigeria and implications for atmospheric ozone pollution: A view from space. *Atmospheric Environment, 99.* https://doi.org/10.1016/j.atmosenv.2014.09.055

Mehmood, K., Mushtag, S., Bao, Y., Sadia-Bibi, S., Yaseen, M., Khan, M. A., Abrar, M. M., Ulhassan, Z., Fahad, S., & Petropoulos, G. P. (2022). The impact of COVID-19 pandemic on air pollution: A global research framework, challenges and future perspectives. *Environmental Science and Pollution.* https://doi.org/10.1007/s11356-022-19484-5 (in press).

Mehmood, K., Bao, Y., Petropoulos, G. P., Abbas, R., Abrar, M. M., Saifullah, Mustafa, A., Shah Saud, A. S., Ahmad, M., Hussain, I., & Fahadl, S. (2021b). Investigating connections between COVID-19 pandemic, air pollution and community interventions for Pakistan employing geoinformation technologies. *Chemosphere, 272*. https://doi.org/10.1016/j.chemosphere.2021.129809

Mehmood, K., Bao, Y., Abbas, R., Saifullah, Petropoulos, G. P., Raza Ahmad, H., Abrar, M. M., Mustafa, A., Abdalla, A., Lasaridi, K., & Fahad, S. (2021a). Pollution characteristics and human health risk assessments of toxic metals and particle pollutants via soil and air using geoinformation in urbanised city of Pakistan. *Environmental Science and Pollution*. https://doi.org/10.1007/s11356-021-14436-x (in press).

Ngo, B. T., Nguyen, A. T., Vu, N. Q., Chu, H., & Cao, M. Q. (2012). Management and monitoring of air and water pollution by using GIS technology. *Journal of Vietnamese Environment, 3*(1). https://doi.org/10.13141/jve.vol3.no1.pp50-54

Pandey, M., Singh, V., Vaishya, R. C., & Shukla, A. K. (2013). Analysis and application of GIS based air quality monitoring-state of art. *International Journal of Engineering Research and Technology, 02*(12).

Perra, V. M., Sdoukopoulos, A., & Latinopulou, P. M. (2017). *Evaluation of sustainable urban mobility in the city of Thessaloniki Vol. 24.* (pp. 329–336) Transportation Research Procedia. https://doi.org/10.1016/j.trpro.2017.05.103.

Popa, A. M., Onose, D. A., Sandric, I. C., Dosiadis, E. A., Petropoulos, G. P., Gavrilidis, A. A., & Faka, A. (2022). Using GEOBIA and vegetation indices to assess small urban green areas in two climatic regions. *Remote Sensing, 14*(19), 4888. https://doi.org/10.3390/rs14194888

Rosmino, C. (2017). *Scientists at Bremen University tracking air pollution in our citied. 16 March, Available at: Scientists at Bremen University tracking air pollution in our cities.* Euronews (Last Access on 19/03/2023).

Shao, M. Y. Bao, Petropoulos, G. P., & Zhang, H. (2019). A two-season impact study of radiative forced tropospheric response to stratospheric initial conditions inferred from satellite radiance assimilation. *Climate MDPI, 7*(114), 1–11. https://doi.org/10.3390/cli7090114 (IF: 1.950).

Srivastava, K. A. (2021). Air pollution: Facts, causes and impact. In P. R. Singh (Ed.), *Asian atmospheric pollution: Sources, characteristics and impacts* (pp. 1–25). Elsevier Science. https://doi.org/10.1016/B978-0-12-816693-2.00020-2

Su, Z., Li, X., Liu, Y., & Deng, B. (2021). The multi-time scale changes in air pollutant concentrations and its mechanism before and during the COVID-19 periods: A case study from Guiyang, Guizhou Province. *Atmosphere, 12*(11). https://doi.org/10.3390/atmos12111490

Tselka, I., Detsikas, S. E., Petropoulos, G. P., & Demertzi, I. I. (2023). Google earth engine and machine learning classifiers for obtaining burnt area cartography: A case study from a Mediterranean setting. Chapter X, pp: Xx-xx. In N. Stathopoulos, A. Tsatsaris, & K. Kalogeropoulos (Eds.), *Geoinformatics for geosciences. Advanced geospatial analysis using RS, GIS and soft computing.* Publisher Elsevier, ISBN 9780323989831 (accepted).

Valanidis, A., Vlachogianni, T., Loridas, S., & Fiotakis, C. (2015). *Atmospheric Pollution in Urban Areas of Greece and Economic Crisis: Trends in air quality and atmospheric pollution data, research and adverse health effects* (pp. 1–21). Dpt. of Chemistry, University of Athens.

Wang, H., Zhou, J., Li, X., Ling, Q., Wei, H., Gao, L., He, Y., Zhu, M., Xiao, X., Liu, Y., Li, S., Chen, C., Duan, G., Peng, Z., Zhou, P., Duan, Y., Wang, J., Yu, T., Yang, Y., … Ding, Y. (2022). Review on recent progress in on-line monitoring technology for atmospheric pollution source emissions in China. *Journal of Environmental Sciences, 123*. https://doi.org/10.1016/j.jes.2022.06.043

Wojciechowski, J., Wisniewska-Danek, K., Friel, C., Coelho, j., Soblet, F., Niemenmaa, V., Happach, B., Kubat, J., Otto, J., Pirelli, L., Simeonova, R., Zalega, A., & O'Doherty, R. (2018). *Air pollution: Our health still insufficiently protected, special report num. 23, 2018.* European Court of Auditors, Luxembourg: Publications Office. https://data.europa.eu/doi/10.2865/170737.

Yang, L., Driscol, J., Sarigai, S., Wu, Q., Chen, H., & Lippitt, C. D. (2022). Google earth engine and artificial intelligence (AI): A comprehensive review. *Remote Sensing, 14*(14), 3253. https://www.mdpi.com/2072-4292/14/14/3253.

Yassi, A., Kjellström, T., De Kok, T., Guidotti, T. L., & World Health Organization. (2001). *Basic environmental health* (pp. 1–51). Oxford, U.K.: Oxford University. https://doi.org/10.1093/acprof:oso/9780195135589.003.0001

Natural disasters

CHAPTER 23

Recent advances and future trends in operational burned area mapping using remote sensing

Alexandra Stefanidou, Dimitris Stavrakoudis and Ioannis Z. Gitas
Laboratory of Forest Management and Remote Sensing, School of Forestry and Natural Environment, Aristotle University of Thessaloniki, Thessaloniki, Greece

1. Introduction

Wildfires constitute a rather crucial threat to forests worldwide with direct and severe ecological, economic, and social impacts (Lin et al., 2023). Although they have always been an integral process of many ecosystems, wildfires are becoming increasingly uncontrollable and intense in the last few decades, leading to major environmental damage, loss of properties, and human casualties (Chuvieco et al., 2019). The devastating fire events of Canada in 2023, Spain in 2022, Australia in 2020, and Greece in 2021 are only some of the prominent examples where not only thousands of forest hectares were destroyed but also human lives were lost as well.

Climate change impacts—including elevated global temperature and reduced precipitation—are considered some of the main causes of extreme wildfire events, whereas burning forest fuels release aerosols and greenhouse gases affecting the atmospheric chemistry (Mouillot et al., 2014; Szpakowski & Jensen, 2019). The forecasts of the Intergovernmental Panel on Climate Change (IPCC) suggest further future increase of temperature along with the emergence of new fire-prone areas, which were not considered vulnerable in the past (IPCC, 2023). In fact, these climate change consequences have already began to materialize, considering the results of a recent study that identified an increasing global trend of forest loss due to wildfires, even in tropical forests of Latin America and Africa where wildfires rarely occur (Tyukavina et al., 2022).

Accurate and timely mapping of burned areas is a vital tool for forest and fire managers, providing essential information for wildfire prevention (e.g., prediction of fire risk), postfire rehabilitation planning (e.g., environmental wildfire impacts assessment), statistics compilation, and design of impact mitigation measures (e.g., prevention of soil erosion) (Knopp et al., 2020; Shah et al., 2022; Stavrakoudis et al., 2019). The large spatial extent and, often, inaccessibility of fire-affected areas render traditional burned area mapping methods (i.e., via field visits) particularly costly and labor-intensive. The use of remote sensing technology has been constituting an alternative solution for several decades, which has been successfully employed in burned area mapping, providing

Geographical Information Science
ISBN 978-0-443-13605-4
https://doi.org/10.1016/B978-0-443-13605-4.31001-3

systematic information over extended areas in a timely and cost-efficient manner (Goodwin & Collett, 2014; Hawbaker et al., 2017; Hosseini & Lim, 2023; Koutsias & Karteris, 1998; Ressl et al., 2009; Roy et al., 2002).

Since the relationship between wildfires and climate is considered bidirectional, fire disturbance (including burned area) has been identified by the Global Climate Observing System (GCOS) as one of the Essential Climate Variables (ECVs) (GCOS, 2016). As a result, the need for operational monitoring of forest fires from Earth Observation (EO) data has become even more prevalent and gained much more attention by the scientific community.

The aim of this chapter is to review the role of remote sensing in operational burned area mapping and the different sensors and methods employed, highlighting their strengths and limitations, as well as to present existing operational burned area products and identify future trends.

2. Remote sensing in operational burned area mapping

2.1 Sensors for mapping burned areas

2.1.1 Passive sensors

Remote sensing application in burned area mapping has a long history of more than 4 decades and still constitutes an active research topic, examining data from advanced sensors and contemporary techniques (Chuvieco et al., 2019). The sensors being widely employed in burned area mapping include both passive and active technology (Filipponi, 2019; Nolde et al., 2019; Pepe et al., 2022; Prabowo et al., 2022; Radman et al., 2023; Tselka et al., 2023). Passive sensors record the sun's energy that is reflected on land surfaces, typically in multiple narrow zones of the electromagnetic spectrum. In contrast, active sensors such as synthetic aperture radar (SAR) and Light Detection and Ranging (LiDAR) emit their own energy (electromagnetic radiation) and record the reflected signal, a process which has some comparative advantages, for example, SAR is relatively unaffected by cloudy conditions or smoke presence (Radman et al., 2023). Nevertheless, the application of active sensors in burned area mapping is still limited due to the difficulty in interpreting the SAR signal on the one hand and the costly acquisition and limited availability of LiDAR data on the other hand (Meng & Zhao, 2017).

All burned area mapping products operationally produced, either on global or regional/national scale, have been derived thus far from passive sensors. In particular, the reduction in the near-infrared (NIR) reflectance and the increase in the shortwave-infrared (SWIR) and middle-infrared (MIR) reflectance of burned vegetation in most passive sensors facilitates the identification and delineation of burned areas (Kurbanov et al., 2022). Moreover, global and regional operational burned area products require information of very high and moderate temporal resolution, respectively, which is provided by various passive sensors, whereas their coarse and medium spatial resolution are considered adequate for delivering reliable results (Chuvieco et al., 2019).

The Système Probatoire d'Observation de la Terre (SPOT)-VEGETATION (SPOT-VGT) imagery was one of the first remote satellite data employed for the generation of a global burned area product, namely the Global Burned Area 2000 (GBA2000) (Grégoire et al., 2003; Tansey et al., 2004). Specifically, the VGT-S1 product was employed, providing calibrated, georeferenced, and atmospherically corrected daily information. The use of the four spectral bands (i.e., blue, red, NIR, and SWIR) of the dataset at 1 km spatial resolution resulted in the production of GBA2000, comprising monthly global burned area maps over the period of five years (i.e., 1999–2003) (Tansey et al., 2004). SPOT-VGT data were also employed later on, by Tansey et al. (2008, 2012), for the improvement of the global classification algorithm. At the same time period, another coarse resolution sensor used for operational burned area mapping at a global scale was the Along Track Scanning Radiometer-2 (ATSR-2) onboard the European Remote Sensing-2 (ERS-2) satellite of the European Space Agency (ESA) (Simon, 2004). ATRS-2 comprised seven bands, that is, one middlewave infrared (MWIR), two thermal IR (TIR), two visible (VIS), one NIR, and one SWIR band, providing information at 1 km spatial resolution. ATRS-2 data analysis led to the generation of a global burned area product with significant omission but generally few commission errors (Simon, 2004).

Heretofore, the satellite data most widely employed for operational burned area mapping were the Moderate Resolution Imaging Spectroradiometer (MODIS) land surface reflectance imagery. The MODIS instrument was launched aboard Terra and Aqua satellites in 1999 and 2002, respectively, providing daily observations of fire activity at spatial resolutions of 250 m, 500 m, and 1 km. The very high temporal resolution and almost complete global coverage has offered scientists the possibility to operationally produce global burned area maps for almost 2 decades (Giglio et al., 2018; Nolde et al., 2019; Roy et al., 2005; Roy et al., 2008; San-Miguel-Ayanz et al., 2012). Either using MODIS imagery from one satellite (Giglio et al., 2013; Ressl et al., 2009), both Terra and Aqua satellites (Giglio et al., 2018; San-Miguel-Ayanz et al., 2012) or in combination with data provided by different passive sensors (Nolde et al., 2019), organizations, and agencies such as ESA and the National Aeronautics and Space Administration (NASA), have produced burned area maps of high overall accuracy.

In terms of regional or national scales, global products are not always considered suitable for the identification of the precise geographical location and extent of burned areas. The public release of moderate resolution satellite data and advances in digital data storage has opened the way not only for the generation of spatially more accurate burned area maps but also for the examination of long-term spatiotemporal trends of burned areas. In particular, free access to the Landsat data archive of 50 years has offered new opportunities and has been widely applied in regional/national burned area mapping at an operational level with a higher spatial resolution of 30 m compared to the daily spectral reflectance data, using time-series approaches (Goodwin & Collett, 2014; Hawbaker et al., 2017, 2020; Neves et al., 2023). Nevertheless, Landsat's temporal resolution of 16 days can sometimes hinder rapid mapping of burned areas, especially in cases of frequent cloud coverage over the fire-affected regions.

The launch of the Copernicus Sentinel-2A and Sentinel-2B satellites in 2014 and 2016, respectively, has facilitated the production of burned area maps on a more frequent and consistent basis. Sentinel-2 data provide spectral information in 13 bands with high spatial (10–20 m depending on the band) and temporal resolution (5 days), rendering them suitable for operational burned area mapping at regional or national level. Since Sentinel-2 data are relatively new, their examination in the context of operational fire research still constitutes a current research topic (Amos et al., 2020).

2.1.2 Combined approaches

Except for the use of single-source satellite data, new approaches have also been recently developed based on different sensors with the aim to operationally produce more accurate burned area maps, reducing commission and omission errors. More specifically, one of the most employed and effective approaches includes the combined use of spectral surface reflectance and thermal anomalies information. Thermal anomalies indicate the presence of fire, whereas the postfire spectral reflectance enables the delineation of the burned area boundaries. Hence, commission errors can be reduced, compared to the single-source data approach, since thermal anomalies discriminate burned pixels from unburned, while the surface reflectance information contributes to the delineation of the total fire-affected areas, avoiding omission errors.

A prominent and recent example of applying this hybrid approach in burned area mapping for operational purposes is the work of Lizundia-Loiola et al. (2022). In particular, the Sentinel-3 Synergy was employed along with Visible Infrared Imaging Radiometer Suite (VIIRS) active fire products resulting in a global burned area product, namely the FireCCIS310 (Lizundia-Loiola et al., 2022). This product is actually an improved version of the C3SBA10 operational burned area map, which was produced using the Sentinel-3 Ocean and Land Colour Instrument (OLCI) NIR reflectance and the MODIS thermal anomaly data (Lizundia-Loiola et al., 2021). Former global burned area maps have been also produced through the hybrid approach in the framework of the Fire Disturbance (Fire_CCI) project (https://climate.esa.int/en/projects/fire/, last accessed in July 21, 2023). Specifically, the combined use of Medium Resolution Imaging Spectrometer (MERIS) and MODIS thermal anomalies data was performed in 2015 and 2016 by Alonso-Canas and Chuvieco and Chuvieco et al., respectively, for the generation of Fire CCI burned area products (Alonso-Canas & Chuvieco, 2015; Chuvieco et al., 2016). Last but not least, Sentinel-3 has been also employed in combination with

MODIS data for the derivation of burned areas for the European Forest Fire Information System (EFFIS), which provides wildland fire information at the European level since 1998 (Nolde et al., 2019).

2.2 Operational burned area mapping methods

2.2.1 Change detection

The change detection approach mainly involves the use of prefire and postfire multispectral images, and the identification of abrupt changes due to fire events (Tansey et al., 2004). The latter is performed through the application of a threshold value to discriminate burned from unburned areas and generate the final change detection product (i.e., burned area map). The selection of the most suitable threshold value can be performed either in an empirical or automatic manner (Pulvirenti et al., 2020). In fact, the threshold value is highly related to a variety of environmental and satellite system factors, which may render empirical threshold selection sensor- and/or site-specific. Therefore, operational thresholding is usually based on automation, which may alleviate the limitation of the empirical method by accounting for the variability arising from different environment or employed satellite sensor (Pulvirenti et al., 2020). Spectral indices, such as the Normalized Burn Index (NBR) and the Normalized Difference Vegetation Index (NDVI), are often employed as ancillary sources of spectral separability information (Pulvirenti et al. 2020, 2023; Tansey et al., 2008).

Change detection has been operationally employed mostly for regional/national than global mapping of burned areas (Eidenshink et al., 2007; Pulvirenti et al., 2023; Tansey et al., 2004). Considering that global burned area maps involve a variety of environmental conditions and, consequently, different vegetation types and phenology, a single change detection algorithm may not be able to capture the entire spectral variation among the burned areas on a global scale. As a result, some approaches in global-scale burned area mapping have been based on the generation of multiple regional algorithms, which—among others—apply change detection, so that burned areas of different ecosystems and vegetation types can be accurately delineated (Grégoire et al., 2003; Tansey et al., 2004). Nevertheless, successful endeavors of generic change detection algorithms development have also been performed for global delineation of burned areas (Goodwin & Collett, 2014).

2.2.2 Multitemporal image analysis

The operational development of global burned area products is mainly based on multitemporal image analysis (Alonso-Canas & Chuvieco, 2015; Chuvieco et al., 2016; Giglio et al., 2018; Roy et al., 2005). Temporal image composition selects the clearest spectral information along a predefined time period, aiming to utilize the most unambiguous information over burned pixels and, thus, to increase the separability between burned and unburned areas (Chuvieco et al., 2005). Therefore, the multitemporal approach has the advantage of reducing commission errors occurring under the presence of dark soils, water bodies, or shaded areas caused by intense topography and/or cloud coverage, which can be mistakenly characterized as burned by algorithms based solely on single postfire images (Chuvieco et al., 2019; Lizundia-Loiola et al., 2022). The resulting temporal composites are subsequently subject to further analysis for the delineation of burned areas.

2.2.3 Classification techniques

A variety of supervised classification techniques have been applied for the operational development of burned area products on regional or national scales. In 2004, Tansey et al. employed different regional burned area mapping methods for the final generation of a global burned area product (Tansey et al., 2004). One of these methods has been developed in a part of Africa and includes the application of the multilayer perceptron (MLP) neural network, which uses the spatial and temporal information of daily images for the production of daily burned area maps (Brivio et al., 2002). Gradient boosting regression models were developed by Hawbaker et al. (2017) for the classification of burn probability, which was then utilized for mapping the burned areas of the conterminous United States. Object-based approaches have also been applied in regional burned area mapping, avoiding common misclassification errors derived from pixel-based methods (Weih & Riggan, 2010). For instance, Stavrakoudis et al. (2019) employed an object-based classification using support vector machines (SVMs) algorithm for regional operational burned area mapping in Greece, while Goodwin and Collett (2014) performed object-based analysis by means of tree-decision approach.

3. Operational burned area mapping products

3.1 Global products

One of the first operationally generated global burned area products was the Global Burned Area (GBA2000), produced by the Joint Research Centre (JRC) of the European Union, and based on daily SPOT Vegetation (VGT) of 1 km spatial resolution (Tansey et al., 2004). For the same year (i.e., 2000), ESA released the GLOBSCAR product derived from ERS-2 ATRS-2 data providing burned area information with 1 km spatial resolution as well. The L3JRC and the Globcarbon products have been also produced by JRC and ESA for the time period 2000–07 and 1998–2007, respectively (Plummer et al., 2006; Tansey et al., 2008).

NASA has been generating for several years global burned area products, which constitute one of the components of the Global Fire Emissions Database (GFED4) (http://www.globalfiredata.org/, last accessed in July 2023). The Collection 6 MCD64A1 product constitutes the result of the third and latest reprocessing of the MODIS timeseries with 26% more global burned areas identified compared to the previous collection (i.e., MCD45A1 product) (Giglio et al., 2018). The new MCD64A1 was derived from MODIS Terra and Aqua data, providing global monthly information on burned areas for the period 2000–22.

Additional recent global burned area products include the new version (3.1) of the Burned Area 300m (BA300) v.3.1 dataset, released by the Copernicus Global Land Service (https://land.copernicus.eu/global/products/ba, last accessed in July 2023). The dataset has been generated with the use of the OLCI and the Sea and Land Surface Temperature Radiometer (SLSTR) onboard the Sentinel-3 platform and is composed of two products, namely the near-real time (NRT) and nontime critical (NTC), providing daily and monthly composites, respectively.

The Global Wildfire Information System (GWIS) is a Web Map Service (https://gwis.jrc.ec.europa.eu/), providing near real-time updated EO information for comprehensive viewing and evaluation of fire regimes and effects worldwide. GWIS constitutes a joint initiative of the GEO and the Copernicus Work Programs and builds upon ongoing activities (e.g., the EFFIS), complementing existing ones around the world with respect to wildfire information gathering. GWIS is currently composed of five applications, namely Current Situation Viewer (Fig. 23.1), Current Statistics Portal, Country Profile, Long-term fire weather forecast, and Data and Services.

One of the most recently generated global burned area products is the FireCCI310, which was developed within the framework of ESA's Climate Change Initiative (CCI) under the Fire Disturbance project (FireCCI) (Lizundia-Loiola et al., 2022). The Fire-CCI310 product is based on the Copernicus Sentinel-3 Synergy (SYN) and VIIRS active fires data, providing information on burned areas for the year 2019 at approximately 300 m spatial resolution at the equator (https://climate.esa.int/es/odp/#/project/fire, last accessed in July 2023). Some of the precursor burned area products, also developed

Figure 23.1 The current situation viewer of the global wildfire information system (GWIS).

within the FireCCI project, include the C3SBA10 product derived from Sentinel-3 OLCI and MODIS thermal anomaly data (Lizundia-Loiola et al., 2021), the FireCCI51 product from MODIS hotspots and NIR reflectance, the FireCCILT11 product based on Advanced Very High Resolution Radiometer (AVHRR) data (Otón et al., 2021), and the Fire_cci v5.0 product derived from MODIS reflectance and thermal anomalies data (Chuvieco et al., 2018).

3.2 Regional/national products

A variety of burned area products have been operationally developed on a regional/national scale worldwide. The National Commission for the Knowledge and Use of the Biodiversity (CONABIO) of Mexico established in 1999 an operational fire monitoring system, in the context of which a burned area algorithm was developed and is being employed for the monitoring of fire-affected areas in Mexico and Central America (Ressl et al., 2009). The burned areas are derived from MODIS Aqua data using spectral indices, namely NDVI and NBR values.

Operational burned area products are also being generated for different parts of northern Australia. The Queensland state of Australia has been producing burned area products on an annual basis using the automated algorithm developed by Goodwin and Collett (2014). Furthermore, the North Australia and Rangelands Fire Information (NAFI) system has been operationally providing MODIS-based information on burned area to land managers since 2000 (Fisher & Edwards, 2015, pp. 57–72). For other Australian regions, burned area mapping is performed only for specific major fire events. To address this limitation, the Australian National University (ANU) and Geoscience Australia have developed an automated workflow for monitoring burned areas on a continental level using the Landsat data archive in GA's Digital Earth Australia data cube (http://wald.anu.edu.au/challenges/bushfires/burn-mapping/, last access in July 2023).

In 2017, the U.S. Geological Survey (USGS) created a Burned Area Essential Climate Variable (BAECV) algorithm for the operational production of regional burned area maps using a dense Landsat timeseries (Hawbaker et al., 2017). The BAECV algorithm was used to delineate burned areas with a minimum extent of 4 ha over the conterminous United States for the period 1984–2015. The BAECV algorithm was updated in 2020, providing up-to-date burned area products and the possibility to routinely process new Landsat data (Hawbaker et al., 2020). All products are available for download from USGS's EarthExplorer platform (https://earthexplorer.usgs.gov). USGS along with the U.S. Department of Agriculture Forest Service (USDA-FS) have also been operationally mapping burned areas within the framework of the Monitoring Trends in Burn Severity (MTBS) project (Eidenshink et al., 2007). The MTBS algorithm employs Landsat data and the difference NBR (dNBR) for delineating fire-affected areas along with the associated fire severity for all events from 1984 onwards.

In Greece, an operational burned area mapping service has been developed by the Aristotle University of Thessaloniki (AUTh) in cooperation with the Hellenic Ministry of Environment and Energy (Stavrakoudis et al., 2019; Tompoulidou et al., 2016). In particular, the Operational Burned Area Mapping (OBAM) service has been established in the framework of the National Observatory of Forest Fires (NOFFi) project under the financial support of the Greece's Green Fund. NOFFi-OBAM is based on object–based image analysis and can be implemented on any satellite data, such as Landsat 8 OLI and Sentinel-2 images. Since 2016, each fire season, the produced burned area maps (an example of which is shown in Fig. 23.2) are provided to the National Forest Service, the General Secretariat for Civil Protection, local forest service departments, and other interested stakeholders (e.g., research institutes, NGOs).

The Forest Research Centre in the School of Agriculture of the University of Lisbon (Portugal) has recently generated a monthly fire atlas covering the time period from 1984 to 2021 over the mainland of Portugal (Neves et al., 2023). The Monthly Fire Atlas product is based on Landsat timeseries and constitutes an improved version of the Portuguese Annual Fire Atlas (Oliveira et al., 2011). The latter was also developed with the use of Landsat data and a semiautomated supervised classification process. The new product is

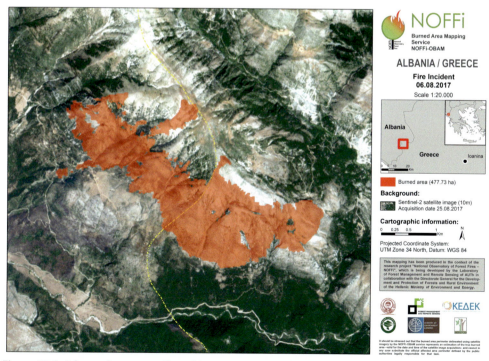

Figure 23.2 A burned area map (fire of August 6, 2017 at the borders between Albania and Greece) produced in the framework of the operational burned area mapping service (NOFFi-OBAM).

publicly available at the PANGAEA repository (https://doi.pangaea.de/10.1594/PANGAEA.955012).

Finally, one of the most widely known fire monitoring systems is the EFFIS, the applications of which are based on remote sensing data and Geographic Information Systems (GIS). EFFIS produces different fire-related modules, namely the Fire Danger Assessment, the Rapid Damage Assessment (active fire detection, fire severity, and land cover damage assessment), the Emissions Assessment and Smoke Dispersion, the Potential Soil Loss Assessment, the Vegetation Regeneration and the Fire News module (https://effis.jrc.ec.europa.eu/, last accessed in July 2023). The system releases—among others—daily updated burned area maps for 43 European, Middle East, and North African countries (Fig. 23.3) and, since 2015, constitutes one of the components of the Emergency Management Services (EMS) in the European Copernicus program (https://www.copernicus.eu/en/copernicus-services/emergency, last accessed in July 2023).

3.3 Validation of burned area products

As described in the previous sections (i.e., 3.1, 3.2), a variety of operational, global, and regional/national burned area products have been made publicly available. Although the methods applied for the generation of these products have significantly improved over time, quality uncertainties, caused by input data characteristics and employed algorithms, still remain present (Chuvieco et al., 2019; Katagis & Gitas, 2022). As such, knowledge regarding a product's quality is necessary for end-users to decide upon whether and how to use it (Boschetti et al., 2016; Mouillot et al., 2014).

Figure 23.3 The current situation viewer providing daily information related to—among others—the burned areas at a regional level (https://effis.jrc.ec.europa.eu/apps/effis_current_situation/index.html).

The term validation describes "the process of assessing, by independent means, the quality of the data products derived from the system outputs," as defined by the Committee on Earth Observing Satellites (CEOS) Working Group on Calibration and Validation (https://ceos.org/ourwork/workinggroups/wgcv/, last accessed in July 2023). CEOS has identified four stages of validation based on its comprehensiveness (https://lpvs.gsfc.nasa.gov/, last accessed in July 2023), each of which are described in Table 23.1.

According to the International Global Burned Area Satellite Product Validation Protocol, the validation of global products requires the use of independent reference data with minimal or no error (Boschetti et al., 2009, 2019). In fact, the reference data are recommended to be derived from satellite images of higher spatial resolution (e.g., Landsat or Sentinel-2 when mapping burned areas via MODIS) and encompass the same time period as the burned area map under validation (Boschetti et al., 2019). The development of reference data is usually performed by stratified spatial and temporal sampling (Katagis & Gitas, 2022). Nevertheless, a standard sampling method for representative reference data generation has not yet been established (Boschetti et al., 2016, 2019; Giglio et al., 2018; Roy et al., 2008).

In recent years, several studies have focused on the accuracy assessment of operational burned area mapping products. The L3JRC, GlobCarbon, and MCD45A1 products were validated in 2009 across southern Africa using independent reference data derived from 11 Landsat images (Roy & Boschetti, 2009). A regional MODIS-derived burned area product covering Latin America was validated by Chuvieco et al. (2008), based on visual analysis of Landsat and CBERS (China–Brazil Earth Resources Satellite) images. In the United States, Giglio et al. (2009) assessed the accuracy of the MCD45A1 product of Collection 5 using high-resolution Landsat imagery in three different regions.

Table 23.1 CEOS validation hierarchy.

Validation stage	Description
1	Product accuracy is assessed from a small (typically <30) set of locations and time periods by comparison with in situ or other suitable reference data.
2	Spatial and temporal consistency of the product, and its consistency with similar products, has been evaluated over globally representative locations and time periods. Results are published in the peer-reviewed literature.
3	Uncertainties in the product and its associated structure are well quantified over a significant (typically >30) set of locations and time periods representing global conditions by comparison with reference in situ or other suitable reference data. Validation procedures follow community-agreed-upon good practices.
4	Validation results for stage 3 are systematically updated when new product versions are released or as the interannual timeseries expands.

From https://lpvs.gsfc.nasa.gov/.

Later on, Giglio et al. (2018) quantified the accuracy of the MCD64A1 Collection 6 product on a regional basis using 108 Landsat images as reference data. The most recent validation of the MCD64A1 Collection 6 is the one performed by Katagis and Gitas (2022) for Greece using Sentinel-2 covering four fire seasons.

The free and open release of Landsat data archive by the USGS in 2008 provided the opportunity to validate an entire global product, instead of just a regional part of it. The first such endeavor was performed by Padilla et al., in 2015, who evaluated the accuracy of six burned area products, including the two MODIS (MCD64 and MCD45) and three Fire_CCI ones (based on MERIS, SPOT VGT, and MERGED) using reference data produced by stratified random sampling on Landsat TM/ETM+ imagery. The results fell rather short of expectations because, although the overall accuracy of the products was very high (i.e., 99%), the commission and omission errors were also high (i.e., above 40% and 65%, respectively) (Padilla et al., 2015).

The validation of global burned area products is usually hindered by the unavailability of public databases with systematically collected reference burned area perimeters. In effort to address this limitation, Franquesa et al. (2020) developed and delivered to the scientific community the first publicly available database created by compiling existing reference BA (burned area) datasets from different international projects. The developed database is called Burned Area Reference Database (BARD) and includes 2661 reference files derived from Landsat and Sentinel-2 data (Franquesa et al., 2020). The datasets are freely available at https://doi.org/10.21950/BBQQU7 and can serve either the generation or validation of new global or regional burned area products (Franquesa et al., 2020).

4. Challenges and future trends

The use of satellite remote sensing in burned area mapping has offered significant opportunities for acquiring knowledge on the geographical location and extent of burned areas worldwide. Nevertheless, despite the rapid technological and processing advancements which led the way for the development of various operational burned area mapping products, inherent limitations still remain unsolved (Chuvieco et al., 2019; Katagis & Gitas, 2022).

On the one hand, satellite data of coarse spatial resolution are characterized by very high temporal resolution, which is an important requirement in operational burned area mapping. Conversely, their low spatial resolution renders the detection of small and/or fragmented fires nearly impossible, resulting in the underestimation of fire-affected areas (Chuvieco et al., 2019; Roteta et al., 2019). This limitation can be alleviated using high-resolution data for global burned area mapping. In fact, studies focusing on the use of high spatial resolution satellite data, such as Landsat and Sentinel-2, have already emerged in the international literature presenting lower commission and omission errors compared to existing operational products (Roteta et al., 2021). Yet, several challenges need to be addressed for converting such approaches into global-scale operational products, with the most important one arguably being the overly increased computational, storage, and data transfer costs compared to coarse-resolution-based algorithms, due to the increased spatial resolution. The rapid advancement of cloud computing observed in recent years can help in this direction, with EO-specific initiatives such as the Copernicus Data Space Ecosystem (https://dataspace.copernicus.eu, last accessed in October 2023) and Google Earth Engine (https://earthengine.google.com, last accessed in October 2023) boosting this effort.

Along with the multispectral sensors already employed for operational burned area mapping, SAR data such as the ones derived from the Sentinel-1A and Sentinel-1B (launched in 2014 and 2016, respectively), may have the potential to be synergistically employed, alleviating current problems related to the detection of burned areas under cloudy or smoke conditions. In the last years, several studies have examined Sentinel-1 data (individually or in combination with optical data), focusing on the development of automated algorithms for operational global and regional/national burned area mapping presenting promising results (Chuvieco et al., 2019; Engelbrecht et al., 2017; Tariq et al., 2023).

The development of new satellite constellations with very- or ultrahigh spatial resolution may also substantially contribute to the operational burned area mapping in the future. Data derived from the new PlanetScope nanosatellite constellation have been recently examined in burned area mapping (Cho et al., 2023; Francini et al., 2020; Huang et al., 2022). The PlanetScope mission uses groups of nanosatellites, called "Doves," equipped with four band sensors in the visible and NIR channels (https:// developers.planet.com/, last accessed in July 2023). The successful employment of the very high spatial and temporal resolution (i.e., 3 m and almost one day, respectively) PlanetScope data may provide the scientific community with the possibility of developing accurate near real-time burned area mapping systems (Francini et al., 2020). Nevertheless, radiometric inconsistencies between different nanosatellites have been highlighted, either due to sensor-specific spectral response functions or to orbital configuration variations (Houborg & McCabe, 2018; Latte & Lejeune, 2020; Leach et al., 2019). Therefore, some methods have been recently developed for radiometric normalization based on larger conventional satellite constellations, such as MODIS (Houborg and McCabe 2016, 2018; Leach et al., 2019). Furthermore, in 2024, the EarthDaily Constellation is scheduled to be launched by EarthDaily Analytics, a satellite imaging company active in the field of remote sensing technologies and processing systems (https://earthdaily.com/company/, last accessed in July 2023). The EarthDaily Constellation will provide high-resolution EO data for forest resources inventory and monitoring as well as climate change mitigation and adaptation, which are expected to effectively contribute to forestry and sustainable forest management (https://earthdaily. com/industry-use-cases/forestry/, last accessed in July 2023). The very high spatial resolution of these constellations aggravates the problem of increased computational resources required (already observed with Sentinel-2 data or similar), possibly challenging even today's cloud infrastructure systems if employed on global scale.

Except for the issues related to the spatiotemporal resolution of sensors, a global and reliable method for operationally producing burned area maps is also under investigation, taking into consideration the boom that the artificial intelligence algorithms have experienced the last few years. Recently, several novel approaches have been introduced to the scientific community, which—although have not yet been operationally examined—showcase promising results. Deep learning methods for burned area mapping have emerged in the past few years and currently constitute a hot research topic in the remote sensing field. Deep learning technology provides the possibility of discriminating burned from unburned areas in a rather time-efficient and automatic manner, which is what operational applications require. Perhaps even more importantly, it can inherently address the problems of reflectance data inconsistencies or even lack of surface reflectance processing, typically observed in recent nanosatellites constellations. A recent example is Sudiana et al. (2023), who employed a hybrid convolutional neural network using optical (Sentinel-2) and SAR (Sentinel-1) data for burned area identification in Indonesia (Sudiana et al., 2023). A deep learning approach was also examined in 2023 by Cho et al. using PlanetScope imagery (Cho et al., 2023). Seydi et al. (2022) mapped burned areas in six European countries (i.e., Portugal, Spain, France, Greece, Turkey, Cyprus) using Sentinel-2 imagery and a deep learning morphological neural network (Seydi et al., 2022). In all these cases, the results of the deep learning approaches applications led to accurate results. Nevertheless, the main challenge of using them on an operational basis lies in the fact that the developed models have not been examined in multiple areas, which is a prerequisite for any operational burned area mapping algorithm (regional or global). Even Seydi et al. (2022), who applied their model to six study areas, cannot yet guarantee its' full transferability. This is why systematic and continuously updated databases of reference wildfire perimeters need to be created and maintained, since most deep learning algorithms require a substantial amount of training data.

Another challenge is related to the validation of operational burned area products. More specifically, the use of higher resolution data for burned area products accuracy assessment, as suggested by the International Global Burned Area Satellite Product Validation Protocol, may not be often possible, especially when the burned area products are derived from high-resolution data such as Sentinel-2. Although commercial satellite images with very high spatial and temporal resolution (such as PlanetScope) are being released, their high costs and lack of historical archives severely limits the possibility of validating extended parts and timeseries of burned area maps (Chuvieco et al., 2019). It also remains an open issue how to validate burned area mapping products produced from very-high-resolution imagery (e.g., those provided by the recent nanosatellites projects), since all internationally accepted validation protocols suggest using higher resolution images for validating the mapping product.

5. Conclusions

This chapter presents an overview of the recent advances and future trends in operational burned area mapping, both global and regional/national, using remote sensing. A variety of burned area products have been developed over the years based on remote sensing data. Global products are mainly based on coarse spatial resolution satellite data, whereas regional or national burned area products usually rely on medium- or high-resolution data. Despite the remote sensing technological progress, certain limitations persist presently, such as product accuracy that could be improved or systematic validation of the products. Contemporary satellite constellations and advanced processing algorithms could have the potential of alleviating current limitations, achieving more accurate and near-real-time burned area products, generated on an operational basis.

References

Alonso-Canas, I., & Chuvieco, E. (2015). Global burned area mapping from ENVISAT-MERIS and MODIS active fire data. *Remote Sensing of Environment, 163*, 140–152. https://doi.org/10.1016/j.rse.2015.03.011

Amos, C., Ferentinos, K. P., Petropoulos, G. P., & Srivastava, P. K. (2020). Assessing the use of Sentinel-2 in burnt area cartography. In *Techniques for disaster risk management and mitigation* (pp. 141–150). John Wiley & Sons, Ltd. https://doi.org/10.1002/9781119359203.ch11

Boschetti, L., Roy, D. P., Giglio, L., Huang, H., Zubkova, M., & Humber, M. L. (2019). Global validation of the collection 6 MODIS burned area product. *Remote Sensing of Environment, 235*, 111490. https://doi.org/10.1016/j.rse.2019.111490

Boschetti, L., Roy, D. P., & Justice, C. O. (2009). *International Global Burned Area Satellite Product Validation Protocol. Part I—production and standardization of validation reference data* (pp. 1–11). Maryland, MD, USA: Committee on Earth Observation Satellites.

Boschetti, L., Stehman, S. V., & Roy, D. P. (2016). A stratified random sampling design in space and time for regional to global scale burned area product validation. *Remote Sensing of Environment, 186*, 465–478. https://doi.org/10.1016/j.rse.2016.09.016

Brivio, P. A., Maggi, M., Binaghi, E., Gallo, I., & Gregoire, J.-M. (2002). Exploiting spatial and temporal information for extracting burned areas from time series of SPOT-VGT data. In *Analysis of multi-temporal remote sensing images* (pp. 132–139). Italy: University of Trento. https://doi.org/10.1142/9789812777249_0013. World Scientific.

Cho, A. Y., Park, S., Kim, D., Kim, J., Li, C., & Song, J. (2023). Burned area mapping using Unitemporal PlanetScope imagery with a deep learning based approach. *IEEE Journal of Selected Topics in Applied Earth Observations and Remote Sensing, 16*, 242–253. https://doi.org/10.1109/JSTARS.2022.3225070

Chuvieco, E., Lizundia-Loiola, J., Pettinari, M. L., Ramo, R., Padilla, M., Tansey, K., Mouillot, F., Laurent, P., Storm, T., Heil, A., & Plummer, S. (2018). Generation and analysis of a new global burned area product based on MODIS 250 m reflectance bands and thermal anomalies. *Earth System Science Data, 10*(4), 2015–2031. https://doi.org/10.5194/essd-10-2015-2018

Chuvieco, E., Mouillot, F., van der Werf, G. R., San Miguel, J., Tanase, M., Koutsias, N., García, M., Yebra, M., Padilla, M., Gitas, I., et al. (2019). Historical background and current developments for mapping burned area from satellite Earth observation. *Remote Sensing of Environment, 225*, 45–64. https://doi.org/10.1016/j.rse.2019.02.013

Chuvieco, E., Opazo, S., Sione, W., del Valle, H., Anaya, J., Bella, C. D., Cruz, I., Manzo, L., López, G., Mari, N., et al. (2008). Global burned-land estimation in Latin America using Modis composite data. *Ecological Applications, 18*(1), 64–79. https://doi.org/10.1890/06-2148.1

Chuvieco, E., Ventura, G., Martín, M. P., & Gómez, I. (2005). Assessment of multitemporal compositing techniques of MODIS and AVHRR images for burned land mapping. *Remote Sensing of Environment, 94*(4), 450–462. https://doi.org/10.1016/j.rse.2004.11.006

Chuvieco, E., Yue, C., Heil, A., Mouillot, F., Alonso-Canas, I., Padilla, M., Pereira, J. M., Oom, D., & Tansey, K. (2016). A new global burned area product for climate assessment of fire impacts. *Global Ecology and Biogeography, 25*(5), 619–629. https://doi.org/10.1111/geb.12440

Eidenshink, J., Schwind, B., Brewer, K., Zhu, Z.-L., Quayle, B., & Howard, S. (2007). A project for monitoring trends in burn severity. *Fire Ecology, 3*(1), 3–21. https://doi.org/10.4996/fireecology.0301003

Engelbrecht, J., Theron, A., Vhengani, L., & Kemp, J. (2017). A Simple normalized difference approach to burnt area mapping using multi-polarisation C-band SAR. *Remote Sensing, 9*(8), 764. https://doi.org/10.3390/rs9080764

Filipponi, F. (2019). Exploitation of Sentinel-2 time series to map burned areas at the national level: A case study on the 2017 Italy wildfires. *Remote Sensing, 11*(6), 622. https://doi.org/10.3390/rs11060622

Fisher, R., & Edwards, A. C. (2015). Fire extent and mapping: Procedures, validation and website application. In *Carbon accounting and savanna fire management*.

Francini, S., McRoberts, R. E., Giannetti, F., Mencucci, M., Marchetti, M., Scarascia Mugnozza, G., & Chirici, G. (2020). Near-real time forest change detection using PlanetScope imagery. *European Journal of Remote Sensing, 53*(1), 233–244. https://doi.org/10.1080/22797254.2020.1806734

Franquesa, M., Vanderhoof, M. K., Stavrakoudis, D., Gitas, I. Z., Roteta, E., Padilla, M., & Chuvieco, E. (2020). Development of a standard database of reference sites for validating global burned area products. *Earth System Science Data, 12*(4), 3229–3246. https://doi.org/10.5194/essd-12-3229-2020

GCOS. (2016). *The global observing system for climate: Implementation needs*. Geneva, Switzerland: World Meteorological Organization.

Giglio, L., Boschetti, L., Roy, D. P., Humber, M. L., & Justice, C. O. (2018). The Collection 6 MODIS burned area mapping algorithm and product. *Remote Sensing of Environment, 217*, 72–85. https://doi.org/10.1016/j.rse.2018.08.005

Giglio, L., Loboda, T., Roy, D. P., Quayle, B., & Justice, C. O. (2009). An active-fire based burned area mapping algorithm for the MODIS sensor. *Remote Sensing of Environment, 113*(2), 408–420. https://doi.org/10.1016/j.rse.2008.10.006

Giglio, L., Randerson, J. T., & van der Werf, G. R. (2013). Analysis of daily, monthly, and annual burned area using the fourth-generation global fire emissions database (GFED4). *Journal of Geophysical Research: Biogeosciences, 118*(1), 317–328. https://doi.org/10.1002/jgrg.20042

Goodwin, N. R., & Collett, L. J. (2014). Development of an automated method for mapping fire history captured in Landsat TM and ETM + time series across Queensland, Australia. *Remote Sensing of Environment, 148*, 206–221. https://doi.org/10.1016/j.rse.2014.03.021

Grégoire, J.-M., Tansey, K., & Silva, J. M. N. (2003). The GBA2000 initiative: Developing a global burnt area database from SPOT-VEGETATION imagery. *International Journal of Remote Sensing, 24*(6), 1369–1376. https://doi.org/10.1080/0143116021000044850

Hawbaker, T. J., Vanderhoof, M. K., Beal, Y.-J., Takacs, J. D., Schmidt, G. L., Falgout, J. T., Williams, B., Fairaux, N. M., Caldwell, M. K., Picotte, J. J., et al. (2017). Mapping burned areas using dense time-series of Landsat data. *Remote Sensing of Environment, 198*, 504–522. https://doi.org/10.1016/j.rse.2017.06.027

Hawbaker, T. J., Vanderhoof, M. K., Schmidt, G. L., Beal, Y.-J., Picotte, J. J., Takacs, J. D., Falgout, J. T., & Dwyer, J. L. (2020). The Landsat Burned Area algorithm and products for the conterminous United States. *Remote Sensing of Environment, 244*, 111801. https://doi.org/10.1016/j.rse.2020.111801

Hosseini, M., & Lim, S. (2023). Burned area detection using Sentinel-1 SAR data: A case study of Kangaroo Island, South Australia. *Applied Geography, 151*, 102854. https://doi.org/10.1016/j.apgeog.2022.102854

Houborg, R., & McCabe, M. F. (2018). A cubesat enabled spatio-temporal enhancement method (CESTEM) utilizing planet, Landsat and MODIS data. *Remote Sensing of Environment, 209*, 211–226. https://doi.org/10.1016/j.rse.2018.02.067

Houborg, R., & McCabe, M. F. (2016). High-resolution NDVI from planet's constellation of earth observing nano-satellites: A new data source for precision agriculture. *Remote Sensing, 8*(9), 768.

Huang, H., Roy, D., Yan, L., Martins, V., & Boschetti, L. (2022). *Demonstration of the suitability of commercial high spatial and temporal resolution PlanetScope imagery for burned area mapping. 2022* pp. IN45B−0364).

IPCC. (2023). *Climate change 2022 − impacts, adaptation and vulnerability: working group II contribution to the sixth assessment report of the intergovernmental panel on climate change* (1st ed.). Cambridge University Press. https://doi.org/10.1017/9781009325844

Katagis, T., & Gitas, I. Z. (2022). Assessing the accuracy of MODIS MCD64A1 C6 and FireCCI51 burned area products in Mediterranean ecosystems. *Remote Sensing, 14*(3), 602. https://doi.org/10.3390/rs14030602

Knopp, L., Wieland, M., Rättich, M., & Martinis, S. (2020). A deep learning approach for burned area segmentation with Sentinel-2 data. *Remote Sensing, 12*(15), 2422.

Koutsias, N., & Karteris, M. (1998). Logistic regression modelling of multitemporal Thematic Mapper data for burned area mapping. *International Journal of Remote Sensing, 19*(18), 3499−3514. https://doi.org/10.1080/014311698213777

Kurbanov, E., Vorobev, O., Lezhnin, S., Sha, J., Wang, J., Li, X., Cole, J., Dergunov, D., & Wang, Y. (2022). Remote sensing of forest burnt area, burn severity, and post-fire recovery: A review. *Remote Sensing, 14*(19), 4714.

Latte, N., & Lejeune, P. (2020). PlanetScope radiometric normalization and Sentinel-2 super-resolution (2.5 m): A straightforward spectral-spatial fusion of multi-satellite multi-sensor images using residual convolutional neural networks. *Remote Sensing, 12*(15), 2366. https://doi.org/10.3390/rs12152366

Leach, N., Coops, N. C., & Obrknezev, N. (2019). Normalization method for multi-sensor high spatial and temporal resolution satellite imagery with radiometric inconsistencies. *Computers and Electronics in Agriculture, 164*, 104893. https://doi.org/10.1016/j.compag.2019.104893

Lin, X., Li, Z., Chen, W., Sun, X., & Gao, D. (2023). Forest fire prediction based on long- and short-term time-series network. *Forests, 14*(4), 778. https://doi.org/10.3390/f14040778

Lizundia-Loiola, J., Franquesa, M., Boettcher, M., Kirches, G., Pettinari, M. L., & Chuvieco, E. (2021). Operational implementation of the burned area component of the Copernicus climate change service: From MODIS 250 m to OLCI300 m data. In *Earth system science data discussions*. https://doi.org/10.5194/essd-2020-399

Lizundia-Loiola, J., Franquesa, M., Khairoun, A., & Chuvieco, E. (2022). Global burned area mapping from Sentinel-3 Synergy and VIIRS active fires. *Remote Sensing of Environment, 282*, 113298. https://doi.org/10.1016/j.rse.2022.113298

Meng, R., & Zhao, F. (2017). Remote sensing of fire effects: A review for recent advances in burned area and burn severity mapping. *Remote Sensing of Hydrometeorological Hazards, 261*−283.

Mouillot, F., Schultz, M. G., Yue, C., Cadule, P., Tansey, K., Ciais, P., & Chuvieco, E. (2014). Ten years of global burned area products from spaceborne remote sensing—a review: Analysis of user needs and recommendations for future developments. *International Journal of Applied Earth Observation and Geoinformation, 26*, 64−79. https://doi.org/10.1016/j.jag.2013.05.014

Neves, A. K., Campagnolo, M. L., Silva, J. M. N., & Pereira, J. M. C. (2023). A Landsat-based atlas of monthly burned area for Portugal, 1984−2021. *International Journal of Applied Earth Observation and Geoinformation, 119*, 103321. https://doi.org/10.1016/j.jag.2023.103321

Nolde, M., Plank, S., & Riedlinger, T. (2019). Derivation of burnt areas from Sentinel-3 and MODIS data for the European forest fire information system using active Contour level sets. *Geophysical Research Abstracts, 21*, 1.

Oliveira, S. L. J., Pereira, J. M. C., Carreiras, J. M. B., Oliveira, S. L. J., Pereira, J. M. C., & Carreiras, J. M. B. (2011). Fire frequency analysis in Portugal (1975−2005), using Landsat-based burnt area maps. *International Journal of Wildland Fire, 21*(1), 48−60. https://doi.org/10.1071/WF10131

Otón, G., Lizundia-Loiola, J., Pettinari, M. L., & Chuvieco, E. (2021). Development of a consistent global long-term burned area product (1982−2018) based on AVHRR-LTDR data. *International Journal of Applied Earth Observation and Geoinformation, 103*, 102473. https://doi.org/10.1016/j.jag.2021.102473

Padilla, M., Stehman, S. V., Ramo, R., Corti, D., Hantson, S., Oliva, P., Alonso-Canas, I., Bradley, A. V., Tansey, K., Mota, B., et al. (2015). Comparing the accuracies of remote sensing global burned area products using stratified random sampling and estimation. *Remote Sensing of Environment, 160*, 114−121. https://doi.org/10.1016/j.rse.2015.01.005

Pepe, A., Sali, M., Boschetti, M., & Stroppiana, D. (2022). Mapping burned areas from Sentinel-1 and Sentinel-2 data. *Environmental Sciences Proceedings, 17*(1), 62. https://doi.org/10.3390/environsciproc 2022017062

Plummer, S., Arino, O., Simon, M., & Steffen, W. (2006). Establishing a earth observation product service for the terrestrial carbon community: The globcarbon initiative. *Mitigation and Adaptation Strategies for Global Change, 11*(1), 97–111. https://doi.org/10.1007/s11027-006-1012-8

Prabowo, Y., Sakti, A. D., Pradono, K. A., Amriyah, Q., Rasyidy, F. H., Bengkulah, I., Ulfa, K., Candra, D. S., Imdad, M. T., & Ali, S. (2022). Deep learning dataset for estimating burned areas: Case study, Indonesia. *Data, 7*(6), 78. https://doi.org/10.3390/data7060078

Pulvirenti, L., Squicciarino, G., Fiori, E., Fiorucci, P., Ferraris, L., Negro, D., Gollini, A., Severino, M., & Puca, S. (2020). An automatic processing chain for near real-time mapping of burned forest areas using Sentinel-2 data. *Remote Sensing, 12*(4), 674. https://doi.org/10.3390/rs12040674

Pulvirenti, L., Squicciarino, G., Fiori, E., Negro, D., Gollini, A., & Puca, S. (2023). Near real-time generation of a country-level burned area database for Italy from Sentinel-2 data and active fire detections. *Remote Sensing Applications: Society and Environment, 29*, 100925. https://doi.org/10.1016/j.rsase.2023.100925

Radman, A., Shah-Hosseini, R., & Homayouni, S. (2023). An unsupervised saliency-guided deep convolutional neural network for accurate burn mapping from Sentinel-1 SAR data. *Remote Sensing, 15*(5), 1184. https://doi.org/10.3390/rs15051184

Ressl, R., Lopez, G., Cruz, I., Colditz, R. R., Schmidt, M., Ressl, S., & Jiménez, R. (2009). Operational active fire mapping and burnt area identification applicable to Mexican Nature Protection Areas using MODIS and NOAA-AVHRR direct readout data. *Remote Sensing of Environment, 113*(6), 1113–1126. https://doi.org/10.1016/j.rse.2008.10.016

Roteta, E., Bastarrika, A., Ibisate, A., & Chuvieco, E. (2021). A preliminary global automatic burned-area algorithm at medium resolution in Google earth engine. *Remote Sensing, 13*(21), 4298. https://doi.org/10.3390/rs13214298

Roteta, E., Bastarrika, A., Padilla, M., Storm, T., & Chuvieco, E. (2019). Development of a Sentinel-2 burned area algorithm: Generation of a small fire database for sub-Saharan Africa. *Remote Sensing of Environment, 222*, 1–17. https://doi.org/10.1016/j.rse.2018.12.011

Roy, D. P., Boschetti, L., Justice, C. O., & Ju, J. (2008). The collection 5 MODIS burned area product — global evaluation by comparison with the MODIS active fire product. *Remote Sensing of Environment, 112*(9), 3690–3707. https://doi.org/10.1016/j.rse.2008.05.013

Roy, D. P., & Boschetti, L. (2009). Southern Africa validation of the MODIS, L3JRC, and GlobCarbon burned-area products. *IEEE Transactions on Geoscience and Remote Sensing, 47*(4), 1032–1044. https://doi.org/10.1109/TGRS.2008.2009000

Roy, D. P., Jin, Y., Lewis, P. E., & Justice, C. O. (2005). Prototyping a global algorithm for systematic fire-affected area mapping using MODIS time series data. *Remote Sensing of Environment, 97*(2), 137–162. https://doi.org/10.1016/j.rse.2005.04.007

Roy, D. P., Lewis, P. E., & Justice, C. O. (2002). Burned area mapping using multi-temporal moderate spatial resolution data—a bi-directional reflectance model-based expectation approach. *Remote Sensing of Environment, 83*(1–2), 263–286. https://doi.org/10.1016/S0034-4257(02)00077-9

San-Miguel-Ayanz, J., Schulte, E., Schmuck, G., Camia, A., Strobl, P., Liberta, G., Giovando, C., Boca, R., Sedano, F., & Kempeneers, P. (2012). Comprehensive monitoring of wildfires in Europe: The European forest fire information system (EFFIS). In *Approaches to managing disaster-Assessing hazards, emergencies and disaster impacts*. InTech.

Seydi, S. T., Hasanlou, M., & Chanussot, J. (2022). Burnt-Net: Wildfire burned area mapping with single post-fire Sentinel-2 data and deep learning morphological neural network. *Ecological Indicators, 140*, 108999. https://doi.org/10.1016/j.ecolind.2022.108999

Shah, S. U., Yebra, M., Van Dijk, A. I., & Cary, G. J. (2022). A new fire danger index developed by random forest analysis of remote sensing derived fire sizes. *Fire, 5*(5), 152.

Simon, M. (2004). Burnt area detection at global scale using ATSR-2: The GLOBSCAR products and their qualification. *Journal of Geophysical Research, 109*(D14), D14S02. https://doi.org/10.1029/2003JD003622

Stavrakoudis, D., Katagis, T., Minakou, C., & Gitas, I. Z. (2019). Towards a fully automatic processing chain for operationally mapping burned areas countrywide exploiting Sentinel-2 imagery. In *Seventh International Conference on Remote Sensing and Geoinformation of the Environment (RSCy2019)* (Vol. 11174). https://doi.org/10.1117/12.2535816

Sudiana, D., Lestari, A. I., Riyanto, I., Rizkinia, M., Arief, R., Prabuwono, A. S., & Sri Sumantyo, J. T. (2023). A hybrid convolutional neural network and random forest for burned area identification with optical and synthetic aperture radar (SAR) data. *Remote Sensing, 15*(3), 728. https://doi.org/10.3390/rs15030728

Szpakowski, D., & Jensen, J. (2019). A review of the applications of remote sensing in fire ecology. *Remote Sensing, 11*(22), 2638. https://doi.org/10.3390/rs11222638

Tansey, K., Bradley, A., Smets, B., van Best, C., & Lacaze, R. (2012). The Geoland2 BioPar burned area product. In *EGU General Assembly Conference Abstracts* (p. 4727).

Tansey, K., GrÉgoire, J.-M., Binaghi, E., Boschetti, L., Brivio, P. A., Ershov, D., Flasse, S., Fraser, R., Graetz, D., Maggi, M., et al. (2004). A global inventory of burned areas at 1 Km resolution for the year 2000 derived from spot vegetation data. *Climate Change, 67*(2), 345−377. https://doi.org/10.1007/s10584-004-2800-3

Tansey, K., Grégoire, J.-M., Defourny, P., Leigh, R., Pekel, J.-F., Van Bogaert, E., & Bartholomé, E. (2008). A new, global, multi-annual (2000−2007) burnt area product at 1 km resolution. *Geophysical Research Letters, 35*(1).

Tariq, A., Jiango, Y., Lu, L., Jamil, A., Al-ashkar, I., Kamran, M., & Sabagh, A. E. (2023). Integrated use of Sentinel-1 and Sentinel-2 data and open-source machine learning algorithms for burnt and unburnt scars. *Geomatics, Natural Hazards and Risk, 14*(1), 2190856. https://doi.org/10.1080/19475705.2023.2190856

Tompoulidou, M., Stefanidou, A., Grigoriadis, D., Dragozi, E., Stavrakoudis, D., & Gitas, I. Z. (2016). The Greek national observatory of forest fires (NOFFi). In *Fourth International Conference on Remote Sensing and Geoinformation of the Environment (RSCy2016)* (Vol. 9688, p. 96880N). International Society for Optics and Photonics.

Tselka, I., Detsikas, S. E., Petropoulos, G. P., & Demertzi, I. I. (2023). Chapter 7 - Google earth engine and machine learning classifiers for obtaining burnt area cartography: A case study from a Mediterranean setting. In N. Stathopoulos, A. Tsatsaris, & K. Kalogeropoulos (Eds.), *Geoinformatics for Geosciences* (pp. 131−148). Elsevier. https://doi.org/10.1016/B978-0-323-98983-1.00008-9

Tyukavina, A., Potapov, P., Hansen, M. C., Pickens, A. H., Stehman, S. V., Turubanova, S., Parker, D., Zalles, V., Lima, A., Kommareddy, I., et al. (2022). Global trends of forest loss due to fire from 2001 to 2019. *Frontiers in Remote Sensing, 3*. https://www.frontiersin.org/articles/10.3389/frsen.2022.825190.

Weih, R. C., & Riggan, N. D. (2010). Object-based classification vs. pixel-based classification: Comparative importance of multi-resolution imagery. *The International Archives of the Photogrammetry, Remote Sensing and Spatial Information Sciences, 38*(4), C7.

CHAPTER 24

Techniques and tools for monitoring agriculture drought: A review

Varsha Pandey[1], Prashant K. Srivastava[2], Anjali Kumari Singh[3], Swati Suman[4] and Swati Maurya[5]

[1]Department of Civil Engineering, Indian Institute of Technology Bombay, Mumbai, Maharashtra, India; [2]Remote Sensing Laboratory, Institute of Environment and Sustainable Development, Banaras Hindu University, Varanasi, Uttar Pradesh, India; [3]Department of Energy and Environment, Faculty of Science and Environment, Mahatma Gandhi Chitrakoot Gramodaya Vishwavidayalya, Satna, Madhya Pradesh, India; [4]Department for Innovations in Biological, Agri-food and Forest System, University of Tuscia, Viterbo, Italy; [5]Department of Environmental Science, Integral University, Lucknow, Uttar Pradesh, India

1. Introduction

Drought is among the foremost natural hazards leading to agricultural, economic, and environmental damage in many parts of the world. Since the last few decades, low to severe drought events have become a worldwide recurrent phenomenon due to the alterations in climate conditions and increased human disturbances (AghaKouchak et al., 2015). Due to the increased frequency and severity and linkages with climate change, droughts have drawn worldwide attention. The developing and underdeveloped countries heavily dependent on agriculture and water resources need to be better adapted, coordinated, and disintegrated in crisis management in response to drought. Drought has a devastating impact on the environmental conditions and socioeconomic status of low-income and marginal farmers. Recurrent and prolonged drought events lead to high water scarcity, loss of natural vegetation, poor groundwater recharge, etc., which adversely affect hydrological balance, crop yields, livestock, allied sectors, and thereby socioeconomic conditions (Glantz, 1994; Bandyopadhyay et al., 2020). Moreover, it causes loss of human and animal lives, leads to migration, and social and political conflicts in various parts of the world. A study conducted in 107 countries indicated about 220 million people are affected by drought annually, wherein the maximum casualty was reported in the developing countries (Samra, 2004). The global risk of drought can be understood from Fig. 24.1, developed by Carrão etal. (2016a,b), as a product of drought hazard, exposure, and vulnerability. The drought risk map depicts the high risk for South and West Asia, including India. Noticeably the risk is high for the areas that are more populated and agriculture-dependent socioeconomic status. According to the Indian Agricultural Research Institute (IARI), 70% of the Indian population depends on agriculture, wherein 68% of the net sown area resides in drought-prone regions. The agricultural drought risk in India is significantly high due to the enormous food demand for above 1.3 billion population (Rosegrant & Cline, 2003; Mishra & Lilhare, 2016).

Geographical Information Science
ISBN 978-0-443-13605-4
https://doi.org/10.1016/B978-0-443-13605-4.00024-2

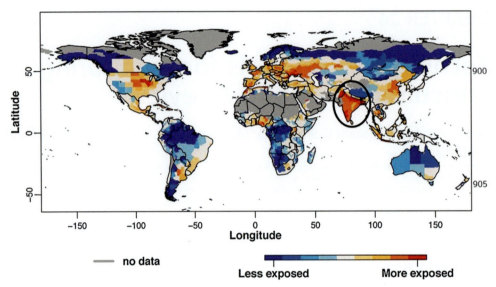

Figure 24.1 Drought risk map produced as a product of hazard, exposure, and vulnerability. *(Map courtesy: Carrão et al. (2016a,b).)*

Meteorological drought triggers agricultural drought, which is directly associated with crop yield reduction, followed by socioeconomic losses. Thus, meteorological and agricultural drought monitoring are prerequisites for socioeconomic well-being assessment. Continuous and long-term observation of soil moisture and vegetation health monitoring make agricultural drought assessment challenging using ground-based data at a large scale. In comparison, the integrated techniques utilizing ground measures, remotely sensed images, and land surface modeled data are widely used to assess agricultural drought employing several parameters, for example, soil moisture (SM), vegetation health, evapotranspiration (ET), crop yield, etc. (Chandrasekar & Sai, 2015; Murthy et al., 2009; Shakya & Yamaguchi, 2010). Ground-sampled *in situ* SM data are accurate, but such data collection is time-consuming and labor-intensive for repetitive analysis at large scale and fine resolution. With limited representative ground-based measurement, the high-resolution remotely sensed satellite data and derived products, reanalysis data products, and hydrological model-simulated data offer more convenient drought assessment and enable spatiotemporal analysis at a large scale. These approaches have become one of the robust approaches for drought mapping to bridge the gaps of spatial contiguity, data unavailability for remote areas, and near real-time data inaccessibility (Shah & Mishra, 2014). The real-time agricultural drought assessment and monitoring is a prerequisite for improved management and prompt activities and to ensure food security. Indices are preferred tools for drought assessment due to their easy and rapid implementation and effectiveness in analyzing, comparing, monitoring, and forecasting. Drought indices, such as Palmer Drought Severity Index (PDSI), Standardized Precipitation Index (SPI), Soil Moisture Deficit Index (SMDI), Soil Water Deficit Index (SWDI), Evapotranspiration Deficit Index (ETDI), etc., are widely used in recent times. Such indices are applied to weather stations collected, hydrological model predicted, and satellite data-derived data, such as precipitation, soil moisture, ET, etc. Various studies have studied agricultural drought using satellite image-derived spectral indices, such as Normalized Difference Vegetation Index (NDVI), Vegetation Health Index (VHI), Crop Moisture Index (CMI), etc. The various indices used for agricultural drought assessment are corroborated with ground measurements. The integration of near-real-time satellite data with ground-based observations allows systematic agricultural drought and its impact assessment, which are helpful for timely mitigation measures.

The chapter is categorized into four major sections. The first section presents a comprehensive overview of agriculture drought assessment and recent approaches. The second section describes different drought types and their propagation. The third section includes the latest approaches for agricultural drought monitoring, including available data sources, followed by the description of widely used indices, usage of advanced tools for agricultural drought assessment, and related case studies. The fourth section concludes the study by highlighting the major challenges and future research needs in agriculture drought assessment.

2. Drought types and their propagation

Drought is a stochastic phenomenon with indistinct onset and ends, characterized by an undefined structure and its slower impact rate that becomes adverse when accumulated over a considerable time period (Yaduvanshi et al., 2015). Drought is a complex process because of its complex relationship with multiple variables starting from the atmospheric components, and hydrologic processes connected *via* moisture exchange with land surface features. Based on the source and impact on other environmental parameters, four types of drought have been categorized into two major groups as physical (meteorological, agricultural, and hydrological) and social (socioeconomic) aspects (Wilhite, 1993, 2000; Keyantash & Dracup, 2002; Mishra & Singh, 2010). The precipitation deficiency initiates the meteorological drought and causes deterioration of soil moisture and streamflow, whereas the agricultural drought results from a deficit in soil moisture and reduced water availability for vegetation growth, and the depletion in streamflow/runoff causes the hydrological drought. Socioeconomic drought is measured based on the supply and demand gap of economic goods caused due to the physical droughts (Wilhite & Glantz, 1985; Wilhite, 2000).

2.1 Meteorological drought

Meteorological drought is considered a primary drought type that occurs due to deficiencies in precipitation for an extended time period (Meze-Hausken, 2004;

Mishra & Singh, 2010). Generally, it is measured when the reduction exceeds 25% of its long-term average value (Meze-Hausken, 2004). The Indian Meteorological Department (IMD) denotes meteorological drought when the mean annual precipitation declines below 75% of long-term average precipitation. IMD further classifies drought conditions into two classes, viz., moderate and severe drought when the precipitation deficiency is between 25% and 50% and exceeds 50%, respectively (Nagarajan, 2010). Meteorological drought is region-specific as the deficiencies in precipitation may vary from place to place. This drought mainly focuses on the physical characteristics as it shows the downswing of precipitation from normal conditions rather than its impacts (Wilhite & Glantz, 1985). The significant consequences of meteorological droughts include reduced soil moisture, decreased surface runoff, groundwater level depletion, and drying up of reservoirs. Precipitation is the primary variable for quantifying meteorological drought (Olukayode Oladipo, 1985). Past studies have assessed meteorological drought conditions by analyzing the precipitation data at various temporal scales (Patel et al., 2007; Smakhtin & Hughes, 2007). Many studies have used indices derived other than precipitation data, such as temperature and evaporation, to determine meteorological drought (Dai et al., 2004b; Li et al., 2012; Paulo et al., 2012).

2.2 Agricultural drought

Agricultural drought refers to inadequate soil moisture availability that results in crop wilting, and, ultimately crop damage or failure (Mishra & Singh, 2010). It is the combined impact of meteorological and hydrological drought. Some researchers concluded that it occurs after a meteorological drought and before a hydrological drought (Hisdal et al., 2000). The agricultural sector is usually at high risk when the drought begins because of its sole dependency on stored soil water. It also occurs due to unsustainable agricultural practices that impact water availability to the crops. Several drought indices, such as SMDI (Calder et al., 1983; Narasimhan & Srinivasan, 2005; Peters et al., 1991), CMI (Keyantash & Dracup, 2002; Walsh, 1987), etc., quantify the agricultural drought based on precipitation, ET, temperature, and soil moisture conditions. Remote sensing image-derived spectral indices, such as NDVI (Peters et al., 2002; Wan et al., 2004), VHI (Rhee et al., 2010; Rojas et al., 2011), SWDI (Moroke et al., 2011; Carrão et al., 2016a,b), etc., denote the vegetation greenness, leaf water content, and deviations from long-term conditions, and effectively highlight the agricultural drought impact on crop yield reduction.

2.3 Hydrological drought

The significant reduction in surface and subsurface water availability causes hydrological drought in a catchment, which leads to water scarcity in various forms (Heim Jr, 2002). Hydrological drought generally results from two successive or long-term meteorological droughts. It is associated with the declines in streamflow as well as lakes, reservoirs, and groundwater levels. Streamflow data is the main variable for analyzing hydrological drought. The relationship of hydrological droughts to the streamflow in a catchment denotes the drought severity, wherein the geological settings also influence the phenomenon. Various indices, such as Standardized Runoff Index (SRI), Streamflow Drought Index (SDI) (Vasiliades et al., 2011; Tabari et al., 2013; Vicente-Serrano et al., 2011), Surface Water Supply Index (SWSI) (Garen, 1993; Keyantash & Dracup, 2004), Standardized Water-Level Index (SWI) (Bhuiyan, 2004; Bhuiyan et al., 2006; Mishra & Nagarajan, 2013), and Groundwater Resource Index (GRI) (Trambauer et al., 2014), are widely used for hydrological drought assessment.

2.4 Socioeconomic drought

Socioeconomic drought occurs due to the combined influence all three physical droughts (meteorological, hydrological, and agricultural). This is associated with an unbalanced demand and supply gap of some economic goods triggered due to the precipitation and water supply deficit. It highlights the availability of food and thereby the food security of the people in the drought-affected area. Therefore, socioeconomic drought is more severe than the physical droughts. The indices used for socioeconomic drought quantification are socioeconomic indicators.

Fig. 24.2 shows the interaction among four drought types and their propagation. The precipitation deficit from its long-term climatology initiates the meteorological drought. Precipitation shortage and irregular precipitation events, or high ET cause the soil moisture deficit and cause agricultural drought, which results in crop damage, vegetation loss, and yield reduction. The hydrological drought triggered by the meteorological and agriculture droughts results into depletion in surface and subsurface runoff or water storage. It develops lastly and slowly because it includes stored water that may be depleted but not replenished (Dai, 2011). The crop failure and disruption of allied sectors led to a deficit of economic goods and increases the gap between demand and supply causes socioeconomic drought. It occurs when the demand exceeds the supply of food and other

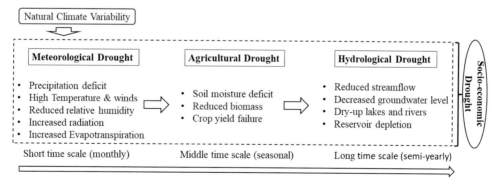

Figure 24.2 Types of droughts and their propagation.

economic goods due to other droughts, that is, meteorological, agricultural, and hydrological droughts. Although the shortage in precipitation triggers agriculture and hydrological drought, other atmospheric factors, and human activities, such as unsustainable land use practices, deforestation, improper water management, and erosion can also cause or aggravate these droughts.

3. Agriculture drought monitoring tools and techniques

The agricultural drought severity is measured primarily by quantifying soil moisture deficit. Periodic soil moisture assessment is a reliable and widely used variable for agricultural drought severity assessment. The agricultural drought onset begins when the soil moisture availability for plants significantly declines, which adversely affects crop health and in the long run decreases the yields. Thus, an accurate and real-time soil moisture measurement is of utmost importance for timely agriculture and water resource management practices. This section describes the soil moisture data available for agricultural drought assessment followed by a few widely used agricultural drought indices, advanced web-based platform for drought assessment, and some case studies showing the application of these datasets for agricultural drought monitoring.

3.1 Available data source

The drought assessment based on the ground collected data constrains accurate monitoring due to its coarse resolution and nonhomogeneous spatial coverage. In contrast, the remotely sensed satellite imageries offer an important input by providing data at different spatial resolutions and available for the last 4–5 decades (Datta et al., 2021; Srivastava et al., 2016). However, remotely sensed data have a few limitations, for example, unavailability of soil moisture data at vegetation root zone depth, lack of long-term reliable data availability, coarse resolution, etc. (Pandey et al., 2021a; Srivastava et al., 2013). On the contrary, the hydrological models employ the long-term climate variables as inputs and provide the soil moisture data at frequent intervals for surface and subsurface soil layers, even at a daily scale. The root zone soil moisture measurement offers antecedent conditions that indicate the potential and available water content (AWC) to the plants.

3.1.1 Satellite-based soil moisture data
Recent remote sensing platforms offer land surface data captured in different ranges of the electromagnetic spectrum for improved soil moisture estimation. Accurate soil moisture estimation requires longer wavelength data, that is, microwave region, which can penetrate the surface vegetation and soil layers. Various past studies have used different optical and thermal remote sensing data to retrieve soil moisture due to the unavailability of microwave data in the public domain. The optical and thermal remote sensing data highlights the reflection and radiation properties of the signal from the surface soil. The dry

soil shows higher reflectance and radiation than wet soil in the near-infrared and thermal bands and allows to estimate soil moisture conditions.

Although optical and thermal remote sensing images have been available for the past 4–5 decades, the main disadvantages of the optical images are the data discontinuity due to the lower penetration power in the soil and the unavailability of cloud free images throughout the year. In comparison, the longer wavelength microwave signal has a higher soil penetration capacity and penetrates through clouds. Thus, microwave remote sensing produces reliable data in all weather conditions and is one of the most suitable alternatives for continuous soil moisture monitoring. Moreover, recent space missions, such as Soil Moisture and Ocean Salinity (SMOS) (Kerr et al., 2010) and Soil Moisture Active and Passive (SMAP) (Entekhabi et al., 2010), provide soil moisture data at a coarse resolution. However, these datasets are widely used for soil moisture estimation and agricultural drought assessment at a larger spatial scale globally (Kedzior & Zawadzki, 2017; Zhu et al., 2019).

However, microwave data products provide a more accurate estimation of soil moisture due to their inherent relationship with dielectric properties of land surface features and water. Thus, the longer wavelength microwave signals provide a reliable soil moisture estimation based on the backscatter signal. The recent Sentinel-1 mission provides the lower wavelength (C-band: 5 cm) microwave data at five days intervals with 5–20 m spatial resolution since 2014. The vertical-vertical (VV) and vertical-horizontal (VH) polarization data have been observed to provide reliable soil moisture data. Various recent studies are using Sentinel-1 data for periodic soil moisture estimation at a high spatial resolution for agricultural drought assessment (Dilip et al., 2023; Shorachi et al., 2022).

3.1.2 Land surface models

Various hydrological models are used to simulate hydrological variables, which often provide more imminent information for hydrological conditions and used for agricultural drought assessment (Sehgal & Sridhar, 2019; Sheffield et al., 2004). The Soil and Water Assessment Tool (SWAT) (Narasimhan & Srinivasan, 2005), Tiled European Centre for Medium-Range Weather Forecast (ECMWF) Scheme for Surface Exchange over Land (TESSEL) (Dutra et al., 2008), One layer Water-Budget (Hao & AghaKouchak, 2013), and the VIC (Sheffield et al., 2004; Sheffield & Wood, 2008; Wang et al., 2011; Wu et al., 2007) are commonly used models for periodic soil moisture simulation. Dash et al. (2019) employed SWAT model to simulate droughts based on SPI in the Kangsabati River basin. Sheffield and Wood (2007) employed VIC model SM data for agriculture drought monitoring over global land areas during 1950–2000. Tang et al. (2009) also employed VIC model-simulated soil moisture data for agriculture drought estimation in the Upper Colorado River Basin, USA, for 50 years and reported the efficiency of the VIC model in simulation for systematic agricultural drought mapping. Kang et al.

(2022) studied the flash drought characteristics in the Mekong River Basin (MRB) using both the SWAT and VIC hydrological models along with multiple climate change scenarios. They identify some disagreement between both model-derived drought indexes due to different approaches followed by models for partitioning the water budget. However, they concluded the usefulness of the models in the identification of various drought perspectives.

3.2 Agricultural drought indices

Various indices and approaches have been adopted for agriculture drought identification and characterization. The use of an index is common in assessing drought and characterization through duration, intensity, severity, and spatial extent (Dai, 2011). A drought index is usually characterized by the standardization of one or more variables that measure the departure of the water availability from its long-term mean. For agricultural drought monitoring, the most widely used factors are soil moisture, vegetation health, and crop yield. Table 24.1 summarizes the widely used indices for quantifying agricultural droughts. Here we have briefly described some widely used indices for agricultural drought assessment.

3.2.1 Palmer drought severity index

For the assessment of the total moisture status of a desired location, Palmer (1965) formulated a comprehensive index as the PDSI. This index uses precipitation, temperature, and AWC as the input parameters for drought severity estimation. The PDSI has been primarily used for determining long-term drought and is not much effective in determining short-term drought conditions. The PDSI is suitable for the categorization of an area prominently affected by wetness and dryness, into various severity levels. It gives standardized moisture conditions (soil water storage) and allows a clear comparison between spatial and temporal components for a better understanding of drought severity. Furthermore, PDSI has made possible comparisons of various climatic zones as it accounts for temperature and soil type (Guttman, 1998).

However, the requirement of substantial inputs of hydrometeorological data and complex computation limits its application in the Asian region, where observational networks are scarce. Moreover, PDSI responds slowly to drought occurrence and its termination. Due to the inherent time scale, PDSI is more suitable for agricultural drought than hydrological droughts. To improve the applicability of PDSI, various studies have modified PDSI and formulated revised indices, such as the Palmer Hydrological Drought Index (PHDI) by Karl (1986).

3.2.2 Crop moisture index

As PDSI responds slowly to detect the onset of drought, Palmer developed CMI to overcome this limitation. CMI is calculated weekly along with PDSI for short-term

Table 24.1 Significant indices for agricultural drought assessment.

Indices	Variables	Formulae	Data used	Authors
Palmer's drought severity index (PDSI)	Evapotranspiration (ET), precipitation, and temperature	$X_i = 0.897 X_{i-1} + \dfrac{z_i}{3}$	Remotely sensed	Palmer (1965)
Standardized precipitation index (SPI)	Precipitation	$SPI = \dfrac{X_{ij} - X_{im}}{\sigma}$	Remotely sensed	McKee et al. (1993)
Standardized precipitation evapotranspiration index (SPEI)	Precipitation and potential evapotranspiration (PET)	Differences between precipitation and potential evapotranspiration	Remotely sensed and modeled	Vicent-Serrano et al. (2010)
Scale drought conditions index (SDCI)	LST, NDVI, and precipitation	$SDCI = (1/4) LST_s$ $+ (2/4) TRMM_s + (1/4) NDVI_s$	Remotely sensed	Rhee et al. (2010)
Vegetation condition index (VCI)	Vegetation indices	$VCI = 100 \times \dfrac{NDVI - NDVI_{min}}{NDVI_{max} - NDVI_{min}}$	Remotely sensed	Kogan (1997)
Temperature condition index (TCI)	Temperature	$TCI = 100 \times \dfrac{TB_{max} - TB}{TB_{max} - TB_{min}}$	Remotely sensed	Kogan (1997)
Vegetation health index (VHIs)	NDVI and TB	$VHI = a VCI + (1 - a) TCI$	Remotely sensed	Kogan (1997)
Crop moisture index (CMI)	ET and soil moisture	Sums of evapotranspiration deficit and soil water recharge	Modeled	Palmer (1968)
Soil moisture anomaly	Soil moisture	$SMA = \dfrac{SMI_t - SMI}{\sigma}$	Remotely sensed and modeled	Bergman Sabol and Miskus (1988)
Soil-adjusted vegetation index (SAVI)	Soil adjustment factor and vegetation density	$SAVI = \dfrac{NIR - RED}{NIR + RED + L} (1 + L)$	Remotely sensed	Huete (1988)

Continued

Table 24.1 Significant indices for agricultural drought assessment.—cont'd

Indices	Variables	Formulae	Data used	Authors
Soil moisture deficit index (SMDI)	Soil moisture	$\text{SMDI} = 0.5 \times \text{SMDI}_{i-1} + \dfrac{\text{SD}_i}{50}$	Modeled	Narasimhan and Srinivasan (2005)
Soil water deficit index (SWDI)	Soil moisture, field capacity, and AWC	$\text{SWDI} = 10 \times \left(\dfrac{\text{SM} - \text{FC}}{\text{AWC}} \right)$	Modeled and remotely sensed	Martínez-Fernández et al. (2015)
Soil moisture agricultural drought index (SMADI)	Soil moisture, LST, and NDVI	$\text{SMADI} = \text{SMCI} \dfrac{\text{MTCI}}{\text{VCI}}$ (here, MTCI is modified TCI where TB is replaced by LST and $\text{SMCI} = \dfrac{\text{SM}_{\max} - \text{SM}}{\text{SM}_{\max} - \text{SM}_{\min}}$)	Modeled and remotely sensed	Sánchez et al. (2016)
Standardized soil moisture index (SSMI)	Soil moisture and precipitation	$\text{SSMI} = \boldsymbol{\phi}^{-1}(p)$	Remotely sensed and modeled	AghaKouchak (2014)

agricultural drought assessment to assess drought impact on crop growth. CMI is sensitive to rapid changes in moisture conditions on vegetation health. CMI used weekly precipitation, temperature, and previous week's CMI value as inputs (Keyantash & Dracup, 2002). Then comparisons are performed between variables from moisture budget and long-term average and are modified by empirical relation to compute the final CMI. CMI is developed for short-term agricultural drought assessment, rather than monitoring the long-term drought conditions. It shows a false sense of recovery for long-term drought events, as short-term improvements hopefully insufficient for offset long-term issues (Svoboda et al., 2002). However, it is an effective tool for measuring agricultural drought during growing seasons. Sensitivity analysis shows that CMI values may increase with increasing potential ET. The increase in potential ET may lead to drier conditions, whereas an increase in the CMI value shows wetter moisture conditions. CMI shows quick vegetation response to changing conditions at a local scale. However, such indices are unsuitable during seed germination or crop growing season (Belal et al., 2014).

3.2.3 Standardized precipitation index

It is a commonly used meteorological drought index derived using long-term precipitation data. SPI can be derived at different time scales (monthly to annual), thus useful for short-term water deficit (soil moisture), as well as long-term water resources deficit (groundwater, stream flow, lake, and reservoir levels) assessment (Vicente-Serrano et al., 2014). Various studies have applied SPI for evaluating agricultural drought stress over different geographical areas and time periods. SPI computation requires current and long-term precipitation data of a region. A probability distribution is applied to the precipitation data for standardized normal data distribution, followed by deriving the mean SPI (set to zero) for the studied location and time period. Thereafter, the SPI is measured by dividing the difference between observed precipitation and long-term mean by the standard deviation (Guttman, 1999).

SPI has limitation when applied for very long time scales, that is, above 24 months as due to bias in the fitted distribution (Mishra & Singh, 2010). It also shows biased values over dry climatic regions where precipitation is very low, and zero values are frequent in a particular season. These climatic regions are prone to high error while computing SPI because it is not normally distributed as precipitation distributions are highly skewed and limitation of the fitting gamma distribution. Moreover, it deals only water supply based on precipitation alone do not involve other factors like ET. Though it has multiple limitations, it is a frequently used index for drought monitoring due to its simplistic and robust approach.

3.2.4 Vegetation health index

Kogan (1995) developed the VHI using remote sensing images derived from Advanced Very High-Resolution Radiometer (AVHRR) satellite. AVHRR satellite is a polar orbiting system, which covers a large area at frequent temporal intervals. For this reason, the AHVRR data are widely used for drought assessment at regional to global scales. VHI is estimated based on three basic environmental laws, minimum value, tolerance, and carrying capacity that define the theoretical basis for calculating the climate vigor (Bhuiyan et al., 2017). VHI is a combination of two indices derived based on greenness (NDVI) and thermal emission. The thermal emission data are converted into the brightness temperature (TB) of the land surface and used to compute Temperature Condition Index (TCI). A brief description of these two subindices is given below.

Vegetation Condition Index (VCI): VCI is derived using AVHRR data-derived greenness indicator (i.e., NDVI). It allows measurement of the onset of drought and its severity and duration based on the vegetation conditions. Jain et al. (2010) describe VCI as vegetation health indicator as a function of NDVI minimum and maximum values for a vegetation ecosystem over a long period. VCI is a better parameter of water stress conditions compared to NDVI as it derives the vegetation health deterioration in a short-term than long-term conditions (Jain et al., 2010). Using the long-term NDVI data, the VCI is computed as:

$$VCI = 100 \times \frac{NDVI - NDVI_{min}}{NDVI_{max} - NDVI_{min}} \qquad (24.1)$$

where NDVI, $NDVI_{max}$, and $NDVI_{min}$ are vegetation conditions in a time, long-term maximum, and minimum vegetation conditions, respectively, for pixel in an image. Various past studies applied VCI for agricultural drought assessment at spatial and temporal scales.

Temperature Condition Index (TCI): TCI was formulated to assess the effect of drought on vegetation health through temperature. TCI is calculated in the same way as VCI but, unlike VCI, TCI reflects the changes in canopy temperature due to drought. The agricultural drought leads to reduced leaf water content, followed by increased canopy temperature. Therefore, a higher canopy temperature denotes lower leaf water content and greater drought. TCI is calculated using the TB data from the AVHRR thermal infrared (TIR) spectral band. Similar to VCI, TCI uses the long-term and immediate temperature conditions to denote the significant change in canopy temperature conditions. Thus, TCI indicates temperature-related vegetation stress than long-term conditions. TCI is computed as Eq. (24.2):

$$TCI = 100 \times \frac{BT_{max} - BT}{BT_{max} - BT_{min}} \qquad (24.2)$$

where TB, TB_{max}, and TB_{min} are canopy temperature in a time, long-term maximum, and minimum canopy temperature, respectively, for each pixel.

VHI is computed using VCI and TCI based on the following equation.

$$VHI = a \times VCI + (1 - a) \times TCI \qquad (24.3)$$

where a is the coefficient that quantifies the contribution of VCI and TCI. All three indices range from 0 to 100 corresponding to changes in vegetation health from extremely unfavorable conditions to normal conditions, respectively. Values below 40 showed drought conditions. VHI measures drought conditions based on the depressed vegetation greenness that indicates a decrease in NDVI and an increase in canopy temperature. This results in a decrease in VCI and TCI, and subsequently, VHI which indicates drought severity. NOAA global area coverage datasets (4 km^2) issued the Global Vegetation Health (GVH) data at a weekly scale from 1981 to the present. For the calculation of VCI and TCI, firstly the AVHRR-derived time-series NDVI and TB are processed to remove outliers through normalization and filtration. Secondly, the values of each pixel are scaled as maximum and minimum values. The NDVI-derived VCI indicates an area of rich or poor vegetation conditions than long-term average conditions, whereas TCI identifies areas which are hotter than usual. Both the subindices are averaged with uniform weighing to generate VHI, which highlights the anomalous vegetation greenness and temperature due to drought. For this reason, VHI has been widely used for agricultural production monitoring, identification of agricultural drought, assessment of irrigated areas, and excessive wetness (Karnieli et al., 2010).

3.2.5 Scaled drought condition index

Rhee et al. (2010) developed a multisensor Scaled Drought Condition Index (SCDI) for monitoring agricultural drought by combining LST and NDVI data from Moderate Resolution Imaging Spectroradiometer (MODIS) and precipitation data from Tropical Rainfall Measuring Mission (TRMM). The SDCI is computed by using the following formula.

$$SDCI = (1/4) \times \text{scaled LST} + (2/4) \times \text{scaled TRMM} + (1/4) \times \text{scaled NDVI}$$

(24.4)

where scaled $LST = (LST_{max} - LST_i)/(LST_{max} - LST_{min})$; scaled $TRMM = (TRMM_i - TRMM_{min})/(TRMM_{max} - TRMM_{min})$, and scaled $NDVI = (NDVI_i - NDVI_{min})/(NDVI_{max} - NDVI_{min})$ and subscripts i, max, and min denote the pixel, maximum, and minimum value of its respective parameters. The SDCI has the flexibility to improve using multiple variables of the multisensor indices according to different drought types.

3.2.6 Soil moisture deficit index

The SMDI is a widely used index that examines the consecutive soil moisture deficit at various soil layers and for a defined time interval (Narasimhan & Srinivasan, 2005). SMDI requires long-term soil moisture datasets for a particular region. First step of the SMDI involves the calculation of long-term median, minimum, and maximum soil moisture for a particular week. Then the weekly percentage of soil moisture deficit ($SD_{k,i}$) for each week i and year k is calculated using the following conditional expression as Eqs. (24.5) and (24.6):

$$SD_{k,i} = \frac{SM_{k,i} - SM_{median,i}}{SM_{median,i} - SM_{min,i}} \times 100 \; \text{If} \quad SM_{k,i} \leq SM_{median,i} \qquad (24.5)$$

$$SD_{k,i} = \frac{SM_{k,i} - SM_{median,i}}{SM_{max,i} - SM_{median,j}} \times 100 \; \text{If} \quad SM_{k,j} > SM_{median,i} \qquad (24.6)$$

where $SM_{k,i}$ is the SM for the current week i and year k, and the $SM_{median,i}$, $SM_{max,i}$ and $SM_{min,i}$ denote the long-term median, maximum, and minimum values of the current week i, respectively. Then the weekly SMDI was computed as Eq. (24.7).

$$SMDI_i = 0.5 \times SMDI_{i-1} + \frac{SD_i}{50} \qquad (24.7)$$

where the $SMDI_{i-1}$ is the SMDI of the previous week and SD_i is the soil moisture deficit in percentage for a current week i. The SMDI was initialized as $SMDI = SD_{i/50}$. SMDI value ranges from -4 to $+4$, indicating extreme drought and wet conditions, respectively.

3.3 Agricultural drought monitoring using satellite and land surface-derived soil moisture product: Case studies

Martínez-Fernández et al. (2016) used the SMOS L2 soil moisture data for calculating SWDI in Spain for 2010—14. The other soil properties information (field capacity and AWC) is computed using pedotransfer function (PTF). They have compared SMOS SWDI ($SWDI_S$) with the two well-established indices (CMI and atmospheric water deficit [AWD]), and *in situ* soil moisture-based SWDI ($SWDI_R$) and observed good agreements (Fig. 24.3). They have concluded that SMOS SWDI can produce reliable soil water stress and appropriately quantifies the agricultural drought. Mishra et al. (2017) used SMAP L3 soil moisture product and soil parameters (field capacity and AWC) to calculate SWDI for Contiguous United States (CONUS). They applied the PTF to derive the soil properties from Harmonized World Soil Database. They have also compared the SMAP-derived SWDI with *in situ* soil moisture and AWD and reported a good agreement. SMAP SWDI shows potential for short-term moisture stress assessment and proved as an effective remotely sensed agricultural drought indicator.

On the other hand, hydrological model-derived agricultural drought studies show its potential at a higher temporal scale and for different (surface and subsurface) soil layers. Pandey and Srivastava (2018) used SMDI for agricultural drought assessment in the Kharif season at weekly scale from 1998 to 2015 in the Bundelkhand region, of Uttar Pradesh, in central subtropical region of India. They used the macroscale VIC

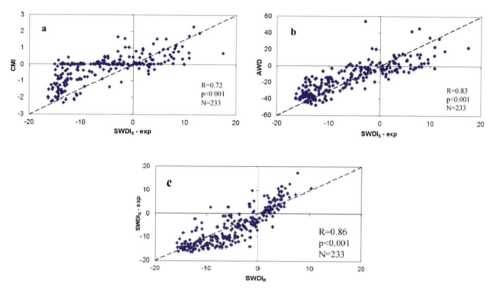

Figure 24.3 Comparison between SWDIS and CMI (a); AWD (b); and SWDIR (c). *(Source: Martínez-Fernández et al. (2016).)*

hydrological model to simulate daily soil moisture at 0.25° spatial resolution. The meteorological forcing, land use land cover (LULC), topography, and vegetation parameters were used as input. The model was calibrated by tuning the soil parameters obtained from the National Bureau of Soil Survey and Land Use Planning (NBSSLUP) and validated the model's performance based on ESA's Climate Change Initiative (ESA-CCI) soil moisture data. The comparison of model-simulated SM indicated a good agreement. The historical agricultural drought affected years were well captured by VIC soil moisture-derived SMDI in the study region. The maximum number of drought weeks were recorded in the year 2001, 2002, 2007, 2008, and 2009, wherein severe drought occurred in the 2007, 2008, 2009, and 2014. The wet (2001) and dry (2002) years are clearly identified by the spatial distribution of SMDI maps in the region, shows 80% of the study area is drought prone during the drought year (Fig. 24.4). The study recommends the suitability of hydrological model for the simulation of soil moisture at large spatial scale with high temporal resolution (1 day), which are useful for reliable agricultural drought assessment.

3.4 Automated and periodic drought assessment using WebGIS platforms

The Google Earth Engine (GEE) is one of the widely used latest cloud-based or WebGIS platforms. This contains a range of remotely sensed imageries, climate and topographic data, and various other spatial layers, which are publicly available. The GEE platform

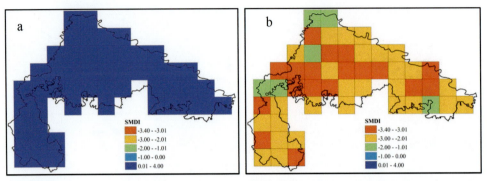

Figure 24.4 Spatiotemporal SMDI maps for 28th week (i.e., July 9—15), which corresponds the sowing period for Kharif season during; (a) normal (2001) and (b) drought (2002) year.

uses Python and JavaScript languages and allows to process large geospatial datasets with limited computation facilities on the user's computer. Thus, the user can perform various analyses without downloading the data, download the input and end products, share the code with others, upload required data and processes, host various data products, and share publicly. Moreover, the platform allows automated data processing and enables near-real-time drought monitoring globally. The platform allows the application of parametric and nonparametric regression and classification tools, including the widely used machine learning models.

Various past studies have used GEE for agriculture drought assessment. For example, the agriculture drought monitoring and hazard assessment through MODIS-based VHI by United Nations (https://un-spider.org/advisory-support/recommended-practices/recommended-practice-agriculture-drought-monitoring). Thilagaraj et al. (2021) used GEE for agriculture drought assessment using the Landsat and Sentinel-2 data–derived VHI and compared with TRMM data-derived SPI (meteorological drought) in the Kodavanar watershed, in Amaravathi basin, India. Sazib et al. (2018) developed the GEE tool by inclusion of global soil moisture products and anomalies in GEE platform for drought assessment and crop health forecasting. The global soil moisture product is developed by the integration of SMOS and SMAP observations into a modified two-layer Palmer model employing 1D data ensemble Kalman filter (EnKF) data assimilation approach. They have assessed the efficacy of GEE tool by comparing the soil moisture products with SPI and NDVI over drought-prone regions in South Africa and Ethiopia. They have recommended soil moisture GEE tool for such studies due to its accessibility and user-friendly interface, and quick processing time. The GEE-based SMAP products used as a proxy for agricultural drought risk assessment studies (Bolten et al., 2010). A sample of this GEE-based anomaly product is shown in Fig. 24.5, which indicates the spatiotemporal distribution of surface soil moisture anomaly over India for the period of 2016—22.

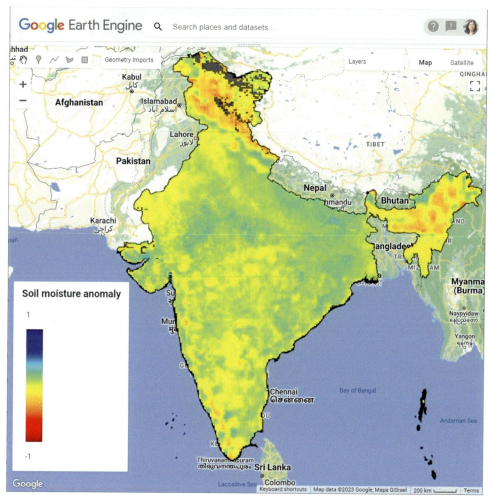

Figure 24.5 Surface soil moisture anomaly derived from NASA-USDA enhanced SMAP global soil moisture product for the period of 2016—22.

4. Challenges and way forward

For agricultural drought assessment, very limited data and products are available on the estimation and assessment of soil moisture and ET, etc., at fine spatial and temporal resolution. However, the long-term observations on soil moisture rely on a few sites, wherein the station data are nonhomogeneous and sparse to support spatially explicit drought analysis (Robock et al., 2000). The International Soil Moisture Network (ISMN) that provides SM data is currently nonfunctional over India, while COsmic-ray Soil Moisture Observing System (COSMOS) is a paid service. A few more Earth observation programs provide real-time soil moisture data such as

ESA CCI (Dorigo et al., 2017; Van der Schalie et al., 2023), Soil Moisture Operational Products System (SMOPS) (Yin et al., 2020), etc., but long-term and root zone information are still lagging. Alternatively, the land surface hydrological models offer a valuable alternative to study soil moisture. Moreover, the assessment of drought hazards and their impact on socioeconomic conditions is less studied, particularly important for securing local and regional food demand and environmental conditions. However, emergence of remote sensing data has overcome these problems from last decades. Moreover, WebGIS platforms such as GEE facilitate large geospatial dataset processing with limited computation facilities on the user's computer. It allows automated data processing and enables near-real-time drought monitoring globally (Khan & Gilani, 2021).

The major concern is increasing frequency and severity of droughts, where one drought event has been reported every three/four years for the last 5 decades (Samra, 2004; Ray et al., 2015). Therefore, early warning systems are necessary to minimize and mitigate the adverse effects of droughts in affected regions. Further, we recommend the assimilation of gauged and remotely sensed data for more robust estimates of drought severity. Significant progress has been found in drought modeling, but still much work remains ahead in this area (Ekmekcioğlu, 2023). Research needs development of suitable models for predicting the onset and termination points of drought and there should be consideration of both statistical and numerical forecasting for better estimation of drought events. Drought forecasting demands a probabilistic approach to enabling the quantification of variation and uncertainty. There should be consideration of both statistical and numerical forecasting for better estimation. The integrated approach must be followed for actual forecasting of drought phenomenon which includes all climatic indicators such as hydrometeorological (precipitation, ET, temperature, soil moisture, etc), climatic representatives [El Niño-Southern Oscillation (ENSO), Sea Surface Temperature (SST), Southern Oscillation Index (SOI), etc.], and parameters-based indices (SPI, SPEI, SWSI, SDI, NDVI, etc.).

References

AghaKouchak, A. (2014). A baseline probabilistic drought forecasting framework using standardized soil moisture index: Application to the 2012 United States drought. *Hydrology and Earth System Sciences, 18*(7), 2485–2492.

AghaKouchak, A., Feldman, D., Hoerling, M., Huxman, T., & Lund, J. (2015). Water and climate: Recognize anthropogenic drought. *Nature News, 524*, 409.

Bandyopadhyay, N., Bhuiyan, C., & Saha, A. (2020). Drought mitigation: Critical analysis and proposal for a new drought policy with special reference to Gujarat (India). *Progress in Disaster Science, 5*, 100049.

Belal, A. A., El-Ramady, H. R., Mohamed, E. S., & Saleh, A. M. (2014). Drought risk assessment using remote sensing and GIS techniques. *Arabian Journal of Geosciences, 7*, 35–53.

Bergman, K. H., Sabol, P., & Miskus, D. (1988). Experimental indices for monitoring global drought conditions. In *Proceedings of the 13th annual climate diagnostics workshop, Cambridge, MA, USA* (Vol. 31, pp. 190–197).

Bhuiyan, C., Singh, R., & Kogan, F. (2006). Monitoring drought dynamics in the Aravalli region (India) using different indices based on ground and remote sensing data. *International Journal of Applied Earth Observation and Geoinformation, 8,* 289–302.

Bhuiyan, C. (2004). *Various drought indices for monitoring drought condition in Aravalli terrain of India* (pp. 12–23). XXth ISPRS Congress.

Bhuiyan, C., Saha, A., Bandyopadhyay, N., & Kogan, F. (2017). Analyzing the impact of thermal stress on vegetation health and agricultural drought—a case study from Gujarat, India. *GIScience and Remote Sensing, 54,* 678–699.

Bolten, J. D., Crow, W. T., Zhan, X., Jackson, T. J., & Reynolds, C. A. (2010). Evaluating the utility of remotely sensed soil moisture retrievals for operational agricultural drought monitoring. *IEEE Journal of Selected Topics in Applied Earth Observations and Remote Sensing, 3*(1), 57–66. https://doi.org/10.1109/JSTARS.2009.2037163

Calder, I., Harding, R., & Rosier, P. (1983). An objective assessment of soil-moisture deficit models. *Journal of Hydrology, 60,* 329–355.

Carrão, H., Naumann, G., & Barbosa, P. (2016a). Mapping global patterns of drought risk: An empirical framework based on sub-national estimates of hazard, exposure and vulnerability. *Global Environmental Change, 39,* 108–124.

Carrão, H., Russo, S., Sepulcre-Canto, G., & Barbosa, P. (2016b). An empirical standardized soil moisture index for agricultural drought assessment from remotely sensed data. *International Journal of Applied Earth Observation and Geoinformation, 48,* 74–84.

Chandrasekar, K., & Sai, M. S. (2015). Monitoring of late-season agricultural drought in cotton-growing districts of Andhra Pradesh state, India, using vegetation, water and soil moisture indices. *Natural Hazards, 75,* 1023–1046.

Dai, A. (2011). Drought under global warming: A review. *Wiley Interdisciplinary Reviews: Climate Change, 2,* 45–65.

Dai, A., Trenberth, K. E., & Qian, T. (2004b). A global dataset of Palmer Drought Severity Index for 1870-2002: Relationship with soil moisture and effects of surface warming. *Journal of Hydrometeorology, 5,* 1117–1130.

Dash, S. S., Sahoo, B., & Raghuwanshi, N. S. (2019). A SWAT-Copula based approach for monitoring and assessment of drought propagation in an irrigation command. *Ecological Engineering, 127,* 417–430.

Datta, S., Das, P., Dutta, D., & Giri, R. K. (2021). Estimation of surface moisture content using Sentinel-1 C-band SAR data through machine learning models. *Journal of the Indian Society of Remote Sensing, 49,* 887–896.

Dilip, T., Kumari, M., Murthy, C. S., Neelima, T. L., Chakraborty, A., & Devi, M. U. (2023). Monitoring early-season agricultural drought using temporal Sentinel-1 SAR-based combined drought index. *Environmental Monitoring and Assessment, 195*(8), 925. https://doi.org/10.1007/s10661-023-11524-y

Dorigo, W., Wagner, W., Albergel, C., Albrecht, F., Balsamo, G., Brocca, L., … Lecomte, P. (2017). ESA CCI Soil Moisture for improved Earth system understanding: State-of-the art and future directions. *Remote Sensing of Environment, 203,* 185–215.

Dutra, E., Viterbo, P., & Miranda, P. (2008). ERA-40 reanalysis hydrological applications in the characterization of regional drought. *Geophysical Research Letters, 35.*

Ekmekcioğlu, Ö. (2023). Drought forecasting using integrated variational mode decomposition and extreme gradient boosting. *Water, 15*(19), 3413. https://doi.org/10.3390/w15193413

Entekhabi, D., Reichle, R. H., Koster, R. D., & Crow, W. T. (2010). Performance metrics for soil moisture retrievals and application requirements. *Journal of Hydrometeorology, 11*(3), 832–840.

Garen, D. C. (1993). Revised surface-water supply index for western United States. *Journal of Water Resources Planning and Management, 119,* 437–454.

Glantz, M. H. (1994). *Drought follows the plow: Cultivating marginal areas.* Cambridge University Press.

Guttman, N. B. (1998). Comparing the palmer drought index and the standardized precipitation index 1. *JAWRA Journal of the American Water Resources Association, 34*(1), 113–121. https://doi.org/10.1111/j.1752-1688.1998.tb05964.x.

Guttman, N. B. (1999). Accepting the standardized precipitation index: A calculation algorithm. *JAWRA Journal of the American Water Resources Association, 35,* 311–322.

Hao, Z., & AghaKouchak, A. (2013). Multivariate standardized drought index: A parametric multi-index model. *Advances in Water Resources, 57,* 12—18.

Heim, R. R., Jr. (2002). A review of twentieth-century drought indices used in the United States. *Bulletin of the American Meteorological Society, 83,* 1149—1166.

Hisdal, H., Tallaksen, L., Peters, E., Stahl, K., & Zaidman, M. (2000). Drought event definition. In *ARIDE technical report No. 6.*

Huete, A. R. (1988). A soil-adjusted vegetation index (SAVI). *Remote Sensing of Environment, 25*(3), 295—309.

Jain, S. K., Keshri, R., Goswami, A., & Sarkar, A. (2010). Application of meteorological and vegetation indices for evaluation of drought impact: A case study for Rajasthan, India. *Natural Hazards, 54,* 643—656.

Kedzior, M., & Zawadzki, J. (2017). SMOS data as a source of the agricultural drought information: Case study of the Vistula catchment, Poland. *Geoderma, 306,* 167—182. https://doi.org/10.1016/j.geoderma.2017.07.018

Kang, H., Sridhar, V., & Ali, S. A. (2022). Climate change impacts on conventional and flash droughts in the Mekong River Basin. *Science of the Total Environment, 838,* 155845. https://doi.org/10.1016/j.scitotenv.2022.155845

Karl, T. R. (1986). The sensitivity of the Palmer Drought Severity Index and Palmer's Z-index to their calibration coefficients including potential evapotranspiration. *Journal of Climate and Applied Meteorology, 25,* 77—86.

Karnieli, A., Agam, N., Pinker, R. T., Anderson, M., Imhoff, M. L., Gutman, G. G., Panov, N., & Goldberg, A. (2010). Use of NDVI and land surface temperature for drought assessment: Merits and limitations. *Journal of Climate, 23,* 618—633.

Kerr, Y. H., Waldteufel, P., Wigneron, J. P., Delwart, S., Cabot, F., Boutin, J., ... Mecklenburg, S. (2010). The SMOS mission: New tool for monitoring key elements of the global water cycle. *Proceedings of the IEEE, 98*(5), 666—687.

Keyantash, J. A., & Dracup, J. A. (2004). An aggregate drought index: Assessing drought severity based on fluctuations in the hydrologic cycle and surface water storage. *Water Resources Research, 40.*

Keyantash, J., & Dracup, J. A. (2002). The quantification of drought: An evaluation of drought indices. *Bulletin of the American Meteorological Society, 83,* 1167—1180.

Khan, R., & Gilani, H. (2021). Global drought monitoring with big geospatial datasets using Google Earth Engine. *Environmental Science and Pollution Research, 28*(14), 17244—17264. https://doi.org/10.1007/s11356-020-12023-0

Kogan, F. N. (1995). Application of vegetation index and brightness temperature for drought detection. *Advances in Space Research, 15,* 91—100.

Kogan, F. N. (1997). Global drought watch from space. *Bulletin of the American Meteorological Society, 78*(4), 621—636. https://doi.org/10.1175/1520-0477(1997)078<0621:GDWFS>2.0.CO;2

Li, W.-G., Yi, X., Hou, M.-T., Chen, H.-L., & Chen, Z.-L. (2012). Standardized precipitation evapotranspiration index shows drought trends in China. *Chinese Journal of Eco-Agriculture, 20,* 643—649.

Martínez-Fernández, J., González-Zamora, A., Sánchez, N., & Gumuzzio, A. (2015). A soil water based index as a suitable agricultural drought indicator. *Journal of Hydrology, 522,* 265—273.

Martínez-Fernández, J., González-Zamora, A., Sánchez, N., Gumuzzio, A., & Herrero-Jiménez, C. M. (2016). Satellite soil moisture for agricultural drought monitoring: Assessment of the SMOS derived Soil Water Deficit Index. *Remote Sensing of Environment, 177,* 277—286. https://doi.org/10.1016/j.rse.2016.02.064

McKee, T. B., Doesken, N. J., & Kleist, J. (1993). The relationship of drought frequency and duration to time scales. In *Proceedings of the 8th conference on applied climatology Boston* (pp. 179—183).

Meze-Hausken, E. (2004). Contrasting climate variability and meteorological drought with perceived drought and climate change in northern Ethiopia. *Climate Research, 27,* 19—31.

Mishra, A. K., & Singh, V. P. (2010). A review of drought concepts. *Journal of Hydrology, 391,* 202—216.

Mishra, A., Vu, T., Veettil, A. V., & Entekhabi, D. (2017). Drought monitoring with soil moisture active passive (SMAP) measurements. *Journal of Hydrology, 552,* 620—632.

Mishra, S. S., & Nagarajan, R. (2013). Hydrological drought assessment in Tel river basin using standardized water level index (SWI) and GIS based interpolation techniques. *International Journal of Civil Engineering, 1*, 01–04.

Mishra, V., & Lilhare, R. (2016). Hydrologic sensitivity of Indian sub-continental river basins to climate change. *Global and Planetary Change, 139*, 78–96.

Moroke, T., Schwartz, R., Brown, K., & Juo, A. (2011). Water use efficiency of dryland cowpea, sorghum and sunflower under reduced tillage. *Soil and Tillage Research, 112*, 76–84.

Murthy, C., Sesha Sai, M., Chandrasekar, K., & Roy, P. (2009). Spatial and temporal responses of different crop-growing environments to agricultural drought: A study in Haryana state, India using NOAA AVHRR data. *International Journal of Remote Sensing, 30*, 2897–2914.

Nagarajan, R. (2010). *Drought assessment*. Springer Science & Business Media.

Narasimhan, B., & Srinivasan, R. (2005). Development and evaluation of soil moisture deficit index (SMDI) and evapotranspiration deficit index (ETDI) for agricultural drought monitoring. *Agricultural and Forest Meteorology, 133*, 69–88.

Olukayode Oladipo, E. (1985). A comparative performance analysis of three meteorological drought indices. *Journal of Climatology, 5*, 655–664.

Palmer, W. C. (1965). *Meteorological drought* (Vol. 30). Weather Bureau: US Department of Commerce.

Palmer, W. C. (1968). *Keeping track of crop moisture conditions, nationwide: The new crop moisture index.*

Pandey, V., Srivastava, P. K., Das, P., & Behera, M. D. (2021a). Irrigation water demand estimation in Bundelkhand region using the variable infiltration capacity model. *Agricultural Water Management, 331*–347.

Pandey, V., & Srivastava, P. K. (2018). Integration of satellite, global reanalysis data and macroscale hydrological model for drought assessment in sub-tropical region of India. *The International Archives of the Photogrammetry, Remote Sensing and Spatial Information Sciences, 42*, 1347–1351.

Patel, N., Chopra, P., & Dadhwal, V. (2007). Analyzing spatial patterns of meteorological drought using standardized precipitation index. *Meteorological Applications, 14*, 329–336.

Paulo, A., Rosa, R., & Pereira, L. (2012). Climate trends and behaviour of drought indices based on precipitation and evapotranspiration in Portugal. *Natural Hazards and Earth System Sciences, 12*, 1481–1491.

Peters, A. J., Rundquist, D. C., & Wilhite, D. A. (1991). Satellite detection of the geographic core of the 1988 Nebraska drought. *Agricultural and Forest Meteorology, 57*, 35–47.

Peters, A. J., Walter-Shea, E. A., Ji, L., Vina, A., Hayes, M., & Svoboda, M. D. (2002). Drought monitoring with NDVI-based standardized vegetation index. *Photogrammetric Engineering and Remote Sensing, 68*, 71–75.

Ray, D. K., Gerber, J. S., Macdonald, G. K., & West, P. C. (2015). Climate variation explains a third of global crop yield variability. *Nature Communication, 6*, 1–9.

Rhee, J., Im, J., & Carbone, G. J. (2010). Monitoring agricultural drought for arid and humid regions using multi-sensor remote sensing data. *Remote Sensing of Environment, 114*, 2875–2887.

Robock, A., Vinnikov, K. Y., Srinivasan, G., Entin, J. K., Hollinger, S. E., Speranskaya, N. A., Liu, S., & Namkhai, A. (2000). The global soil moisture data bank. *Bulletin of the American Meteorological Society, 81*, 1281–1300.

Rojas, O., Vrieling, A., & Rembold, F. (2011). Assessing drought probability for agricultural areas in Africa with coarse resolution remote sensing imagery. *Remote Sensing of Environment, 115*, 343–352.

Rosegrant, M. W., & Cline, S. A. (2003). Global food security: Challenges and policies. *Science, 302*, 1917–1919.

Sánchez, N., González-Zamora, Á., Piles, M., & Martínez-Fernández, J. (2016). A new soil moisture agricultural drought index (SMADI) integrating MODIS and SMOS products: A case of study over the Iberian Peninsula. *Remote Sensing, 8*(4), 287.

Samra, J. (2004). *Review and analysis of drought monitoring, declaration and management in India*. IWMI.

Sazib, N., Mladenova, I., & Bolten, J. (2018). Leveraging the Google Earth Engine for drought assessment using global soil moisture data. *Remote Sensing, 10*(8), 1265.

Sehgal, V., & Sridhar, V. (2019). Watershed-scale retrospective drought analysis and seasonal forecasting using multi-layer, high-resolution simulated soil moisture for Southeastern US. *Weather and Climate Extremes, 23*, 100191.

Shah, R., & Mishra, V. (2014). Evaluation of the reanalysis products for the monsoon season droughts in India. *Journal of Hydrometeorology, 15*, 1575–1591.

Shakya, N., & Yamaguchi, Y. (2010). Vegetation, water and thermal stress index for study of drought in Nepal and central northeastern India. *International Journal of Remote Sensing, 31*, 903–912.

Sheffield, J., Goteti, G., Wen, F., & Wood, E. F. (2004). A simulated soil moisture based drought analysis for the United States. *Journal of Geophysical Research: Atmospheres, 109.*

Sheffield, J., & Wood, E. F. (2007). Characteristics of global and regional drought, 1950–2000: Analysis of soil moisture data from off-line simulation of the terrestrial hydrologic cycle. *Journal of Geophysical Research: Atmospheres, 112.*

Sheffield, J., & Wood, E. F. (2008). Global trends and variability in soil moisture and drought characteristics, 1950–2000, from observation-driven simulations of the terrestrial hydrologic cycle. *Journal of Climate, 21,* 432–458.

Shorachi, M., Kumar, V., & Steele-Dunne, S. C. (2022). Sentinel-1 SAR backscatter response to agricultural drought in The Netherlands. *Remote Sensing, 14*(10), 2435. https://doi.org/10.3390/rs14102435

Smakhtin, V., & Hughes, D. A. (2007). Automated estimation and analyses of meteorological drought characteristics from monthly rainfall data. *Environmental Modelling and Software, 22,* 880–890.

Srivastava, P. K., Han, D., Ramirez, M. A. R., & Islam, T. (2013). Appraisal of SMOS soil moisture at a catchment scale in a temperate maritime climate. *Journal of Hydrology, 498,* 292–304.

Srivastava, P. K., Petropoulos, G. P., & Kerr, Y. H. (2016). *Satellite soil moisture retrieval: Techniques and applications.* Elsevier.

Svoboda, M., LeComte, D., Hayes, M., & Heim, R. (2002). The drought monitor. *Bulletin of the American Meteorological Society, 83,* 1181.

Tabari, H., Nikbakht, J., & Talaee, P. H. (2013). Hydrological drought assessment in Northwestern Iran based on streamflow drought index (SDI). *Water Resources Management, 27,* 137–151.

Tang, C., & Piechota, T. C. (2009). Spatial and temporal soil moisture and drought variability in the Upper Colorado River Basin. *Journal of Hydrology, 379,* 122–135.

Thilagaraj, P., Masilamani, P., Venkatesh, R., & Killivalavan, J. (2021). Google earth engine based agricultural drought monitoring in Kodavanar watershed, Part of Amaravathi Basin, Tamil Nadu, India. *The International Archives of the Photogrammetry, Remote Sensing and Spatial Information Sciences, 43,* 43–49.

Trambauer, P., Maskey, S., Werner, M., Pappenberger, F., Van Beek, L., & Uhlenbrook, S. (2014). Identification and simulation of space-time variability of past hydrological drought events in the Limpopo River basin, southern Africa. *Hydrology and Earth System Sciences Discussions, 11,* 2014.

Van der Schalie, R., Preimesberger, W., Pasik, A., & Scanlon, T. (2023). *ESA climate change initiative plus soil moisture product user guide (PUG) supporting product version v08.1 (V1.0 issue 1.0).* Zenodo. https://doi.org/10.5281/zenodo.8320914

Vasiliades, L., Loukas, A., & Liberis, N. (2011). A water balance derived drought index for Pinios River Basin, Greece. *Water Resources Management, 25,* 1087–1101.

Vicente-Serrano, S. M., López-Moreno, J. I., Beguería, S., Lorenzo-Lacruz, J., Azorin-Molina, C., & Morán-Tejeda, E. (2011). Accurate computation of a streamflow drought index. *Journal of Hydrologic Engineering, 17,* 318–332.

Vicente-Serrano, S. M., Lopez-Moreno, J.-I., Beguería, S., Lorenzo-Lacruz, J., Sanchez-Lorenzo, A., García-Ruiz, J. M., Azorin-Molina, C., Morán-Tejeda, E., Revuelto, J., & Trigo, R. (2014). Evidence of increasing drought severity caused by temperature rise in southern Europe. *Environmental Research Letters, 9,* 044001.

Vicente-Serrano, S. M., Beguería, S., & López-Moreno, J. I. (2010). A multiscalar drought index sensitive to global warming: The standardized precipitation evapotranspiration index. *Journal of Climate, 23*(7), 1696–1718.

Walsh, S. J. (1987). Comparison of NOAA AVHRR data to meteorologic drought indices. *Photogrammetric Engineering and Remote Sensing, 53,* 1069–1074.

Wan, Z., Wang, P., & Li, X. (2004). Using MODIS land surface temperature and normalized difference vegetation index products for monitoring drought in the southern Great Plains, USA. *International Journal of Remote Sensing, 25,* 61–72.

Wang, A., Lettenmaier, D. P., & Sheffield, J. (2011). Soil moisture drought in China, 1950–2006. *Journal of Climate, 24,* 3257–3271.

Wilhite, D. A. (2000). *Drought as a natural hazard: Concepts and definitions.*

Wilhite, D. A., & Glantz, M. H. (1985). Understanding: The drought phenomenon: The role of definitions. *Water International, 10,* 111–120.

Wilhite, D. A. (1993). *The enigma of drought. Drought assessment, management, and planning: Theory and case studies* (pp. 3—15). Springer.

Wu, Z., Lu, G., Wen, L., Lin, C. A., Zhang, J., & Yang, Y. (2007). Thirty-five year (1971—2005) simulation of daily soil moisture using the variable infiltration capacity model over China. *Atmosphere-Ocean, 45,* 37—45.

Yaduvanshi, A., Srivastava, P. K., & Pandey, A. (2015). Integrating TRMM and MODIS satellite with socio-economic vulnerability for monitoring drought risk over a tropical region of India. *Physics and Chemistry of the Earth, Parts A/B/C, 83,* 14—27.

Yin, J., Zhan, X., & Liu, J. (2020). NOAA satellite soil moisture operational product system (SMOPS) version 3.0 generates higher accuracy blended satellite soil moisture. *Remote Sensing, 12*(17), 2861. https://doi.org/10.3390/rs12172861

Zhu, Q., Luo, Y., Xu, Y.-P., Tian, Y., & Yang, T. (2019). Satellite soil moisture for agricultural drought monitoring: Assessment of SMAP-derived soil water deficit index in Xiang River Basin, China. *Remote Sensing, 11*(3), 362. https://doi.org/10.3390/rs11030362

CHAPTER 25

Exploring the use of random forest classifier with Sentinel-2 imagery in flooded area mapping

Cinzia Albertini[1,2], Andrea Gioia[2], Vito Iacobellis[2], Salvatore Manfreda[3] and George P. Petropoulos[4]

[1]Dipartimento di Scienze del Suolo, della Pianta e degli Alimenti, Università degli Studi di Bari Aldo Moro, Bari, Italy;
[2]Dipartimento di Ingegneria Civile, Ambientale, del Territorio, Edile e di Chimica, Politecnico di Bari, Bari, Italy;
[3]Dipartimento di Ingegneria Civile, Edile e Ambientale, Università degli Studi di Napoli Federico II, Naples, Italy;
[4]Department of Geography, Harokopio University of Athens, Athens, Greece

1. Introduction

Floods are the most frequent type of natural disasters occurring worldwide. In 2021 alone, in fact, the majority of the recorded catastrophic events were floods (223 of 432 occurrences), which also dominated in terms of the highest number of disaster–related fatalities according to the Centre for Research on the Epidemiology of Disasters (CRED, Delforge et al., 2022). The reported number of flood occurrences is pointed out to be above the 2001—2020 annual average of 163 floods, raising concerns about the increasing flood risk, already triggered by issues regarding economic growth, land use modifications, and climate change (Blöschl, 2022; Hall et al., 2014; Winsemius et al., 2016). In recent years, there have been numerous large floods caused by strong storm events due to the increasingly alarming ability to retain water of a warmer atmosphere (Blöschl et al., 2019). Concurrently, there has been a proliferation of studies examining the potential quantitative impacts on flood hazard under diverse climate change and land use scenarios (Alfieri et al., 2017; Blöschl, 2022; Blöschl et al., 2019; García-Valdecasas Ojeda et al., 2022; Padulano et al., 2021; Whitfield, 2012; Wilhelm et al., 2022).

In this context, the need for flood hazard assessment, including the development of inundation maps, is of primary importance since they represent effective tools for flood risk management and to assist with decision-making and disaster response planning. To this end, satellite remote sensing plays a crucial role thanks to the ability to monitor the Earth's surface over large areas and through time, thus allowing for studying the spatio-temporal dynamics of flood processes (Kim et al., 2017; Singh et al., 2011; Tripathi et al., 2019; Tsatsaris et al., 2021) and mapping flood risk at large scale and in data-scarce environments (Albano et al., 2020; Pandey et al., 2022). Numerous studies have examined the contribution, strengths, and weaknesses of Earth observation (EO) data for flood mapping and monitoring, including the recent review by Schumann et al. (2022). In

Geographical Information Science
ISBN 978-0-443-13605-4
https://doi.org/10.1016/B978-0-443-13605-4.00017-5

particular, microwave remote sensing techniques offer the possibility to map flood inundation both day and night and under all weather conditions, while optical satellite imagery can be used to easily extract the flood extent through direct interpretation of color composites or applying fully or semiautomatic approaches (Albertini et al., 2022; Schumann et al., 2022). Although microwave sensors can provide more efficient measurements thanks to their good penetration capabilities through the atmosphere (Shen et al., 2019), extensive literature exists that proves the ability of multispectral satellite sensors to map flooded areas and water bodies (Albertini et al., 2022). Application of data derived from such satellites can either exploit surfaces' spectral signature information retrieved from single bands or adopt multiband approaches mainly based on bands ratio or math algebra expressions. Several multispectral indices have been proposed for mapping flooded areas and obtaining water segmentation in contexts of varying degrees of landscape complexity. These include the Normalized Difference Moisture Index (NDMI, Gao, 1996), the Normalized Difference Water Index (NDWI, McFeeters, 1996), the Red and Short-Wave Infra-Red (RSWIR, Memon et al., 2015; Rogers & Kearney, 2004) Index, and the Modified Normalized Difference Water Index (MNDWI, Xu, 2006).

Flood inundation mapping from satellites has gained increasing interest and seen a growth in applications motivated not only by the availability of data from new satellite missions with varying and enhanced spatial and temporal resolutions but also by the development of improved and powerful processing algorithms and image classification techniques. This specifically refers to the advent of machine learning (ML) methods which enable to handling large and complex datasets, such as remote sensing data, and are shown to produce higher performances than traditional approaches (Maxwell et al., 2018). Flood inundation modeling with ML has, in fact, received a lot of attention in recent years and numerous studies have compared the accuracies and robustness of different methods (Bentivoglio et al., 2022; Ghorpade et al., 2021; Karim et al., 2023; Mosavi et al., 2018). Popular algorithms in flood modeling are supervised classification techniques, including random forest (RF), support vector machine (SVM), decision tree (DT), and artificial neural network (ANN), which require information known a priori in order to train the classifier and make generalizations (Serpico et al., 2012; Volpi et al., 2013). Among those methods, RF has proven a powerful classification algorithm. This is due to its ability to make excellent predictions, handle multicollinearity and high-dimensionality, its low sensitivity to overfitting and computational efficiency (Belgiu & Drăgut, 2016).

Different RF models have been developed and used for several applications, including the evaluation of regional flood hazards using multisource datasets, such as precipitation, land use, topographic and soil texture data (e.g., Esfandiari et al., 2020; Wang et al., 2015), modeling urban coastal flood severity using crowd-sourced data (Sadler et al.,

2018), and mapping flood inundation by fusing Sentinel-1 and Sentinel-2 imagery (Billah et al., 2023; Tavus et al., 2020). However, to our knowledge, none or very few studies have specifically investigated the ability of Sentinel-2 spectral bands alone to delineate flooded areas, nor the added value of multispectral indices. For this reason, the current work aims to explore the use of RF with Sentinel-2 imagery and some of the most common spectral indices for mapping the flood event that occurred in October 2020 along the Sesia River (Northern Italy). In detail, this study aims to contribute to:

(1) gain insights into the ability of one of the most common supervised ML algorithms, that is, the RF, to effectively detect flooded areas using Sentinel-2 imagery.

(2) identify the RF model exploiting spectral bands and one or more multispectral indices that allow for improving floodwater detection and reducing data omission.

To this scope, two different RF models were developed using Sentinel-2 multispectral bands at 20 m resolution in combination with one or more spectral indices. The indices include the ratio between the two SWIR bands, the MNDWI, the NDMI, and the RSWIR index. Validation of the proposed models was carried out using as reference the delineation map (first estimate product, FEP) delivered by the Rapid Mapping service of the Copernicus Emergency Management Service (CEMS). The obtained performances and the validation framework pointed out the contribution of multispectral indices in improving flood delineation and reducing omissions as well as the added value of Sentinel-2 and the Copernicus service for cost-effective flood mapping.

2. Methodology

RF is a powerful classification and regression algorithm introduced by Breiman (2001) based on an ensemble of trees each providing a specific prediction, that is, the class membership. The final outcome is determined by a majority rule according to which the final class is assigned based on the most frequent vote received in each tree. In order to grow a tree in the forest, a random subset of the original training dataset is selected according to a procedure called bootstrapping, while to identify the best split randomly selected features among all the predictor variables are used (Breiman, 2001). The forest is therefore built based on a user-adjustable number of predictors (*mtry*) and up to a user-defined number of trees (*Ntree*). These parameters can be fine-tuned within a k-fold cross-validation in order to optimize the performances (see subsection 2.3.2).

Belonging to supervised classification algorithms, labeled data are needed as input to train the classifier and build a learned model. These data instances are generally called regions of interest (ROIs) and can be collected by experienced users through visual interpretation of the scene of interest (see subsection 2.3.1).

In this study, two different supervised classification schemas were built (see Section 2.3) and compared for detecting the Sesia River flood that occurred in October 2020 in Northern Italy, using the Sentinel-2 imagery and derived multispectral indices as predictor variables to train the RF algorithm. In order to develop the RF models and test the accuracy of the final classification, a four-step approach was followed: first, the preprocessing of the Sentinel-2 bands was carried out to mask potential clouds and their shadows and also compute some multispectral indices. In the second step, ROIs were collected to be used for training and testing the classifier. Following this, the RF classification using the training dataset was carried out, which led to the final classification rules. This step also includes the parameters tuning, implemented within a 10-fold cross-validation that allows searching for the best set to optimize the performances. In the final step, the data testing and validation of the final classified flood map with the Copernicus FEP map were carried out to assess the reliability of the classification using some of the most common accuracy and error metrics.

An overview of the methodology implemented to meet the study objectives is presented in Fig. 25.1, while in the following subsections, each step is fully explained.

2.1 Experimental setup

2.1.1 Study site and flood event description

The area of interest (AOI) selected for the investigation is located at the border between the Piedmont and Lombardy regions (Northern Italy) and, as identified by the CEMS

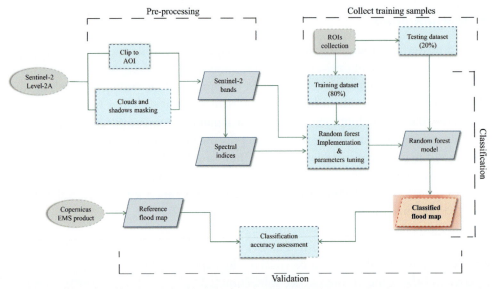

Figure 25.1 Methodological workflow for the implementation of the random forest classifier. The four main steps are highlighted, including the preprocessing of the Sentinel-2 bands in the area of interest (AOI), the collection of the Regions of Interest (ROIs), the algorithm implementation, and the validation using the Copernicus Emergency Management Service (EMS) product.

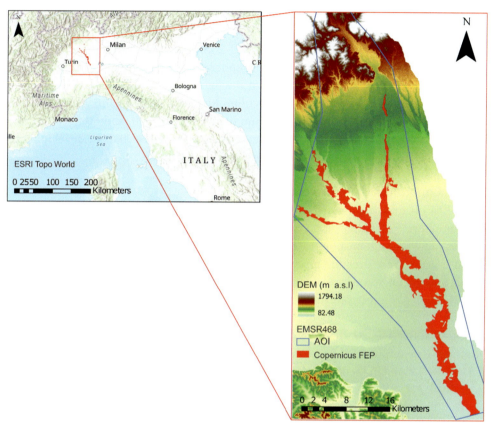

Figure 25.2 Location of the study area and Shuttle Radar Topography Mission (SRTM) digital elevation model (DEM, Farr et al., 2007) clipped at the river basin. The Copernicus Emergency Management Service (EMS) Rapid Mapping products (EMSR468) are also shown: the flood delineation from the First Estimate Product (FEP) in red and the area of interest (AOI) in light blue. *(Map source: ESRI Topo World.)*

(Copernicus Emergency Management Service, 2020b), approximately extends from Quarona to Valmacca municipalities (see Fig. 25.2), respectively, in the Vercelli and Alessandria provinces, for about 1090 km^2. In particular, the study area is situated in the Piedmont part of the flat Padana plain, and it is crossed by the Sesia River, characterized by a torrential regime and mainly used in agriculture for irrigation purposes.

From 2 to 3 October 2020, the North of Italy was interested by heavy precipitation that especially hit the north-western part of the Piedmont region. Precipitation levels reached 325 mm in the Sesia River basin on 2 October, which led to exceptional flooding, particularly in the valley part of the basin with record river discharge levels (above 3000 m^3/s at "Borgosesia" station and above 5000 m^3/s at the "Palestro" station according to the data provided in the report by the Piedmont Agenzia Regionale per la Prevenzione e la Protezione dell'Ambiente, ARPA, 2020).

2.1.2 Datasets
2.1.2.1 Satellite data

The flood event was captured by the Sentinel-2 satellite at the time of the flood peak. Therefore, the Level-2A (Bottom of Atmosphere [BOA] reflectance) scene acquired on 3 October 2020 was downloaded from the Sentinel Scientific Data Hub (Sentinel, 2024) and used for the analyses. It consists of multispectral bands in the visible and thermal regions of the electromagnetic spectrum delivered at 10, 20, and 60 m spatial resolution. In this work, all nine bands at 20 m resolution were selected for developing the RF models. Details about the bands are provided in Table 25.1.

2.1.2.2 Copernicus data for validation

The CEMS Rapid Mapping (Copernicus Emergency Management Service, 2024) provides monitoring of natural disasters and emergencies from EO satellite imagery, as well as delivers free-of-charge mapping services following the evolution of the event. Released products can include pre- and postevent maps, such as first estimates, delineation, monitoring, and grading products, obtained either from active or passive remote sensing data and through photointerpretation, semiautomatic, automatic, or modeling extraction methods.

On October 4, 2020, the CEMS Rapid Mapping started monitoring the situation in Northern Italy to provide updated information following the flood evolution (activation [EMSR468] Flood in Piedmont region, Italy (Copernicus Emergency Management Service, 2020a)).

The first estimate product ([EMSR468] Sesia: FEP Product (Copernicus Emergency Management Service, 2020b)) provided by the CEMS 1 day after the event (4 October) through visual interpretation of the Sentinel-2 scene is available for the considered AOI (see Fig. 25.2). It depicts the flood extent situation as of 3 October affecting the Sesia River and the initial part of the Po River after the confluence.

Table 25.1 Sentinel-2 bands of the scene acquired on 2023-10-03 over the study area.

Sentinel-2 Level-2A acquired on 2023-10-03			
Band name	Band number	Central wavelength (nm)	Spatial resolution (m)
Blue	Band 2	490	20
Green	Band 3	560	20
Red	Band 4	665	20
Vegetation (Red-edge 1)	Band 5	705	20
Vegetation (Red-edge 2)	Band 6	740	20
Vegetation (Red-edge 3)	Band 7	783	20
Vegetation (NIR narrow)	Band 8a	865	20
SWIR	Band 11	1610	20
SWIR2	Band 12	2190	20

2.2 Data preprocessing

2.2.1 Sentinel-2 imagery

For the algorithm implementation, the Sentinel–2 bands at 20 m spatial resolution were selected (see Table 25.1). The preprocessing step includes the clipping of the scene to the AOI and masking the clouds and cloud shadows. No topographic or atmospheric corrections were carried out since the images are derived from the Level-2A product, which provides already terrain and atmospherically corrected BOA surface reflectance images.

The map extent layer [AOI01] provided in the Copernicus vector data package (EMS-R468_AOI01_FEP_PRODUCT_r1_VECTORS_v2_vector) was used to clip the spectral bands, while both the cloud mask layer and the Scene Classification Layer (SCL) map at 20 m spatial resolution contained in the distributed Sentinel-2 Level-2A product were employed for masking clouds. In detail, the cloud mask layer provided in the Quality Indicators data folder (QI_DATA) is representative of opaque and cirrus clouds, while the SCL raster layer provides a classification map consisting of snow/ice, water, vegetation, and bare soil, as well as cirrus, medium and high probability clouds and cloud shadows, which are thus detectable and can be easily removed. It is important to point out that although a 24% cloud coverage in the AOI was registered, it was mainly concentrated in the upper part of the scene and did not affect areas covered by the floodwater. Therefore, once the bands were clipped to the AOI and clouds masked, they were ready to be employed as predictor variables in the supervised algorithm and to be applied in the computation of some multispectral indices to be additionally used in the algorithm implementation.

2.2.2 Copernicus flood map

The CEMS Rapid Mapping delivers data package in vector format and the delineation of the Sesia river flood is provided as an ESRI shapefile. Therefore, before the pixel-per-pixel classification accuracy assessment, the FEP was converted to a raster file with the same spatial resolution of the Sentinel-2 bands (i.e., 20 m).

2.3 Random forest implementation

In this work, two different supervised classification schemes using the R package "caret" (Kuhn et al., 2020) were built and compared adopting the Sentinel-2 bands listed in Table 25.1 and the subsequently specified spectral indices:

(1) **Scenario 1** (hereinafter referred to as S1): It was developed using the Sentinel-2 bands at 20 m resolution and the ratio between the two SWIR bands, expressed as:

$$\text{SWIR ratio} = \frac{\text{SWIR}}{\text{SWIR2}} = \frac{\text{B11}}{\text{B12}} \qquad (25.1)$$

The SWIR ratio was chosen in order to exploit, on the one hand, the sensitivity of the SWIR band to the water and moisture content (Huber et al., 2013; Wolski et al., 2017) and, on the other hand, the ability of the SWIR2 bands to distinguish vegetation from soil and characterize the land surface vegetation (Wang et al., 2018).

(2) **Scenario 2** (hereinafter referred to as S2): It was built considering the Sentinel-2 bands at 20 m resolution and the following spectral indices:

$$\text{MNDWI} = \frac{\text{green} - \text{SWIR}}{\text{green} + \text{SWIR}} = \frac{\text{B3} - \text{B11}}{\text{B3} + \text{B11}} \tag{25.2}$$

$$\text{NDMI} = \frac{\text{nir} - \text{SWIR}}{\text{nir} + \text{SWIR}} = \frac{\text{B8A} - \text{B11}}{\text{B8A} + \text{B11}} \tag{25.3}$$

$$\text{RSWIR} = \frac{\text{red} - \text{SWIR}}{\text{red} + \text{SWIR}} = \frac{\text{B4} - \text{B11}}{\text{B4} + \text{B11}} \tag{25.4}$$

In this case, the aim was to explore the benefits of adding some of the most powerful spectral indices in discriminating water features from other surfaces.

2.3.1 ROIs collection

ROIs collection was performed based on the visual interpretation of Sentinel-2 RGB band combinations, in particular, the true (B4, B3, B2), false (B8a, B4, B3), and SWIR (B12, B8a, B4) color composites, from which it is possible to derive useful information on the floodwater presence in the scene. A minimum of 10–30p training pixels per class is suggested to be used to train the classifier (Mather & Koch, 2011; Petropoulos et al., 2011; Piper, 1992; Van Niel et al., 2005), where p is the number of bands (i.e., the predictors). A total of 737 ROIs (a minimum of 120 pixels per class, if referring to S2 with 12 predictors) were selected for five different land cover classes, namely flooded areas, artificial surfaces, crops, forests, and permanent water, as reported in Table 25.2, identified with the help of the CORINE Land Cover 2018 (European Union Copernicus Land Monitoring Service, 2018) raster data. For the algorithm implementation, we used 80% of the collected ROIs for training and the remaining 20% for testing (see Table 25.2).

Table 25.2 Region of Interest (ROIs) collected to train and test the random forest classifier.

Class	Training dataset	Testing dataset	Total
Flooded areas	145	36	181
Artificial surfaces	111	27	138
Crops	115	28	143
Forests	108	27	135
Permanent water	112	28	140
Total	591	146	737

In the study, the ability of each individual wavelength in the selected ROIs to discriminate between the five classes was examined both visually, comparing the mean spectral signatures of each class, and quantitatively using one of the most common metrics for separability analysis, that is, the Jeffries—Matusita (JM) distance (Bruzzone et al., 1995). Fig. 25.3 illustrates the mean spectra profiles of the bands in the collected ROIs for each considered class. As it can be observed, there is a good separability of floodwater from the other classes, especially from forests and crops. These results are verified by looking at Table 25.3, in which JM values above 1.9 between the flooded areas class and the others are registered, confirming a very good separation (Bindel et al., 2012; Lane et al., 2014; Soubry & Guo, 2021).

2.3.2 Data training and parameters tuning

As mentioned in Section 2, during the algorithm training, some parameters to build the RF can be adjusted by the user, in particular, the number of predictors randomly selected to build each tree (*mtry*) and the number of trees in the forest (*Ntree*). Parameters fine-tuning can, in fact, assist with searching for the best set that allows for optimizing the performances (Bergstra & Bengio, 2012; Duarte & Wainer, 2017; Schratz et al., 2019). This process is usually performed in a *k*-fold cross-validation (e.g., Efron, 1983; Friedl & Brodley, 1997; Kohavi, 1995; Maxwell et al., 2018) consisting in a random split of the original training dataset into *k* smaller sets (or folds) of which *k-1* are used for training and the remaining for testing the algorithm. Thus, the classifier is trained and tested *k* times

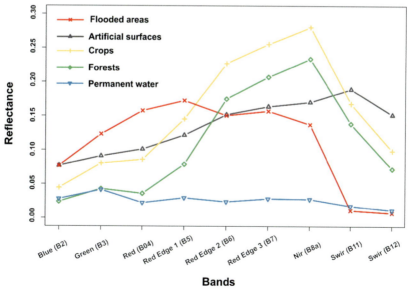

Figure 25.3 Mean spectral signatures of the selected regions of interest in each Sentinel-2 band for all the classes used in the random forest classifier implementation.

Table 25.3 Spectral separability between the collected Regions of Interest (ROIs) as measured by the Jeffries—Matusita distance.

	Flooded areas	Artificial surfaces	Crops	Forests	Permanent water
Flooded areas		1.99	2	2	1.99
Artificial surfaces			1.89	1.95	1.99
Crops				1.82	1.99
Forests					1.99
Permanent water					

and the cross-validation performances are averaged among the values computed in each iteration (Kohavi, 1995; Maxwell et al., 2018).

In this work, we fine-tuned *mtry* and *Ntree* using a 10-fold cross-validation. In particular, the variability of the performances for 100, 200, 300, 400, 500, and 1000 trees and for *mtry* varying from a minimum of two predictors to the maximum number of variables in each scenario (i.e., 10 and 12 variables for S1 and S2, respectively) was investigated.

2.3.3 Data testing and accuracy assessment

Once the RF models were developed, data testing was performed to check the ability of the classification rules to correctly categorize new unseen data. It was carried out on the remaining 20% of the selected samples and the accuracy was evaluated by building the confusion matrix and looking at the overall accuracy (OA), producer accuracy (PA), and user accuracy (UA) metrics. The former refers to the total correctly classified pixels, PA to the number of pixels in each class correctly assigned by the model to class i, while UA, also called precision, refers to the number of pixels of class i in the reference map which are present in the classified map but not necessarily correctly labeled.

As the last step, the RF models were applied to the whole scene in order to derive the final classified maps identifying the five classes. Since the interest was specially directed to detecting floodwater, the flooded areas class was isolated from the other classes and a binary map of flooded and not flooded pixels was derived to be validated with the Copernicus FEP on a pixel-per-pixel—based assessment. A confusion matrix was derived for each scenario and some of the most common accuracy and error metrics in image classification accuracy assessment were derived (Eqs. 25.5—25.12), namely the detected area efficiency (DAE), which corresponds to PA, the skipped area rate (SAR) or omission error, and the false area rate (FAR) or commission error (Ireland et al., 2015; Kontoes et al., 2009). In addition, the OA and F-score metrics were also computed. In the following equations, the expressions of the mentioned metrics are reported:

$$OA\ (\%) = \frac{TP + TN}{T} \cdot 100 \tag{25.5}$$

$$PA\ (\%) = \frac{TP}{R_i} \cdot 100 \tag{25.6}$$

$$UA\ (\%) = \frac{TP}{C_i} \cdot 100 \tag{25.7}$$

$$DAE\ (\%) = PA = \frac{DFA}{DFA + SFA} \cdot 100 \tag{25.8}$$

$$SAR\ (\%) = \frac{SFA}{DFA + SFA} \cdot 100 \tag{25.9}$$

$$FAR\ (\%) = \frac{FFA}{DFA + FFA \cdot 100} \tag{25.10}$$

$$Precision\ (\%) = UA = \frac{TP}{TP + FP} \cdot 100 \tag{25.11}$$

$$F-score\ (\%) = \frac{2 \cdot Precision \cdot DAE}{Precision + DAE} \tag{25.12}$$

where T is the total number of pixels; TP and FP are the true and false positives, respectively; TN represents the true negatives; R_i and C_i indicate the total number of pixels in class i in the reference and classified map, respectively; DFA is the detected flooded area and represent the common area between the generated flooded area and the reference area; SFA is the skipped flooded area representing the area included in the reference area but not in the generated flooded area; FFA is the false flooded area and represents the area included in the generated flooded area but not in the reference area.

3. Results

3.1 Training and testing results

Parameters tuning performed within a 10-fold cross–validation with the training dataset allowed us to identify the best values of *mtry* and *Ntree* giving the highest training accuracy. The results of the parameters optimization are reported in Table 25.4. In

Table 25.4 Optimized parameter values and training accuracy obtained from the 10-fold cross-validation.

Optimized parameters			
Scenario	*Mtry*	*Ntree*	Accuracy (%)
S1	2	1000	85.79
S2	2	500	86.71

particular, the optimized value of *mtry* is two in each scenario, while *Ntree* is set to 1000 and 500 for S1 and S2, respectively.

Once the optimum parameter values were identified, the accuracy of the final RF models was evaluated on the testing dataset of each scenario. Table 25.5 and Table 25.6 show the confusion matrices obtained from the data testing predictions derived from S1 and S2, respectively, together with the most relevant metrics, that is, the PA, UA, and OA. Regarding S1 (see Table 25.5), the highest producer's accuracy, that is the number of pixels correctly assigned to the class *i* as in the reference map, is registered for the flooded areas and artificial surfaces (PA = 100%), while the lowest value for the crops (PA = 85.71%). PA values for the forests and permanent water classes are equal to 96.30% and 92.86%, respectively. Looking at the user's accuracy, the flooded areas and crops classes are characterized by the highest values (UA = 100%), while artificial surfaces, permanent water, and forests have a UA value equal to 96.43%, 92.86%, and 81.25%, respectively. Finally, the OA of the RF classification in S1 is 95.21%.

Regarding S2 (see Table 25.6), the highest producer's accuracy is registered for the flooded areas and artificial surfaces, as in S1, but also for the forests class (PA = 100%), while permanent water and crops are characterized by a PA value of 92.86% and 85.71%, respectively. User's accuracy is equal to 100% for the artificial surfaces, crops, and permanent water classes, while UA for flooded areas and forests is 97.30% and 84.38%, respectively. In this case, the OA of the RF model is 95.89%.

3.2 Detection efficiency and classification errors

For each scenario, the corresponding RF model was applied to the whole scene in order to derive the final classification map containing the five classes. Fig. 25.4 shows the results of the classification for S1 (top panel) and S2 (bottom panel). In particular, in Fig. 25.4a—c,

Table 25.5 Confusion matrix derived from the data testing over Scenario 1 (S1) and accuracy metrics.

S1	References				
Predicted	*Flooded areas*	*Artificial surfaces*	*Crops*	*Forests*	*Permanent water*
Flooded areas	36	0	0	0	0
Artificial surfaces	0	27	0	1	0
Crops	0	0	24	0	0
Forests	0	0	4	26	2
Permanent water	0	0	0	0	26
PA	100%	100%	85.71%	96.30%	92.86%
UA	100%	96.43%	100%	81.25%	92.86%
OA	95.21%				

Table 25.6 Confusion matrix derived from the data testing over Scenario 2 (S2) and accuracy metrics.

S2	References				
Predicted	Flooded areas	Artificial surfaces	Crops	Forests	Permanent water
Flooded areas	36	0	1	0	0
Artificial surfaces	0	27	0	0	0
Crops	0	0	24	0	0
Forests	0	0	3	27	2
Permanent water	0	0	0	0	26
PA	100%	100%	85.71%	100%	92.86%
UA	97.30%	100%	100%	84.38%	100%
OA	95.89%				

the flooded areas, artificial surfaces, crops, forests, and permanent water classes as derived from the final RF models in S1 and S2 are depicted, while in Fig. 25.4b–d, the flooded areas (shown in blue) were isolated from the other classes and compared with the validation map, that is, the Copernicus FEP (shown in red). These allowed us to qualitatively compare the two scenarios and visually assess the presence of omissions and/or commissions. It is possible, in fact, to verify the presence of a higher number of overdetections in S1 compared to S2, but also a lower number of omissions. To better clarify this point, a map showing where commissions, omissions, and truly detected flooded pixels occur, together with some detailed zoom in these areas, is reported in Fig. 25.5. The DFA, SFA, and FFA between the two scenarios are compared in this figure from which it is possible to immediately appreciate the lower number of commissions (depicted in purple) in S2, but also the lowest omissions in S1 (depicted in blue).

In order to quantitatively compare the validation results in the two scenarios, the error and accuracy metrics mentioned in subsection 2.3.3 were computed and reported in Table 25.7. Both scenarios show very high accuracy values (OA above 90%), slightly higher for S2 (OA = 92.59% compared to 91.31% of S1). The same is true for the F-score metric, which in both cases is above 70%, with a slightly higher value for S2 (F-score = 73.86%) than for S1 (F-score = 71.47%). S2 performs better than S1, especially concerning the commission errors, represented by the significantly lower value of the FAR (16.11% against 25.35% for S1), which also affects the UA metric (83.89% for S2 compared to 74.65% for S1). However, as already verified by visually comparing the maps, S1 shows a higher ability in detecting the true flood areas, expressed by the DAE equal to 68.55% compared to S2 (DAE = 65.97%) and, consequently, by the lowest number of omissions (SAR = 31.44% for S1 compared to SAR = 34.02% for S2).

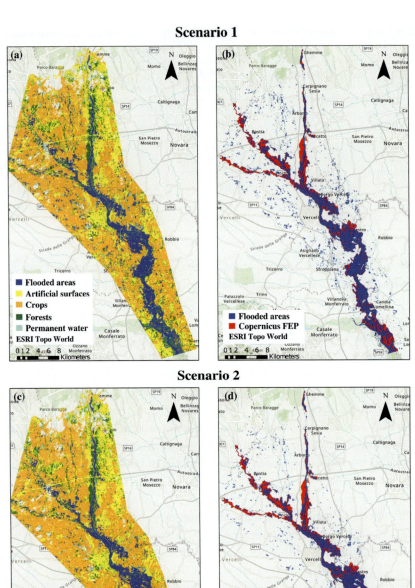

Figure 25.4 Classification results derived from the implementation of the Random Forest (RF) algorithm in Scenario 1 (top panel) and Scenario 2 (bottom panel): (a, c) classified maps showing the five classes as identified by the corresponding RF model and (b, d) binary maps of flooded areas (in blue) compared to the Copernicus first estimate product (FEP, in red) used for the validation. *(Map source: ESRI Topo World.)*

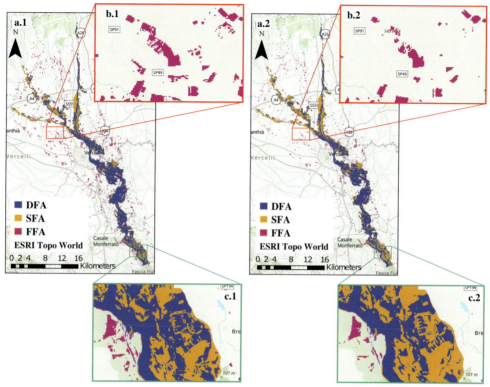

Figure 25.5 Maps showing the Detected Flooded Area (DAE), the Skipped Flooded Area (SFA) and the False Flooded Area (FFA) for (a.1) Scenario 1 and (a.2) Scenario 2. Panels (b.1) and (b.2) show a comparison of commission errors, while (c.1) and (c.2) compare the omissions. *(Map source: ESRI Topo World.)*

Table 25.7 Accuracy metrics derived from the validation of flood maps in Scenario 1 (S1) and Scenario 2 (S2) using the Copernicus first estimate product as reference flood map.

	S1 (%)	S2 (%)
OA	91.31	92.59
PA, DAE	68.55	65.97
UA, Precision	74.65	83.89
SAR	31.44	34.02
FAR	25.35	16.11
F-score	71.47	73.86

4. Discussion

In this work, an investigation on the use of the RF classifier with Sentinel-2 bands and some of the most common multispectral indices for water segmentation was carried out to map the flooding extent of the Sesia River flood in October 2020. The RF algorithm has been successfully applied in several studies, including ecological predictions of tree species distribution under future climate conditions (Prasad et al., 2006), land cover classification from hyperspectral (Ham et al., 2005) and multisource data such as multispectral imagery and topographic features (Gislason et al., 2006), as well as in flood hazard and risk assessment. For example, Wang et al. (2015) using several predictor variables, such as topographic data, soil properties, hydrological variables, and the Normalized Difference Vegetation Index (NDVI), developed a flood risk map to analyze the hazard distribution of the Dongjiang River Basin (China).

In the present study, by considering two different scenarios of predictor variables, we show the ability of the RF classification algorithm to delineate the flooded area and look for the model that leads to the best performance in terms of the number of omissions. In Scenario 1, the ratio between the two SWIR bands was included as one of the predictors to be used in the training process together with the Sentinel-2 bands, as these are valuable indicators of the water presence and moisture conditions, as well as of the vegetation cover (Huber et al., 2013; Wang et al., 2018; Wolski et al., 2017). Scenario 2 analyses the value of jointly using the Sentinel-2 bands with some of the spectral indices that have been recognized as the best performing for detecting floodwater (Albertini et al., 2022; Asmadin et al., 2018; Boschetti et al., 2014; Li et al., 2018; Munasinghe et al., 2018).

First, the training and tuning of the RF parameters were carried out considering five different classes to better discriminate between floodwater and other land surface features. In particular, the classification algorithm was built based on the training ROIs representative of artificial surfaces, forests, crops, permanent water, and flooded areas. Once the optimum parameters were set, the data testing was performed to assess the models' capabilities of correctly labeling new data. The two scenarios showed similar and satisfactory results in terms of OA (above 95% in both cases), with just a few misclassifications. For example, some confusion of the forests class with crops and permanent water occurred in both scenarios (confirmed by the lowest values of the UA metric for forests and PA for crops and permanent water). However, since the present study objective has been to assess the ability of the RF supervised algorithm to detect flooded areas, a binary map of flooded and not flooded pixels for each scenario was obtained by isolating them from the other classes and the final maps were validated using the Copernicus FEP as reference. It is worth noting that validation of the other land cover classes was out of the scope of this study, as they were instead specifically used to improve the discrimination of flooded areas.

The comparison between the reference map and the flood inundation maps showed that, overall, the RF models produced a good delineation of flooded areas (OA above 91% in each scenario), but some noise (i.e., false detections) was observed in S1. Those errors were found to be linked to the characteristics of the study area, mainly flat and well-known for its rice fields as the predominant local crop. Due to the presence of irrigation channels and the frequently flooded conditions of these land parcels, they are likely to be misclassified as flooded areas. At the time of the flooding, in fact, the rice was already waterlogged and had still to be harvested (De Petris et al., 2021, 2022). This confirms the contribution of the SWIR ratio in correctly detecting the surface water and moisture content, also verified by the fact that S1 was the best scenario in terms of the amount of true detected flooded areas and number of omissions compared to S2, but which also leads to the aforementioned errors. On the other hand, S2 exploits the power of different spectral indices employed in combination with the single spectral bands in better discriminating flooded pixels from other features, including vegetation liquid water, built-up areas, and wetland environments (Gao, 1996; Rogers & Kearney, 2004; Xu, 2006), thus largely reducing false alarms.

Worthy of consideration are some assumptions and limitations of the present work. First, the developed classification assumes that multispectral images from Sentinel-2 depicting the flood event are available. This is not always the case, as satellites are characterized by their own revisit time, which can lead to images availability from hours to days from the flood occurrence. It is, indeed, acknowledged that mismatches between the time of the satellite pass and the flood wave peak can influence the delineation accuracies (Ireland et al., 2015; Notti et al., 2018). In addition, a feasible use of multispectral images implies none or limited issues related to cloud cover, which is often dominant during a flood event due to bad weather conditions. Nonetheless, none of these issues were found in our application. It is also worth underlining that in this study, ROIs collected through photointerpretation of Sentinel-2 color composites were used, which may have to some extent influenced the overall quality of the classification. However, the collected samples were trustworthy, as confirmed by the JM values.

Despite some assumptions, our study results clearly demonstrated that the RF classification algorithm employed with data derived from multispectral satellites, that is, the single spectral bands or the common multispectral indices, allows the flooded areas to be reliably extracted. In addition, our RF models achieved very good OAs and compared very well with previous applications found in the literature. Similar studies conducted with different datasets to assess the ability of the RF classifier for flood extent mapping confirmed the good accuracy of this supervised ML algorithm. Notably, an OA of around 87% was achieved by applying the RF classification with Landsat imagery (Ghansah et al., 2021) and unmanned aerial vehicles (Feng et al., 2015).

5. Conclusions

The RF algorithm is a highly reliable classification method widely applied across various studies, recognized as one of the most powerful ML classifiers compared to other techniques.

In this study, the RF algorithm was employed in conjunction with Sentinel-2 imagery to determine the extent of the Sesia River flood that occurred in October 2020 in Northern Italy. The key contributions of this study are as follows:

- RF exhibited exceptional predictive capabilities in identifying flooded areas with very good accuracy (OA above 91% and F-score above 71%).
- The choice of spectral bands, coupled with specific spectral indices, played a crucial role in determining the accuracy of flood predictions. In our study, the model exploiting the SWIR ratio together with the single Sentinel-2 bands (Scenario 2) was found to have very good identification capabilities of areas affected by flood, since omissions were minimized and the true flood areas were better detected. The scenario including also the MNDWI, NDMI, and RSWIR led to improved results in terms of reduction of the false alarms.

Satellite-based flood area extraction can significantly benefit from the integration of ML techniques, as they allow the inclusion of several predictor variables with moderate efforts. Multispectral imagery, offering visual interpretations, facilitates the collection of ROIs for training supervised algorithms. However, challenges persist, such as the confusion between wet vegetation, agricultural fields, and actual floodwater, as well as cloud interference that hampers the usability of optical data. Combining images from passive microwave instruments, active sensors, and geomorphic data can enhance outcomes and minimize false alarms, as demonstrated in prior studies (Albertini et al., 2022; D'Addabbo et al., 2016; Samela et al., 2022).

Future research directions may involve exploring the predictive power of each variable used in RF classification, incorporating multisource data. This investigation could lead to the development of a model that considers only the most relevant variables, thus enhancing classification accuracy and identifying robust features applicable to new study areas. In fact, a pathway that can be followed in taking this work further is the application of the proposed methodology in other research areas, which could confirm, on the one hand, its good predictive capabilities and, on the other hand, if integrating multiple predictor variables (e.g., SAR and optical satellite imagery, morphologic features, etc.), the power of the synergic use of multisource datasets, as well as indicate the most stable predictors.

Acknowledgments

This study was carried out within the RETURN Extended Partnership and received funding from the European Union Next-Generation EU (National Recovery and Resilience Plan—NRRP, Mission 4, Component 2, Investment 1.3—D.D. 1243 August 2, 2022, PE0000005) and within the framework of the technical-scientific collaboration agreement signed on November 25, 2021 between the Southern Apennine District Basin Authority, the Interuniversity Consortium for Hydrology (CINID), and the Department of Engineering for the Environment of the University of Calabria (DIAM).

References

Agenzia Regionale per la Prevenzione e la Protezione dell'Ambiente Piemonte. (2020). *Evento del 2-3 ottobre 2020*.

Albano, R., Samela, C., Crăciun, I., Manfreda, S., Adamowski, J., Sole, A., Sivertun, Å., & Ozunu, A. (2020). Large scale flood risk mapping in data scarce environments: An application for Romania. *Water, 12*, 1834.

Albertini, C., Gioia, A., Iacobellis, V., & Manfreda, S. (2022). Detection of surface water and floods with multispectral satellites. *Remote Sensing, 14*, 6005.

Alfieri, L., Bisselink, B., Dottori, F., Naumann, G., de Roo, A., Salamon, P., Wyser, K., & Feyen, L. (2017). Global projections of river flood risk in a warmer world. *Earth's Future, 5*, 171–182. https://doi.org/10.1002/2016EF000485

Asmadin, Siregar, V. P., Sofian, I., Jaya, I., & Wijanarto, A. B. (2018). Feature extraction of coastal surface inundation via water index algorithms using multispectral satellite on North Jakarta. *IOP Conference Series: Earth and Environmental Science, 176*, 1–10. https://doi.org/10.1088/1755-1315/176/1/012032

Belgiu, M., & Drăgut, L. (2016). Random forest in remote sensing: A review of applications and future directions. *ISPRS Journal of Photogrammetry and Remote Sensing, 114*, 24–31.

Bentivoglio, R., Isufi, E., Jonkman, S. N., & Taormina, R. (2022). Deep learning methods for flood mapping: A review of existing applications and future research directions. *Hydrology and Earth System Sciences, 26*, 4345–4378.

Bergstra, J., & Bengio, Y. (2012). Random search for hyper-parameter optimization. *Journal of Machine Learning Research, 13*.

Billah, M., Islam, A. K. M. S., Mamoon, W., & Bin, Rahman, M. R. (2023). Random forest classifications for landuse mapping to assess rapid flood damage using Sentinel-1 and Sentinel-2 data. *Remote Sensing Applications: Society and Environment, 30*, 100947.

Bindel, M., Hese, S., Berger, C., & Schmullius, C. (2012). Feature selection from high resolution remote sensing data for biotope mapping. *The International Archives of the Photogrammetry, Remote Sensing and Spatial Information Sciences, 38*, 39–44.

Blöschl, G., Hall, J., Viglione, A., Perdigão, R. A. P., Parajka, J., Merz, B., Lun, D., Arheimer, B., Aronica, G. T., & Bilibashi, A. (2019). Changing climate both increases and decreases European river floods. *Nature, 573*, 108–111.

Blöschl, G. (2022). Three hypotheses on changing river flood hazards. *Hydrology and Earth System Sciences, 26*, 5015–5033.

Boschetti, M., Nutini, F., Manfron, G., Brivio, P. A., & Nelson, A. (2014). Comparative analysis of normalised difference spectral indices derived from MODIS for detecting surface water in flooded rice cropping systems. *PLoS One, 9*, e88741.

Breiman, L. (2001). Random forests. *Machine Learning, 45*, 5–32.

Bruzzone, L., Roli, F., & Serpico, S. B. (1995). An extension of the Jeffreys-Matusita distance to multiclass cases for feature selection. *IEEE Transactions on Geoscience and Remote Sensing, 33*, 1318–1321.

Copernicus Emergency Management Service (© 2020 European Union), 2020a EMSR468 https://emergency.copernicus.eu/mapping/list-of-components/EMSR468.

Copernicus Emergency Management Service (© 2020 European Union), 2020b [EMSR468] Sesia: FEP product https://emergency.copernicus.eu/mapping/list-of-components/EMSR468/ALL/EMSR468_AOI01.

Copernicus Emergency Management Service (n.d.). 2024. Retrieved July 1, 2023 from https://emergency.copernicus.eu/

D'Addabbo, A., Refice, A., Pasquariello, G., Lovergine, F. P., Capolongo, D., & Manfreda, S. (2016). A Bayesian network for flood detection combining SAR imagery and ancillary data. *IEEE Transactions on Geoscience and Remote Sensing, 54*, 3612–3625.

De Petris, S., Ghilardi, F., Sarvia, F., & Borgogno-Mondino, E. (2022). A simplified method for water depth mapping over crops during flood based on Copernicus and DTM open data. *Agricultural Water Management, 269*, 107642.

De Petris, S., Sarvia, F., & Borgogno-Mondino, E. (2021). Multi-temporal mapping of flood damage to crops using sentinel-1 imagery: A case study of the Sesia River (October 2020). *Remote Sensing Letters, 12*, 459—469.

Delforge, D., Below, R., Wathelet, V., Jones, R., Tubeuf, S., & Speybroek, N. (2022). *2021 disasters in numbers*. Brussels.

Duarte, E., & Wainer, J. (2017). Empirical comparison of cross-validation and internal metrics for tuning SVM hyperparameters. *Pattern Recognition Letters, 88*, 6—11.

Efron, B. (1983). Estimating the error rate of a prediction rule: Improvement on cross-validation. *Journal of the American Statistical Association, 78*, 316—331.

Esfandiari, M., Jabari, S., McGrath, H., & Coleman, D. (2020). Flood mapping using random forest and identifying the essential conditioning factors; A case study in fredericton, new brunswick, Canada. In *ISPRS annals of the photogrammetry, remote sensing and spatial information sciences* (pp. 609—615). https://doi.org/10.5194/isprs-Annals-V-3-2020-609-2020

European union Copernicus land monitoring service. (2018). European Environment Agency.

Farr, T. G., Rosen, P. A., Caro, E., Crippen, R., Duren, R., Hensley, S., Kobrick, M., Paller, M., Rodriguez, E., & Roth, L. (2007). The shuttle radar topography mission. *Reviews of Geophysics, 45*.

Feng, Q., Liu, J., & Gong, J. (2015). Urban flood mapping based on unmanned aerial vehicle remote sensing and random forest classifier—A case of Yuyao, China. *Water, 7*, 1437—1455.

Friedl, M. A., & Brodley, C. E. (1997). Decision tree classification of land cover from remotely sensed data. *Remote Sensing of Environment, 61*, 399—409.

Gao, B. (1996). NDWI—A normalized difference water index for remote sensing of vegetation liquid water from space. *Remote Sensing of Environment, 58*, 257—266. https://doi.org/10.1016/S0034-4257(96)00067-3

García-Valdecasas Ojeda, M., Di Sante, F., Coppola, E., Fantini, A., Nogherotto, R., Raffaele, F., & Giorgi, F. (2022). Climate change impact on flood hazard over Italy. *Journal of Hydrology, 615*, 128628. https://doi.org/10.1016/j.jhydrol.2022.128628

Ghansah, B., Nyamekye, C., Owusu, S., & Agyapong, E. (2021). Mapping flood prone and Hazards Areas in rural landscape using landsat images and random forest classification: Case study of Nasia watershed in Ghana. *Cogent Engineering, 8*, 1923384.

Ghorpade, P., Gadge, A., Lende, A., Chordiya, H., Gosavi, G., Mishra, A., Hooli, B., Ingle, Y. S., & Shaikh, N. (2021). Flood forecasting using machine learning: A review. In *2021 8th international conference on smart computing and communications (ICSCC)* (pp. 32—36). IEEE.

Gislason, P. O., Benediktsson, J. A., & Sveinsson, J. R. (2006). Random forests for land cover classification. *Pattern Recognition Letters, 27*, 294—300.

Hall, J., Arheimer, B., Borga, M., Brázdil, R., Claps, P., Kiss, A., Kjeldsen, T. R., Kriaučiūnienė, J., Kundzewicz, Z. W., & Lang, M. (2014). Understanding flood regime changes in Europe: A state-of-the-art assessment. *Hydrology and Earth System Sciences, 18*, 2735—2772.

Ham, J., Chen, Y., Crawford, M. M., & Ghosh, J. (2005). Investigation of the random forest framework for classification of hyperspectral data. *IEEE Transactions on Geoscience and Remote Sensing, 43*, 492—501.

Huber, C., Battiston, S., Yésou, H., Tinel, C., Laurens, A., & Studer, M. (2013). Synergy of VHR pleiades data and SWIR spectral bands for flood detection and impact assessment in urban areas: Case of Krymsk, Russian Federation, in July 2012. In *2013 IEEE international geoscience and remote sensing symposium-IGARSS* (pp. 4538—4541). IEEE.

Ireland, G., Volpi, M., & Petropoulos, G. P. (2015). Examining the capability of supervised machine learning classifiers in extracting flooded areas from landsat TM imagery: A case study from a Mediterranean flood. *Remote Sensing, 7*, 3372—3399.

Karim, F., Armin, M. A., Ahmedt-Aristizabal, D., Tychsen-Smith, L., & Petersson, L. (2023). A review of hydrodynamic and machine learning approaches for flood inundation modeling. *Water, 15*, 566.

Kim, D., Lee, H., Laraque, A., Tshimanga, R. M., Yuan, T., Jung, H. C., Beighley, E., & Chang, C.-H. (2017). Mapping spatio-temporal water level variations over the central Congo River using PALSAR ScanSAR and Envisat altimetry data. *International Journal of Remote Sensing, 38*, 7021—7040. https://doi.org/10.1080/01431161.2017.1371867

Kohavi, R. (1995). A study of cross-validation and bootstrap for accuracy estimation and model selection. In *IJCAI* (pp. 1137–1145). Montreal: Canada.

Kontoes, C. C., Poilvé, H., Florsch, G., Keramitsoglou, I., & Paralikidis, S. (2009). A comparative analysis of a fixed thresholding vs. a classification tree approach for operational burn scar detection and mapping. *International Journal of Applied Earth Observation and Geoinformation, 11*, 299–316.

Kuhn, M., Wing, J., Weston, S., Williams, A., Keefer, C., Engelhardt, A., Cooper, T., Mayer, Z., Kenkel, B., & Team, R. C. (2020). Package 'caret.' *R J., 223*.

Lane, C. R., Liu, H., Autrey, B. C., Anenkhonov, O. A., Chepinoga, V. V., & Wu, Q. (2014). Improved wetland classification using eight-band high resolution satellite imagery and a hybrid approach. *Remote Sensing, 6*, 12187–12216.

Li, J., Yang, X., Maffei, C., Tooth, S., & Yao, G. (2018). Applying independent component analysis on Sentinel-2 imagery to characterize geomorphological responses to an extreme flood event near the non-vegetated Río Colorado terminus, Salar de Uyuni, Bolivia. *Remote Sensing, 10*, 725.

Mather, P. M., & Koch, M. (2011). *Computer processing of remotely-sensed images: An introduction*. John Wiley and Sons.

Maxwell, A. E., Warner, T. A., & Fang, F. (2018). Implementation of machine-learning classification in remote sensing: An applied review. *International Journal of Remote Sensing, 39*, 2784–2817.

McFeeters, S. K. (1996). The use of the Normalized Difference Water Index (NDWI) in the delineation of open water features. *International Journal of Remote Sensing, 17*, 1425–1432. https://doi.org/10.1080/01431169608948714

Memon, A. A., Muhammad, S., Rahman, S., & Haq, M. (2015). Flood monitoring and damage assessment using water indices: A case study of Pakistan flood-2012. *Egyptian Journal of Remote Sensing and Space Science, 18*, 99–106. https://doi.org/10.1016/j.ejrs.2015.03.003

Mosavi, A., Ozturk, P., & Chau, K. (2018). Flood prediction using machine learning models: Literature review. *Water, 10*, 1536.

Munasinghe, D., Cohen, S., Huang, Y., Tsang, Y., Zhang, J., & Fang, Z. (2018). Intercomparison of satellite remote sensing-based flood inundation mapping techniques. *JAWRA Journal of the American Water Resources Association, 54*, 834–846.

Notti, D., Giordan, D., Caló, F., Pepe, A., Zucca, F., & Galve, J. P. (2018). Potential and limitations of open satellite data for flood mapping. *Remote Sensing, 10*(11), 1673. https://doi.org/10.3390/rs10111673

Padulano, R., Costabile, P., Costanzo, C., Rianna, G., Del Giudice, G., & Mercogliano, P. (2021). Using the present to estimate the future: A simplified approach for the quantification of climate change effects on urban flooding by scenario analysis. *Hydrological Processes, 35*, e14436.

Pandey, A. C., Kaushik, K., & Parida, B. R. (2022). Google Earth Engine for large-scale flood mapping using SAR data and impact assessment on agriculture and population of Ganga-Brahmaputra basin. *Sustainability, 14*, 4210.

Petropoulos, G. P., Kontoes, C., & Keramitsoglou, I. (2011). Burnt area delineation from a uni-temporal perspective based on Landsat TM imagery classification using Support Vector Machines. *International Journal of Applied Earth Observation and Geoinformation, 13*, 70–80.

Piper, J. (1992). Variability and bias in experimentally measured classifier error rates. *Pattern Recognition Letters, 13*, 685–692.

Prasad, A. M., Iverson, L. R., & Liaw, A. (2006). Newer classification and regression tree techniques: Bagging and random forests for ecological prediction. *Ecosystems, 9*, 181–199.

Rogers, A. S., & Kearney, M. S. (2004). Reducing signature variability in unmixing coastal marsh Thematic Mapper scenes using spectral indices. *International Journal of Remote Sensing, 25*, 2317–2335. https://doi.org/10.1080/01431160310001618103

Sadler, J. M., Goodall, J. L., Morsy, M. M., & Spencer, K. (2018). Modeling urban coastal flood severity from crowd-sourced flood reports using Poisson regression and random forest. *Journal of Hydrology, 559*, 43–55.

Samela, C., Coluzzi, R., Imbrenda, V., Manfreda, S., & Lanfredi, M. (2022). Satellite flood detection integrating hydrogeomorphic and spectral indices. *GIScience and Remote Sensing, 59*, 1997–2018.

Schratz, P., Muenchow, J., Iturritxa, E., Richter, J., & Brenning, A. (2019). Hyperparameter tuning and performance assessment of statistical and machine-learning algorithms using spatial data. *Ecological Modelling, 406*, 109—120.

Schumann, G., Giustarini, L., Tarpanelli, A., Jarihani, B., & Martinis, S. (2022). Flood modeling and prediction using Earth observation data. *Surveys in Geophysics*, 1—26.

Sentinel Scientific Data Hub 2024 (n.d.). Available online: https://scihub.copernicus.eu/ (Accessed on 4 February 2022).

Serpico, S. B., Dellepiane, S., Boni, G., Moser, G., Angiati, E., & Rudari, R. (2012). Information extraction from remote sensing images for flood monitoring and damage evaluation. *Proceedings of the IEEE, 100*, 2946—2970.

Shen, X., Wang, D., Mao, K., Anagnostou, E., & Hong, Y. (2019). Inundation extent mapping by synthetic aperture radar: A review. *Remote Sensing, 11*(7), 879. https://doi.org/10.3390/rs11070879

Singh, S. K., Pandey, A. C., & Nathawat, M. S. (2011). Rainfall variability and spatio temporal dynamics of flood inundation during the 2008 Kosi flood in Bihar State, India. *Asian Journal of Earth Sciences, 4*, 9—19.

Soubry, I., & Guo, X. (2021). Identification of the optimal season and spectral regions for shrub cover estimation in Grasslands. *Sensors, 21*, 3098.

Tavus, B., Kocaman, S., Nefeslioglu, H. A., & Gokceoglu, C. (2020). A fusion approach for flood mapping using Sentinel-1 and Sentinel-2 datasets. *The International Archives of the Photogrammetry, Remote Sensing and Spatial Information Sciences, 43*, 641—648.

Tripathi, G., Parida, B. R., & Pandey, A. C. (2019). Spatio-temporal rainfall variability and flood prognosis analysis using satellite data over North Bihar during the August 2017 flood event. *Hydrology, 6*, 38.

Tsatsaris, A., Kalogeropoulos, K., Stathopoulos, N., Louka, P., Tsanakas, K., Tsesmelis, D. E., Krassanakis, V., Petropoulos, G. P., Pappas, V., & Chalkias, C. (2021). Geoinformation technologies in support of environmental hazards monitoring under climate change: An extensive review. *ISPRS International Journal of Geo-Information, 10*, 94.

Van Niel, T. G., McVicar, T. R., & Datt, B. (2005). On the relationship between training sample size and data dimensionality: Monte Carlo analysis of broadband multi-temporal classification. *Remote Sensing of Environment, 98*, 468—480.

Volpi, M., Petropoulos, G. P., & Kanevski, M. (2013). Flooding extent cartography with Landsat TM imagery and regularized kernel Fisher's discriminant analysis. *Computers and Geosciences, 57*, 24—31.

Wang, B., Jia, K., Liang, S., Xie, X., Wei, X., Zhao, X., Yao, Y., & Zhang, X. (2018). Assessment of Sentinel-2 MSI spectral band reflectances for estimating fractional vegetation cover. *Remote Sensing, 10*, 1927.

Wang, Z., Lai, C., Chen, X., Yang, B., Zhao, S., & Bai, X. (2015). Flood hazard risk assessment model based on random forest. *Journal of Hydrology, 527*, 1130—1141.

Whitfield, P. H. (2012). Floods in future climates: A review. *Journal of Flood Risk Management, 5*, 336—365.

Wilhelm, B., Rapuc, W., Amann, B., Anselmetti, F. S., Arnaud, F., Blanchet, J., Brauer, A., Czymzik, M., Giguet-Covex, C., & Gilli, A. (2022). Impact of warmer climate periods on flood hazard in the European Alps. *Nature Geoscience, 15*, 118—123.

Winsemius, H. C., Aerts, J. C. J. H., Van Beek, L. P. H., Bierkens, M. F. P., Bouwman, A., Jongman, B., Kwadijk, J. C. J., Ligtvoet, W., Lucas, P. L., & Van Vuuren, D. P. (2016). Global drivers of future river flood risk. *Nature Climate Change, 6*, 381—385.

Wolski, P., Murray-Hudson, M., Thito, K., & Cassidy, L. (2017). Keeping it simple: Monitoring flood extent in large data-poor wetlands using MODIS SWIR data. *International Journal of Applied Earth Observation and Geoinformation, 57*, 224—234.

Xu, H. (2006). Modification of normalised difference water index (NDWI) to enhance open water features in remotely sensed imagery. *International Journal of Remote Sensing, 27*, 3025—3033.

CHAPTER 26

Landslide susceptibility assessment and mapping in Breaza town, Romania

Ionut Sandric[1], Zenaida Chitu[2] and Radu Irimia[1]
[1]Faculty of Geography, University of Bucharest, Bucharest, Romania; [2]National Meteorological Administration, Bucharest, Romania

1. Introduction

Understanding natural phenomena and their spatial distribution is a complex task fraught with uncertainties, which can manifest randomly, locally, or on a large scale. Random uncertainties relate to the precision of instruments used for landslide mapping, local systematic uncertainties stem from specific space–time characteristics, and large-scale uncertainties are associated with spatial data discretization and harmonization. Researchers have grappled with these uncertainties over the past 3 decades, with a notable increase in research efforts in the last 10 years (Stefan Steger et al., 2016; Steger et al., 2017; Wu et al., 2015).

Some approaches have centered on the spatial accuracy of mapped landslides and their influence on the final landslide susceptibility maps (Carrara, 1993; Guzzetti, 2005; Zêzere et al., 2009). Others have delved into the impact of elevation precision from digital elevation models (DEMs) on terrain modeling (Qin et al., 2013; Schlögel et al., 2018) or the effects of their spatial resolution on the primary landslide predisposing factors (Arnone et al., 2016; Claessens et al., 2005; Schlögel et al., 2018). While random simulations for positional accuracy of landslide bodies have been explored (Stefan Steger et al., 2016; Steger et al., 2017), they often neglect the random errors originating from elevation sources. A more intricate approach, involving simulations of positional accuracy of landslide bodies, terrain attributes, and categorical data using a bootstrap technique, was undertaken by Rossi & Reichenbach (2016), although it did not model spatial autocorrelation.

The primary objective of this study is to evaluate the cumulative impact of uncertainties stemming from landslide inventory, DEMs, and their derived products on landslide susceptibility maps. The study's secondary goal is to develop a tool that can be utilized to replicate similar analyses in different regions worldwide and provide scientists with estimates of uncertainties in their landslide susceptibility maps.

Geographical Information Science
ISBN 978-0-443-13605-4
https://doi.org/10.1016/B978-0-443-13605-4.00003-5

2. Materials and methods

2.1 Regional setting

The research is conducted within the geographical scope of Breaza town, situated in the Curvature Subcarpathians of Romania. Breaza encompasses 10 settlements, covering an expanse of approximately 50 km^2 (refer to Fig. 26.1). Over the past few decades, Breaza has garnered recognition in Romania as a region susceptible to landslide occurrences, which have inflicted damage on roads and residential structures. The topography of the area primarily comprises hills, with elevations of around 400 m in the southern region and approximately 800 m in the northwestern part, characterized by an average slope gradient of approximately 20°. The Prahova River bisects the area, marking a more dynamic and active right side.

In terms of geology, the area lies within a syncline characterized by a south-east to north-west orientation. Another critical aspect is the region's lithology, which exhibits significant heterogeneity. It includes layers of clay and marl, often interrupted by sandstone, conglomerates, gypsum, gravel, and sand (Damian et al., 2003). These varied lithologies are intersected by a dense network of streams, with values ranging from 0.3 to 1 km^2, featuring intricate gullies and torrential channels. Additionally, a complex

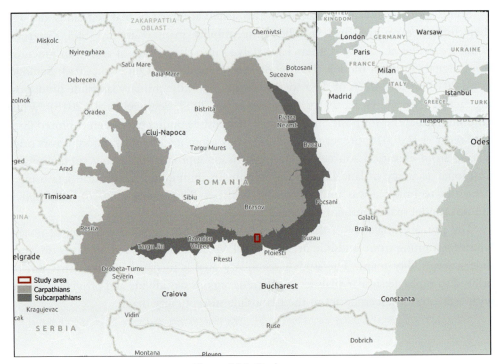

Figure 26.1 Study area.

network of faults and strike-slip faults criss-crosses the area. Notably, in the last 25 years, the Prahova River has incised a deep and steep channel, exerting a substantial influence on the activity of landslides situated along the terrace scarps. The Prahova River and its valley are currently undergoing extensive engineering projects in preparation for the future Bucuresti—Brasov highway. As a result of these endeavors, a section of the river now flows underground, and the primary channel has been redirected from its old course. Consequently, due to these engineering modifications, the relationship between landslides and river erosion is expected to no longer hold true.

Over the centuries, the natural vegetation has been largely replaced by meadows and orchards, predominantly cultivating apple and plum trees (Chitu, 2010). Orchards are primarily situated on areas prone to landslides, where the low rainfall interception rate, combined with favorable lithology, increases their susceptibility to landslide occurrences. Similar conditions apply to meadows, where the presence of domestic animals sometimes enhances the infiltration rate. The climate conditions in the region favor landslide occurrences, with intense precipitation occurring during spring and autumn. The rainfall threshold that triggers landslides in this area is approximately 100 mm within 24 h (Sandric, 2008), or 300 mm over an extended period, leading to deep-seated landslides (Chitu, 2010; Chitu et al., 2016).

A comprehensive landslide inventory was established in 2009 (Şandric et al., 2009) and has been regularly updated each year until 2017. The classification system for the landslide inventory aligns with that of Cruden and Varne (1996). Over 90% of the landslides in the area manifest as deep-seated rotational slides, with the remainder comprising translational slides and earth flows. For this particular study, our focus is solely on the deep-seated rotational slides, amounting to a total of 310. Most of these landslides are situated on marl and clay, forming hill slopes with gradients ranging from 10 to 25 degrees.

The dataset utilized in this study (refer to Table 26.1) was manually digitized from topographical maps, geological maps, and ortho-images. The land cover was categorized based on a user-defined classification system, carefully designed to best represent its role as a predisposing factor. A portion of the study area is covered by geological maps at a scale of 25,000, while another portion is mapped at a scale of 50,000. Consequently, boundary adjustments and on-site observations were conducted to ensure a consistent spatial distribution of the lithological units.

Landslide susceptibility assessment typically relies on various factors, including the first and second derivatives of DEMs, lithology, land use, and land cover. DEMs provide an estimate of the elevation distribution on a pixel-by-pixel basis, but they do not represent precise elevation values. As a result, data derived from DEMs, including those obtained from technologies like drones or LIDAR, are susceptible to errors. While technologies like LIDAR, InSAR, and close-range photogrammetry have reduced these uncertainties, they may not be ideal when working at a regional scale.

Table 26.1 Datasets with predisposing factors used in landslide susceptibility assessment.

Class No.	Land-cover—Land-use	Lithology	Slope (degrees)	Plan curvature	Profile curvature
1	Hardwood	Tuff	<3	<−3	<−3
2	Built-up area	Marl with sand	3—6	−3−−2.5	−3−−2.5
3	Bare soil	Marl	6—9	−2.5−−2.0	−2.5−−2.0
4	Meadow	Sand	9—12	−2.0−−1.5	−2.0−−1.5
5	Shrubs	Gypsum and marl	12—15	−1.5−−1.0	−1.5−−1.0
6	Warbler meadow	Marl and clay	15—18	−1.0−−0.5	−1.0−−0.5
7	Orchard	Clay	18—21	−0.5−0.0	−0.5−0.0
8	Road	Gravel	21—24	0.0—0.5	0.0—0.5
9	Conifers	Sandstone	24—27	0.0	0.0
10	Gardens	Clay and marl	27—30	0.5—1.0	0.5—1.0
11	Prahova river	Conglomerate	30—33	1.0—1.5	1.0—1.5
12		Sandstone, marl, and clay	33—36	1.5—2.0	1.5—2.0
13		Gravel and sand	36—39	2.0—2.5	2.0—2.5
14		Marl and limestone	39—42	2.5—3.0	2.5—3.0
15		Lower flych horizon	42—45	>3.0	>3.0
16		Gura Beliei marl	45—48		
17		Flych and marl	>48		

3. Landslide susceptibility assessment

The Weight of Evidence (WofE) method is grounded in Bayesian conditional probability theory, with a key distinction being the log transformation of odds derived from Bayesian probability theory (Bonham-Carter, 1994; Bonham-Carter et al., 1989; Regmi et al., 2010). This transformation simplifies the interpretation of each factor's contribution to the process's evolution. Results yielding negative values signify unfavorable factors, while positive weights indicate favorable ones (Bonham-Carter, 1994; Regmi et al., 2010).

WofE draws upon knowledge gathered from diverse information sources, which inherently makes the WofE analysis, like all data-driven methods, more susceptible to errors and uncertainties compared to other approaches. WofE utilizes ground truth data concerning the presence and absence of phenomena along with explanatory variables, referred to as evidence, to calculate landslide susceptibility values on a pixel basis. The known natural phenomena locations can be recorded as either point data or polygon data. In the case of point data, these are typically situated on the landslide's scarp, which closely resembles the initial state. However, a concern with point data lies in the spatial discretization of the polygonal shape into a dimensionless representation.

On the other hand, when working with polygon data, the location is represented by the entire body of the landslide. Both cases are equally susceptible to uncertainties in Bayesian inference. WofE employs a ratio between the presence and absence of phenomena and the presence and absence of evidence. Polygon data, due to covering a significantly larger area associated with the presence of phenomena, encompass a greater number of possible cases in which landslides may have occurred. Representing the landslide body as polygons introduces additional uncertainties (Sandric, 2008):

- Misinterpretation of aerial or satellite imagery, leading to incorrect delineation of the landslide body.
- The precision of landslide body delineation in the field is dependent on the expertise of the personnel and the accuracy of the topographical equipment. This aspect significantly influences prior, conditional, and posterior probability calculations, particularly when dealing with very high-resolution datasets.

A conceptual model used for assessing uncertainty propagation in landslide susceptibility is presented below in Fig. 26.2.

3.1 Generation of noisy DEMs

Considering that both elevation values and mapped landslides inherently contain errors (Fisher & Tate, 2006; Fisher et al., 2004), we developed an uncertainty propagation model implemented as a Python script within ArcGIS for Desktop using the ArcPy API (ESRI, 2013). The underlying conceptual model (see Fig. 26.3) relies on Monte Carlo simulations to introduce noise into both elevation data and landslide locations. Since this noise is generated based on the assumption of a normal distribution, it is advisable to employ a statistical test such as Shapiro—Wilk or Kolmogorov—Smirnov to verify if the user's data adheres to a normal distribution. If the data does not conform to a

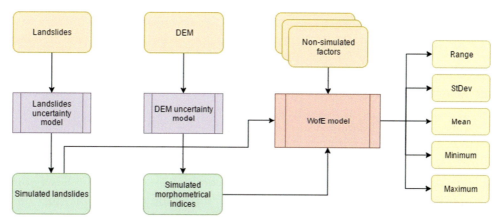

Figure 26.2 Landslide susceptibility propagation model. It comprises the landslide location uncertainty model, the elevation uncertainty model, and its propagation into terrain analysis.

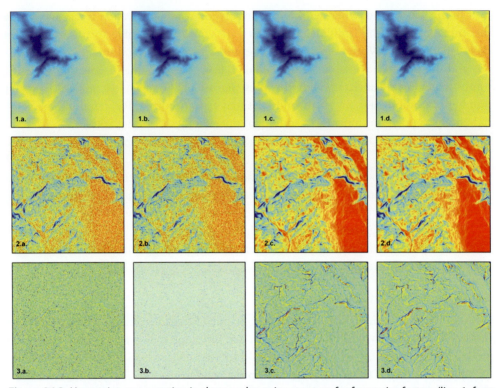

Figure 26.3 Uncertainty propagation in slope and terrain curvature for four noise factors (line 1, from a to d are four different noisy DEMs: 0, 1, 5, and 10), from left to right: high noise, moderate noise, small noise, and no noise. The noise propagation is less present in the slope images (line 2, from a to d) than the noise present in the terrain curvature (line 3, from a to d).

normal distribution, a transformation test, like the Box–Cox method (Csillik et al., 2015), should be employed.

By utilizing the mean and standard deviation of the errors, a random value, normally distributed, is generated for each pixel. Subsequently, this noise is subject to spatial smoothing and then reintegrated into the original DEM. As a result, a novel, simulated, and plausible elevation dataset is derived (Karssenberg & De Jong, 2005; Temme et al., 2009).

Smoothing the elevation noise serves to eliminate DEM artifacts (Tarekegn & Sayama, 2013), while retaining the integrity of second-order terrain derivatives. When noise is directly added to the original elevations, it leads to excessive noise in all terrain derivatives, rendering them impractical for use. Our tool offers a noise factor parameter, with a recommended value falling within the range of 4 to 7 on a scale from 0 to 10, where 0 represents the highest noise level and 10 signifies the lowest noise level.

3.2 Uncertainty propagation in landslide location

The origins of these uncertainties can be categorized as either random, local, or large-scale systematic. Random uncertainties are linked to the precision of instruments utilized for landslide mapping. Local systematic uncertainties are influenced by specific space-time characteristics, while large-scale uncertainties stem from the spatial discretization of phenomena and data harmonization. Researchers have addressed these uncertainties over the past 3 decades, with a noticeable upward trend in the last 10 years (Stefan Steger et al., 2016; Steger et al., 2017; Wu et al., 2015).

Moreover, uncertainties arise from issues such as the incorrect classification of landslide types, the mapping of nonexistent landslides (particularly common in cases of visual interpretation of aerial images), the accuracy of DEMs used for terrain analyses, and errors in predisposing factor mapping like land cover and lithology. Conversions between vector and raster datasets also contribute to uncertainties, making it essential to select an appropriate method (Arnone et al., 2016). These combined uncertainties significantly impact the final susceptibility map. As a result, it is imperative to conduct studies on how uncertainties propagate into the final results to provide more accurate estimates of marginal landslide susceptibility values (Rossi & Reichenbach, 2016; Schlögel et al., 2018; Stefan Steger et al., 2016; Steger et al., 2017).

The underlying of the uncertainty propagation model employed for generating noisy landslide data involves breaking down each landslide at the vertex level and then randomly displacing each vertex within an estimated mapping error (Brown & Heuvelink, 2007). This mapping error should be measured in distance units and documented. Using these values, mean and standard deviation values are computed to introduce noise. This noise is applied to the original vertex locations, effectively generating new landslide locations through random displacement (Brown & Heuvelink, 2007; Sandric, 2008; Stefan Steger et al., 2016; Steger et al., 2017).

3.3 Uncertainty propagation modeling weight of evidence

The Weight of Evidence with Uncertainty Propagation model is grounded in Bayesian conditional probability theory, with minor adaptations (Bonham-Carter, 1994; Bonham-Carter et al., 1989). In this model, the prior probability of landslide occurrence in a specific area is calculated as the total area or the total number of landslides for the n simulation. The conditional probability for the n simulation is determined by the ratio of the overlapping area between the landslides and each factor. As each simulation alters both the landslide area and factor area, the conditional probability changes accordingly. The posterior probability, reflecting the likelihood of a landslide occurring when a factor is present, is calculated based on the conditional probability, denoted as P(Tn), for each simulation.

Assessing the correlation between the presence or absence of predictive factors and the presence of landslides involves the examination of negative and positive weights, their variances, as well as the contrast and studentized contrast (Bonham-Carter et al., 1988, 1989). A negative weight signifies the absence of a predictive factor, while a positive weight indicates the factor's presence. A negative contrast points to a negative association between the predictive factor and landslide presence, while a positive contrast suggests a positive association. The studentized contrast serves as a metric of the uncertainty associated with the weights and helps determine if the contrast significantly deviates from 0 (Bonham-Carter et al., 1989). A contrast value close to 0 indicates the absence of either a negative or positive association between the predictive factor and landslides.

The landslide susceptibility map is computed as the sum of contrasts from all the factors. After completing all simulations, each susceptibility map is normalized, and statistical analyses are performed on the entire stack of maps. This analysis yields measures such as mean, minimum, maximum, range, and standard deviation per pixel. The final landslide susceptibility maps are presented as a collection of these maps, rather than a single map. The minimum and maximum values establish the lower and upper boundaries for landslide susceptibility, while the range and standard deviations reflect its sensitivity. In regions where the range and standard deviation values are high, the landslide susceptibility value is highly sensitive, compared to areas where the range and standard deviation values are lower, indicating a more reliable estimated susceptibility.

4. Results and discussion

4.1 Model application

A DEM was created through interpolation from contour lines and elevation points using the ANUDEM method (Hutchinson et al., 2011). Elevation errors were determined by cross-referencing the elevation points gathered in the field with the base DEM. All elevation points were collected using a Trimble GeoXH 2008 DGPS (Differential Global Positioning Systems) system, featuring horizontal and vertical accuracy of less than 20 cm. For Breaza Town, the mean and standard deviation values were found to be 4.50 and 3.67 m, respectively. These points were uniformly distributed across the entire study area, and no discernible correlation with specific terrain features was identified.

A total of five thousand five hundred noisy DEMs were generated based on the conceptual model depicted in Fig. 26.3. This entailed creating 500 noisy DEMs for each noise factor ranging from 0 to 10. For each DEM, a WofE analysis was conducted.

In Fig. 26.3, several DEMs and their derivatives are showcased for four distinct noise values: 0, 1, 5, and 10. A noise value of 0 and 1 signifies a considerable amount of noise introduced into the base DEM. The slope gradient and plan curvature obtained from this DEM illustrate the absence of spatial autocorrelation when excessive noise is present, particularly noticeable in the plan curvature. Gradually increasing the noise value, denoting a reduced level of noise added to the base DEM, stabilizes spatial autocorrelation. Notably, from a value of 5 up to 10, there are no significant alterations in the results.

4.2 Landslide susceptibility for Breaza town

In the context of Breaza town, we examined uncertainty propagation in the WofE analysis with a combination of predisposing factors, which included lithology, land cover, slope, and terrain curvatures (plan and profile). To ensure sensitivity to the smallest variations in landslide susceptibility predisposing factors, we classified the first and second derivatives of the DEM using extremely narrow equal intervals. A DEM noise value of 5 was deemed optimal for this analysis. The noise factor plays a vital role in controlling the smoothness of the simulated elevation model, thereby influencing the level of noise integrated into the analysis. An optimal noise value of 5 was chosen because higher values did not significantly impact the results (Sandric, 2008; Micu et al., 2017).

While no specific simulations were conducted on land use, land cover, and lithology, their contributions as predisposing factors were assessed through the landslide simulations. For each simulated landslide, new weights were computed for each class within these predisposing factors.

In the case of slope classification, an equal interval of 3 degrees was considered optimal for measuring changes in slope. This interval struck a balance between being too small and too large, resulting in the creation of 17 classes. Smaller and larger intervals led to excessively noisy slope contributions with no statistically significant correlation with landslides.

As for terrain curvature, values with an equal interval of 0.5 radians were selected and distributed within an open interval spanning 3 radians.

The use of a normal distribution for generating noise prompted the application of the Shapiro–Wilk normality test (Shapiro & Wilk, 1965) to both elevation and landslide errors. The results revealed that, with a 95% confidence interval, a normal distribution was maintained.

The study generated three landslide susceptibility maps (Fig. 26.4) for Breaza, illustrating the maximum, minimum, and average susceptibility values per pixel. As anticipated, lower values were observed in flat areas featuring sand and gravel deposits along the Prahova River bed and terraces. Moderate and high susceptibility regions were predominantly located on gentle to steep slopes housing orchards, bare soil, pastures, and lithological units comprising clay, marls, and mixtures.

An additional map was generated, showing the disparity between the maximum and minimum susceptibility per pixel across all simulations. This map notably identifies areas, particularly on terraces and in the river valley, as being highly sensitive to elevation changes and errors in landslide mapping.

Surprisingly, terrain curvature exhibited minimal sensitivity to elevation uncertainties. All classes of terrain curvature (plan or profile) exhibited a consistently positive contribution to landslide susceptibility. Only high values of convexity had a negative impact on landslide presence (Stefan Steger et al., 2016; Steger et al., 2017).

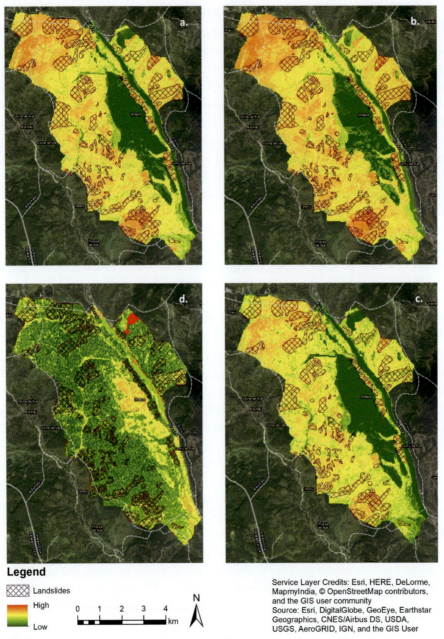

Legend

⬚⬚⬚ Landslides

High

Low

0 1 2 3 4 km

N

Service Layer Credits: Esri, HERE, DeLorme,
MapmyIndia, © OpenStreetMap contributors,
and the GIS user community
Source: Esri, DigitalGlobe, GeoEye, Earthstar
Geographics, CNES/Airbus DS, USDA,
USGS, AeroGRID, IGN, and the GIS User

Figure 26.4 Landslide susceptibility assessment in Breaza town. The figures present: (a) mean land-slide susceptibility, (b) maximum landslide susceptibility, (c) landslide susceptibility uncertainty given by difference between the minimum and maximum values recorded per pixel, (d) minimum landslide susceptibility.

As expected, slope demonstrated a clear contribution to landslide susceptibility within the range of 7 −30 degrees. This contribution was distinctly divided into positive and negative weights, without any instances of sign reversal. The highest sensitivity was observed on steep slopes, specifically those exceeding 35 degrees (Camilo et al., 2017; Rossi & Reichenbach, 2016; Zêzere et al., 2009).

Lithology exhibited positive weights, as anticipated, especially in clay, marls, and their mixtures. Throughout the uncertainty propagation analysis, the weights and contrasts for lithological units remained unaltered.

The factor of land use and land cover displayed positive weights, particularly in orchards and meadows, and negative values in forests. The fluctuations within land use and land cover classes remained consistent across all classes, with slight exceptions for bare soil, meadows, and gardens. In these cases, the marginal contrast values were higher, but still maintained the same sign.

The absence of a correlation between slope curvature and landslide presence may be attributed to the widespread distribution of landslides across the entire slope, overlapping on both types of curvature (Guzzetti et al., 2006; Stefan Steger et al., 2016).

While we did not conduct simulations for land use, land cover, and lithology, these predisposing factors were indirectly influenced by the landslide inventory simulations. Among these, bare soil and meadows within the vegetation category displayed the most substantial impact, with probability values ranging approximately 0.2 between the highest and lowest values (Cai et al., 2017; Gullà et al., 2008).

Regarding lithology, the most sensitive categories exhibited either a significantly positive or negative contrast. This sensitivity can be readily explained by the simulations of landslide bodies (Damian et al., 2003).

We believe that the models presented in this study can be applied on a global scale without specific limitations. They offer a reliable estimation of how landslide susceptibility uncertainty is distributed spatially. The primary limitation lies in the absence of simulations for the lithology and land-use/land-cover factors. These limitations are not solely linked to mapping errors but are also influenced by the contextual classification of these factors.

5. Conclusions

In conclusion, incorporating uncertainty propagation from terrain analyses and landslide bodies enhances our ability to provide a more accurate representation of the spatial distribution of landslide susceptibility. This approach aids in defining susceptibility boundaries, facilitating more effective planning for mitigation strategies.

It's worth noting that employing extremely high noise values in the simulation of DEMs resulted in the complete removal of spatial autocorrelation and yielded unrealistic first and second-order terrain derivatives, such as plan and profile curvatures. Conversely, very low noise values introduced excessive spatial autocorrelation and reduced uncertainty propagation. Optimal outcomes were achieved with a noise factor falling within the range of 4 to 7.

Acknowledgments

The study was financed by National Research Council, Romania, Post-doctoral research project: "Modelling the Uncertainty from Spatial-Temporal Data for Geomorphological Hazards Assessment," and by the project GPU framework for geospatial analysis, Young Researchers Grant (ICUB-University of Bucharest). Code and sample tools are available at https://github.com/sandricionut/Landslide-Analysis.

References

Arnone, E., Francipane, A., Scarbaci, A., Puglisi, C., & Noto, L. V. (2016). Effect of raster resolution and polygon-conversion algorithm on landslide susceptibility mapping. *Environmental Modelling & Software, 84*(October), 467—481. https://doi.org/10.1016/J.ENVSOFT.2016.07.016

Bonham-Carter, G. F., Agterberg, F. P., & Wright, D. (1988). *Integration of geological datasets for gold exploration in Nova Scotia.*

Bonham-Carter, G. F., Agterberg, F. P., & Wright, D. (1989). Weights of evidence modelling: A new approach to map mineral potential. In F. P. Agterberg, & G. F. Bonham-Carter (Eds.), *Statistical applications in the earth sciences.* Geological Survey of Canada.

Bonham-Carter, G. F. (1994). *Geographic information systems for geoscientists, modelling with GIS.* Pergamon.

Brown, J. D., & Heuvelink, G. B. M. (2007). The data uncertainty engine (DUE): A software tool for assessing and simulating uncertain environmental variables. *Computers & Geosciences, 33*(2), 172—190. https://doi.org/10.1016/j.cageo.2006.06.015

Cai, J.-S., Yeh, T.-C. J., Yan, E.-C., Hao, Y.-H., Huang, S.-Y., & Wen, J.-C. (2017). Uncertainty of rainfall-induced landslides considering spatial variability of parameters. *Computers and Geotechnics, 87*(July), 149—162. https://doi.org/10.1016/j.compgeo.2017.02.009

Camilo, D. C., Lombardo, L., Mai, P. M., Dou, J., & Huser, R. (2017). Handling high predictor dimensionality in slope-unit-based landslide susceptibility models through LASSO-penalized generalized linear model. *Environmental Modelling & Software, 97*(November), 145—156. https://doi.org/10.1016/J.ENVSOFT.2017.08.003

Carrara, A. (1993). Potentials and pitfalls of GIS technology in assessing natural hazards. In P. Reichenbach, F. Guzzetti, & A. Carrara (Eds.), *Abstracts, proceed. Int. Workshop GIS in assess. Nat. Hazards, Perugia, Sept. 20-22, 1993* (pp. 128—137).

Chitu, Z., Bogaard, T., Busuioc, A., Burcea, S., Sandric, I., & Adler, M.-J. (2016). Identifying hydrological pre-conditions and rainfall triggers of slope failures at catchment scale for 2014 storm events in the Ialomita Subcarpathians, Romania. *Landslides,* 1—16.

Chitu, Z. (2010). Predictia Spatio-Temporala a Hazardului La Alunecari de Teren Utilizand Tehnici SIG. *Studiu de Caz Arealul Subcarpatic Cuprins Intre Valea Prahovei Si Valea Ialomitei.*

Claessens, L., Heuvelink, G. B. M., Schoorl, J. M., & Veldkamp, A. (2005). DEM resolution effects on shallow landslide hazard and soil redistribution modelling. *Earth Surface Processes and Landforms, 30*(4), 461—477. https://doi.org/10.1002/esp.1155

Cruden, D. M., & Varnes, D. J. (1996). Landslide types and processes. In A. K. Turner, & R. L. Schuster (Eds.), *Landslide investigation and mitigation* (pp. 36—71). Washington DC: National Academy of Sciences. Transportation research board special report.

Csillik, O., Evans, I. S., & Drăgut, L. (2015). Transformation (normalization) of slope gradient and surface curvatures, automated for statistical analyses from DEMs. *Geomorphology, 232*(March), 65—77. https://doi.org/10.1016/j.geomorph.2014.12.038

Damian, R., Armas, I., Sandric, I., & Osaci-Costache, G. (2003). *Vulnerabilitatea Versantilor La Alunecari de Teren in Sectorul Subcarpatic Al Vaii Prahova*. Ed. România de Mâine.

ESRI. (2013). Introduction in ArcPy. http://resources.arcgis.com/en/help/main/10.1/index.html#//000v00000001000000%23GUID-4EC90E5F-F497-4FC0-99FB-7703ED4C8F77.

Fisher, P. F., & Tate, N. J. (2006). Causes and consequences of error in digital elevation models. *Progress in Physical Geography, 30*(4), 467–489. https://doi.org/10.1191/0309133306pp492ra

Fisher, P., Wood, J., & Cheng, T. (2004). Where is Helvellyn? Fuzziness of multi-scale landscape morphometry. *Transactions of the Institute of British Geographers, 29*(1), 106–128. https://doi.org/10.1111/j.0020-2754.2004.00117.x

Gullà, G., Antronico, L., Iaquinta, P., & Terranova, O. (2008). Susceptibility and triggering scenarios at a regional scale for shallow landslides. *Geomorphology, 99*(1–4), 39–58. https://doi.org/10.1016/j.geomorph.2007.10.005

Guzzetti, F. (2005). *Landslide hazard and risk assessment*. http://geomorphology.irpi.cnr.it/Members/fausto/PhD-dissertation.

Guzzetti, F., Reichenbach, P., Ardizzone, F., Cardinali, M., & Galli, M. (2006). Estimating the quality of landslide susceptibility models. *Geomorphology, 81*(1–2), 166–184. https://doi.org/10.1016/j.geomorph.2006.04.007

Karssenberg, D., & De Jong, K. (2005). Dynamic environmental modelling in GIS: 2. Modelling error propagation. *International Journal of Geographical Information Science, 19*(6), 623–637. https://doi.org/10.1080/13658810500104799

Micu, M., Jurchescu, M., Şandric, I., Mărgărint, M. C., Chiţu, Z., Micu, D., Ciurean, R., Ilinca, V., & Vasile, M. (2017). Mass movements. In Maria Radoane, & Alfred Vespremeanu-Stroe (Eds.), *Landform dynamics and evolution in Romania* (pp. 765–820). Cham: Springer International Publishing. https://doi.org/10.1007/978-3-319-32589-7_32

Qin, C.-Z., Bao, L.-L., Zhu, A.-X., Wang, R.-X., & Hu, X.-M. (2013). Uncertainty due to DEM error in landslide susceptibility mapping. *International Journal of Geographical Information Science, 27*(7), 1364–1380. https://doi.org/10.1080/13658816.2013.770515

Regmi, N. R., Giardino, J. R., & Vitek, J. D. (2010). Modeling susceptibility to landslides using the weight of evidence approach: Western Colorado, USA. *Geomorphology, 115*(1–2), 172–187. https://doi.org/10.1016/j.geomorph.2009.10.002

Rossi, M., & Reichenbach, P. (2016). LAND-SE: A software for statistically based landslide susceptibility Zonation, version 1.0. *Geoscientific Model Development, 9*(10), 3533–3543. https://doi.org/10.5194/gmd-9-3533-2016

Şandric, I., & Chitu, Z. (2009). Landslide inventory for the administrative area of Breaza, curvature Subcarpathians, România. *Journal of Maps, 5*(1), 75–86. https://doi.org/10.4113/jom.2009.1051

Sandric, I. (2008). *Sistem Informational Geografic Pentru Evaluarea Hazardelor Naturale. O Abordare Bayesiana Cu Propagare de Erori*. University of Bucharest.

Schlögel, R., Marchesini, I., Alvioli, M., Reichenbach, P., Rossi, M., & Malet, J.-P. (2018). Optimizing landslide susceptibility zonation: Effects of DEM spatial resolution and slope unit delineation on logistic regression models. *Geomorphology, 301*(January), 10–20. https://doi.org/10.1016/j.geomorph.2017.10.018

Steger, S., Brenning, A., Bell, R., & Glade, T. (2017). The influence of systematically incomplete shallow landslide inventories on statistical susceptibility models and suggestions for improvements. *Landslides, 14*(5), 1767–1781. https://doi.org/10.1007/s10346-017-0820-0

Steger, S., Brenning, A., Bell, R., & Glade, T. (2016). The propagation of inventory-based positional errors into statistical landslide susceptibility models. *Natural Hazards and Earth System Sciences, 16*(12), 2729–2745. https://doi.org/10.5194/nhess-16-2729-2016

Tarekegn, T. H., & Sayama, T. (2013). Correction of SRTM DEM artefacts by Fourier Transform for flood inundation modeling. *Journal of Japan Society of Civil Engineers, Ser. B1 (Hydraulic Engineering), 69*(4), I_193–I_198. https://doi.org/10.2208/jscejhe.69.I_193

Temme, A. J. A. M., Heuvelink, G. B. M., Schoorl, J. M., & Claessens, L. (2009). Chapter 5 geostatistical simulation and error propagation in geomorphometry. In geomorphometry concepts, software, applications. In Tomislav Hengl, & Hannes I. Reuter (Eds.), *Developments in soil science* (Vol 33, pp. 121–140). Elsevier. https://doi.org/10.1016/S0166-2481(08)00005-6

Wu, X., Chen, X., Zhan, F. B., & Hong, S. (2015). Global research trends in landslides during 1991–2014: A bibliometric analysis. *Landslides, 12*(6), 1215–1226. https://doi.org/10.1007/s10346-015-0624-z

Zêzere, J. L., Henriques, C. S., Garcia, R. A. C., Oliveira, S. C., Piedade, A., & Neves, M. (2009). Effects of landslide inventories uncertainty on landslide susceptibility modelling. *Landslide Processes: From Geomorphologic Mapping to Dynamic Modelling*, 81–86. http://eost.u-strasbg.fr/omiv/Landslide_Processes_Conference/Zezere_et_al.pdf

Index

Note: 'Page numbers followed by *f* indicate figures, *t* indicates tables.'

Printed in the United States
by Baker & Taylor Publisher Services